Contributions to Economics

The series *Contributions to Economics* provides an outlet for innovative research in all areas of economics. Books published in the series are primarily monographs and multiple author works that present new research results on a clearly defined topic, but contributed volumes and conference proceedings are also considered. All books are published in print and ebook and disseminated and promoted globally. The series and the volumes published in it are indexed by Scopus and ISI (selected volumes).

Bruce C. Dieffenbach

Conjugate Duality in Economic Analysis

Springer

Bruce C. Dieffenbach
Independent author
Sarasota, FL, USA

ISSN 1431-1933 ISSN 2197-7178 (electronic)
ISBN 978-3-032-21395-2 ISBN 978-3-032-21396-9 (eBook)
https://doi.org/10.1007/978-3-032-21396-9

This Springer imprint is published by the registered company Springer Nature Switzerland AG.
The registered company address is: Gewerbestrasse 11, 6330 Cham, Switzerland

Bruce's wife June Dieffenbach grew up in England during World War II and graduated in biology from the University of Liverpool. She was very English in her attitude and behavior. Her love of shopping was matched by her distinctive fashion sense, a reflection of her creativity and vibrant personality. June adored entertaining in our mansion Holland House, on the terrace overlooking the lake and the formal garden. She relished our house exchange-vacations to many countries. Many times she crossed the Atlantic Ocean on a ship and once flew around the world for a birthday party. She enjoyed ballroom dancing in two formal-dress clubs and organized a splendid dance at Belcourt Castle in Newport.

June taught biology and chemistry in middle school and high school, in Asia and America. A motivating teacher, June successfully got the students interested in the subject matter, in an amusing way. June excelled at well-prepared talks to groups. She described the growing of lavender in

Provence and in England, showing distilling equipment, a lavender bottle, a model building, slides, photos, and books. The writing of this book took a very long time, and June believed resolutely that it would be published and well-received. June died tragically in an automobile accident, October 19, 2024.

Preface

Dorfman, Samuelson, and Solow present important economic concepts via linear primal and the dual optimizations. Defined as an optimization, the Fenchel conjugate of a convex function generalizes and extends their work. Conjugate duality asserts that under general conditions the conjugate of the conjugate is the original function. A Fenchel factor is a convex function plus its conjugate less the inner product of the arguments, which must be nonnegative.

Given a "primal" optimization problem minimizing a convex function, augment the objective function by introducing a perturbation variable, but keeping convexity. The optimum value of the primal is a convex function of the perturbation. Rockafellar explains how calculating its conjugate obtains a "dual" optimization problem maximizing a concave function.

Necessarily the optimum value of the primal is greater than or equal to the optimum value of the dual, and under general conditions both are equal. Solve the primal and the dual simultaneously—an algebraic solution—by setting the primal equal to the dual. This expression factors into a sum of Fenchel factors. That each factor be zero constitute necessary and sufficient conditions for the simultaneous solution. A significant task is to understand the mathematical and economic meaning of these conditions.

Invoking the Fenchel factors in reverse furnishes an effective way to calculate the dual. Rearrange a sum of Fenchel factors to obtain both the primal and the dual, simultaneously. This book contains bountiful applications of primal and dual optimization.

"Proof by optimization" via duality furnishes a unified development of economic theory. To discover and prove an economic theorem, set up a primal that embodies some economic condition. Calculate a dual and interpret the economic meaning of the dual. Figuring out the economic meaning of the simultaneous solution to the primal and the dual both discovers and proves an economic theorem.

A theorem of the alternative asserts that either of two conditions holds, but not both simultaneously. Calculate an alternative by calculating a dual. A cone alternative generalizes a traditional theorem of the alternative, which involves linear inequalities. The efficiency alternative reveals the close linkage between efficiency and weak efficiency. A major economic application is the derivation and proof of first and second theorems of efficiency. The primal models efficiency, whereas the dual models market equilibrium.

An allocation is in the core if no coalition of consumers can trade their endowments among themselves so at least one member gains and no member loses. For an economy having many consumers, the intuitive idea is that an allocation lies in the core if and only if it is attainable as a general market equilibrium. A primal/dual convex optimization models whether an allocation is in the core of a large economy. The primal is a maximization choosing simultaneously across both allocations and coalitions, to see whether any coalition can do better. For a core allocation, no coalition can do better and the optimum value is zero. If the allocation is not in the core, then some coalition can do better, and the optimum value is positive. The dual is a minimization, looking for an equilibrium price vector. For a core allocation, the dual solution solves for a price vector for which the allocation is a general equilibrium. Conversely, if the allocation is not in the core, then the dual has no solution.

Jordan, von Neumann, and Wigner axiomatize a Euclidean Jordan algebra by adding algebraic structure to a Euclidean space. The axioms generalize the real numbers, but not the complex numbers. The key example is the self-adjoint linear transformations, pervasive in statistical analysis. The Lorentz Euclidean Jordan algebra has application to finance theory and portfolio choice. A Hermitian Euclidean Jordan algebra yields a theorem of the alternative to analyze the stability of linear differential and difference equations.

Moment space is a Euclidean space of random variables that embodies the first and second moments. The cross-moment transformation between subspaces is the unique linear transformation such that the inner product of two vectors is the inner product of one vector and the transformation of the other. Consider the explanatory- and instrumental-variable subspaces in an instrumental-variables framework. For an identified equation, the explanatory variable is the dependent variable transformed by the generalized inverse of the cross-moment transformation.

Competing Interests The author has no competing interests to declare that are relevant to the content of this manuscript.

Optimization by Duality

Discovery and Proof via Duality

Convex optimization via duality furnishes a unified development of economic theory, and this book applies this mathematical technique to a wide variety of problems in economic and econometric theory. Given a convex function f on a Euclidean space, an optimization problem defines its conjugate f^*, another convex function:

$$f^*(\boldsymbol{x}^*) := \sup_{\boldsymbol{x}} \left[\langle \boldsymbol{x}^*, \boldsymbol{x} \rangle - f(\boldsymbol{x})\right].$$

Here $\langle \boldsymbol{x}^*, \boldsymbol{x} \rangle$ is the inner product. Under general conditions, "conjugate duality" holds: the conjugate of the conjugate is the original function. The conjugate is the building block of convex optimization by Fenchel and Rockafellar.

Given a convex optimization problem (the "primal"), one can associate with it a "dual" optimization problem. Perturbing the primal makes the problem slightly different, and this perturbation determines the dual. One calculates the dual from the primal via the conjugate. If the primal is a minimization, then the dual is a maximization. Under general conditions, there is "no duality gap": the minimum of the primal is the maximum of the dual.

No Duality Gap

Consider the primal

$$V(\boldsymbol{y}) = \inf_{\boldsymbol{x}} F(\boldsymbol{x}, \boldsymbol{y}).$$

Here $\boldsymbol{y}$ is the perturbation,, and the dual is

$$V^{**}(\boldsymbol{y}) := \sup_{\boldsymbol{y}^*} [\langle \boldsymbol{y}^*, \boldsymbol{y} \rangle - V^*(\boldsymbol{y}^*)] .$$

Under genderal conditions, there is no duality gap: $V(\boldsymbol{y}) = V^{**}(\boldsymbol{y})$. Furthermore, if the optimum value is finite, then it is attained by a dual solution.

"Proof by optimization" via duality furnishes a unified development of economic theory. To discover and prove an economic theorem, set up a primal that embodies some economic condition. Calculate a dual and interpret it. What is the economic meaning of a solution to the dual? Our standard method is to solve the primal and the dual simultaneously—an algebraic solution—by setting the primal equal to the dual. To figure out the economic meaning of this simultaneous solution both discovers and proves an economic theorem.

Fenchel Factors

A "Fenchel factor" is $f(\boldsymbol{x}) + f^*(\boldsymbol{x}^*) - \langle \boldsymbol{x}^*, \boldsymbol{x} \rangle$, which is nonnegative by the definition of the conjugate. The primal less the dual is nonnegative, and one can factor this difference into a sum of Fenchel factors. Because each is greater than or equal to zero, each factor must be zero when there is no duality gap. Together these conditions are necessary and sufficient for a simultaneous solution.

To reverse this procedure furnishes an effective and simple way to calculate the dual. Start with Fenchel inequalities. Add them together and rearrange to obtain an inequality with the primal objective function on the left and the dual objective function on the right. In applications, reversal is a simple and direct way to derive the dual. Valuable for both derivation and exposition, it facilitates the analysis of complicated primal/dual pairs. The reversal methodology is the technique mostly used throughout this book.

Conjugate Duality

Fenchel proves the conjugate-duality theorem: for a convex function, under general conditions $f^{**} = f$. That a closed, convex set is the intersection of its containing, closed, half spaces is key to proving conjugate duality. The effective domain $\text{dom}\, f$ is the vectors where the function is finite.

Support

The indicator function of a set is zero for vectors in the set, and infinity for vectors outside the set. The support to a set is the conjugate of its indicator.

We study efficiency or weak efficiency by maximizing the efficiency function or the weak efficiency function. These two functions enable us to distinguish sharply and effectively between the two concepts. The maximization is especially fruitful in primal/dual convex optimization. The primal analyzes whether a particular vector of goals is efficient or weakly efficient, and this vector is the perturbation variable. One interprets the dual choice variable as a shadow price for the goals.

Epigraph

The epigraph is the points $(\boldsymbol{x}, y)$ above the function. "Epi-multiplication" is the support to the epigraph of the conjugate:

$$(y \star f)(\boldsymbol{x}) := \delta^*_{\text{epi}\, f^*}(\boldsymbol{x}, -y),$$

in which y is the "epi-multiplier." Epi-multiplication by zero is the "horizon function":

$$f^\infty(\boldsymbol{x}) := (0 \star f)(\boldsymbol{x}) = \delta^*_{\text{dom}\, f^*}(\boldsymbol{x}),$$

the support to the effective domain of the conjugate.

Subdifferential

Generalizing the derivative, the subdifferential ∂f describes the local behavior of a convex function. Define the subdifferential via the nonvertical hyperplanes supporting the epigraph. Whereas differential calculus deals with derivatives, subdifferential calculus deals with subdifferentials. In subdifferential calculus, one works with rules for subdifferentials, which form only a subset of the rules for differential calculus. For example, whereas the chain rule is a key tool of differential calculus, unfortunately this rule does not hold for subdifferential calculus. In general the composition of two convex functions need not be convex, so the chain rule cannot apply in general.

For the value function

$$V(\boldsymbol{y}) = \inf_{\boldsymbol{x}} F(\boldsymbol{x}, \boldsymbol{y}),$$

the following envelope inclusion holds: if choice $\boldsymbol{x}$ is minimizing for perturbation $\boldsymbol{y}$, then under general conditions

$$\partial V(\boldsymbol{y}) \subset \partial_Y F(\boldsymbol{x}, \boldsymbol{y}),$$

a generalization of the envelope theorem.

Lagrangian

Economists commonly introduce a Lagrangian to analyze an optimization problem with a constraint. Generalizing the traditional *ad hoc* definition employed by economists, define the Lagrangian as the value of an optimization. Given the perturbation function for a problem, taking the conjugate in the perturbation variable while holding the choice variable fixed obtains the Lagrangian. It is a saddle-value function of the choice variables in the primal and the dual, and its value lies between the values of the primal and dual objective functions. A basic result is the saddle-point theorem that the choice variables form a saddle point of the Lagrangian if and only if they solve the primal and the dual.

Karush-Kuhn-Tucker

For concave g and convex f, consider the Karush-Kuhn-Tucker problem

$$\sup_{\boldsymbol{x}} \left[g(\boldsymbol{x}) - \delta_{f(\boldsymbol{x}) \leq z} \right],$$

in which the perturbation base value $z = 0$. The Lagrangian is

$$L(\boldsymbol{x}, z^*) = g(\boldsymbol{x}) - z^* f(\boldsymbol{x}) - \delta_{\mathrm{dom}\, f}(\boldsymbol{x}) + \delta_{z^* \geq 0}.$$

For $(\boldsymbol{x}, z^*)$ to be a saddle point of the Lagrangian is necessary and sufficient for $\boldsymbol{x}$ to solve the Karush-Kuhn-Tucker problem and for z^* to solve its dual. The "Lagrange multiplier" is the dual solution set.

Compared to the traditional *ad hoc* Lagrangian analyzed by economists, this expression contains the extra term $\delta_{\mathrm{dom}\, f}(\boldsymbol{x})$, which comes into play for the exceptional case $z^* = 0$. It can be that expenditure minimization does not imply utility maximization: possibly compensated demand does not maximize utility among consumption costing the same or less. We analyze efficiency by maximizing the efficiency function among vectors in a feasible set. In the Karush-Kuhn-Tucker form of the efficiency dual, the exceptional case is inherent: a vector is efficient if and only if the unique Lagrange multiplier is zero.

Linear Algebra

Linear Transformation

In line with the modern mathematical point-of-view, we see linear algebra as coordinate-free. A vector in a Euclidean space exists independently of any coordinates used to describe it. A linear transformation from one Euclidean space to another maps one vector to another, independently of any coordinate system.

In contrast, in matrix algebra, one works with a basis. A matrix describes how a linear transformation maps one vector to another. In economic theory and econometrics, typically vectors are not seen as coordinate-free. A particular basis is singled out, and one works with coordinates. Despite this tradition, the coordinate-free point-of-view is superior. Not using coordinates reduces the use of subscripts and makes expressions simpler, and theorems are easier to state and to prove.

Given a linear transformation $A : \mathbf{X} \to \mathbf{Y}$, then there exists a unique linear transformation (the adjoint) $A^\top : \mathbf{Y} \to \mathbf{X}$ that preserves the inner product:

$$\langle y^*, Ax \rangle = \left\langle A^\top y^*, x \right\rangle.$$

Coordinate-free and not the same as the transpose of a matrix representation, the adjoint is a fundamental component of duality. If the primal optimization problem involves a linear transformation, then the dual optimization problem involves the adjoint.

Let $\mathrm{R}(\cdot)$ and $\mathrm{N}(\cdot)$ denote the range and the null space of a linear transformation. Perturbation duality furnishes a straightforward proof of the "fundamental theorem of linear algebra": the orthogonal complement of the null space of a linear transformation is the range of the adjoint. Given a linear transformation $A : \mathbf{X} \to \mathbf{Y}$, the restricted linear transformation $A : \mathrm{R}\left(A^\top\right) \to \mathrm{R}(A)$ is invertible. Extending this inverse obtains the Moore-Penrose generalized inverse $A^+ : \mathbf{Y} \to \mathbf{X}$.

Decomposition

The spectral decomposition of a self-adjoint linear transformation expresses it via its real eigenvalues and orthonormal eigenvectors. We derive the singular-value decomposition of a linear transformation (not necessarily square) from the spectral decomposition of a Jordan-Wielandt partitioned linear transformation. A square linear transformation transforms area by the product of its singular values.

In this book an idempotent is a self-adjoint linear transformation equal to its square. There is a one-to-one correspondence between a subspace and an idempotent having rank equal to the dimension of the subspace. To find an optimizing subspace, recast the problem as seeking an optimizing idempotent.

Linear Equation

Using perturbation duality, we calculate the general formula for the solution set to a linear equation. We express the solution as the optimum value of a convex optimization, find the reciprocal dual, and calculate the value function. Its subdifferential is the solution set. To calculate the solution set contrasts the traditional approach, which is not a calculation. Instead, one conjectures the solution and then verifies it via the fundamental theorem of linear algebra.

Economic Applications

"Economic Applications" employs conjugate duality to explain traditional concepts in mathematical economics.

Benefit and Cost

The primal chooses the quantity to maximize consumer benefit minus cost, and the dual chooses the price to minimize the sum of consumers' surplus plus profit. As a solution to the primal and the dual is on the demand curve and the supply curve, market equilibrium maximizes benefit minus cost.

Diet

Stigler's diet problem is a classic primal/dual application. The problem for the consumer is to minimize the cost of buying food to achieve nutrition targets. Interpret the dual choice variable as the price (value) of nutrition. Choose this price to maximize the value of nutrition, subject to food value less than or equal to food cost.

Activity Analysis

Koopmans studies efficient production in activity analysis, and we present his ideas via perturbation duality. An activity produces output and uses inputs. A production equilibrium is a price/quantity combination such that profit is maximized. The key results are first and second theorems of production equilibrium, analogous to the first and second theorems of welfare economics. The first theorem is that production equilibrium implies that production is efficient. The second theorem is that efficient production can be realized as a production equilibrium: for chosen prices, the production maximizes profit.

Zero-Sum Game

Nash equilibrium is the cornerstone for game theory. For a two-player, zero-sum game, a Nash equilibrium is a saddle point of the payoff as a function of the strategies of the two players. This function is the Lagrangian of a primal/dual pair. In the primal one player maximizes the minimum payoff across his pure strategies, and in the dual the other player minimizes the maximum payoff across his pure strategies. The game has a Nash equilibrium, and one can calculate the optimum strategies by solving a convex optimization.

Assignment

We analyze the problem of assigning n people to n jobs. The assignment problem was important in the historical development of the theory of linear programming and duality. At first glance, it might seem that the problem is not a convex optimization. Choosing the optimum perturbation from among the $n!$ possible permutations is a choice among discrete alternatives, not a convex problem. Nevertheless one can cast the problem as an interesting convex optimization, having a dual with an insightful economic interpretation.

Constant Elasticity of Substitution

We study constant-returns-to-scale production functions having constant elasticity of substitution between any two inputs. Solving the dual of a Karush-Kuhn-Tucker problem obtains the cost function. The cost function resembles the production function: it has the same constant elasticity of substitution functional form, but with the elasticity of substitution inverted. When the elasticity of substitution is high, then a small change in the input price causes a strong substitution of one input for another, but the change in cost is small.

Cone Representation

One can replace the analysis of a closed, convex set by an equivalent analysis of a closed, convex cone. That a closed, convex set is the intersection of its containing closed half spaces epi δ_C^* is the key: the cone representation of the set is the polar of the epigraph of the support. For a closed, convex set C,

$$\text{cone } C = \left\{ (\boldsymbol{x}, -y) \;\middle|\; \begin{array}{l} \boldsymbol{x} \in yC \text{ if } y > 0 \\ \boldsymbol{x} \in C^\infty \text{ if } y = 0 \end{array} \right\}.$$

The barrier cone is the effective domain of the support, the directions in which the set is bounded. The horizon cone C^∞ is the polar of the barrier cone and describes the set "at infinity," the directions in which the set is unbounded. The cone representation of a convex function is the cone representation of its epigraph.

Local Analysis

Directional Derivative and Subdifferential

The directional derivative of a convex function is the derivative in each direction from a vector—a positively homogeneous, convex function. The conjugate of the directional derivative is the indicator of the subdifferential. We study the local behavior of a convex function by translating its epigraph so $f(\boldsymbol{0}) = 0$ and studying the local behavior at the origin. The conjugate is nonnegative, and the subdifferential is the vectors such that the conjugate is zero.

Because the function is convex, one can express the derivative as the value function of a perturbed convex optimization: the derivative in direction x is

$$\inf_y (y \star f)(\boldsymbol{x}).$$

The objective function is the epi-multiplication of the function. The perturbation is the direction, and the choice variable is the multiplying factor. One obtains the directional derivative as the choice variable rises to infinity.

Normal Cone

Define the projection $X : (\boldsymbol{x}, y) \longmapsto \boldsymbol{x}$. Under general conditions the subdifferential of the indicator of a contour set is the projection of the normal cone of the epigraph,

$$\partial\delta_{f \le 0} = X\,\partial\delta_{\text{epi } f}.$$

This subdifferential is also

$$\partial\delta_{f\leq 0} = \mathbf{R}_{+}\partial f \cup \partial\delta_{\mathrm{dom}\, f},$$

in which the two terms derive from the non-vertical and vertical supporting hyperplanes, respectively.

Subdifferential Calculus

To derive results of subdifferential calculus directly from the directional derivative is a powerful technique. Duality leads naturally to the discovery of an alternate formula for the subdifferential. For example, let the primal calculate the conjugate of the directional derivative of the sum of two functions; the value function is the indicator of the subdifferential of the sum. In the dual, the value function is the indicator of the sum of the subdifferentials. Under general conditions there is no duality gap, so the subdifferential of a sum is the sum of the subdifferentials.

Polyhedra

A polyhedron is the intersection of a finite number of closed half spaces, the solutions to linear inequalities. A cone is finitely generated if its elements are the nonnegative linear combinations of a finite set of generators. The Minkowski-Weyl theorem that a cone is finitely-generated if and only if it is polyhedral is intuitive geometrically, but the proof is subtle. No simple, general formula relates a finitely-generated representation and a polyhedral representation. For this reason, sophisticated algorithms are put forward for numerical polyhedral optimization.

Euclidean Jordan Algebra

Euclidean Jordan Algebra

Although the setting for conjugate and perturbation duality is Euclidean space, many important applications involve additional structure. Whereas applied optimization theory commonly adds this extra structure informally, we proceed formally, by working in a "Euclidean Jordan algebra," a Euclidean space having additional algebraic structure. In mathematics, an algebra is a vector space with a bilinear, multiplicative product. A Euclidean Jordan algebra is a finite-dimensional real algebra.

The prime example of a Euclidean Jordan algebra is the self-adjoint linear transformations, in which the Jordan product

$$X \bullet Y := \frac{1}{2}(XY + YX).$$

That the product is commutative but not associative is a key property, the reverse of standard matrix multiplication.

We present the original axioms of Jordan, von Neumann, and Wigner, which capture very well how a Euclidean Jordan algebra generalizes the real numbers, but not the complex numbers. By definition, $\boldsymbol{x} \succeq \boldsymbol{y}$ means that $\boldsymbol{x} - \boldsymbol{y}$ is a square. The classification theorem states that there are five distinct types of Euclidean Jordan algebras.

Spectral Decomposition

A Euclidean Jordan algebra admits a spectral decomposition that generalizes the spectral decomposition of the self-adjoint linear transformations. Each vector has unique real eigenvalues, and the spectral decomposition expresses the vector as the sum of the eigenvalues times idempotents. The trace is the sum of the eigenvalues.

Making the algebra into a Euclidean space, the inner product of two vectors is by definition the trace of their Jordan product. The trace inner product is less than or equal to the inner product of the ordered eigenvalues:

$$\langle \boldsymbol{x}^*, \boldsymbol{x} \rangle \leq \langle \boldsymbol{\lambda}(\boldsymbol{x}^*), \boldsymbol{\lambda}(\boldsymbol{x}) \rangle .$$

Equality holds if and only if the two vectors have a simultaneous ordered spectral decomposition.

Spectral Conjugate

A spectral function is one expressible as the composition of a symmetric function and the ordered eigenvalues: $f[\boldsymbol{\lambda}(\boldsymbol{x})]$. The conjugate $(f \circ \boldsymbol{\lambda})^* = f^* \circ \boldsymbol{\lambda}$.

Quadratic Function

We define and analyze the quadratic function, a generalization of the quadratic $\frac{1}{2}\langle \boldsymbol{x}, \boldsymbol{x} \rangle$. The quadratic set having indicator $\delta_Q(\boldsymbol{x}, y) = \delta_{y \geq \frac{1}{2}\boldsymbol{x}^\top \boldsymbol{x}}$ is the epigraph of the quadratic $\frac{1}{2}\langle \boldsymbol{x}, \boldsymbol{x} \rangle$. The quadratic function $\delta_Q^*(\boldsymbol{x}^*, -y^*)$ is the epi-multiplication of the conjugate $\frac{1}{2}\langle \boldsymbol{x}^*, \boldsymbol{x}^* \rangle$ by y^*. We generalize this concept. Let $(A, S) \in \mathbf{T}^{\mathbf{X} \to \mathbf{Y}} \oplus \mathbf{S}^{\mathbf{X}}$. The quadratic set has indicator $\delta_Q(A, S) = \delta_{S \succeq \frac{1}{2} A^\top A}$, and the quadratic function is the support $\delta_Q^*(A^*, -S^*)$. This book applies the quadratic function to least-squares linear regression and maximum-likelihood linear regression in econometrics.

Lorentz Euclidean Jordan Algebra

One of the five possible types of Euclidean Jordan algebra, the Lorentz Euclidean Jordan algebra $\mathbf{L}^{\mathbf{X}}$ consists of pairs $(\boldsymbol{x}, y)$ in $\mathbf{X} \times \mathbf{R}$ such that

$$(\boldsymbol{x}^*, y^*) \bullet (\boldsymbol{x}, y) = (y\boldsymbol{x}^* + y^*\boldsymbol{x}, \langle \boldsymbol{x}^*, \boldsymbol{x} \rangle + y^* y).$$

The nonnegative cone is the $(\boldsymbol{x}, y)$ such that $y \geq |\boldsymbol{x}|$, commonly called the "ice-cream" cone. In the application to finance theory, $\boldsymbol{x}$ is a portfolio, and the optimum y is standard deviation.

Hermitian Euclidean Jordan Algebra

The complexification of a Euclidean space $\mathbf{X}$ is its natural extension to a complex vector space having a complex-valued inner product. The complexification expresses a vector as the sum of its real and complex parts, $\boldsymbol{x} = \boldsymbol{y} + \boldsymbol{z}\mathrm{i}$. The complex inner product is

$$\langle \boldsymbol{x}^*, \boldsymbol{x} \rangle = \langle \boldsymbol{y}^* + \boldsymbol{z}^*\mathrm{i}, \boldsymbol{y} + \boldsymbol{z}\mathrm{i} \rangle := \left(\boldsymbol{y}^{*\top} - \boldsymbol{z}^{*\top}\mathrm{i}\right)(\boldsymbol{y} + \boldsymbol{z}\mathrm{i}).$$

The square linear transformation $A + B\mathrm{i}$ is self-adjoint if and only if A is self-adjoint and B is skew. The Hermitian Euclidean Jordan algebra $\mathbf{H}^{\mathbf{X}}$ is the real vector space of self-adjoint linear transformations on the complexification.

Mathematical Programming

Hahn-Banach

Using perturbation duality, we prove the Hahn-Banach theorem in Euclidean space. The theorem generalizes basic separation. The primal models a simple condition: for a convex set and an affine set, either the sets have a vector in common (optimum value zero) or they do not (optimum value infinity). For the latter condition, one obtains the Hahn-Banach separation theorem. For the former condition, one obtains the Hahn-Banach extension theorem. That the same primal/dual pair obtains both theorems reveals their intrinsic interconnection.

Ekeland Variational Principle

The standard variational principle is the first-order condition for an optimum. If the optimum is not attained, however, then the first-order condition fails to hold. At an approximate minimum, an approximate first-order condition need not hold. Given an approximate minimum, the Ekeland variational principle considers a slight variation of the original problem. This variational problem has a minimum that is an appproximate minimum for the original problem. It satisfies an approximate first-order condition for the original problem, and its location is near the approximate minimum. This chapter uses perturbation duality to prove the Ekeland variational principle.

Homogeneous Fractional Programming

We analyze a "standard homogeneous fractional program," the maximization of the ratio of a concave positively homogeneous function to a convex positively homogeneous function:

$$\sup_{\boldsymbol{x}\neq\mathbf{0}} \frac{g(\boldsymbol{x})}{f(\boldsymbol{x})} = \sup_{\boldsymbol{x}\neq\mathbf{0}} \frac{-\delta_C^*(-\boldsymbol{x})}{\delta_D^*(\boldsymbol{x})},$$

for convex sets C and D

Because in general the ratio is not a concave function of $\boldsymbol{x}$, reduce it to a Karush-Kuhn-Tucker maximization:

$$\sup_{\boldsymbol{x}\neq\mathbf{0}} \left[\frac{g(\boldsymbol{x})}{f(\boldsymbol{x})}\right] = \sup_{\boldsymbol{x}} \left[g(\boldsymbol{x}) - \delta_{f(\boldsymbol{x})\leq z}\right],$$

in which the perturbation base value $z = 1$. Under general conditions, the zero-maximum condition

$$0 = \max_{\boldsymbol{x}\neq\mathbf{0}} \left[g(\boldsymbol{x}) - z^* f(\boldsymbol{x})\right] \text{ for some } z^* > 0.$$

is attained by a choice $\boldsymbol{x}$ if and only if $\boldsymbol{x}$ solves the homogeneous fractional program and z^* is the optimum value.

Fractional Programming

We generalize the analysis of a homogeneous fractional program to the case in which the denominator and numerator are not positively homogeneous. For the concave and nondecreasing value function of the following Karush-Kuhn-Tucker problem,

$$V(z) := \sup_{x} \left[g(\boldsymbol{x}) - \delta_{f(x)\le z} \right].$$

we show

$$\sup_{x} \frac{g(\boldsymbol{x})}{f(\boldsymbol{x})} = \sup_{z>0} \frac{V(z)}{z}.$$

The optimum value is the unique z^* such that $V_*(z^*) = 0$.

Alternatively, even if the numerator and the denominator in the fractional program are not positively homogeneous, nevertheless one can reduce the analysis to the positively homogeneous case, via epi multiplication.

The Alternative

A theorem of the alternative asserts that either of two conditions holds, but not both simultaneously. Theorems of the alternative have played a major role in the development of mathematical economics. See a theorem of the alternative as an existence theorem: the failure of one condition to hold implies the other condition.

This chapter develops the concept of the alternative via perturbation duality. The primal/dual pair has no duality gap. The optimum value is zero if the first alternative holds and the second does not, and is infinity if the second alternative holds and the first does not. In the historic development of the alternative, the mathematician posits the alternative out of a hat, without explanation or motivation, and then proves the result. In contrast, the duality approach arrives at the alternative by calculating the dual.

In this chapter the starting point is a cone alternative. Express one alternative as a closed, convex set. One derives the second alternative from the polar of the cone representation of this set.

Efficiency

We seek conditions for the feasible set to contain an efficient vector or a weakly-efficient vector. For a closed, convex cone K in a Euclidean Jordan algebra, the "efficiency alternative" states that exactly one of the following has a solution:

1. Some $\boldsymbol{x} \gg \mathbf{0}$ belongs to K;
2. Some $\boldsymbol{x}^* > \mathbf{0}$ belongs to K°.

The efficiency alternative leads to both efficiency duality and weak-efficiency duality. Does a vector maximize value in the feasible set? Alternatively, does a particular Karush-Kuhn-Tucker problem have a zero Lagrange multiplier?

Production

Domestic Product and Income

One can model production equilibrium via a domestic product and income duality. The primal maximizes domestic product: choose output to maximize its value, for given input. The dual minimizes the domestic income: choose the input price to minimize the input value, among all prices for which the maximum profit is zero. A price/quantity combination is a production equilibrium if and only if it solves the primal and the dual.

Shephard Duality

Shephard's book *Cost and Production Functions* explains and proves the equivalence between the production function and the cost function: either determines the other. Furthermore, the relationship is symmetric: the relationship of cost to product is the same as the relationship of product to cost. This chapter uses conjugate and perturbation duality to develop Shephard's ideas.

Heckscher-Ohlin via Shephard Duality

Heckscher-Ohlin international-trade theory develops the theory of comparative advantage, by focussing on how factor endowments determine the direction of trade. A country having a high ratio of capital to labor endowment exports capital-intensive goods and imports labor-intensive goods. The natural modelling framework takes the world price of outputs and the factor endowments as given, and one solves for the output quantities and the factor prices. "Factor-price equalization" applies: There is no economic incentive for capital or labor to flow between countries.

The Stolper-Samuelson theorem asserts that imposing a tariff on the labor-intensive good raises the real wage. One models this idea by raising the output price of the labor-intensive good. For each sector, the capital/labor ratio rises. In each sector the marginal product of labor rises, so the real wage rises, measured in terms of either good.

Factor-Intensive Production Function

Solow models production by the factor-intensive production function: output per worker is a nondecreasing, concave function of the capital/labor ratio. Its conjugate is the factor-price frontier, the combinations of capital cost and labor cost such that the minimum cost of production is one. The conjugate evaluated at the capital cost is the negative of the labor cost.

An application is the golden rule of saving, the saving rate that maximizes long-run per capita consumption in the neoclassical one-sector growth model. Choosing the optimum saving rate reduces to the evaluation of the conjugate of the factor-intensive production function. Per capita consumption is the negative of the conjugate evaluated at the rate of population growth, and the steady-state capital/labor ratio is a subgradient.

For a competitive economy in market equilibrium, that the conjugate is the factor-price frontier links the golden rule of saving to the real interest rate and the real wage. The golden rule of saving is achieved if the real interest rate equals the rate of population growth and the real wage equals per capita consumption. Saving equals profit.

Heckscher-Ohlin via Factor-Intensive Production Function

We develop further the Heckscher-Ohlin international-trade theory. The domestic-product primal maximizes the output value, subject to the technology and input constraints. Because the theory focuses on factor endowments and the capital/labor ratio in production, it is natural to set up the primal via factor-intensive production functions. The domestic-income dual then involves the factor-price frontier, the conjugate of the factor-intensive production function. The dual chooses input prices to minimize the value of the resource endowments, subject to the constraint that the prices lie on or above the factor-price frontiers.

Consumption

Hicks and Marshall Decompositions

We contrast two alternative decompositions of the calculation of the conjugate of a concave function—the "Hicks decomposition" and the "Marshall decomposition."

One obtains an economic interpretation by viewing the function as the utility of consumption. Economists refer to expenditure-minimizing consumption as Hicks demand (compensated demand) and to utility-maximizing consumption as Marshall demand (consumer demand).

The expenditure function is the minimum cost of obtaining utility. The indirect-utility function is the maximum utility possible, subject to the budget constraint that expenditure is less than or equal to income. The relationship between the expenditure function and the indirect-utility function is simple: reversing the axes in the epigraph of the expenditure function obtains the epigraph of the indirect-utility function. We prove this axis-reversal theorem via the duals of expenditure minimization and utility maximization.

Consumption Demand

The fundamental research of Hildenbrand and Kannai justifies the use of concave utility, but what economists have not done is to develop consumer demand theory via concave utility. Because indifference curves are ordinal, the concept of the concave conjugate of utility is not part of economic thinking. Concave utility allows analysis by duality, which makes possible the derivation of results and the discovery of relationships by algebraic calculation.

For a "classical consumer," we contrast expenditure minimization and utility maximization via the directional derivative and the subdifferential of utility. Axis reversal explains the situation. If income/expenditure is positive, then expenditure is minimized if and only if utility is maximized. If income/expenditure is zero, then necessarily expenditure is minimized, because it cannot be nonnegative. However perhaps utility is not maximized: to consume more free goods perhaps raises utility.

Indirect Utility

By the Marshall decomposition applied to a consumer, indirect utility—the maximum utility for given income and price—is equivalent to utility; either suffices to determine the other. Consumer demand maximizes utility subject to the budget constraint set by price and income. Alternatively, see this same condition as constraining price for a given quantity and income, such that spending is less than or equal to income. A parallelogram theorem holds: quantity maximizes utility for a given price if and only if this price minimizes indirect utility for this quantity. Maximum utility equals minimum indirect utility. A parallelogram furnishes a geometric interpretation.

Benefit

Luenberger introduces and applies "benefit" for a consumer, as a novel alternative to utility. We define and analyze benefit and relate it to compensated demand. Because benefit is a dollar value, it is meaningful to consider benefit minus cost. Take the price of consumption goods as given. If consumption maximizes benefit minus cost and benefit is zero, then it is the compensated demand. Conversely, the compensated demand maximizes benefit minus cost, and benefit is zero.

Welfare Economics

Resource Utilization

We study resource utilization in an exchange economy. Is an allocation of consumption to consumers weakly efficient? Could a positive amount of each good be discarded, and the remaining supply reallocated to attain the same levels of preference? If not, then the utilization of resources is weakly efficient. We model this issue by three alternative optimizations.

The resource-utilization primal solves for the quantity such that this quantity of every good can be discarded, and the remaining supply is sufficient to attain the same levels of preference. We calculate an expenditure dual, which chooses prices to minimize the value of supply less the sum of the expenditure functions for all consumers. Using this duality, we prove the first and second theorems of resource utilization, akin to the first and second theorems of welfare economics.

From the expenditure dual we derive the equivalent formulation by Debreu. We restate the expenditure dual as a positively homogeneous fractional program, and find the equivalent concave maximization and its Fenchel dual. The dual interprets the fractional program as Debreu's coefficient of resource utilization: its optimum value is the fraction of resources that is just sufficient to attain the same levels of preference.

Whether the resource allocation is benefit-efficient is a third formulation. The dual of the benefit-efficiency problem reduces to the expenditure dual. That the utilization of resources is weakly efficient is thus equivalent to benefit efficiency. We restate the first and second theorems of resource utilization in terms of benefit efficiency.

Welfare and Equilibrium

Working from the concave utility of consumers, this chapter derives first and second theorems of welfare economics for an exchange economy. General equilibrium refers to simultaneous demand and supply equilibrium in all markets. The Hicks and Marshall decompositions for utility contrast expenditure-minimizing compensated demand and utility-maximizing consumer demand. Always consumer demand is expenditure-minimizing, but compensated demand is perhaps not utility-maximizing. Compensated equilibrium is general equilibrium for compensated demand, whereas competitive equilibrium is general equilibrium for consumer demand. A competitive equilibrium is necessarily a compensated equilibrium, but not conversely.

Perturbation duality leads to first and second theorems of welfare economics. Starting from a weak-efficiency primal, we derive a compensated-equilibrium dual. The primal/dual analysis yields first and second theorems of welfare economics for compensated equilibrium. Starting from an efficiency primal, we derive a competitive-equilibrium dual, resembling the compensated-equilibrium dual, but slightly different. The primal/dual analysis leads to first and second theorems of welfare economics for competitive equilibrium.

Core

Core and Equilibrium

Debreu and Scarf analyze the core of a large economy, inspired by Edgeworth. Consider an exchange economy in which each consumer has a utility function and an endowment of goods. An allocation of goods across the consumers is in the core if no coalition of consumers can trade their endowments among themselves so at least one member of the coalition gains and no member loses. For an economy with many consumers, the intuitive idea is that an allocation of goods lies in the core if and only if it is attainable as a general market equilibrium. For an allocation in the core, there exists a price vector such that it is a general market equilibrium. Conversely, for a general market equilibrium, the allocation is in the core.

We develop their analysis via primal/dual convex optimization modelling whether an allocation is in the core of a large economy. The primal is a maximization choosing simultaneously across both allocations and coalitions, to see whether any coalition can do better. For a core allocation, no coalition can do better and the optimum value is zero. If the allocation is not in the core, then some coalition can do better, and the optimum value is positive. The dual is a general-equilibrium minimization, looking for an equilibrium price vector. For a core allocation, the dual solution solves for a price vector for which the allocation is a general equilibrium. Conversely, if the allocation is not in the core, then the dual has no solution.

Bondareva-Shapley

We study the core of a cooperative multi-person game with transferable utility. A payoff specifies the utility received by each player. By cooperating, a coalition of players can guarantee a certain utility—its *worth*—to be divided somehow among its members. The core is the set of payoffs such that no coalition can do better for itself. The worth of the grand coalition of all players is the value of the game. The primal models which payoffs are in the core. Furnishing the Bondareva-Shapley condition for the core to be non-empty, the dual finds the support to the "set of balanced weights," evaluated at the worth. Whether the core is non-empty or empty is a matter of whether the support to the set of balanced weights evaluated at the worth equals the value or is greater, respectively.

Econometrics

Moment Space

Considerable statistical analysis and probabilistic economic modelling deals just with the first and second moments of the random variables. The full probability distribution of the variables plays no role. Moment space is a Euclidean space of random variables that embodies the first and second moments. A vector in the space represents a random variable. The inner product of random variables is the expected value of their product (the second noncentral cross-moment). The mean is a linear function of the random variable. To work in moment space enables the application of convex optimization and duality theory.

Moment space is convenient for modelling random variables in a problem depending only on the first and second moment. We apply moment space to econometrics and finance. Least-squares linear regression involves only the first and second moments of the random variables, so moment space is applicable. Maximum-likelihood estimation of a normal distribution involves only the first and second moments. In finance, using moment space to study mean/variance choice under uncertainty is effective. To embed the second moment in the inner product simplifies the derivation of relationships and the statement of results.

The cross-moment transformation between subspaces is the unique linear transformation $A : M \to N$ such that the inner product of $\boldsymbol{x} \in M$ and $\boldsymbol{y} \in N$ is $\langle \boldsymbol{y}, A\boldsymbol{x} \rangle$. A second interpretation is valuable. Let M and N denote the idempotents corresponding to these subspaces. Alternatively see the cross-moment transformation as the square linear transformation NM, such that the domain and range are restricted to M and N.

Least-Squares Linear Regression

Let $\boldsymbol{y}$ denote the dependent random variable, and M is the subspace of independent variables. The least-squares linear regression of $\boldsymbol{y}$ on M is the $\boldsymbol{x} \in M$ such that the quadratic $\langle \boldsymbol{y} - \boldsymbol{x}, \boldsymbol{y} - \boldsymbol{x} \rangle$ is minimized: The following three conditions are necessary and sufficient for solutions to the primal and the dual:

$$
\begin{aligned}
\boldsymbol{y}^* &= \boldsymbol{y} - \boldsymbol{x} \\
\boldsymbol{x} &\in M \\
\boldsymbol{y}^* &\in M^{\perp}.
\end{aligned}
$$

These conditions are symmetric between the fitted value $\boldsymbol{x}$ and the residual $\boldsymbol{y}^*$.

Taking the dual of a transformation of the Fenchel dual obtains another fundamental result: the fitted values maximize the correlation with the dependent variable. We interpret least-squares linear regression as a fractional program.

Multivariate Least-Squares Linear Regression

We analyze multivariate least-squares linear regression in moment space. For the multivariate dependent variable Y, choose fitted values X to minimize $\langle Y - X, Y - X \rangle$. One expresses the minimizing choice via the cross-moment transformation between the subspaces of the dependent variables and the independent variables. The multivariate least-squares linear regression is the best linear regression, in that any other choice X makes larger the matrix $(Y - X)^{\top}(Y - X)$ of the second moments of the residual $Y - X$.

We extend the analysis to reduced-rank least-squares linear regression, in which the fitted values X are constrained to having reduced rank l. The primal embodies the reduced-rank requirement. We present primal and dual solutions such that the primal and dual values are the same. A solution X is the truncation of the singular value decomposition of $A^{\top} Y$ at l terms.

Instrumental Variables

Working in moment space, we develop instrumental-variables estimation via duality, in a coordinate-free way. In a structural equation, the explanatory variable causes the dependent variable. The unexplained component is the dependent variable minus the explanatory variable. An estimate of the unknown explanatory variable is a choice from the explanatory variable subspace. If one believes that the explanatory variable and the unexplained component are correlated, then estimation poses a problem, because least-squares linear regression of the dependent variable on the explanatory variable subspace will incorrectly estimate the explanatory variable.

To overcome this problem, one approach is instrumental-variables estimation. One assumes that, in the population moment space, the subspace of instrumental random variables is orthogonal to the unexplained component. In the sample moment space, the estimation chooses the explanatory variable to achieve this orthogonality as well as possible.

We first set up the instrumental-variables framework. The solution to a linear equation determines necessary and sufficient conditions for identification. The analysis revolves around the cross-moment transformation linking the explanatory- and instrumental-variable subspaces. For an identified equation, the explanatory variable is the dependent variable transformed by the generalized inverse of the cross-moment transformation.

One also obtains the analog estimate as the solution to minimization problems. A quadratic minimization implements the generalized method of moments approach to estimating the explanatory variable. Solving this problem and its dual finds that the primal solution is the analog estimate. The two-stage least-squares estimate is also the analog estimate.

Likelihood Statistical Inference

Maximum-likelihood estimation permits elegant testing of the null hypothesis that the unknown parameters satisfy a linear constraint. The context is asymptotic statistical inference when the sample size is large. The Wald test checks directly how closely the maximum-likelihood estimate satisfies the constraint. The Wilks likelihood-ratio test maximizes the likelihood subject to the constraint and checks how much it reduces the likelihood. Rao's score test looks at the derivative of the log-likelihood at the constrained maximum-likelihood estimate and checks whether this derivative is close to zero. The Aitchison and Silvey Lagrange-multiplier test checks whether the Lagrange multiplier in the constrained maximum-likelihood is close to zero. Asymptotically these four tests are equivalent.

We model the estimation in Euclidean space, in which the inner product embodies the negative of the matrix of second derivatives, in a coordinate-free formulation simplifying the mathematics. Because the sample size is large, approximate the log-likelihood by a simple quadratic in the maximum-likelihood estimate minus the vector of parameters. Each test reduces to a least-squares linear regression.

Conjugate Maximum Likelihood

In some important cases, maximum-likelihood estimation amounts to evaluating a conjugate, in which the data vector of sufficient statistics is the dual perturbation for the vector of parameters. In a large sample, the variance of the parameter estimate is the derivative of the estimate with respect to the sufficient statistics. If the parameter estimate is sensitive to the data, then the variance of the estimate is large.

Maximum-Likelihood Estimation of Linear Regression

We perform maximum-likelihood estimation of linear regression by perturbation duality. Transforming the parameters and deploying the quadratic function obtains a maximization concave in the parameters. To carry out a comparative-statics calculation of parameter variances and covariances, adding a choice variable obtains a conjugate maximum-likelihood problem, in which the inner product multiplies the second moments of the data by the parameters. We find the large-sample variances of the parameter estimates via the comparative-statics approach, by calculating how a change in the data affects the estimates.

Finance

Stochastic Discount Factor

A stochastic discount factor generalizes the concept of discounted present value. The component of a stochastic discount factor for a state is the value per unit of probability of a one-dollar payoff in that state. The inner product of a stochastic discount factor and the payoff vector for an asset is the asset price. In a primal/dual analysis in finance theory, commonly the choice variable in the dual defines a stochastic discount factor.

The law of one price states that a portfolio having payoff zero must have cost zero. Invoking the Fredholm alternative establishes a basic result: there exists a stochastic discount factor for asset markets if and only if the law of one price holds. The "pricing payoff" is the unique payoff that is a stochastic discount factor.

The fundamental theorem of finance asserts that there is no opportunity for arbitrage if and only if there exists a positive stochastic discount factor. There exists a cost/mean/variance efficient portfolio if and only if there is no risk-free portfolio providing an arbitrage opportunity.

"Cost and mean positivity" is the condition that the inner product of the pricing payoff and the mean vector in the payoff space is positive. The cost of the mean vector is positive, and the expected value of the pricing payoff is positive. See positivity as a requirement for asset-market equilibrium, a condition distinct from the requirement that there should be no opportunity for arbitrage.

Expected Utility

We analyze portfolio choice by an investor who maximizes the expected utility of the payoff. The conjugate of the expected utility of a payoff is the expected value of the conjugate of utility evaluated at the marginal utility. The analysis parallels and generalizes the standard theory of a consumer maximizing utility subject to a budget constraint, except that the budget constraint involves the set of stochastic discount factors, not just a single price vector. The dual chooses a particular stochastic discount factor.

Price Bounds

The fundamental theorem of finance asserts that there is no arbitrage if and only if there exists a positive stochastic discount factor. As a corollary, one obtains upper and lower price bounds for an asset having an arbitrary payoff. Given asset payoffs and prices not allowing any arbitrage, there exists a non-empty set of positive stochastic discount factors. For an extra asset having an arbitrary payoff, to rule out arbitrage its price must be the value of the payoff discounted by one of the positive stochastic discount factors. The supremum of these prices sets an upper bound for the price, and the infimum of these prices sets a lower bound.

Interpreting the bounds, the "upper-hedging" and "lower-hedging" duals furnish alternative expressions for the bounds. The upper bound is the minimum price among all larger payoffs in the payoff space, whereas the lower bound is the maximum price among smaller payoffs.

Sharpe Ratio

The Sharpe ratio is the maximum value of the ratio of mean to standard deviation in a moment space of random variables. The Sharpe ratio for excess returns plays an important role in optimum portfolio choice and in the capital-asset pricing model. The Sharpe ratio is attained at the mean vector. The inverse-Sharpe-ratio theorem is that the Sharpe ratio in a subspace is the inverse of the ratio in the complementary subspace.

The Hansen-Jagannathan bound in a financial market is an application of this theorem. Under cost and mean positivity, the maximum ratio of mean to standard deviation among all stochastic discount factors is the inverse of the Sharpe ratio for excess returns.

Efficient Portfolios Risk-Free Asset

We solve an efficiency problem to determine the cost/mean/variance efficient portfolios for asset markets having a risk-free asset. The portfolio-choice separation theorem describes the efficient portfolios: invest a nonnegative amount in a particular risky portfolio and an arbitrary positive or negative amount in the risk-free asset. The calculations for the Sharpe ratio—the maximum ratio of mean to standard deviation—confirm Hansen-Jagannathan duality.

Efficient Portfolios No Risk-Free Asset

Working with an asset space in which there is no risk-free asset, we solve for the portfolios that are cost/mean/variance efficient. Setting the problem in Lorentz space obtains a (nonpolyhedral) linear program. The efficient portfolios consist of the cone generated by the mean portfolio and the negative of the cost portfolio, a result correcting traditional thinking. Cost and mean positivity is necessary and sufficient for the existence of efficient portfolios having a positive cost.

Capital-Asset Pricing

The analysis here focuses on the allocation of risk and the implication of risk for asset pricing. Working with a simple mean/variance utility, we obtain Sharpe's capital-asset pricing model. Via primal and dual efficiency problems, we obtain first and second theorems of efficient risk sharing, a variant of the first and second theorems of welfare economics. Asset-market equilibrium shares risk efficiently. Conversely, an efficient allocation of risk can be obtained as a market equilibrium.

Arbitrage Pricing Theory

Ross puts forward the *arbitrage pricing theory*. The idea is that a factor structure of asset returns leads naturally to a model of risk premia, because otherwise there would exist an opportunity for arbitrage. After summarizing the argument of Ross, we develop the idea via a moment space of portfolios, which captures the factor structure by introducing a basis. The moment space decomposes into orthogonal complements: a systematic-risk subspace and a no-systematic-risk subspace. Calculating the Sharpe ratio in each subspace obtains a formulation of the arbitrage pricing theory.

Dynamics

Stability

We investigate the stability of constant coefficients, linear difference and differential equations via the von Neumann-Morgenstern theorem of the alternative, applied in the Hermitian Euclidean Jordan algebra $\mathbf{H}^{\mathbf{X}}$ of self-adjoint linear transformations. For a square linear transformation A on $\mathbf{H}^{\mathbf{X}}$, exactly one of the following has a solution:

Stable Alternative $AX \ll \mathbf{0}, X \gg \mathbf{0}$;
Not Stable Alternative $A^{\top}X^{*} \geq \mathbf{0}, X^{*} > \mathbf{0}$.

We apply this theorem to study stability. For the difference equation, one choice of A applies. For the differential equation, a different choice applies. However the same two alternatives apply to either. In each case, the stable alternative states that there exists a Lyapunov transformation, so the dynamic system is stable. We interpret the not stable alternative in terms of eigenvectors and eigenvalues.

Balanced Growth

In the von Neumann balanced growth model, an input of goods this period produces an output of goods next period. The inputs and outputs are the same goods, and there is no consumption of goods. Each period all inputs and outputs grow at the same rate. There exists a maximum ratc of balanced growth, and furthermore this balanced growth is attainable as a production equilibrium.

The total input of goods is $A\boldsymbol{x}$, and the total output is $B\boldsymbol{x}$, for activity levels $\boldsymbol{x}$. We analyze the saddle-point problem

$$\min_{y^*}\left(\max_{x}\frac{\langle y^*, B\boldsymbol{x}\rangle}{\langle y^*, A\boldsymbol{x}\rangle}\right),$$

the output value divided by the input value. We establish that a saddle point $\langle \boldsymbol{x}, \boldsymbol{y}^*\rangle$ exists. Activity $\boldsymbol{x}$ attains the maximum feasible balanced growth, and the saddle value is the growth factor. Activity $\boldsymbol{x}$ is a production equilibrium for price $\boldsymbol{y}^*$, in which the interest factor equals the growth factor. We analyze the saddle-point problem as two standard fractional programs. Together the two zero-conjugate conditions amount to finding a Nash equilibrium for a two-person zero-sum game.

Contents

Part X Consumption

Part XI Welfare Economics

Part XII Core

Part I
Optimization by Duality

Chapter 1
Discovery and Proof Via Duality

Abstract Convex optimization via duality furnishes a unified development of economic theory, and this book applies this mathematical technique to a wide variety of problems in economic and econometric theory. Given a convex function f, an optimization problem defines its conjugate f^*, another convex function:

$$f^*(\boldsymbol{x}^*) := \sup_{\boldsymbol{x}} [\langle \boldsymbol{x}^*, \boldsymbol{x}\rangle - f(\boldsymbol{x})].$$

Under general conditions, “conjugate duality” holds: the conjugate of the conjugate is the original function. The conjugate is the building block of convex optimization by Fenchel and Rockafellar. Given a convex optimization problem (the “primal”), one can associate with it a “dual” optimization problem. Perturbing the primal makes the problem slightly different, and this perturbation determines the dual. One calculates the dual from the primal via the conjugate. If the primal is a minimization, then the dual is a maximization. Under general conditions, there is “no duality gap”: the minimum of the primal is the maximum of the dual. “Proof by optimization” via duality furnishes a unified development of economic theory. To discover and prove an economic theorem, set up a primal that embodies some economic condition. Calculate a dual and interpret it. What is the economic meaning of a solution to the dual? Our standard method is to solve the primal and the dual simultaneously—an algebraic solution—by setting the primal equal to the dual. To figure out the economic meaning of this simultaneous solution both discovers and proves an economic theorem.

Convex optimization via duality furnishes a unified development of economic theory, and this book applies this mathematical technique to a wide variety of problems in economic and econometric theory. The building block of convex optimization is the “conjugate,” put forward and developed by the German mathematician Fenchel (1905–1988) Fenchel (1949, 1953). Given a convex function, an optimization problem defines its conjugate, another convex function. Under general conditions, “conjugate duality” holds: the conjugate of the conjugate is the original function. Thus a convex function and its conjugate are equivalent, in that either determines

B. C. Dieffenbach, *Conjugate Duality in Economic Analysis*, Contributions to Economics,
https://doi.org/10.1007/978-3-032-21396-9_1

the other. The conjugate is the basis for the development of convex analysis and convex optimization Rockafellar (1970, 1974).

Given a convex optimization problem (the "primal"), one can associate with it a "dual" optimization problem. Perturbing the primal makes the problem slightly different, and this perturbation determines the dual. One calculates the dual from the primal via the conjugate. See the two problems as equivalent but having different points-of-view. If the primal is a minimization, then the dual is a maximization. Under general conditions, there is "no duality gap": the minimum of the primal is the maximum of the dual. In many economic applications, the primal and the dual each have an interesting economic interpretation.

"Proof by optimization" via duality furnishes a unified development of economic theory. To discover and prove an economic theorem, set up a primal that embodies some economic condition. Calculate a dual and interpret it. What is the economic meaning of a solution to the dual? Embodying the same condition as the primal, the dual sees it from a different direction, in a different light. Our standard method is to solve the primal and the dual simultaneously—an algebraic solution—by setting the primal equal to the dual. To figure out the economic meaning of this simultaneous solution both discovers and proves an economic theorem.

Contrast this method with traditional mathematical economics. By a "stroke of genius," an economist may conjecture that two seemingly different conditions are in fact equivalent. A clever argument proves the equivalence. In contrast, duality can discover the relationship directly by algebraic calculation, without requiring any stroke of genius or clever argument.

This book contains many economic applications of perturbation duality. Of course it is standard to analyze certain economic theories as optimizations. The basic examples are the theory of the consumer, analyzed by utility maximization, and the theory of the firm, analyzed by profit maximization.

However many aspects of economic theory are traditionally not analyzed as optimizations. Market equilibrium and its properties are not developed via optimization. Economic efficiency—the idea of a trade-off between objectives—pervades economic thinking. Intuitively, efficiency is a sort of optimization, as no superior possibility exists. Yet typically the analysis of efficiency is not carried out in an explicit optimization context. An important application of duality is to analyze efficiency explicitly via primal and dual optimizations. An efficiency problem, the primal chooses quantities; interpret the dual as choosing prices that attain market equilibrium.

The primal and dual choice variables need not be quantity and price. In two-person game theory, the strategy of one player is the choice variable of the primal, and the strategy of the other player is the choice variable of the dual. In finance, the choice variables might be the portfolio payoff for the primal and the stochastic discount factor for the dual. In statistical least-squares estimation, the choice variables are the fitted values for the primal and the forecast errors for the dual.

The mathematical theory of the conjugate has developed over a long historical period. It derives from the Legendre transformation (Adrien-Marie Legendre 1752–1833), which focuses on the subgradient invertibility presented below.

Physicists apply invertibility in classical mechanics and thermodynamics to obtain alternate presentations. Extending the theory of the Legendre transformation, Fenchel defines the conjugate and proves the conjugate-duality theorem Fenchel (1949). Fenchel establishes how the conjugate is fundamental to convex optimization Fenchel (1953). The foundation for the development of convex analysis by Rockafellar is the conjugate (Rockafellar 1970). There and also in Rockafellar (1974), he defines the dual via the conjugate of the value function; the optimum values of the dual choice variables are Lagrange multipliers for the perturbations.

Strang explains how many theories in physics pair a function and its conjugate (Strang 1986, p. 732). For kinetic energy depending on velocity, the conjugate is kinetic energy depending on momentum. For strain energy depending on strain, the conjugate is complementary energy depending on stress. For Gibbs free energy depending on pressure, the conjugate is Helmholtz free energy depending on volume. For electrical networks, the potential difference and current are paired.

Even though the conjugate is the building block of optimization, most economists have never heard of it, and it has scarcely been used in economic research. An application in economics has been production theory, to show that different representations of technology are equivalent: if a convex function describes technology, then its conjugate is an equivalent description. An incisive use of the conjugate in economics is the encyclopedia article by Newman, who applies it to prove economic results (Newman 1987).

Studying a wide range of economic applications, the classic work of Dorfman, Samuelson, and Solow explains the economic insight afforded by primal and dual optimizations (Dorfman et al. 1958). In the terminology of this book, their problems are polyhedral (traditional linear programming). Their methodology does not employ the conjugate.

Luenberger proves important economic theorems in a creative way, by posing and analyzing innovative primal and dual optimizations (Luenberger 1995). We derive his results by conjugate and perturbation duality.

1.1 Euclidean Space

We set our analysis in Euclidean space, a finite-dimensional real vector space having an inner product $\langle \boldsymbol{x}^*, \boldsymbol{x} \rangle$, a real-valued function of two vectors. The inner product is bilinear—linear in one vector, while holding the other fixed—and symmetric between them. It is positive: $\langle \boldsymbol{x}, \boldsymbol{x} \rangle > 0$ unless $\boldsymbol{x} = \mathbf{0}$. Two vectors are orthogonal if their inner product is zero. One defines closed and open sets via the inner product. Whereas economists and econometricians typically work with a basis and basis coordinates for a vector, we see a vector as coordinate-free, in line with the modern mathematical point-of-view.

In applications, to embody important aspects of the problem into the inner product simplifies both derivation and understanding. For Euclidean geometry in three-dimensions, the inner product captures the concept of distance and orientation. The norm $|\boldsymbol{x}| = \sqrt{\langle \boldsymbol{x}, \boldsymbol{x} \rangle}$ is the length of a vector. Two vectors are perpendicular if their inner product is zero.

In microeconomics, Debreu sees the multiplication of price by quantity in the context of the dual space (Debreu 1959). A quantity vector $\boldsymbol{x}$ lies in the Euclidean space, and a price vector $\boldsymbol{x}^*$ lies in the dual space. The inner product $\langle \boldsymbol{x}^*, \boldsymbol{x} \rangle$ is the dollar value of the price vector multiplied by the quantity vector.

In econometrics, one can analyze a Euclidean space of random variables, in which the inner product of two random variables is their covariance. The norm is the standard deviation. Two random variables are uncorrelated if their inner product is zero.

Given Euclidean spaces $\mathbf{X}$ and $\mathbf{Y}$, their direct sum $\mathbf{X} \oplus \mathbf{Y}$ is another Euclidean space, which one can model via an ordered pair $(\boldsymbol{x}, \boldsymbol{y})$ of a vector from each space. The sum of two pairs is the pair of the sum:

$$(\boldsymbol{x}^*, \boldsymbol{y}^*) + (\boldsymbol{x}, \boldsymbol{y}) := (\boldsymbol{x}^* + \boldsymbol{x}, \boldsymbol{y}^* + \boldsymbol{y}).$$

The inner product of two pairs is the sum of the inner products:

$$\langle (\boldsymbol{x}^*, \boldsymbol{y}^*), (\boldsymbol{x}, \boldsymbol{y}) \rangle := \langle \boldsymbol{x}^*, \boldsymbol{x} \rangle + \langle \boldsymbol{y}^*, \boldsymbol{y} \rangle.$$

The simplest Euclidean space is the real numbers $\mathbf{R}$, in which the inner product is the multiplication of one number by the other. The Euclidean space $\mathbf{R}^n = \mathbf{R} \oplus \cdots \oplus \mathbf{R}$ is direct sum of $\mathbf{R}$ taken n times.

In a figure in this book, a point denotes a vector. The position of the point relative to the origin shows the length and direction of the vector. In contrast, a line segment having an arrowhead denotes a vector having the given length and direction.

1.2 Convex Function

We analyze convex functions. An extended real-valued function f on a Euclidean space $\mathbf{X}$ is one whose values are either real or $\pm\infty$. Its epigraph epi f is all vectors $(\boldsymbol{x}, y)$ vertically above the function. The indicator function $\delta_C(\boldsymbol{x})$ is 0 on set C and ∞ elsewhere, so epi $\delta_C = C \times \mathbf{R}_+$. Hence

$$f(\boldsymbol{x}) = \inf_y \left[y + \delta_{\operatorname{epi} f}(\boldsymbol{x}, y) \right] = \inf_y \left[y + \delta_{f(\boldsymbol{x}) \le y} \right].$$

In a convex set, any vector on the line segment connecting any two vectors must also lie in the set.

Definition 1 (Convex Function) A function is convex if its epigraph is a convex set.

The framework of this book is convex optimization, which refers to the minimization of a convex function Rockafellar (1970). (We do not maximize a convex function.) The problem

$$\inf_{\boldsymbol{x}} f(\boldsymbol{x}) \tag{1.1}$$

seeks to choose $\boldsymbol{x} \in \mathbf{X}$ to minimize f. We write inf rather than min, to allow for the possibility that the infimum is not attained, but is only approached in the limit by a sequence of choices $\boldsymbol{x}$. In general a minimum need not be unique. To model constraints on the choice variable, set $f = \infty$ wherever the constraints are violated. Whereas a century ago mathematical economists commonly assumed differentiability, to work with a convex function seems more appropriate in many economic models and applications. A convex function can have kinks and not be differentiable there. In particular, in constrained optimization, the objective function is not differentiable where constraints are binding.

We use the following terminology.

Definition 2 (Effective Domain) For a convex function, its effective domain dom f is all $\boldsymbol{x}$ such that $f(\boldsymbol{x}) < \infty$.

Definition 3 (Closure) A function is closed if its epigraph is closed. The epigraph of the closure is the closure of the epigraph:

$$\operatorname{epi}(\operatorname{cl} f) := \operatorname{cl}(\operatorname{epi} f).$$

Definition 4 (Proper) A convex function is proper if it is finite somewhere and nowhere $-\infty$.

1.3 The Conjugate

An optimization problem defines the conjugate f^* of a function f.

Definition 5 (Conjugate) The conjugate

$$f^*(\boldsymbol{x}^*) := \sup_{\boldsymbol{x}} [\langle \boldsymbol{x}^*, \boldsymbol{x} \rangle - f(\boldsymbol{x})]. \tag{1.2}$$

(A function is concave if its negative is convex. "Convex optimization" can also refer to the maximization of a concave function. The conjugate of a concave function f is $f_*(\boldsymbol{x}^*) := \inf_{\boldsymbol{x}} [\langle \boldsymbol{x}^*, \boldsymbol{x} \rangle - f(\boldsymbol{x})]$.)

An economic example is the cost function $f(x)$, the minimum cost of producing x units of output. Revenue x^*x is price x^* times quantity. The conjugate $f^*(x)$ is the profit function, the maximum profit as a function of the price.

By the definition (1.2) of the conjugate, the "Fenchel inequality"

$$f(\boldsymbol{x}) + f^*(\boldsymbol{x}^*) - \langle \boldsymbol{x}^*, \boldsymbol{x} \rangle \geq 0$$

holds for all $\boldsymbol{x}$ and $\boldsymbol{x}^*$. The three terms on the left constitute a "Fenchel factor." There is a "Fenchel equality" if the Fenchel factor is zero.

In mathematics, the term "conjugate" means that the conjugate of the conjugate is the original. For example, for a complex number, the complex conjugate of the complex conjugate is the original number. Under general conditions, this relationship also holds for a convex function.

Theorem 1 (Conjugate Duality) *For a proper, closed, convex function,* $f^{**} = f$

Fenchel (1949). In this chapter we prove conjugate duality throughout the effective domain, by a simple, algebraic calculation that directly relates f^{**} and f via an optimization. The chapter "Conjugate Duality" proves conjugate duality throughout the entire space via basic separation and also derives the geometric relationship between the epigraphs of a convex function and its conjugate.

A supporting hyperplane is a fundamental concept in economic theory.

Definition 6 (Hyperplane) For some nonzero $\boldsymbol{x}^*$, a hyperplane is the set of vectors $\boldsymbol{x}$ such that $\langle \boldsymbol{x}^*, \boldsymbol{x} \rangle$ is constant.

Here $\boldsymbol{x}^*$ is "normal" to the hyperplane.

Definition 7 (Half Space) A half space is all $\boldsymbol{x}$ such that $\langle \boldsymbol{x}^*, \boldsymbol{x} \rangle \leq y$, for some $\boldsymbol{x}^* \neq \boldsymbol{0}$ and y. Its boundary is the hyperplane $\langle \boldsymbol{x}^*, \boldsymbol{x} \rangle = y$. The vector $\boldsymbol{x}^*$ is a normal to the hyperplane.

Definition 8 (Support) The support to a set C is the conjugate $\delta_C^*(\boldsymbol{x}^*)$ of its indicator.

The halfspace of vectors $\boldsymbol{x}$ such that $\langle \boldsymbol{x}^*, \boldsymbol{x} \rangle \leq \delta_C^*(\boldsymbol{x}^*)$ contains C. The support is nonnegative if and only if the set contains the origin.

Definition 9 (Supporting Hyperplane) The hyperplane $\langle \boldsymbol{x}^*, \boldsymbol{x} \rangle = \delta_C^*(\boldsymbol{x}^*)$ supports C at $\boldsymbol{y} \in C$ if $\boldsymbol{y}$ is in the hyperplane.

Figure 1.1 graphs a hyperplane supporting a closed, convex set C at $\boldsymbol{y}$. The normal $\boldsymbol{x}^*$ points *away* from the set.

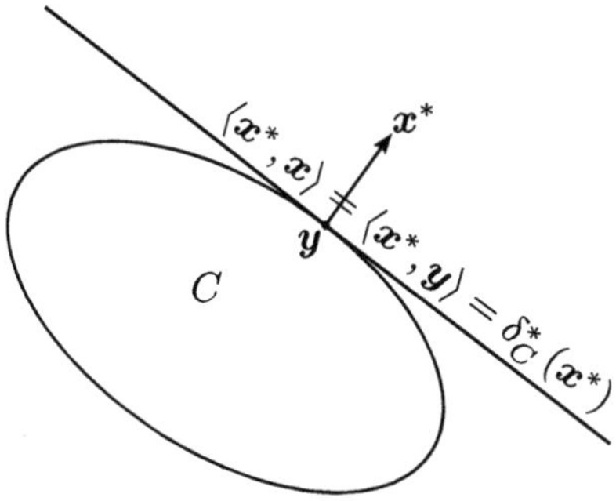

Fig. 1.1 Supporting hyperplane

Expressing the conjugate via the support to the epigraph provides a geometric interpretation of the conjugate:

$$\begin{aligned} f^*(\boldsymbol{x}^*) &= \sup_{\boldsymbol{x}} \left[\langle \boldsymbol{x}^*, \boldsymbol{x}\rangle - f(\boldsymbol{x})\right] \\ &= \sup_{\boldsymbol{x}} \left\{\langle \boldsymbol{x}^*, \boldsymbol{x}\rangle - \inf_{y}\left[y + \delta_{\operatorname{epi} f}(\boldsymbol{x}, y)\right]\right\} \\ &= \sup_{\boldsymbol{x},y} \left[\langle \boldsymbol{x}^*, \boldsymbol{x}\rangle - 1y - \delta_{\operatorname{epi} f}(\boldsymbol{x}, y)\right] \\ &= \delta^*_{\operatorname{epi} f}(\boldsymbol{x}^*, -1). \end{aligned}$$

If the value of conjugate is attained at $\overline{\boldsymbol{x}}$, then $f^*(\boldsymbol{x}^*) = \langle \boldsymbol{x}^*, \overline{\boldsymbol{x}}\rangle - f(\overline{\boldsymbol{x}})$.

In Fig. 1.2, the nonvertical hyperplane

$$\langle \boldsymbol{x}^*, \boldsymbol{x}\rangle - 1y = f^*(\boldsymbol{x}^*) \tag{1.3}$$

supports the epigraph at $(\boldsymbol{x}, y) = [\overline{\boldsymbol{x}}, f(\overline{\boldsymbol{x}})]$, and its vertical intercept is $-f^*(\boldsymbol{x}^*)$.

The conjugate is closed and convex. In (1.2), see $\langle \boldsymbol{x}^*, \boldsymbol{x}\rangle - f(\boldsymbol{x})$ as a function of $\boldsymbol{x}^*$ parametrized by $\boldsymbol{x}$, the linear function $\langle \boldsymbol{x}^*, \boldsymbol{x}\rangle$ minus the scalar $f(\boldsymbol{x})$. Its epigraph is the half space of $(\boldsymbol{x}^*, y^*) \in \mathbf{X}\oplus\mathbf{R}$ having indicator $\delta_{\langle \boldsymbol{x}^*, \boldsymbol{x}\rangle - f(\boldsymbol{x}) \le y^*}$. By the definition (1.2), the epigraph of the conjugate $f^*(\boldsymbol{x}^*)$ is the intersection of these epigraphs, which is a closed and convex set.

The conjugate of a proper function is proper (Rockafellar 1970, Theorem 12.2, p. 104).

To compare $f(\boldsymbol{x})$ and $f^{**}(\boldsymbol{x})$ for a particular $\boldsymbol{x} \in \operatorname{dom} f$, translate $[\boldsymbol{x}, f(\boldsymbol{x})]$ to the origin: define $\vec{f}$ by the translation $\operatorname{epi}\vec{f} := \operatorname{epi} f - \{[\boldsymbol{x}, f(\boldsymbol{x})]\}$. Its value at zero is zero: $\vec{f}(\mathbf{0}) = 0$. The general relation $\delta^*_{C-\{z\}}(\boldsymbol{x}^*) = -\langle \boldsymbol{x}^*, z\rangle + \delta^*_C(\boldsymbol{x}^*)$ implies

$$f^*(\boldsymbol{x}^*) = \delta^*_{\operatorname{epi} f}(\boldsymbol{x}^*, -1) = \langle \boldsymbol{x}^*, \boldsymbol{x}\rangle - f(\boldsymbol{x}) + \delta^*_{\operatorname{epi}\vec{f}}(\boldsymbol{x}^*, -1).$$

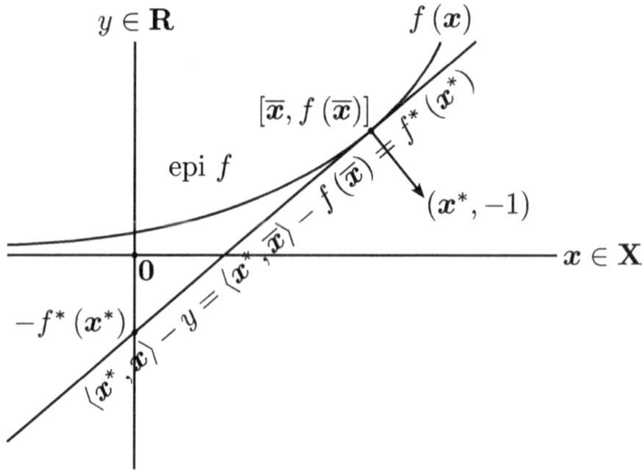

Fig. 1.2 Conjugate as supporting hyperplane

Taking the conjugate again yields

$$\begin{aligned} f^{**}(\boldsymbol{x}) &= \sup_{\boldsymbol{x}^*} [\langle \boldsymbol{x}^*, \boldsymbol{x} \rangle - f^*(\boldsymbol{x}^*)] \\ &= \sup_{\boldsymbol{x}^*} \left\{ \langle \boldsymbol{x}^*, \boldsymbol{x} \rangle - \left[\langle \boldsymbol{x}^*, \boldsymbol{x} \rangle - f(\boldsymbol{x}) + \delta^*_{\mathrm{epi}\, \vec{f}} (\boldsymbol{x}^*, -1) \right] \right\} \\ &= f(\boldsymbol{x}) - \inf_{\boldsymbol{x}^*} \vec{f}^*(\boldsymbol{x}^*). \end{aligned}$$

Since $\mathbf{0} \in \mathrm{epi}\, \vec{f}$, the conjugate $\vec{f}^*(\boldsymbol{x}^*) = \delta^*_{\mathrm{epi}\, \vec{f}}(\boldsymbol{x}^*, -1) \geq 0$, so $f^{**}(\boldsymbol{x}) \leq f(\boldsymbol{x})$.

Figure 1.3 shows that for conjugate duality to hold is a general condition. To simplify the notation, we omit the arrow overscore on the translated function and work with a function such that $f(\mathbf{0}) = 0$. In the figure, conjugate duality holds because

$$\inf_{\boldsymbol{x}^*} f^*(\boldsymbol{x}^*) = 0. \tag{1.4}$$

For the $\boldsymbol{x}^*$ in the figure, $f^*(\boldsymbol{x}^*) = 0$, and this $\boldsymbol{x}^*$ is indeed minimizing. For any other $\boldsymbol{x}^*$, the conjugate is positive, because the hyperplane supporting epi f intersects the vertical axis below the origin.

Conjugate duality fails where the function is not closed. For a proper, convex function, always $f^{**}(\boldsymbol{x}) = \mathrm{cl}\, f(\boldsymbol{x})$. Where the function is not closed, $f^{**}(\boldsymbol{x}) < f(\boldsymbol{x})$. The optimum value in the minimization (1.4) is positive.

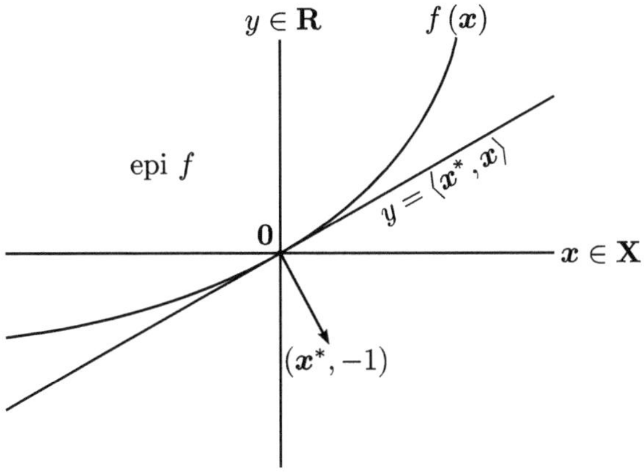

Fig. 1.3 Conjugate duality

1.4 Subdifferential

Marginal conditions are fundamental in economic theory (Samuelson 1947). Economists work with marginal cost, marginal product, marginal utility, etc. Traditional mathematical economics models a marginal value by a derivative. Unfortunately, to assume differentiability seems inappropriate in many economic models and applications. For example, a convex function can have a kink, where the derivative is not defined. In particular, in constrained optimization, the problem is nondifferentiable when constraints are binding. In many economic models and applications, convex analysis seems appropriate, even if differentiability is not appropriate. Yet to abandon the concept of slope and derivative would be a mistake, as marginal conditions are important. In convex analysis, the subdifferential generalizes the derivative.

The nonvertical supporting hyperplanes (1.3) define the subdifferential.

Definition 10 (Subgradient) If $(\boldsymbol{x}^*, -1)$ is a normal of a hyperplane supporting the epigraph, then $\boldsymbol{x}^*$ is a *subgradient*.

Definition 11 (Subdifferential) The *subdifferential* $\partial f(\boldsymbol{x})$ is the set of subgradients at $\boldsymbol{x}$.

Here "sub" means "almost," not "under."

The subdifferential is a closed, convex set. When a function is translated to the origin (Fig. 1.4), the conjugate f^* is nonnegative, and the subdifferential is all vectors such that $f^*(\boldsymbol{x}^*) = 0$. Thus the subdifferential is all vectors such that $f^*(\boldsymbol{x}^*) \leq 0$, a closed, convex set.

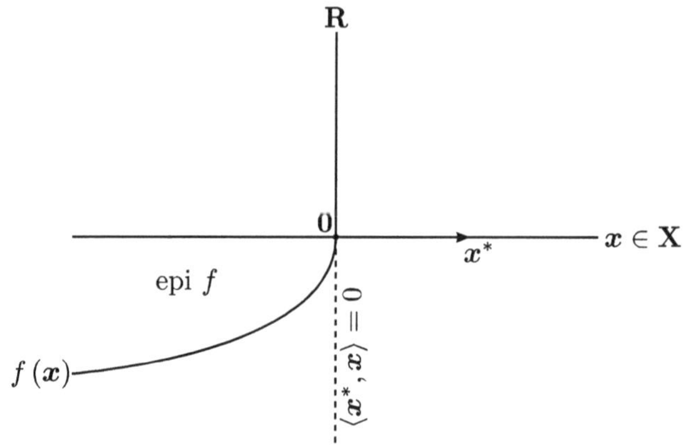

Fig. 1.4 Empty subdifferential

A convex function is differentiable at a vector if and only if there exists a unique subgradient. The subgradient is the derivative. In Fig. 1.2, the subdifferential $\partial f(\overline{x}) = \{x^*\}$. At a kink, there are multiple subgradients.

Possibly the subdifferential is empty. In Fig. 1.4, the hyperplane supporting the epigraph at the origin is vertical. There is no nonvertical supporting hyperplane, so the subdifferential is empty. Conjugate duality holds; in the minimization (1.4) the optimum value zero is not attained but is only approached as an infimum. Outside the effective domain the subdifferential is empty. If the subdifferential is not empty, one says that it is subdifferentiable.

The subgradient invertibility theorem explains the sense in which the subdifferentials of a function and its conjugate are inverses. The initial condition (1.5) is a Fenchel equality, and the two subgradient conditions (1.6) and (1.7) interpret it.

Theorem 2 (Subgradient Invertibility) *The following two conditions are equivalent and imply the conjugate duality $f^{**}(x) = f(x)$:*

$$f(x) + f^*(x^*) - \langle x^*, x \rangle = 0 \tag{1.5}$$

$$x^* \in \partial f(x). \tag{1.6}$$

Either condition implies the inverse relation

$$x \in \partial f^*(x^*). \tag{1.7}$$

Conversely, if conjugate duality holds, then the inverse relation implies the two conditions.

The first two conditions are equivalent, by the definition of the subgradient. In the evaluation (1.2) of the conjugate, the Fenchel equality holds if and only if the supremum is attained for some $\boldsymbol{x}$. Conjugate duality holds, as $f^{**}(\boldsymbol{x}) \geq \langle \boldsymbol{x}^*, \boldsymbol{x} \rangle - f^*(\boldsymbol{x}) = f(\boldsymbol{x})$, and always $f^{**}(\boldsymbol{x}) \leq f(\boldsymbol{x})$. Furthermore, $\boldsymbol{x} \in \partial f^*(\boldsymbol{x}^*)$. Conversely, the inverse relation (1.7) is equivalent to $f^{**}(\boldsymbol{x}) + f^*(\boldsymbol{x}^*) - \langle \boldsymbol{x}^*, \boldsymbol{x} \rangle = 0$. If $f^{**}(\boldsymbol{x}) = f(\boldsymbol{x})$, then (1.5) and (1.6) follow.

For the cost function, the subdifferential $\partial f(x)$ is the marginal cost as a function of output (a closed interval at a kink). For the profit function, the subdifferential $\partial f^*(x)$ is the output supply of the firm as a function of the output price, the set of outputs that maximize the profit.

Convex optimization generalizes the first-order condition of differential calculus. For the minimization (1.1),

$$\inf_x f(\boldsymbol{x}) = -\sup_x f(\boldsymbol{x}) = -f^*(\boldsymbol{0}).$$

Thus $\boldsymbol{x}$ minimizes f if and only if the Fenchel equality

$$f(\boldsymbol{x}) + f^*(\boldsymbol{0}) - \langle \boldsymbol{0}, \boldsymbol{x} \rangle = 0$$

holds.

Theorem 3 (First-Order Condition) *The first-order condition $\boldsymbol{0} \in \partial f(\boldsymbol{x})$ is necessary and sufficient for the minimization (1.1).*

In contrast to traditional optimization, this book makes little use of the first-order condition. Instead the simultaneous solution of a primal/dual pair has supplanted it, and one should see the first-order condition as obsolete.

1.5 Perturbation Duality

Fenchel defines a dual for a general primal and proves that under general conditions there is no duality gap (Fenchel 1953). Extending this work, Rockafellar defines the dual via perturbation of the primal (Rockafellar 1970, 1974).

Problem 1 (Perturbed Primal) The perturbed primal is the minimization

$$V(\boldsymbol{y}) = \inf_x F(\boldsymbol{x}, \boldsymbol{y}). \tag{1.8}$$

Defined on $\mathbf{X} \oplus \mathbf{Y}$, the perturbation function $F(\boldsymbol{x}, \boldsymbol{y})$ is a jointly convex function of the choice variable $\boldsymbol{x}$ and the perturbation variable $\boldsymbol{y}$ (rather than just convex in each variable individually). A change in $\boldsymbol{y}$ perturbs the problem and typically affects the optimum value of the choice variable and the value $V(\boldsymbol{y})$. The base value $\boldsymbol{y}$ refers to the particular value selected.

A calculation proves that the value function is convex: the value of a convex combination is less than or equal to the convex combination of the value.

Theorem 4 (Convex Value Function) *If the perturbation function is proper and convex, then the value function V is convex.*

For a proper perturbation function, for any $0 < \lambda < 1$,

$$\begin{aligned} &V\left[\lambda \boldsymbol{y}_1 + (1-\lambda)\boldsymbol{y}_2\right] \\ &\quad = \inf_{\boldsymbol{x}} F\left[\boldsymbol{x}, \lambda \boldsymbol{y}_1 + (1-\lambda)\boldsymbol{y}_2\right] \\ &\quad = \inf_{\boldsymbol{x}_1,\boldsymbol{x}_2} F\left[\lambda \boldsymbol{x}_1 + (1-\lambda)\boldsymbol{x}_2, \lambda \boldsymbol{y}_1 + (1-\lambda)\boldsymbol{y}_2\right] \\ &\quad \le \inf_{\boldsymbol{x}_1,\boldsymbol{x}_2} \left[\lambda F\left(\boldsymbol{x}_1,\boldsymbol{y}_1\right) + (1-\lambda) F\left(\boldsymbol{x}_2,\boldsymbol{y}_2\right)\right] \\ &\quad = \lambda \inf_{\boldsymbol{x}_1} F\left(\boldsymbol{x}_1,\boldsymbol{y}_1\right) + (1-\lambda) \inf_{\boldsymbol{x}_2} F\left(\boldsymbol{x}_2,\boldsymbol{y}_2\right) \\ &\quad = \lambda V\left(\boldsymbol{y}_1\right) + (1-\lambda) V\left(\boldsymbol{y}_2\right). \end{aligned}$$

The inequality applies because F is jointly convex in $\boldsymbol{x}$ and $\boldsymbol{y}$.

The value function sets the dual. Changing how $\boldsymbol{y}$ perturbs the primal alters the dual.

Definition 12 (Dual) The dual of the perturbed primal is the concave maximization

$$V^{**}(y) = \sup_{y^*} \left[\langle y^*, y\rangle - V^*(y^*)\right].$$

That $V \geq V^{**}$ means that the infimum of the primal is necessarily greater than or equal to the supremum of the dual. If conjugate duality holds, they are equal; one says that there is "no duality gap."

That the solutions to the dual show how a change in the perturbation variable affects the optimum value follows immediately from the subgradient invertibility theorem.

Theorem 5 (Solution Set) *If there is no duality gap, then the solution set to the dual is the subdifferential $\partial V(\boldsymbol{y})$.*

This subdifferential is empty if the supremum in the dual is not attained.

Thinking of how traditional economic theory focuses on marginal conditions, one might expect that our subsequent analysis would focus on the subdifferential, via the latter two relationships (1.6) and (1.7) in the subgradient invertibility theorem. However instead we focus on the first condition (1.5), the Fenchel equality. Our use of the subdifferential is surprisingly minimal.

Let us summarize the standard methodology used in this book. We study a convex optimization in which the perturbation function is the sum of one or more proper, closed, convex functions. An algebraic calculation finds the dual from the primal. We want there to be no duality gap between the primal and the dual, and we want the optimum value to be attained in both. We solve the primal and the dual simultaneously, by setting the objective function of the primal equal to the objective function of the dual.

We express the primal less the dual as a Fenchel factorization, a sum of Fenchel factors. If there is no duality gap, then for any solution to the primal and the dual the sum must be zero. Because each factor is nonnegative, every factor must be zero, necessary and sufficient for primal and dual solutions. To obtain economic and geometric understanding, interpret the Fenchel equality connecting a function and its conjugate by figuring out what it means for pairs $\boldsymbol{x}$ and $\boldsymbol{x}^*$ to satisfy it. The chapter "Fenchel Factors" illustrates this factorization in simple examples.

An advantage of working with a Fenchel equality rather than the equivalent subdifferential relation is that one works with functions, rather than sets. Many economic applications deal with nondifferentiable functions. Since the subdifferential is a set, to figure out and to write down the subdifferential at every point is far more difficult and complicated than to figure out and to state the conjugate.

For a perturbed optimization (1.8), one can define a Lagrangian $L(\boldsymbol{x}, \boldsymbol{y}^*)$, a convex function of the primal choice variable $\boldsymbol{x}$ and a concave function of the dual choice variable $\boldsymbol{y}^*$. The value of the Lagrangian lies between the value of the primal and dual objective functions. If there is no duality gap and the optimum value is attained in both the primal and the dual, then a pair $(\boldsymbol{x}, \boldsymbol{y}^*)$ solves the primal and the dual if and only if it is a saddle point of the Lagrangian. The chapter "Lagrangian" defines and analyzes the Lagrangian.

A goal is to show that duality has a plethora of applications. Because various proof techniques employed in mathematical economics can themselves be proved via duality, there is no necessity to use these alternatives. Instead duality is the method for nearly every proof, and applying the same methodology repeatedly is unifying. Duality has numerous advantages. That its algebraic calculation breaks down a complex problem into simple, distinct parts is a key advantage. The Fenchel equalities represent concisely the necessary and sufficient conditions. Seeing the same problem from different points-of-view—primal, dual, Lagrangian—can be informative and can yield insightful economic intuition. The wide-ranging applications reveal the power of the methodology.

A practical issue is how to perturb of the primal. In most problems, a certain perturbation is obvious. If the problem contains a vector as a parameter, commonly this vector should be the perturbation. In a least-squares linear regression, the perturbation is the dependent variable; and the dual choice variable is the residual. If the primal has little structure, then there may be only one natural way to perturb it. On the other hand, if the primal has much structure, then various perturbations are feasible; and an effective selection is valuable.

If the selected perturbation in the primal fails to produce a closed-form expression for the dual, add further perturbation. Conversely, perhaps the solution for a dual choice variable is automatically a particular value, such as zero or one. One can then simplify the analysis by removing the perturbation for this variable.

Each alternative viewpoint on a convex optimization is potentially valuable. For example, the primal choice variable might be quantity, and one might interpret the dual choice variable as price. To understand the primal and the dual as independent optimizations is important. What is the economic meaning for quantity to solve the primal and for price to solve the dual? What is the meaning of each first-order condition? Perhaps one can interpret the dual as choosing the price to equate demand and supply. Perhaps one can interpret a saddle point of the Lagrangian as a market equilibrium of quantity and price. Even though these different viewpoints are essentially equivalent, together they can provide deeper insight.

In the history of science and mathematics, calculus and linear algebra have been fundamental. These techniques permit the mechanical calculation of complex results, in which the calculation is broken down into many small steps. Discovery and proof via duality shares this advantage, as one calculates the results in small steps, via the conjugate. The primary challenge is to express an economic condition as a convex optimization, and one then solves it mechanically.

A common use of the term "duality" by economists is signify the equivalence of two different representations of a concept. For example, the cost and profit functions are equivalent; either determines the other. An economist might say that the two functions are "dual" to each other. See Diewert (1974, 1982) for examples. This book does not employ this terminology, however, and here the term "duality" refers to conjugate duality or perturbation duality, always in the context of optimization. In many cases conjugate duality or perturbation duality is a simple and effective way to show that two representations are equivalent, such as the equivalence of the cost and profit functions.

Books studying convex optimization from various points of view have been valuable in the writing of this book. In their "concise book" (Borwein and Zhu 2005, p. 137), Borwein and Lewis present a remarkable variety of theorems and applications Borwein and Lewis (2006). Luenberger covers many topics, with great clarity, and goes beyond Euclidean space to infinite dimensions Luenberger (1969). Hiriart-Urruty and Lemaréchal (1993a, b) derive many relationships among conjugates. Boyd and Vandenberghe (2004) and Brinkhuis and Tikhomirov (2005) present convex optimization generally from the Karush-Kuhn-Tucker Lagrangian point-of-view. Rockafellar and Wets (1998) extend optimization theory in new ways.

This book does not prove fundamental results of convex analysis, but it does invoke them to prove in detail the fundamental theorems and techniques of duality. A "proposition" is a fundamental result stated here without proof, but then used in applications. A "theorem" is a something proved here, usually via conjugate or perturbation duality. Many terms are defined via a conjugate. Because one can understand convex analysis well by geometric intuition, this procedure works well.

Writing about the reaction to Hilbert's challenge, Gray contends

> that for at least 30 years pure mathematicians won the battle to define the soul of mathematics. They did so in terms that were strikingly reminiscent of Hilbert's overall philosophy: an active and articulate version of a wide variety of mathematical topics, but one that stops short of solving genuine applied problems. Some of it is applicable, and even intended to be applied, but the implication is that will be done by others.

(Gray 2000, p. 215). The same situation holds for convex optimization and duality. Although mathematicians have developed the abstract theory, they have been disinclined to apply it. This book applies the mathematics to economic theory and econometrics. In doing so the mathematical theory is extended, because specific applications call for new tricks.

In his preface, Rockafellar explains that

> an attempt has been made to keep the exposition on a relatively elementary technical level, and details have been supplied which, in a work aimed only at a mathematical in-group, might merely have been alluded to as "exercises."

(Rockafellar 1970, p. vii). This book seeks to follow his example. The mathematical prerequisite is a good knowledge of linear algebra and an introductory knowledge of convex sets. The use of differential calculus is minimal.

This book contains no application to numerical computation of the solution to a convex optimization.

References

Borwein JM, Lewis AS (2006) Convex analysis and nonlinear optimization: theory and examples, 2nd edn. Springer-Verlag, New York

Borwein JM, Zhu QJ (2005) Techniques of variational analysis, Vol 20. Springer, New York

Boyd S, Vandenberghe L (2004) Convex optimization. Cambridge University Press, Cambridge

Brinkhuis J, Tikhomirov V (2005) Optimization: insights and applications. Princeton University Press, Princeton, NJ

Debreu G (1959) Theory of value: an axiomatic analuysis of economic equilibrium. Yale University Press, New Haven, CT

Diewert WE (1974) Applications of duality theory. In: Intriligator M, Kendrick D (eds) Frontiers of quantitative economics, Vol II. North-Holland, Amsterdam, pp 106–171

Diewert WE (1982) Duality approaches in microeconomic theory. In: Arrow KJ, Intriligator MD (eds) Handbook of mathematical economics, Vol II. North-Holland, Amsterdam, pp 535–599

Dorfman R, Samuelson PA, Solow RM (1958) Linear programming and economic analysis. McGraw-Hill, New York

Fenchel W (1949) On conjugate convex functions. Canad J Math 1:73–77

Fenchel W (1953 Sept) Convex cones, sets, and functions. Department of Mathematics [University Microfilms, Ann Arbor (typeset 2010 by Kim C Border, California Institute of Technology)]

Gray JJ (2000) The Hilbert challenge. Oxford University Press, Oxford

Hiriart-Urruty JB, Lemaréchal C (1993a) Convex analysis and minimization algorithms I: fundamentals, Vol 305. Springer-Verlag, Berlin

Hiriart-Urruty JB, Lemaréchal C (1993b) Convex analysis and minimization algorithms II: advanced theory and bundle methods, Vol 306. Springer-Verlag, Berlin

Luenberger DG (1969) Optimization by vector space methods (Howard RA, ed). Wiley, New York

Luenberger DG (1995) Microeconomic theory. McGraw-Hill, New York

Newman P (1987) Duality. In Eatwell J, Milgate M, Newman P (eds) The New Palgrave, a dictionary of economics, Vol 1. Macmillan Press Limited, London, pp 924–934

Rockafellar RT (1970) Convex analysis. Princeton University Press, Princeton, NJ

Rockafellar RT (1974) Conjugate duality and optimization. Society for Industrial and Applied Mathematics, Philadelphia

Rockafellar RT, Wets RJB (1998) Variational analysis, Vol 317. Springer, Berlin
Samuelson PA (1947) Foundations of economic analysis, Vol 80. Harvard University Press, Cambridge, MA
Strang G (1986) Introduction to applied mathematics. Wellesley-Cambridge Press, Wellesley, MA (QA37.2S87 1986)

Chapter 2
No Duality Gap

Abstract We present the relative and polyhedral conditions for no duality gap. Consider the primal

$$V(y) := \inf_{x} F(x, y)$$

Its dual is

$$V^{**}(y) = \sup_{y^*} [\langle y^*, y\rangle - V^*(y^*)].$$

For the perturbation function F to be proper, closed, and convex is not sufficient to ensure that the value function satisfies conjugate duality: there might be a duality gap, $V(y) > V^{**}(y)$. The relative and polyhedral conditions both imply that there is no duality gap. Furthermore, if the optimum value is finite, then it is attained by a dual solution. The relative condition for no duality gap is that there is some $\boldsymbol{x}$ such that $(\boldsymbol{x}, \boldsymbol{y})$ is in the relative interior of dom F. The polyhedral condition for no duality gap is that F is polyhedral and there is some $\boldsymbol{x}$ such that $(\boldsymbol{x}, \boldsymbol{y})$ is in dom F.

We present the relative and polyhedral conditions for no duality gap. Consider the primal

$$V(y) := \inf_{x} F(x, y).$$

in which $F(\boldsymbol{x}, \boldsymbol{y})$ is defined on $\mathbf{X} \oplus \mathbf{Y}$, the direct sum of two Euclidean spaces. For the perturbation function to be proper, closed, and convex is not sufficient to ensure that the value function satisfies conjugate duality: there might be a duality gap, $V(y) > V^{**}(y)$. The relative and polyhedral conditions both imply that there is no duality gap. Furthermore, if the optimum value is finite, then it is attained by a dual solution $\boldsymbol{y}^*$.

B. C. Dieffenbach, *Conjugate Duality in Economic Analysis*, Contributions to Economics,
https://doi.org/10.1007/978-3-032-21396-9_2

2.1 Relative Condition

A vector is in the interior int C of a set C if a sufficiently small movement in any direction stays in the set. The interior of a set can be empty. For example, a line segment in a space of dimension two has an empty interior.

An affine space is a subspace translated by a vector. For example, a line or hyperplane containing the origin is a subspace. A line or hyperplane not containing the origin is an affine space. A singleton set is an affine space, a space of dimension zero. The affine hull of a set is the smallest affine space containing it, the intersection of all affine spaces containing it. For example, the affine hull of a line segment is the line containing it. The affine hull of an affine space is itself.

Viewing a set as a subset of its affine hull, rather than as a subset of the entire space, one can speak of its properties relative to its affine hull. For example, a set is relatively open if it is an open set relative to its affine hull. A line segment excluding its two endpoints is relatively open.

Definition 1 (Relative Interior) The relative interior rint C of a set C is its interior relative to its affine hull.

Definition 2 (Relative Boundary) The relative boundary rb C of a set C is its boundary relative to its affine hull.

A nonempty convex set has a nonempty relative interior. The relative interior of an affine space is itself. When viewed as defined on the affine hull of its effective domain, a convex function is relatively continuous throughout the relative interior of its effective domain.

The relative condition for a perturbation function pertains to its effective domain.

Definition 3 (Relative Condition) There exists some $\boldsymbol{x}$ such that $(\boldsymbol{x}, \boldsymbol{y}) \in$ rint (dom F) .

Theorem 1 (Relative Condition for No Duality Gap) *If the relative condition holds, then there is no duality gap. If the optimum value is finite, then it is attained in the dual.*

This theorem is a result of Fenchel (1953). Below the chapter "Relative Condition" proves it via best separation. For there to exist some $\boldsymbol{x}$ such that $(\boldsymbol{x}, \boldsymbol{y}) \in$ dom F ensures that the optimum value of the primal is not ∞, but there might be a duality gap.

2.2 Polyhedral Condition

A set is polyhedral if it is the intersection of a finite number of closed half-spaces. A function is polyhedral if its epigraph is a polyhedral set.

Definition 4 (Polyhedral Condition) The perturbation function F is polyhedral, and there exists some $\boldsymbol{x}$ such that $(\boldsymbol{x}, \boldsymbol{y}) \in \text{dom } F$.

If the polyhedral condition holds, then the optimum value of the primal is not ∞.

Theorem 2 (Polyhedral Condition for No Duality Gap) *If the polyhedral condition holds, then there is no duality gap. If the optimum value is finite, then it is attained in the dual.*

The chapter "Polyhedra" invokes the Minkowski-Weyl theorem to prove this result.

See both the relative condition and the polyhedral condition as existence conditions. Each implies not only that there is no duality gap, but furthermore that the dual has a solution, if the optimum value is finite. That there is a solution constitutes an existence theorem: the economic conditions embodied in the dual are satisfied by the solution.

2.3 Reciprocal Dual

Given a dual, one can perturb it to obtain a dual of the dual, a third optimization perhaps different from both the primal and the dual. However an important case is reciprocal duality, in which the dual of the dual reverts to the primal.

Theorem 3 (Reciprocal Dual) *For perturbation* $\boldsymbol{y}$, *the primal*

$$\inf_{\boldsymbol{x}} [\langle \boldsymbol{x}^*, \boldsymbol{x} \rangle + F(\boldsymbol{x}, \boldsymbol{y})] \tag{2.1}$$

has the reciprocal dual

$$\sup_{\boldsymbol{y}^*} [\langle \boldsymbol{y}^*, \boldsymbol{y} \rangle - F^*(-\boldsymbol{x}^*, \boldsymbol{y}^*)]. \tag{2.2}$$

For perturbation $\boldsymbol{x}^*$, *its dual is the primal.*

In each problem, the first term is the inner product of the choice variable and the perturbation variable for its dual.

An important role for the reciprocal dual can be to establish that there is no duality gap. For example, the reciprocal dual may satisfy the relative condition, while the primal does not. The reciprocal dual also interprets the primal, as a primal solution $\boldsymbol{x}$ shows how a change in $\boldsymbol{x}^*$ affects the optimum value.

The chapter "Polyhedra" shows that the conjugate of a polyhedral function is polyhedral, so a polyhedral primal has a polyhedral dual. The only way for a polyhedral primal/dual pair to have a dual gap is when the optimum value of the primal is ∞, and the optimum value of the dual is $-\infty$. Otherwise either the primal or the reciprocal dual satisfies the polyhedral condition for no duality gap.

We calculate the dual of the primal. Defining

$$V(\boldsymbol{y}) := \inf_{\boldsymbol{x}} [\langle \boldsymbol{x}^*, \boldsymbol{x} \rangle + F(\boldsymbol{x}, \boldsymbol{y})],$$

then the conjugate

$$\begin{aligned} V^*(\boldsymbol{y}^*) &= \sup_{\boldsymbol{y}} [\langle \boldsymbol{y}^*, \boldsymbol{y} \rangle - V(\boldsymbol{y})] \\ &= \sup_{\boldsymbol{y}} \left\{ \langle \boldsymbol{y}^*, \boldsymbol{y} \rangle - \inf_{\boldsymbol{x}} [\langle \boldsymbol{x}^*, \boldsymbol{x} \rangle + F(\boldsymbol{x}, \boldsymbol{y})] \right\} \\ &= \sup_{\boldsymbol{x}, \boldsymbol{y}} [\langle -\boldsymbol{x}^*, \boldsymbol{x} \rangle + \langle \boldsymbol{y}^*, \boldsymbol{y} \rangle - F(\boldsymbol{x}, \boldsymbol{y})] \\ &= F^*(-\boldsymbol{x}^*, \boldsymbol{y}^*). \end{aligned}$$

The dual is

$$\sup_{\boldsymbol{y}^*} [\langle \boldsymbol{y}^*, \boldsymbol{y} \rangle - V^*(\boldsymbol{y}^*)] = \sup_{\boldsymbol{y}^*} [\langle \boldsymbol{y}^*, \boldsymbol{y} \rangle - F^*(-\boldsymbol{x}^*, \boldsymbol{y}^*)].$$

The dual of the reciprocal dual is the primal. The concave value function for the dual is

$$V(\boldsymbol{x}^*) := \sup_{\boldsymbol{y}^*} [\langle \boldsymbol{y}^*, \boldsymbol{y} \rangle - F^*(-\boldsymbol{x}^*, \boldsymbol{y}^*)].$$

Taking $\boldsymbol{x}^*$ as the perturbation, calculate its dual. The concave conjugate

$$\begin{aligned} V_*(\boldsymbol{x}) &= \inf_{\boldsymbol{x}^*} [\langle \boldsymbol{x}^*, \boldsymbol{x} \rangle - V(\boldsymbol{x}^*)] \\ &= \inf_{\boldsymbol{x}^*} \left(\langle \boldsymbol{x}^*, \boldsymbol{x} \rangle - \left\{ \sup_{\boldsymbol{y}^*} [\langle \boldsymbol{y}^*, \boldsymbol{y} \rangle - F^*(-\boldsymbol{x}^*, \boldsymbol{y}^*)] \right\} \right) \\ &= -\sup_{\boldsymbol{x}^* \boldsymbol{y}^*} [-\langle \boldsymbol{x}^*, \boldsymbol{x} \rangle + \langle \boldsymbol{y}^*, \boldsymbol{y} \rangle - F^*(-\boldsymbol{x}^*, \boldsymbol{y}^*)] \\ &= -\sup_{\boldsymbol{x}^* \boldsymbol{y}^*} [\langle \boldsymbol{x}^*, \boldsymbol{x} \rangle + \langle \boldsymbol{y}^*, \boldsymbol{y} \rangle - F^*(\boldsymbol{x}^*, \boldsymbol{y}^*)] \\ &= -F(\boldsymbol{x}, \boldsymbol{y}). \end{aligned}$$

The dual of the dual is

$$\inf_{\boldsymbol{x}} [\langle \boldsymbol{x}^*, \boldsymbol{x} \rangle - V_*(\boldsymbol{x})] = \inf_{\boldsymbol{x}} [\langle \boldsymbol{x}^*, \boldsymbol{x} \rangle + F(\boldsymbol{x}, \boldsymbol{y})],$$

which is back to the primal (2.2).

If there is no duality gap, let $V(\boldsymbol{x}^*, \boldsymbol{y})$ denote the value function common to the primal and the reciprocal dual. Here V is a "saddle-value function," concave in the dual perturbation $\boldsymbol{x}^*$ and convex in the primal perturbation $\boldsymbol{y}$. Given $\boldsymbol{x}^*$, the subdifferential of V with respect to $\boldsymbol{y}$ is the solution set to the reciprocal dual. Given $\boldsymbol{y}$, the subdifferential of V with respect to $\boldsymbol{x}^*$ is the solution set of the primal.

Subtracting the dual from the primal obtains the Fenchel inequality

$$F(\boldsymbol{x},\boldsymbol{y}) + F^*(-\boldsymbol{x}^*,\boldsymbol{y}^*) - \langle -\boldsymbol{x}^*,\boldsymbol{x}\rangle - \langle \boldsymbol{y}^*,\boldsymbol{y}\rangle \geq 0 \tag{2.3}$$

If there is no duality gap, then for the Fenchel factor to be zero is necessary and sufficient for a simultaneous solution $\boldsymbol{x}$ and $\boldsymbol{y}^*$ to the primal and the dual. A simple way to derive a primal/dual pair is to reverse this sequence. Starting with the Fenchel inequality (2.3), rearrange it as

$$\langle \boldsymbol{x}^*,\boldsymbol{x}\rangle + F(\boldsymbol{x},\boldsymbol{y}) \geq \langle \boldsymbol{y}^*,\boldsymbol{y}\rangle - F^*(-\boldsymbol{x}^*,\boldsymbol{y}^*).$$

The objective function for the primal is on the left, and the objective function for the reciprocal dual is on the right. This reversal technique is employed throughout this book, because it is so simple.

For the Fenchel inequality (2.3), variable interchange admits a second primal/dual pair. Rearrange the Fenchel inequality (2.3) as

$$\langle \boldsymbol{x}^*,\boldsymbol{x}\rangle + F^*(-\boldsymbol{x}^*,\boldsymbol{y}^*) \geq \langle \boldsymbol{y}^*,\boldsymbol{y}\rangle - F(\boldsymbol{x},\boldsymbol{y}).$$

Interchanging the choice and perturbation variables, define the primal

$$\inf_{\boldsymbol{x}^*}[\langle \boldsymbol{x}^*,\boldsymbol{x}\rangle + F^*(-\boldsymbol{x}^*,\boldsymbol{y}^*)]$$

and its reciprocal dual

$$\sup_{\boldsymbol{y}}[\langle \boldsymbol{y}^*,\boldsymbol{y}\rangle - F(\boldsymbol{x},\boldsymbol{y})].$$

In applications, occasionally—but not commonly—this variable interchange has an interesting economic interpretation.

Reference

Fenchel W (1953 Sept) Convex cones, sets, and functions. Department of Mathematics. (University Microfilms, Ann Arbor (typeset 2010 by Kim C. Border, California Institute of Technology))

Chapter 3
Fenchel Factors

Abstract The methodology of this book is to solve the primal and the dual simultaneously. The primal less the dual is nonnegative, and one can factor this difference into a sum of Fenchel factors. Each is greater than or equal to zero, so when there is no duality gap, each factor must be zero. Together these conditions are necessary and sufficient for a simultaneous solution. To reverse this procedure furnishes an effective and simple way to calculate the dual. Start with Fenchel inequalities. Add them together and rearrange to obtain an inequality with the primal objective function on the left and the dual objective function on the right. This chapter presents two examples of this reversal methodology. The reversed sequence conforms with how economists treat this topic. Each firm maximizes profit by producing where the output price equals marginal cost. Starting our analysis with the Fenchel factors means starting with profit maximization by firms. Deriving the primal and the dual from the Fenchel factors discovers how profit maximization implies industry-cost minimization. In applications, reversal is a simple and direct way to derive the dual. Valuable for both derivation and exposition, it facilitates the analysis of complicated primal/dual pairs. The reversal methodology is the technique mostly used throughout this book.

3.1 Cost and Profit

In the first example, the primal is output-cost minimization, for an industry of firms producing the same output. This simple example illustrates how a primal/dual analysis can discover a fundamental economic result. Interpret the dual choice variable as the output price. Solving the primal and the dual simultaneously expresses industry supply as a function of the output price. Each firm maximizes its profit by choosing output to make its marginal cost equal to the output price. Decentralization in the market is cost-minimizing: profit maximization implies that total industry cost is minimized.

Let $f_i(x_i)$ denote the cost function for firm i, its minimum cost of producing x_i units of output. The cost function is proper, closed, and convex. In its effective domain—an interval beginning at zero and continuing higher—cost increases as

B. C. Dieffenbach, *Conjugate Duality in Economic Analysis*, Contributions to Economics,
https://doi.org/10.1007/978-3-032-21396-9_3

output rises. The conjugate

$$f_i^*(x_i^*) = \sup_{x_i} [x_i^* x_i - f_i(x_i)]$$

is the maximum profit of the firm for output price x_i^*, revenue $x_i^* x_i$ minus cost $f_i(x_i)$. The subdifferential $\partial f_i(x_i)$ is the marginal cost (at a kink, a closed interval). The subdifferential $\partial f_i^*(x_i^*)$ is the output supply of the firm, the set of outputs that maximize profit when output price is x_i^*.

For an industry having n firms, we employ perturbation duality to analyze the industry-cost function, the minimum cost of producing the total output of the industry.

Problem 1 (Industry-Cost Minimization)

$$V(y) := \inf_{\boldsymbol{x}} \left[\sum_i f_i(x_i) + \delta_{y=\sum_i x_i} \right] = \inf_{\boldsymbol{x}} F(\boldsymbol{x}, y).$$

The value function is the minimum cost to produce industry output y. The perturbation function $F(\boldsymbol{x}, y)$ is defined on $\mathbf{X} \oplus \mathbf{Y} = \mathbf{R}^n \oplus \mathbf{R}$. For any positive industry output less than its least upper bound, the primal satisfies the polyhedral/relative condition, so there is no duality gap.[1]

We calculate the cost dual. The conjugate of the value function

$$\begin{aligned} V^*(y^*) &= \sup_y [y^* y - V(y)] \\ &= \sup_y \left\{ y^* y - \inf_{\boldsymbol{x}} \left[\sum_i f_i(x_i) + \delta_{y=\sum_i x_i} \right] \right\} \\ &= \sup_{\boldsymbol{x}, y} \left[y^* y - \sum_i f_i(x_i) - \delta_{\{0\}} \left(y - \sum_i x_i \right) \right] \\ &= \sup_{\boldsymbol{x}, z} \left[y^* \left(z + \sum_i x_i \right) - \sum_i f_i(x_i) - \delta_{\{0\}}(z) \right] \\ \left(z := y - \sum_i x_i \right) & \end{aligned}$$

[1] That the reciprocal dual (defined below) satisfies the relative condition also implies no duality gap.

$$
\begin{aligned}
&= \sum_i \sup_{x_i} [y^* x_i - f_i(x_i)] + \sup_z [y^* z - \delta_{\{0\}}(z)] \\
&= \sum_i f_i^*(y^*) + \delta_{\{0\}}^*(y^*) \\
&= \sum_i f_i^*(y^*),
\end{aligned}
$$

the sum of the profit functions.

Problem 2 (Industry-Cost Dual)

$$
\sup_{y^*} [y^* y - V^*(y^*)] = \sup_{y^*} \left[y^* y - \sum_i f_i^*(y^*) \right].
$$

The dual chooses price y^* to maximize industry revenue minus firm profits.

To solve the primal and the dual simultaneously, set the primal equal to the dual; and factor the expression into a sum of Fenchel factors. The primal less the dual is necessarily greater than or equal to zero:

$$
\begin{aligned}
0 &\le \left[\sum_i f_i(x_i) - \delta_{\{0\}} \left(y - \sum_i x_i \right) \right] - \left[y^* y - \sum_i f_i^*(y^*) - \delta_{\{0\}}^*(y^*) \right] \\
&= \sum_i [f_i(x_i) + f_i^*(y^*) - y^* x_i] \\
&\quad + \left[\delta_{\{0\}} \left(y - \sum_i x_i \right) + \delta_{\{0\}}^*(y^*) - y^* \left(y - \sum_i x_i \right) \right],
\end{aligned}
$$

the sum of n Fenchel factors for the firm cost functions and a Fenchel factor for output. Because there is no duality gap, at a solution the primal must equal the dual. Each Fenchel factor is nonnegative, so for their sum to be zero requires that each factor be zero. Thus we obtain $n+1$ necessary and sufficient conditions for simultaneous solution to the primal and the dual.

Interpret the dual choice y^* as the output price. For the Fenchel factor to be zero,

$$
f_i(x_i) + f_i^*(y^*) - y^* x_i = 0,
$$

the firm chooses output x_i to maximize profit for price y^*. Price equals marginal cost, $y^* \in \partial f_i(x_i)$. Output supply for the firm is the subdifferential $\partial f_i^*(y^*)$, and industry supply is the sum of the supply across firms. Decentralization in the market is cost-minimizing: profit maximization implies that total industry cost is minimized.

The industry cost function $V(y)$ has the properties of the firm cost function. Its subdifferential is industry marginal cost. Its conjugate is industry profit, and its subdifferential is industry supply, the output that maximizes industry profit for a given output price. The epigraph of the value function is the sum of the epigraphs of the firm cost functions:

$$
\text{epi}\, V = \sum_i \text{epi}\, f_i. \tag{3.1}
$$

A calculation shows this property:

$$V^*(y^*) = \sum_i f_i^*(y^*) = \sum_i \delta^*_{\text{epi } f_i}(y^*, -1) = \delta^*_{\sum_i \text{epi } f_i}(y^*, -1),$$

as the sum of supports is the support of the sum. But also

$$V^*(y^*) = \delta^*_{\text{epi } V}(y^*, -1).$$

The sets involved here are closed, and (3.1) follows.

In applications, a powerful analytical shortcut is to reverse this sequence. Rather than setting the primal equal to the dual and factoring, one starts with Fenchel factors and quickly derives the primal and the dual from them.

The cost-minimization primal contains the cost functions f_i, so the dual must contain their conjugates f_i^*. Start with the Fenchel inequalities for the f_i, and add a Fenchel inequality for industry output as the sum of firm output:

$$f_i(x_i) + f_i^*(y^*) - y^* x_i \geq 0 \text{ for all } i$$

$$\delta_{\{0\}}\left(y - \sum_i x_i\right) + \delta^*_{\{0\}}(y^*) - y^*\left(y - \sum_i x_i\right) \geq 0$$

Adding and rearranging obtains

$$\sum_i f_i(x_i) + \delta_{\{0\}}\left(y - \sum_i x_i\right) \geq \langle y^*, y\rangle - \sum_i f_i^*(y^*) - \delta^*_{\{0\}}(y^*).$$

The primal objective function is the left-hand side, and the dual objective function is the right-hand side. The key to this addition and rearrangement is that the cross terms $y^* x_i$ must cancel.

One places plus or minus signs on the choice and perturbation variables to bring about this cancellation. Furthermore, this placement affects the signs of the choice variables in the solution. Like traditional Lagrange multiplier optimization, changing the sign on a perturbation variable reverses the sign of the corresponding choice variable. In the cost-minimization problem, one wants the price y^* to be positive.

This reversed sequence conforms with how economists treat this topic. Each firm maximizes profit by producing where the output price equals marginal cost. Production is cost-minimizing, because the marginal cost is the same for all firms. Starting our analysis with the Fenchel factors means starting with profit maximization by firms. Deriving the primal and the dual from the Fenchel factors discovers how profit maximization implies industry-cost minimization.

3.2 Fenchel Duality

The concept of a tradeoff is fundamental in economics: more of one goal implies less of another. A simple model of a tradeoff is to minimize the sum of two proper, closed, convex functions. Fenchel duality captures this concept.

Set the perturbation function on $\mathbf{X} \oplus \mathbf{Y}$ by defining $\mathbf{Y}$ as a copy of $\mathbf{X}$, just like it but distinct from it. The copy map $C : \mathbf{X} \to \mathbf{Y}$ is an invertible linear transformation mapping a vector to the corresponding vector in the copy. Let the notation $(\boldsymbol{x},\boldsymbol{y})$ and $(\boldsymbol{x}^*,\boldsymbol{y}^*)$ denote such pairings, in which $\boldsymbol{y} = C\boldsymbol{x}$ and $\boldsymbol{y}^* = C\boldsymbol{x}^*$. For a linear transformation, the copy of a sum is the sum of the copies. A linear transformation preserves scalar multiplication: $C(\lambda \boldsymbol{x}) = \lambda (C\boldsymbol{x})$. In addition, we require that the copy map preserves inner products: $\langle \boldsymbol{x}^*,\boldsymbol{x}\rangle = \langle \boldsymbol{y}^*,\boldsymbol{y}\rangle$ (the first inner product is in $\mathbf{X}$, whereas the second is in the copy $\mathbf{Y}$).[2]

The tradeoff duality is Fenchel duality. Here f and g are proper, closed, convex functions, but f is defined on $\mathbf{X}$, whereas g is defined on its copy $\mathbf{Y}$. The choice variable $\boldsymbol{x} \in \mathbf{X}$, whereas the perturbation variable $\boldsymbol{y} \in \mathbf{Y}$.

Problem 3 (Tradeoff Primal)

$$\inf_{\boldsymbol{x}\in\mathbf{X}} F(\boldsymbol{x},\boldsymbol{y}) = \inf_{\boldsymbol{x}\in\mathbf{X}} [f(\boldsymbol{x}) + g(C\boldsymbol{x}+\boldsymbol{y})],$$

in which the perturbation base value $\boldsymbol{y} = \mathbf{0}$.

Theorem 1 (Tradeoff Dual) *The dual of the tradeoff primal is*

$$\sup_{\boldsymbol{y}^*\in\mathbf{Y}} \left[-f^*\left(-C^{-1}\boldsymbol{y}^*\right) - g^*(\boldsymbol{y}^*)\right].$$

We obtain the dual by calculating the conjugate of the value function:

$$\begin{aligned}
V^*(\boldsymbol{y}^*) &= \sup_{\boldsymbol{y}} [\langle \boldsymbol{y}^*,\boldsymbol{y}\rangle - V(\boldsymbol{y})] \\
&= \sup_{\boldsymbol{y}} \left\{\langle \boldsymbol{y}^*,\boldsymbol{y}\rangle - \inf_{\boldsymbol{x}\in\mathbf{X}} [f(\boldsymbol{x}) + g(C\boldsymbol{x}+\boldsymbol{y})]\right\} \\
&= \sup_{\boldsymbol{x},\boldsymbol{y}} [\langle \boldsymbol{y}^*,\boldsymbol{y}\rangle - f(\boldsymbol{x}) - g(C\boldsymbol{x}+\boldsymbol{y})] \\
&= \sup_{\boldsymbol{x}\in\mathbf{X}} [-\langle \boldsymbol{y}^*, C\boldsymbol{x}\rangle - f(\boldsymbol{x})] + \sup_{\boldsymbol{z}\in\mathbf{Y}} [\langle \boldsymbol{y}^*,\boldsymbol{z}\rangle - g(\boldsymbol{z})] \quad (\boldsymbol{z} := C\boldsymbol{x}+\boldsymbol{y}) \\
&= \sup_{\boldsymbol{x}\in\mathbf{X}} \left[-\left\langle C^{-1}\boldsymbol{y}^*,\boldsymbol{x}\right\rangle - f(\boldsymbol{x})\right] + \sup_{\boldsymbol{z}\in\mathbf{Y}} [\langle \boldsymbol{y}^*,\boldsymbol{z}\rangle - g(\boldsymbol{z})] \\
&\quad (C \text{ preserves the inner product}) \\
&= f^*\left(-C^{-1}\boldsymbol{y}^*\right) + g^*(\boldsymbol{y}^*).
\end{aligned}$$

[2] The copy map is a Euclidean-space isomorphism of $\mathbf{X}$ and the copy. The inverse of the copy map is its adjoint—the copy map is an orthogonal linear transformation. Extended Fenchel duality generalizes Fenchel duality by replacing the copy map with an arbitrary linear transformation.

In calculations more simply one can identify **X** and **Y**, without any explicit reference to the copy map. View both f and g as defined on **X**.

Corollary 1 (Fenchel Duality) *For perturbation base value* $\mathbf{y} = \mathbf{0}$, *the Fenchel primal*

$$\inf_{x \in \mathbf{X}} [f(\boldsymbol{x}) + g(\boldsymbol{x} + \boldsymbol{y})] \tag{3.2}$$

has the Fenchel dual

$$\sup_{y^* \in \mathbf{X}} [-f^*(-\boldsymbol{y}^*) - g^*(\boldsymbol{y}^*)]. \tag{3.3}$$

Fenchel (1953, Chap. III, Sect. 6)

We solve the primal and the dual simultaneously, by setting the primal equal to the dual, and factoring the expression into a sum of Fenchel factors. The primal less the dual is necessarily greater than or equal to zero:

$$\begin{aligned} 0 &\leq [f(\boldsymbol{x}) + g(\boldsymbol{x} + \boldsymbol{y})] - [-f^*(-\boldsymbol{y}^*) - g^*(\boldsymbol{y}^*)] \\ &= [f(\boldsymbol{x}) + f^*(-\boldsymbol{y}^*) - \langle -\boldsymbol{y}^*, \boldsymbol{x} \rangle] + [g(\boldsymbol{x} + \boldsymbol{y}) + g^*(\boldsymbol{y}^*) - \langle \boldsymbol{y}^*, \boldsymbol{x} + \boldsymbol{y} \rangle]. \end{aligned}$$

the sum of two Fenchel factors, for f and for g. If there is no duality gap, at a solution the primal must equal the dual. Each Fenchel factor is nonnegative, so for their sum to be zero requires that each factor be zero. We obtain two necessary and sufficient conditions for simultaneous solution to the primal and the dual.

Presenting the Fenchel primal and dual graphically, Fig. 3.1 plots g and the concave function $-f$. The primal (3.2) chooses $\boldsymbol{x}$ to minimize the vertical distance between them. In the figure, $g+f > 0$, and the optimum value is positive. The minimum is where the vertical line connects the two epigraphs.

See the choice of $\boldsymbol{y}^*$ in the dual as setting parallel hyperplanes supporting epi g and epi $(-f)$. The hyperplane supporting epi g is

$$\langle \boldsymbol{y}^*, \boldsymbol{x} \rangle + \langle -1, z \rangle = \delta^*_{\text{epi } g}(\boldsymbol{y}^*, -1) = g^*(\boldsymbol{y}^*),$$

for $(\boldsymbol{x}, z) \in \mathbf{X} \oplus \mathbf{R}$. The hyperplane supporting epi $(-f)$ is

$$\langle -\boldsymbol{y}^*, \boldsymbol{x} \rangle + \langle 1, z \rangle = \delta^*_{\text{epi}(-f)}(-\boldsymbol{y}^*, 1) = \delta^*_{\text{epi } f}(-\boldsymbol{y}^*, -1) = f^*(-\boldsymbol{y}^*).$$

The hyperplanes have vertical intercepts $-g^*(\boldsymbol{y}^*)$ and $f^*(-\boldsymbol{y}^*)$, so the Fenchel dual chooses $\boldsymbol{y}^*$ to maximize the vertical distance between the intercepts. (For a different choice $\boldsymbol{y}^*$, the vertical distance between supporting hyperplanes is less.)

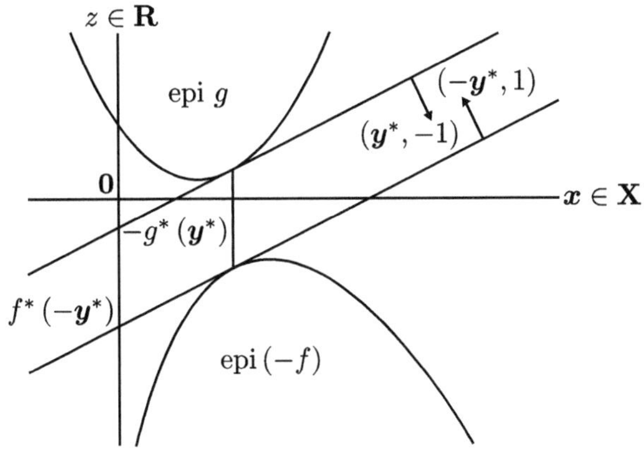

Fig. 3.1 Fenchel duality

Reversal is a quick way to derive the dual. The Fenchel primal (3.2) contains both f and g, so the dual must contain their conjugates f^* and g^*. Hence start with the two Fenchel inequalities for f and g:

$$f(x) + f^*(-y^*) - \langle -y^*, x \rangle \geq 0$$
$$g(x+y) + g^*(y^*) - \langle y^*, x+y \rangle \geq 0$$

Adding and rearranging obtains

$$f(x) + g(x+y) \geq \langle y^*, y \rangle - f^*(-y^*) - g^*(y^*).$$

The objective function of the primal is on the left, and the objective function of the dual is on the right. That the cross terms $\langle y^*, x \rangle$ cancel is key to this derivation.

A key property of primal and dual solutions is "independence". In general, either the primal or the dual might have multiple solutions. *Any* primal solution x and *any* dual solution y^* must make every Fenchel factor zero. The independence property is that primal and dual solutions can be paired together arbitrarily, and the Fenchel equalities must hold. In an application, independence may establish that the primal or dual solution is unique.

The situation in the examples above is general. The theoretical concept is to calculate the dual via the conjugate of the value function. One can factor the primal less the dual as a sum of Fenchel factors, each nonnegative. The primal equals the dual if and only if each factor is zero. If there is no duality gap, these conditions are necessary and sufficient for simultaneous solution of the primal and the dual.

One understands and explains the solutions by interpreting the Fenchel factors in some meaningful way. The interpretation might be economic, geometric, or mathematical. As illustrated by the industry cost-minimization problem, there can be many factors. For example, in a problem for an exchange economy, there might be one

Fenchel factor for each consumer, plus additional Fenchel factors. (In general, the Fenchel factorization is not unique.)

From this viewpoint, to solve a problem in the traditional way via the first-order condition is obsolete. Because the subdifferential is a set, to write this set down for each possible value of the choice variable can be extremely complicated. Furthermore, a complicated first-order condition may be difficult to interpret.

In applications, reversal is a simple and direct way to derive the dual. Valuable for both derivation and exposition, it facilitates the analysis of complicated primal/dual pairs. The reversal methodology is the technique mostly used throughout this book.

Reference

Fenchel W (1953 Sept) Convex cones, sets, and functions. Department of Mathematics. (University Microfilms, Ann Arbor (typeset 2010 by Kim C. Border, California Institute of Technology))

Chapter 4
Conjugate Duality

Abstract Fenchel proves the conjugate-duality theorem: for a proper, closed, convex function, $f^{**} = f$. That a closed, convex set is the intersection of its containing, closed, half spaces is key to proving conjugate duality. Conjugate duality for an indicator function is equivalent to basic separation. To prove that a closed, convex set is the intersection of its containing, closed, half spaces, we calculate the support to the indicator of these half spaces. Applying this result proves conjugate duality. For a proper, closed, convex function, its epigraph is the intersection of all nonvertical containing, closed, half spaces. A calculation establishes epi f^{**} = epi f. The containing half-space symmetry derived below furnishes a geometric interpretation of conjugate duality: the nonvertical containing half spaces of the epigraph define the epigraph of the conjugate.

Fenchel proves the conjugate-duality theorem (Fenchel 1949).

Theorem 1 (Conjugate Duality) *For a proper, closed, convex function,* $f^{**} = f$.

Below we analyze a proper, closed, convex function f. That a closed, convex set is the intersection of its containing, closed, half spaces is key to proving conjugate duality.

Conjugate duality for an indicator function is equivalent to basic separation. To prove that a closed, convex set is the intersection of its containing, closed, half spaces, we calculate the support to the indicator of these half spaces. Applying this result proves conjugate duality. For a proper, closed, convex function, its epigraph is the intersection of all nonvertical containing, closed, half spaces. A calculation establishes epi f^{**} = epi f.

The containing half-space symmetry derived below furnishes a geometric interpretation of conjugate duality: the nonvertical containing half spaces of the epigraph define the epigraph of the conjugate.

B. C. Dieffenbach, *Conjugate Duality in Economic Analysis*, Contributions to Economics,
https://doi.org/10.1007/978-3-032-21396-9_4

4.1 Basic Separation

Propostion 1 (Basic Separation) *If C is a nonempty, closed, convex set such that $\mathbf{0} \notin C$, then there exists some $\boldsymbol{x}^*$ such that $\delta_C^*(\boldsymbol{x}^*) < 0$.*

Figure 4.1 illustrates basic separation.

Basic separation proves conjugate duality for an indicator. The indicator of a set is a proper, closed, convex function if and only if the set is a nonempty, closed, convex set.

Theorem 2 (Conjugate Duality for an Indicator) *The indicator of a nonempty, closed, convex set C satisfies conjugate duality: $\delta_C^{**} = \delta_C$.*

The proof is by contradiction. Because the support δ_C^* is positively homogeneous, $\delta_C^{**} = \delta_D$ for some closed, convex set D. That $\delta_C^{**} \le \delta_C$ means $D \supset C$. If $\delta_C^{**}(z) < \delta_C(z)$ for some z, then $z \in D$ but $z \notin C$. We have

$$0 = \delta_{D-\{z\}}(\mathbf{0}) = \delta_{C-\{z\}}^{**}(\mathbf{0}) = \sup_{x^*} \left[-\delta_{C-\{z\}}^*(\boldsymbol{x}^*) \right].$$

By basic separation, $\delta_{C-\{z\}}^*(\boldsymbol{x}^*) < 0$ for some $\boldsymbol{x}^*$, so the supremum is ∞, a contradiction.

Conversely, the conjugate duality for an indicator function implies basic separation. Hence anything that one can prove via basic separation can also be proved via conjugate duality.

Theorem 3 (Set Membership) *For a closed, convex set C, the following three conditions are equivalent:*

- $\mathbf{0} \in C$;
- $\delta_C \le \delta_{\{0\}}$;
- $\delta_C^* \ge \delta_{\{0\}}^* = 0$.

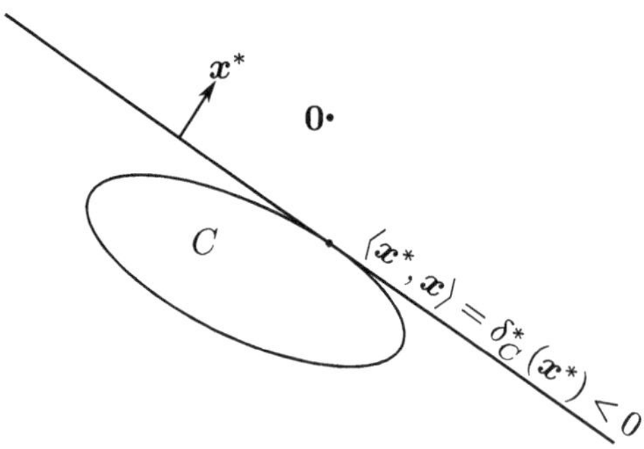

Fig. 4.1 Basic separation

The first and second statements are equivalent. The second statement implies the third, because taking the conjugate reverses the inequality. By conjugate duality, the third statement implies the second, again because taking the conjugate reverses the inequality. Basic separation follows immediately from the set-membership theorem, as a contrapositive: if C is a closed, convex set such that $\mathbf{0} \notin C$, then there exists some $\boldsymbol{x}^*$ such that $\delta_C^*(\boldsymbol{x}^*) < 0$.

4.2 Containing Closed Half-Space Representation

The containing closed half-space representation is fundamental, the basis for conjugate duality.

Theorem 4 (Containing Half-Space Representation) *A closed, convex set is the intersection of its containing, closed half spaces.*

Figure 4.2 illustrates the theorem, by graphing three containing half spaces.

Definition 1 (Cone) A cone K is a set such that $\lambda \boldsymbol{x} \in K$ for all $\boldsymbol{x} \in K$ and $\lambda > 0$.

A cone need not contain the origin, but a nonempty closed cone does contain it. The following is immediate from the definition of the support.

Summary 1 (Containing Half Spaces) The closed half space defined by $\langle \boldsymbol{x}^*, \boldsymbol{x} \rangle \leq y^*$ contains a set C if and only if

$$(\boldsymbol{x}^*, y^*) \in \operatorname{epi} \delta_C^*,$$

a closed, convex cone.

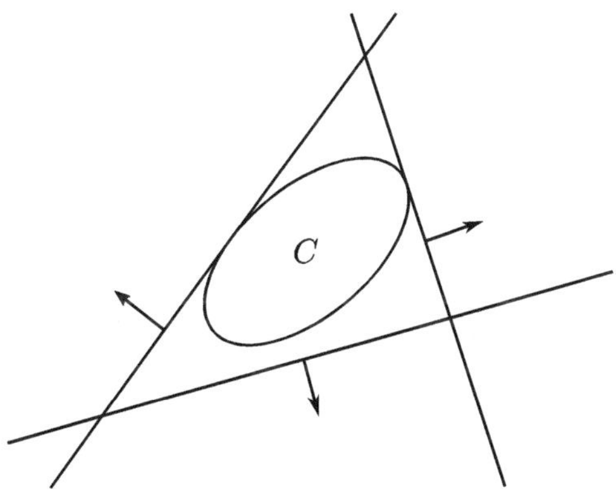

Fig. 4.2 Containing half spaces

The containing half spaces constitute a cone. Here we admit $(\boldsymbol{x}^*, y^*) = (\boldsymbol{0}, 0)$ as a half space.

We prove the containing half-space representation theorem by calculating a support. The indicator of the intersection of the closed half spaces containing C is the support

$$\delta^*_{\text{epi}\ \delta^*_C}(\boldsymbol{x}, -1) = \sup_{\boldsymbol{x}^*, y^*} \left[\langle \boldsymbol{x}^*, \boldsymbol{x} \rangle - y^* - \delta_{\text{epi}\ \delta^*_C}(\boldsymbol{x}^*, y^*)\right].$$

The objective function is positively homogeneous in $(\boldsymbol{x}^*, y^*)$, so the optimum value is either zero or infinity. If $\boldsymbol{x}$ belongs to the containing, closed half-spaces, the maximum is zero. If $\boldsymbol{x}$ fails to belong to some containing, closed half-space $(\boldsymbol{x}^*, y^*)$, then $\langle \boldsymbol{x}^*, \boldsymbol{x} \rangle - y^* > 0$, so the supremum is ∞.

We evaluate

$$\begin{aligned}
&\sup_{\boldsymbol{x}^*, y^*} \left[\langle \boldsymbol{x}^*, \boldsymbol{x} \rangle - y^* - \delta_{\text{epi}\ \delta^*_C}(\boldsymbol{x}^*, y^*)\right] \\
&\quad = \sup_{\boldsymbol{x}^*} \left[\langle \boldsymbol{x}^*, \boldsymbol{x} \rangle - \inf_{y^*}\left[y^* + \delta_{\text{epi}\ \delta^*_C}(\boldsymbol{x}^*, y^*)\right]\right] \\
&\quad = \sup_{\boldsymbol{x}^*} \left[\langle \boldsymbol{x}^*, \boldsymbol{x} \rangle - \delta^*_C(\boldsymbol{x}^*)\right] \\
&\quad = \delta^{**}_C(\boldsymbol{x}) \\
&\quad = \delta_C(\boldsymbol{x}),
\end{aligned}$$

by conjugate duality for an indicator function.[1]

4.3 Conjugate Duality

We apply the containing half-space representation to the epigraph of a proper, closed, convex function f. The closed half spaces $\langle \boldsymbol{x}^*, \boldsymbol{x} \rangle + y^* y \leq z^*$ containing $(\boldsymbol{x}, y) \in \text{epi}\, f$ have indicator $\delta_{\text{epi}\ \delta^*_{\text{epi}\, f}}(\boldsymbol{x}^*, y^*, z^*)$. The epigraph is the intersection of these half spaces:

$$\delta_{\text{epi}\, f}(\boldsymbol{x}, y) = \sup_{\boldsymbol{x}^*, y^*, z^*} \left[\langle \boldsymbol{x}^*, \boldsymbol{x} \rangle + y^* y - z^* - \delta_{\text{epi}\ \delta^*_{\text{epi}\, f}}(\boldsymbol{x}^*, y^*, z^*)\right] \tag{4.1}$$

for closed and convex f, as the intersection of all closed half spaces containing epi f.

Because the set is an epigraph, necessarily $y^* \leq 0$ for each containing half space. The nonvertical containing half spaces are those such that $y^* < 0$, and the vertical containing half spaces are those such that $y^* = 0$. Because the function is proper, the epigraph contains no vertical lines, so there exist nonvertical containing half spaces.

[1] The theorem also holds for $C = \varnothing$, even though its indicator is improper.

The vertical containing half spaces are those such that $y^* = 0$, so each is the limit of a sequence of nonvertical containing half spaces. Consequently epi f is also the intersection of just the nonvertical containing half spaces.

Equivalently, the intersection of the containing half spaces is $(\boldsymbol{x}, y)$ such that

$$0 \geq \sup_{\boldsymbol{x}^*, y^*, z^*} \left[\langle \boldsymbol{x}^*, \boldsymbol{x} \rangle + y^* y - z^* - \delta_{\text{epi } \delta^*_{\text{epi } f}} (\boldsymbol{x}^*, y^*, z^*) \right].$$

Because epi f is the intersection of just the nonvertical containing half spaces, set $y^* = -1$; and epi f is also all $(\boldsymbol{x}, y)$ such that

$$0 \geq \sup_{\boldsymbol{x}^*, z^*} \left[\langle \boldsymbol{x}^*, \boldsymbol{x} \rangle - y - z^* - \delta_{\text{epi } \delta^*_{\text{epi } f}} (\boldsymbol{x}^*, -1, z^*) \right],$$

Because $f^* (\boldsymbol{x}^*) = \delta^*_{\text{epi } f} (\boldsymbol{x}^*, -1)$,

$$\begin{aligned} 0 &\geq -y + \sup_{\boldsymbol{x}^*, z^*} \left[\langle \boldsymbol{x}^*, \boldsymbol{x} \rangle - z^* - \delta_{\text{epi } \delta^*_{\text{epi } f}} (\boldsymbol{x}^*, -1, z^*) \right] \\ &= -y + \sup_{\boldsymbol{x}^*, z^*} \left[\langle \boldsymbol{x}^*, \boldsymbol{x} \rangle - z^* - \delta_{\text{epi } f^*} (\boldsymbol{x}^*, z^*) \right] \\ &= -y + \delta^*_{\text{epi } f^*} (\boldsymbol{x}, -1) \\ &= -y + f^{**} (\boldsymbol{x}), \end{aligned}$$

so $(\boldsymbol{x}, y) \in$ epi f^{**}. Thus epi $f^{**} =$ epi f.

These calculations imply a symmetric geometric relationship between epi f and epi f^*: each epigraph consists of the nonvertical containing half spaces of the other.

Summary 2 (Containing Half-Space Symmetry) The nonvertical half spaces $\langle \boldsymbol{x}^*, \boldsymbol{x} \rangle - y \leq z^*$ containing $(\boldsymbol{x}, y) \in$ epi f are the $(\boldsymbol{x}^*, z^*) \in$ epi f^*. The nonvertical half spaces $\langle \boldsymbol{x}^*, \boldsymbol{x} \rangle - z^* \leq y$ containing $(\boldsymbol{x}^*, z^*) \in$ epi f^* are the $(\boldsymbol{x}, y) \in$ epi f.

Rather than proving conjugate duality via basic separation, Fenchel proves it by demonstrating this containing half-space symmetry, via a geometric argument (Fenchel 1949).

The calculation above breaks down if f is closed and convex but not proper. Its value is $-\infty$ in dom f, a closed, convex set, and ∞ elsewhere. The epigraph is the set dom $f \times \mathbf{R}$ of vertical lines. Apart from the case dom $f = \varnothing$, no nonvertical half space contains epi f. Every containing half space is vertical and thus is not the limit of a sequence of nonvertical containing half spaces. Apart from the cases dom $f = \varnothing$ or dom $f = \mathbf{X}$, then $f^{**} \neq f$.

4.4 Concave Function

Given a result for convex function, there exists an analogous result for a concave function. Typically this book states the analogous result, without proof. The concave conjugate is the support

$$f_*(\boldsymbol{x}^*) = -\delta^*_{\mathrm{epi}\, f}(-\boldsymbol{x}^*, 1).$$

The Fenchel inequality

$$f(\boldsymbol{x}) + f_*(\boldsymbol{x}^*) - \langle \boldsymbol{x}^*, \boldsymbol{x} \rangle \le 0$$

holds for all $\boldsymbol{x}$ and $\boldsymbol{x}^*$.

Suppose that the concave function $f(\boldsymbol{x})$ is the negative of the convex function $g(\boldsymbol{x})$. The concave conjugate

$$\begin{aligned} f_*(\boldsymbol{x}^*) &:= \inf_{\boldsymbol{x}} [\langle \boldsymbol{x}^*, \boldsymbol{x} \rangle - f(\boldsymbol{x})] \\ &= \inf_{\boldsymbol{x}} [\langle \boldsymbol{x}^*, \boldsymbol{x} \rangle + g(\boldsymbol{x})] \\ &= -\sup_{\boldsymbol{x}} [-\langle \boldsymbol{x}^*, \boldsymbol{x} \rangle - g(\boldsymbol{x})] \\ &= -g^*(-\boldsymbol{x}^*). \end{aligned} \tag{4.2}$$

There are *two* minus signs on the right-hand side; the concave conjugate f_* is *not* the negative of g^*. Given a result for convex function, it is possible to obtain the analogous result for a concave function, by carefully invoking the double minus sign property (4.2). However, the many sign reversals required make this procedure error prone, so it is more straightforward to obtain the analogous result by deriving it in parallel to the original derivation for a convex function.

Reference

Fenchel W (1949) On conjugate convex functions. Canad J Math 1:73–77

Chapter 5
Support

Abstract A proper, closed, convex function is positively homogeneous if and only if it is the support to a non-empty, closed, convex set. We interpret the Fenchel equality for the indicator and the support as a supporting hyperplane. The Cauchy-Schwarz inequality shows that the absolute value of the cosine of the angle between two vectors is less than or equal to one. In a statistical application in which the inner product is the covariance, the correlation between two random variables is less than or equal to one in absolute value. We study efficiency or weak efficiency by maximizing the efficiency function or the weak efficiency function. These two functions enable us to distinguish sharply and effectively between the two concepts. The maximization is especially fruitful in primal/dual convex optimization. The primal analyzes whether a particular vector of goals is efficient or weakly efficient, and this vector is the perturbation variable. One interprets the dual choice variable as a shadow price for the goals.

5.1 Supporting Hyperplane

For a nonempty, closed, convex set, below we interpret the Fenchel equality for the indicator and the support as a supporting hyperplane (defined in the chapter "Discovery and Proof via Duality"). A support is positively homogeneous.

Definition 1 (Positive Homogeneity) A function f is positively homogeneous if

$$f(\lambda \boldsymbol{x}) = \lambda f(\boldsymbol{x})$$

for every $\boldsymbol{x}$ for every $\lambda > 0$.

The conjugate of a positively homogeneous function is the indicator of a set, so conjugate duality implies the following.

Theorem 1 (Positively Homogeneous Function) *A proper, closed, convex function is positively homogeneous if and only if it is the support to a non-empty, closed, convex set.*

B. C. Dieffenbach, *Conjugate Duality in Economic Analysis*, Contributions to Economics,
https://doi.org/10.1007/978-3-032-21396-9_5

Interpret the Fenchel equality for a support as characterizing a supporting hyperplane.

Summary 1 (Supporting Hyperplane) For a closed, convex set C and $\boldsymbol{x}^* \neq \boldsymbol{0}$, the Fenchel equality

$$\delta_C(\boldsymbol{x}) + \delta_C^*(\boldsymbol{x}^*) - \langle \boldsymbol{x}^*, \boldsymbol{x} \rangle = 0 \tag{5.1}$$

holds if and only if $\langle \boldsymbol{x}^*, \boldsymbol{y} \rangle = \delta_C^*(\boldsymbol{x}^*)$ *is a supporting hyperplane at* $\boldsymbol{x}$.

If the equality holds, then $\langle \boldsymbol{x}^*, \boldsymbol{y} \rangle = \delta_C^*(\boldsymbol{x}^*)$ is a supporting hyperplane at $\boldsymbol{x}$. Conversely, a hyperplane $\langle \boldsymbol{x}^*, \boldsymbol{y} \rangle = \delta_C^*(\boldsymbol{x}^*)$ supporting the set at $\boldsymbol{x}$ satisfies the Fenchel equality.

The normal cone is a geometric concept.

Definition 2 (Normal Cone) Given a closed, convex set C, the normal cone at $\boldsymbol{x}$ is the cone of normals to the hyperplanes (plus $\boldsymbol{0}$) supporting C at $\boldsymbol{x}$.

The normal cone is the subdifferential $\partial\delta_C(\boldsymbol{x})$, so it is closed and convex. The normal cone is also all $\boldsymbol{x}^*$ satisfying (5.1). If $\boldsymbol{x} \in \text{int}\, C$, then $\delta_C^*(\boldsymbol{x}^*) > \langle \boldsymbol{x}^*, \boldsymbol{x} \rangle$ for any $\boldsymbol{x}^* \neq \boldsymbol{0}$, so the normal cone is $\{\boldsymbol{0}\}$. If $\boldsymbol{x} \notin C$, then (5.1) cannot be satisfied, so the normal cone is empty.

The indicator of a nonempty, closed, convex cone K is positively homogeneous, so its support is the indicator of a set. Because the support is also positively homogeneous, it is the indicator of a cone, the "polar cone" K°. The normal cone at the origin is the polar cone.

5.2 Geometric Interpretation of the Support

We obtain a geometric interpretation of the support. In Fig. 5.1, the hyperplane $\langle \boldsymbol{x}^*, \boldsymbol{x} \rangle = \delta_C^*(\boldsymbol{x}^*)$ supports the closed, convex set C at $\boldsymbol{z}$. Its normal $\boldsymbol{x}^*$ points away from the set. Calculate the vector $\boldsymbol{x}$ on the hyperplane closest to the origin:

$$\inf_{\boldsymbol{x}} \left[\frac{1}{2} \langle \boldsymbol{x}, \boldsymbol{x} \rangle + \delta_{\langle \boldsymbol{x}^*, \boldsymbol{x} \rangle = y} \right].$$

The perturbation base value $y = \delta_C^*(\boldsymbol{x}^*)$.

Calculate the dual by adding the Fenchel inequalities

$$\frac{1}{2} \langle \boldsymbol{x}, \boldsymbol{x} \rangle + \frac{1}{2} \langle \boldsymbol{x}^* y^*, \boldsymbol{x}^* y^* \rangle - \langle \boldsymbol{x}^* y^*, \boldsymbol{x} \rangle \geq 0 \tag{5.2}$$

$$\delta_{\{0\}}(\langle \boldsymbol{x}^*, \boldsymbol{x} \rangle - y) + \delta_{\{0\}}^*(-y^*) - (-y^*)(\langle \boldsymbol{x}^*, \boldsymbol{x} \rangle - y) \geq 0 \tag{5.3}$$

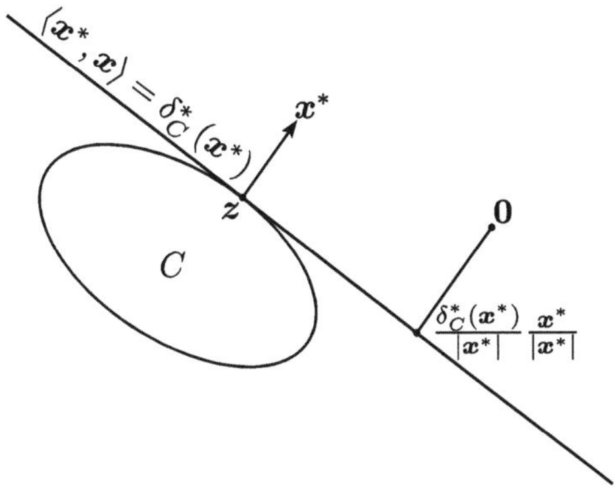

Fig. 5.1 Supporting hyperplane

and rearranging, to obtain

$$\frac{1}{2}\langle x, x\rangle + \delta_{\{0\}}(\langle x^*, x\rangle - y) \geq y^* y - \frac{1}{2}\langle x^* y^*, x^* y^*\rangle - \delta^*_{\{0\}}(-y^*).$$

The primal satisfies the polyhedral condition, so there is no duality gap.

For the two Fenchel factors (5.2)–(5.3) to be zero means

$$x = x^* y^*$$
$$\langle x^*, x\rangle = y.$$

Because the problem is quadratic, the necessary and sufficient conditions are linear in the primal and dual choice variables. Thus $x \propto x^*$ such that $\langle x^*, x\rangle = \delta^*_C(x^*)$.

Theorem 2 (Closest Vector) *The vector on the hyperplane closest to the origin is*

$$x = \frac{\delta^*_C(x^*)}{|x^*|}\frac{x^*}{|x^*|}.$$

The norm $|x|$ is the absolute value of $\delta^*_C(x^*) / |x^*|$. Whether x points in the same or opposite direction to x^* depends on the sign of $\delta^*_C(x^*)$. In the figure, $\delta^*_C(x^*) < 0$, so they point in opposite directions.

5.3 Norm

The Cauchy-Schwarz inequality is a basic property of Euclidean space. For two vectors $\boldsymbol{x}^*$ and $\boldsymbol{x}$ each of norm one,

$$\langle \boldsymbol{x}^* - \boldsymbol{x}, \boldsymbol{x}^* - \boldsymbol{x} \rangle = 2\,(1 - \langle \boldsymbol{x}^*, \boldsymbol{x} \rangle)$$
$$\langle \boldsymbol{x}^* + \boldsymbol{x}, \boldsymbol{x}^* + \boldsymbol{x} \rangle = 2\,(1 + \langle \boldsymbol{x}^*, \boldsymbol{x} \rangle)$$

Unless $\boldsymbol{x}^* = \boldsymbol{x}$, the first inner product is positive, so $\langle \boldsymbol{x}^*, \boldsymbol{x} \rangle < 1$. Unless $\boldsymbol{x}^* = -\boldsymbol{x}$, the second inner product is positive, so $\langle \boldsymbol{x}^*, \boldsymbol{x} \rangle > -1$. That the inner product is bilinear then yields the Cauchy-Schwarz inequality.

Theorem 3 (Cauchy-Schwarz Inequality) *For two vectors* $\boldsymbol{x}^*$ *and* $\boldsymbol{x}$, $|\langle \boldsymbol{x}^*, \boldsymbol{x} \rangle| \le |\boldsymbol{x}^*|\,|\boldsymbol{x}|$, *with equality only if one vector is a multiple of the other.*

In a geometric application, the cosine $\langle \boldsymbol{x}^*, \boldsymbol{x} \rangle / |\boldsymbol{x}^*|\,|\boldsymbol{x}|$ of the angle between the two vectors is less than or equal to one. In a statistical application in which the inner product is the covariance, the correlation between two random variables is less than or equal to one in absolute value.

The unit ball is fundamental in applications involving the norm.

Definition 3 (Unit Ball) The unit ball B is the closed, convex set having norm less than or equal to one,

$$\delta_B(\boldsymbol{x}) = \delta_{|\boldsymbol{x}| \le 1}. \tag{5.4}$$

Theorem 4 (Norm as Support) *The support to the unit ball is the norm,*

$$\delta_B^*(\boldsymbol{x}^*) = |\boldsymbol{x}^*|.$$

The unit ball is closed and convex, by its definition via the convex inequality (5.4). As a conjugate, the norm is closed and convex.

If $\boldsymbol{x}^* \neq \boldsymbol{0}$, then

$$\delta_B^*(\boldsymbol{x}^*) = \sup_{\boldsymbol{x}}\,[\langle \boldsymbol{x}^*, \boldsymbol{x} \rangle - \delta_B(\boldsymbol{x})] = |\boldsymbol{x}^*| \sup_{\boldsymbol{x}}\,[\langle \boldsymbol{x}^* / |\boldsymbol{x}^*|, \boldsymbol{x} \rangle - \delta_B(\boldsymbol{x})] = |\boldsymbol{x}^*|;$$

by Cauchy-Schwarz the second maximum is one, attained at $\boldsymbol{x} = \boldsymbol{x}^* / |\boldsymbol{x}^*|$. If $\boldsymbol{x}^* = \boldsymbol{0}$, then the support is zero, and the solution set is $\boldsymbol{x} \in B$. Thus the Fenchel equality for the unit ball and its support amounts to a Cauchy-Schwarz equality.

The subdifferential of the norm is important later in this book.

Theorem 5 (Subdifferential of the Norm) *The subdifferential*

$$\partial\,|\boldsymbol{x}^*| = \begin{cases} \{\boldsymbol{x}^* / |\boldsymbol{x}^*|\} \textit{ if } \boldsymbol{x}^* \neq \boldsymbol{0} \\ B \textit{ if } \boldsymbol{x}^* = \boldsymbol{0}. \end{cases}$$

Inverting this subdifferential obtains the subdifferential (the normal cone) of the unit ball.

Theorem 6 (Subdifferential of the Unit Ball) *The subdifferential of the indicator of the unit ball is*

$$\partial\delta_B(\boldsymbol{x}) = \begin{cases} \{\mathbf{0}\} \text{ if } |\boldsymbol{x}| < 1 \\ \mathbf{R}_+\{\boldsymbol{x}\} \text{ if } |\boldsymbol{x}| = 1 \\ \varnothing \text{ if } |\boldsymbol{x}| > 1. \end{cases}$$

5.4 Partial Order

In economics, a fundamental concept is order. For example, for one market basket of goods the quantity of each good might be greater than or equal to the quantity in a second market basket. The concept of a partial ordering is basic to economic efficiency: one outcome can be better than another, or perhaps one cannot say which outcome is better.

The axioms defining Euclidean space involve no concept of order, but one can define a partial ordering on a Euclidean space by imposing additional structure. A simple way is to single out a particular orthonormal basis $\boldsymbol{e}_i$, and to express a vector in terms of basis coefficients: $\boldsymbol{x} = \sum_i x_i \boldsymbol{e}_i$. Define a partial ordering via the basis coefficients:

Nonnegative	$\boldsymbol{x} \geq \mathbf{0}$	$x_i \geq 0$ for all i
Greater than zero	$\boldsymbol{x} > \mathbf{0}$	$x_i \geq 0$ for all i and $x_i > 0$ for some i
Positive	$\boldsymbol{x} \gg \mathbf{0}$	$x_i > 0$ for all i.

For example, one says that $\boldsymbol{x} \geq \boldsymbol{y}$ if $\boldsymbol{x} - \boldsymbol{y} \geq \mathbf{0}$. The nonnegative vectors form the "nonnegative cone" $\mathbf{R}^n_+$, and the positive vectors form the "positive cone" $\mathbf{R}^n_{++}$. The polar of the nonnegative cone is its negative.

"Complementary slackness" refers to the conditions

$$\boldsymbol{x} \geq \mathbf{0}$$
$$\boldsymbol{x}^* \geq \mathbf{0}$$
$$\langle \boldsymbol{x}^*, \boldsymbol{x} \rangle = 0.$$

Every x_i and x_i^* is nonnegative. That $x_i^* x_i = 0$ means that either one component or the other must be zero. Observe that complementary slackness is equivalent to the Fenchel equality for the indicator of the nonnegative cone,

$$\delta_{\mathbf{R}^n_+}(\boldsymbol{x}) + \delta^*_{\mathbf{R}^n_+}(-\boldsymbol{x}^*) - \langle -\boldsymbol{x}^*, \boldsymbol{x} \rangle = 0. \tag{5.5}$$

5.5 Efficiency

The concept of a tradeoff is fundamental in economics. For $\boldsymbol{x} = \sum_i x_i \boldsymbol{e}_i$ in $\mathbf{R}^n$, see the x_i as the values of n desired goals. For each goal, a higher value is better. "Efficiency" means that more of one goal can be attained only by settling for less of another. (In economics, "Pareto efficiency" means efficiency.) Suppose that $\mathbf{0}$ satisfies specified constraints. For $\mathbf{0}$ to be efficient means that no $\boldsymbol{x} \succ \mathbf{0}$ satisfies the constraints.

We study efficiency by maximizing the efficiency function (5.6). Ecker and Kouada (1975) and Benson (1978) study efficiency via an equivalent formulation.

Definition 4 (Efficiency Function) The efficiency function is the concave function

$$\sum_i \left(x_i - \delta_{x_i \geq 0}\right) = \langle \mathbf{1}, \boldsymbol{x} \rangle - \delta_{\boldsymbol{x} \geq \mathbf{0}} = -\delta^*_{\mathbf{R}^n_+ + \{\mathbf{1}\}}(-\boldsymbol{x}). \tag{5.6}$$

Its conjugate is

$$\begin{aligned} &\inf_{\boldsymbol{x}} \left[\langle \boldsymbol{x}^*, \boldsymbol{x} \rangle - \left(\langle \mathbf{1}, \boldsymbol{x} \rangle - \delta_{\boldsymbol{x} \geq \mathbf{0}}\right)\right] \\ &= -\sup_{\boldsymbol{x}} \left[\langle \mathbf{1} - \boldsymbol{x}^*, \boldsymbol{x} \rangle - \delta_{\boldsymbol{x} \geq \mathbf{0}}\right] \\ &= -\delta_{\boldsymbol{x}^* \geq \mathbf{1}} \\ &= -\delta_{\mathbf{R}^n_+ + \{\mathbf{1}\}}(\boldsymbol{x}^*). \end{aligned} \tag{5.7}$$

Every x_i must be greater than or equal to one. Conjugate duality then implies the final formula in (5.6).

To test for the efficiency of $\mathbf{0}$, maximize the efficiency function subject to the constraints. If the optimum value is positive, then $\mathbf{0}$ is not efficient. The maximizing choice attains a positive value for at least one goal, with all goals greater than or equal to zero. If the optimum value is zero, then $\mathbf{0}$ is efficient.

Maximizing the efficiency function is especially fruitful in primal/dual convex optimization. The primal analyzes whether a particular vector of goals is efficient, and this vector is the perturbation variable. One interprets the dual choice variable as a shadow price for the goals. The dual involves the conjugate (5.7). In applications, that the price be greater than or equal to $\mathbf{1}$ boils down to requiring that the price be positive.

We interpret the Fenchel equality for the efficiency function as complementary slackness between $\boldsymbol{x}$ and $\boldsymbol{x}^* - \mathbf{1}$:

$$\begin{aligned} 0 &= \left[\langle \mathbf{1}, \boldsymbol{x} \rangle - \delta_{\boldsymbol{x} \geq \mathbf{0}}\right] - \delta_{\boldsymbol{x}^* \geq \mathbf{1}} - \langle \boldsymbol{x}^*, \boldsymbol{x} \rangle \\ &= -\left\{\delta_{\mathbf{R}^n_+}(\boldsymbol{x}) + \delta^*_{\mathbf{R}^n_+}\left[-(\boldsymbol{x}^* - \mathbf{1})\right] - \langle -(\boldsymbol{x}^* - \mathbf{1}), \boldsymbol{x} \rangle\right\}, \end{aligned}$$

which expresses the complementary slackness

$$\begin{aligned} \boldsymbol{x} &\geq \mathbf{0} \\ \boldsymbol{x}^* - \mathbf{1} &\geq \mathbf{0} \\ \langle \boldsymbol{x}^* - \mathbf{1}, \boldsymbol{x} \rangle &= 0. \end{aligned}$$

In particular, $x_i^* > 1$ requires $x_i = 0$.

5.6 Weak Efficiency

A related but distinct concept is "weak efficiency." Weak efficiency means that no other outcome is superior for every goal, whereas efficiency means that no other outcome is superior in some goal and not inferior in any. Efficiency implies weak efficiency, but not the converse.

Suppose that **0** satisfies specified constraints. The "weak efficiency" of **0** means that no $\boldsymbol{x} \gg \mathbf{0}$ satisfies the constraints. No vector is better for every goal.

Definition 5 (Weak-Efficiency Function) The weak-efficiency function is the concave function $\min(\boldsymbol{x}) = \min_i [x_i]$.

To test for weak efficiency, maximize $\min(\boldsymbol{x})$ subject to the constraints. If the optimum value is positive, then **0** is not weakly efficient. If the optimum value is zero, then **0** is weakly efficient. Economists know the min function in the context of fixed-coefficients production, in which production of an output requires fixed minimum quantities of each input.

Let us study the convex function $\max(\boldsymbol{x}) := \max[x_i]$. In terms of its epigraph,

$$\max(\boldsymbol{x}) = \inf_y \left[y + \delta_{\text{epi max}}(\boldsymbol{x}, y) \right] = \inf_y \left(y + \delta_{\boldsymbol{x} \leq \mathbf{1}y} \right).$$

As the value function of a convex minimization, max is convex. As it is closed and positively homogeneous, it is the support to a set P. By reversing the order of optimization,

$$\begin{aligned} \delta_P(\boldsymbol{x}^*) &= \max{}^*(\boldsymbol{x}^*) \\ &= \sup_{\boldsymbol{x}} [\langle \boldsymbol{x}^*, \boldsymbol{x} \rangle - \max(\boldsymbol{x})] \\ &= \sup_{\boldsymbol{x}} \left[\langle \boldsymbol{x}^*, \boldsymbol{x} \rangle - \inf_y \left(y + \delta_{\boldsymbol{x} \leq \mathbf{1}y} \right) \right] \\ &= \sup_{\boldsymbol{x}, y} \left(\langle \boldsymbol{x}^*, \boldsymbol{x} \rangle - y - \delta_{\boldsymbol{x} \leq \mathbf{1}y} \right) \end{aligned}$$

$$
\begin{aligned}
&= \sup_y \left[-y + \sup_x \left(\langle \boldsymbol{x}^*, \boldsymbol{x} \rangle - \delta_{\boldsymbol{x} \le \mathbf{1}y} \right) \right] \\
&= \delta_{\boldsymbol{x}^* \ge \mathbf{0}} + \sup_y \left[-y + \langle \boldsymbol{x}^*, \mathbf{1}y \rangle \right] \\
&= \delta_{\boldsymbol{x}^* \ge \mathbf{0}} + \delta_{\langle x^*, \mathbf{1} \rangle = 1}.
\end{aligned}
$$

In $\mathbf{R}^n$, the set P is the unit simplex and is the convex hull of the $\boldsymbol{e}_i$.

We interpret the Fenchel equality for the max function via complementary slackness. The Fenchel equality

$$
\begin{aligned}
0 &= \max(\boldsymbol{x}) + \max^*(\boldsymbol{x}^*) - \langle \boldsymbol{x}^*, \boldsymbol{x} \rangle \\
&= \max(\boldsymbol{x}) + \delta_{\max(\boldsymbol{x})\mathbf{1} \ge \boldsymbol{x}} + \delta_{\boldsymbol{x}^* \ge \mathbf{0}} + \delta_{\langle x^*, \mathbf{1} \rangle = 1} - \langle \boldsymbol{x}^*, \boldsymbol{x} \rangle \\
&= \delta_{\max(\boldsymbol{x})\mathbf{1} \ge \boldsymbol{x}} + \delta_{\boldsymbol{x}^* \ge \mathbf{0}} - \langle -\boldsymbol{x}^*, \max(\boldsymbol{x})\,\mathbf{1} - \boldsymbol{x} \rangle + \delta_{\langle x^*, \mathbf{1} \rangle = 1} \\
&= \left\{ \delta_{\mathbf{R}^n_+} \left[\max(\boldsymbol{x})\,\mathbf{1} - \boldsymbol{x} \right] + \delta^*_{\mathbf{R}^n_+}(-\boldsymbol{x}^*) - \langle -\boldsymbol{x}^*, \max(\boldsymbol{x})\,\mathbf{1} - \boldsymbol{x} \rangle \right\} + \delta_{\langle x^*, \mathbf{1} \rangle = 1}.
\end{aligned}
$$

The term in braces in greater than or equal to zero, a Fenchel inequality expressing complementary slackness. Hence the Fenchel equality for the max function is equivalent to

$$
\begin{aligned}
\max(\boldsymbol{x})\,\mathbf{1} - \boldsymbol{x} &\ge \mathbf{0} \\
\boldsymbol{x}^* &\ge \mathbf{0} \\
\langle \boldsymbol{x}^*, \max(\boldsymbol{x})\,\mathbf{1} - \boldsymbol{x} \rangle &= 0 \\
\langle \boldsymbol{x}^*, \mathbf{1} \rangle &= 1.
\end{aligned}
$$

The first three conditions are complementary slackness. If $x_i < \max(\boldsymbol{x})$, necessarily $x_i^* = 0$. If $x_i^* > 0$, necessarily $x_i = \max(\boldsymbol{x})$.

The weak-efficiency function $\min(\boldsymbol{x}) = -\max(-\boldsymbol{x})$, and its concave conjugate $\min_*(\boldsymbol{x}^*) = -\delta_P(\boldsymbol{x}^*)$. To maximize it in a primal/dual convex optimization is analogous to the study of efficiency. The dual minimization involves its conjugate, and again one interprets the dual choice variable as a shadow price for the goals. That the price belongs to P boils down to requiring that it be greater than zero. Commonly in economic theory for convenience a price vector is normalized to sum to one; in contrast here this normalization arises naturally via an optimization.

In $\mathbf{R}^n$, advantageously both the efficiency function and the weak-efficiency function are polyhedral.

That the effective domains of the efficiency function and the weak-efficiency function are different is significant. For the former it is the nonnegative cone, whereas for the latter it is the entire space. If $\mathbf{0}$ satisfies specified constraints, for it to be efficient the constraint set must just "touch" the nonnegative cone at the origin but cannot contain any vectors greater than zero. This technical difference can make the analysis of efficiency more challenging than the analysis of weak efficiency. Even though typically the efficiency dual satisfies the relative condition, the efficiency primal does not. There is no duality gap, but in special cases the optimum value zero is not attained in the dual. In contrast, for weak efficiency typically both the primal and the dual satisfy the relative condition.

Even though in general weak efficiency does not imply efficiency, commonly economic theorists blur the distinction between them by imposing additional conditions on the model, to guarantee that a weakly-efficient vector is necessarily efficient. They proceed to analyze weak efficiency. That their methods cannot readily handle efficiency is a key motive for this procedure. In contrast, the weak-efficiency and efficiency functions enable us to distinguish sharply and effectively between the two concepts, and we do so.

5.7 Euclidean Jordan Algebra

Economic applications commonly augment the Euclidean space $\mathbf{R}^n$ with the standard partial ordering defined above, in an informal way. However there are other useful partial orderings of a Euclidean space. This book develops and describes these partial orderings in a formal way, by working in a "Euclidean Jordan algebra," a Euclidean space having additional algebraic structure, in which the partial ordering derives from the formal structure (Jordan et al. 1934). The Euclidean space $\mathbf{R}^n$ augmented with the standard partial ordering is an example, and this book refers to it as the Euclidean Jordan algebra $\mathbf{R}^n$.

In abstract algebra, "isomorphic" means "the same." If two objects of a given type (group, ring, vector space, Euclidean space, algebra, etc.) are isomorphic, then they are "the same," when considered as objects of that type. An "isomorphism" is a one-to-one and onto mapping from one space to the other that "preserves" all properties defining the space. Any n-dimensional Euclidean space is isomorphic to the Euclidean space $\mathbf{R}^n$. In contrast, two different Euclidean Jordan algebras of having the same dimension are isomorphic as Euclidean spaces, but perhaps not as Euclidean Jordan algebras.

Although the Euclidean Jordan algebra $\mathbf{R}^n$ is the mainstream application, the classification theorem asserts that there are five distinct types of Euclidean Jordan algebras. This book employs three types, in applications to econometrics, finance, and stability. As the five types share many properties, later we develop and describe their common properties at length. Euclidean Jordan algebra is a valuable structure for convex optimization and duality.

In this book, applications not involving a partial ordering are set in Euclidean space, whereas applications involving a partial ordering are set in a Euclidean Jordan algebra. This difference affects the graphical presentation in a Euclidean space. Viewing the space as coordinate-free and not involving a basis, a figure plots the origin but not axes. Orthogonal vectors are plotted as perpendicular. On the other hand, for the direct sum of two Euclidean spaces, in a figure the horizontal axis represents one space, and the vertical axis represents the other space. The axes are perpendicular, because vectors in the two spaces are orthogonal.

References

Benson HP (1978 Dec) Existence of efficient solutions for vector maximization problems. J Optim Theor Appl 26(4):569–580

Ecker JG, Kouada IA (1975) Finding efficient points for linear multiple objective programs. Math Program 8:375–377

Jordan P, Neumann JV, Wigner E (1934 Jan) On an algebraic generalization of the quantum mechanical formalism. Ann Math 35(1):29–64

Chapter 6
Epigraph

Abstract The epigraph is the points $(\boldsymbol{x}, y)$ above the function,

$$\delta_{\mathrm{epi}\, f}(\boldsymbol{x}, y) = \delta_{f(\boldsymbol{x}) \leq y}.$$

The conjugate is the support to the epigraph, $f^*(\boldsymbol{x}^*) = \delta^*_{\mathrm{epi}\, f}(\boldsymbol{x}^*, -1)$.

"Epi-multiplication" is the support to the epigraph of the conjugate:

$$(y \star f)(\boldsymbol{x}) := \delta^*_{\mathrm{epi}\, f^*}(\boldsymbol{x}, -y).$$

Here y is the "epi-multiplier." That epi-multiplication corresponds to multiplying the epigraph by y is the source of the terminology.

Epi-multiplication by zero is the "horizon function":

$$f^\infty(\boldsymbol{x}) := (0 \star f)(\boldsymbol{x}) = \delta^*_{\mathrm{dom}\, f^*}(\boldsymbol{x}),$$

the support to the effective domain of the conjugate.

We study the epigraph and its support for a proper, closed, convex function $f(\boldsymbol{x})$. The epigraph is the points $(\boldsymbol{x}, y)$ above the function,

$$\delta_{\mathrm{epi}\, f}(\boldsymbol{x}, y) = \delta_{f(\boldsymbol{x}) \leq y}.$$

The conjugate is the support to the epigraph, $f^*(\boldsymbol{x}^*) = \delta^*_{\mathrm{epi}\, f}(\boldsymbol{x}^*, -1)$. The relative interior of the epigraph is the $(\boldsymbol{x}, y)$ such that $\boldsymbol{x} \in \mathrm{rint}(\mathrm{dom}\, f)$ and $f(\boldsymbol{x}) < y$ (Rockafellar 1970, Lemma 7.3, p. 54).

B. C. Dieffenbach, *Conjugate Duality in Economic Analysis*, Contributions to Economics,
https://doi.org/10.1007/978-3-032-21396-9_6

6.1 Support

The support to the epigraph has three cases.

Theorem 1 (Support to the Epigraph)

$$\delta^*_{\text{epi } f}(\boldsymbol{x}^*, -y^*) = \begin{cases} y^* f^*(\boldsymbol{x}^*/y^*) \ \textit{for}\ y^* > 0 \\ \delta^*_{\text{dom } f}(\boldsymbol{x}^*) \ \textit{for}\ y^* = 0 \\ \infty \ \textit{for}\ y^* < 0. \end{cases}$$

Because the support is positively homogeneous, for $y^* > 0$

$$\delta^*_{\text{epi } f}(\boldsymbol{x}^*, -y^*) = y^* \delta^*_{\text{epi } f}(\boldsymbol{x}^*/y^*, -1) = y^* f^*(\boldsymbol{x}^*/y^*).$$

For $y^* = 0$, the support to epi f is the support to dom f (Rockafellar 1966):

$$\begin{aligned} \delta^*_{\text{epi } f}(\boldsymbol{x}^*, 0) &= \sup_{\boldsymbol{x},y} \left[\langle \boldsymbol{x}^*, \boldsymbol{x} \rangle - \delta_{\text{epi } f}(\boldsymbol{x}, y) \right] \\ &= \sup_{\boldsymbol{x}} \left\{ \left[\langle \boldsymbol{x}^*, \boldsymbol{x} \rangle + \sup_{y} \left[-\delta_{\text{epi } f}(\boldsymbol{x}, y) \right] \right] \right\} \\ &= \sup_{\boldsymbol{x}} \left[\langle \boldsymbol{x}^*, \boldsymbol{x} \rangle - \inf_{y} \delta_{\text{epi } f}(\boldsymbol{x}, y) \right] \\ &= \sup_{\boldsymbol{x}} \left[\langle \boldsymbol{x}^*, \boldsymbol{x} \rangle - \delta_{\text{dom } f}(\boldsymbol{x}) \right] \\ &= \delta^*_{\text{dom } f}(\boldsymbol{x}^*). \end{aligned}$$

We refer to $y^* = 0$ as the "exceptional case," and many exceptional cases in economic theory correspond to this case.

6.2 Epi-Multiplication

"Epi-multiplication" is the support to the epigraph of the conjugate.[1]

Definition 1 (Epi-Multiplication) The epi-multiplication

$$(y \star f)(\boldsymbol{x}) := \delta^*_{\text{epi } f^*}(\boldsymbol{x}, -y).$$

[1] Whereas here we define epi-multiplication via the support to the epigraph of the conjugate, Rockafellar and Wets define it via multiplication of the epigraph, so for the exceptional case their definition of epi-multiplication differs (Rockafellar and Wets 1998).

Here y is the "epi-multiplier." For $y > 0$, that epi-multiplication corresponds to multiplying the epigraph by y is the source of the terminology. Define a function $g(\boldsymbol{x})$ having epigraph

$$\text{epi}\, g := y\, \text{epi} f.$$

We have

$$\begin{aligned} g(\boldsymbol{x}) &= \inf_z \left[z + \delta_{\text{epi}\, g}(\boldsymbol{x}, z)\right] \\ &= \inf_z \left[z + \delta_{y\, \text{epi}\, f}(\boldsymbol{x}, z)\right] \\ &= \inf_z \left[z + \delta_{\text{epi}\, f}(\boldsymbol{x}/y, z/y)\right] \\ &= y \inf_w \left[w + \delta_{\text{epi}\, f}(\boldsymbol{x}/y, w)\right] \ (w := z/y) \\ &= yf(\boldsymbol{x}/y) \\ &= (y \star f)(\boldsymbol{x}). \end{aligned}$$

Epi-multiplication by zero is the "horizon function," the support to the effective domain of the conjugate.

Definition 2 (Horizon Function) The horizon function

$$f^{\infty}(\boldsymbol{x}) := (0 \star f)(\boldsymbol{x}) = \delta^*_{\text{dom}\, f^*}(\boldsymbol{x}).$$

The horizon function shows the behavior of f at infinity. Because the support to the epigraph is closed, the vectors $(\overline{\boldsymbol{x}}, \overline{z}) \in \text{epi} f^{\infty}$ constitute the limit points $(\overline{\boldsymbol{x}}, 0, \overline{z})$ of the sequences $(\boldsymbol{x}, y, z)$ such that $(\boldsymbol{x}/y, z/y) \in \text{epi} f$ and $y \searrow 0$. The horizon function is positively homogeneous.

Summary 1 (Epi-Multiplication)

$$(y \star f)(\boldsymbol{x}) = \begin{cases} yf(\boldsymbol{x}/y) \ \text{for}\ y > 0 \\ f^{\infty}(\boldsymbol{x}) \ \text{for}\ y = 0 \\ \infty \ \text{for}\ y < 0. \end{cases}$$

6.3 Fenchel Equality

We interpret the Fenchel equality connecting the epigraph and its support:

$$\delta_{\text{epi}\, f}(\boldsymbol{x}, y) + \delta^*_{\text{epi}\, f}(\boldsymbol{x}^*, -y^*) - \langle \boldsymbol{x}^*, \boldsymbol{x} \rangle - (-y^*)\, y = 0. \tag{6.1}$$

Consider first the positive epi-multiplier case. Dividing the epigraph Fenchel equality (6.1) by $y^* > 0$ obtains

$$\begin{aligned} 0 &= \delta_{\text{epi}\, f}(\boldsymbol{x}, y) + f^*(\boldsymbol{x}^*/y^*) - \langle \boldsymbol{x}^*/y^*, \boldsymbol{x}\rangle + y \\ &= [f(\boldsymbol{x}) + f^*(\boldsymbol{x}^*/y^*) - \langle \boldsymbol{x}^*/y^*, \boldsymbol{x}\rangle] + \left[y - f(\boldsymbol{x}) + \delta_{\text{epi}\, f}(\boldsymbol{x}, y)\right] \end{aligned}$$

Each term in brackets is nonnegative, so the sum is zero if and only if each is zero.

Theorem 2 (Positive Epi-Multiplier) *For $y^* > 0$, the epigraph Fenchel equality (6.1) is equivalent to the Fenchel equality*

$$f(\boldsymbol{x}) + f^*(\boldsymbol{x}^*/y^*) - \langle \boldsymbol{x}^*/y^*, \boldsymbol{x}\rangle = 0$$

together with the condition

$$f(\boldsymbol{x}) = y.$$

For exceptional case $y^* = 0$, the Fenchel equality (6.1) is

$$\begin{aligned} 0 &= \delta_{\text{epi}\, f}(\boldsymbol{x}, y) + \delta^*_{\text{epi}\, f}(\boldsymbol{x}^*, 0) - \langle \boldsymbol{x}^*, \boldsymbol{x}\rangle - 0y \\ &= \delta_{f(\boldsymbol{x})\le y} + \delta_{\text{dom}\, f}(\boldsymbol{x}) + \delta^*_{\text{dom}\, f}(\boldsymbol{x}^*) - \langle \boldsymbol{x}^*, \boldsymbol{x}\rangle \\ &= \left[\delta_{\text{dom}\, f}(\boldsymbol{x}) + \delta^*_{\text{dom}\, f}(\boldsymbol{x}^*) - \langle \boldsymbol{x}^*, \boldsymbol{x}\rangle\right] + \delta_{f(\boldsymbol{x})\le y}. \end{aligned}$$

The term in brackets and the final term are each nonnegative, so the sum is zero if and only if each is zero.

Theorem 3 (Exceptional Case) *For the exceptional case $y^* = 0$, the Fenchel equality (6.1) is equivalent to*

$$\delta_{\text{dom}\, f}(\boldsymbol{x}) + \delta^*_{\text{dom}\, f}(\boldsymbol{x}^*) - \langle \boldsymbol{x}^*, \boldsymbol{x}\rangle = 0 \tag{6.2}$$

together with the inequality

$$f(\boldsymbol{x}) \le y.$$

Here dom f is perhaps not closed. The same equivalence holds if one replaces dom f by its closure, and (6.2) becomes a Fenchel equality. The exceptional case is important for the analysis of efficiency and weak efficiency.

6.4 Concave Function

We state the corresponding results for a concave function. The epigraph is the points $(\boldsymbol{x}, y)$ below the function,

$$\delta_{\text{epi}\, f}(\boldsymbol{x}, y) = \delta_{f(\boldsymbol{x}) \geq y}.$$

The concave conjugate is

$$f_*(\boldsymbol{x}^*) = -\delta^*_{\text{epi}\, f}(-\boldsymbol{x}^*, 1).$$

Definition 3 (Epi-Multiplication) For a concave function

$$(y \star f)(\boldsymbol{x}) := -\delta^*_{\text{epi}\, f_*}(-\boldsymbol{x}, y).$$

Definition 4 (Horizon Function) For a concave function

$$f_\infty(\boldsymbol{x}) := (0 \star f)(\boldsymbol{x}) = -\delta^*_{\text{dom}\, f_*}(-\boldsymbol{x}).$$

Summary 2 (Concave Epi-Multiplication)

$$(y \star f)(\boldsymbol{x}) = \begin{cases} yf(\boldsymbol{x}/y) \ \text{for } y > 0 \\ f_\infty(\boldsymbol{x}) \ \text{for } y = 0 \\ -\infty \ \text{for } y < 0. \end{cases}$$

We interpret the Fenchel equality connecting the epigraph and its support:

$$0 = \delta_{\text{epi}\, f}(\boldsymbol{x}, y) + \delta^*_{\text{epi}\, f}(-\boldsymbol{x}^*, y^*) - \langle -\boldsymbol{x}^*, \boldsymbol{x} \rangle - y^* y. \tag{6.3}$$

Theorem 4 (Positive Epi-Multiplier) *For $y^* > 0$, the Fenchel equality (6.3) is equivalent to the Fenchel equality*

$$f(\boldsymbol{x}) + f_*(\boldsymbol{x}^*/y^*) - \langle \boldsymbol{x}^*/y^*, \boldsymbol{x} \rangle = 0$$

together with the equality

$$f(\boldsymbol{x}) = y.$$

Theorem 5 (Zero Epi-Multiplier) *For the exceptional case $y^* = 0$, the Fenchel equality (6.3) is equivalent to*

$$\delta_{\text{dom}\, f}(\boldsymbol{x}) + \delta^*_{\text{dom}\, f}(-\boldsymbol{x}^*) + \langle \boldsymbol{x}^*, \boldsymbol{x} \rangle = 0$$

together with the inequality

$$f(\boldsymbol{x}) \geq y.$$

References

Rockafellar RT (1966) Level sets and continuity of conjugate convex functions. Trans Am Math Soc 123:46–63
Rockafellar RT (1970) Convex analysis. Princeton University Press, Princeton, NJ
Rockafellar RT, Wets RJB (1998) Variational analysis, Vol 317. Springer, Berlin

Chapter 7
Subdifferential

Abstract The subdifferential describes the local behavior of a convex function. Previously we defined the subdifferential via the nonvertical hyperplanes supporting the epigraph. The subgradient inequality is an equivalent, traditional definition. For a convex function $f(\boldsymbol{x})$, a change in $\boldsymbol{x}$ changes a subgradient $\boldsymbol{x}^*$ in a like direction, rather than in the opposite direction. Whereas differential calculus deals with derivatives, subdifferential calculus deals with subdifferentials. In subdifferential calculus, one works with rules for subdifferentials, which form only a subset of the rules for differential calculus. For example, whereas the chain rule is a key tool of differential calculus, unfortunately this rule does not hold for subdifferential calculus. In general the composition of two convex functions need not be convex, so the chain rule cannot apply in general. For convex F, define the value function

$$V(\boldsymbol{y}) = \inf_{\boldsymbol{x}} F(\boldsymbol{x}, \boldsymbol{y}),$$

Under general conditions, the following envelope inclusion holds: if choice $\boldsymbol{x}$ is minimizing for perturbation $\boldsymbol{y}$, then

$$\partial V(\boldsymbol{y}) \subset \partial_Y F(\boldsymbol{x}, \boldsymbol{y}),$$

a generalization of the envelope theorem.

The subdifferential describes the local behavior of a convex function. Previously we defined the subdifferential via the nonvertical hyperplanes supporting the epigraph. If a hyperplane supporting the epigraph at $[\boldsymbol{x}, f(\boldsymbol{x})]$ has a normal $(\boldsymbol{x}^*, -1)$, then $\boldsymbol{x}^*$ is a subgradient at $\boldsymbol{x}$, and the subdifferential $\partial f(\boldsymbol{x})$ is the set of subgradients.

B. C. Dieffenbach, *Conjugate Duality in Economic Analysis*, Contributions to Economics,
https://doi.org/10.1007/978-3-032-21396-9_7

7.1 Subgradient Inequality

The equivalent traditional definition is via the subgradient inequality. We prove the equivalence below.

Theorem 1 (Subgradient Inequality) *A subgradient* $\boldsymbol{x}^* \in \partial f(\boldsymbol{x})$ *if and only if* $f(z) \geq f(\boldsymbol{x}) + \langle \boldsymbol{x}^*, z - \boldsymbol{x}\rangle$, *the subgradient inequality.*

For an indicator, the subdifferential $\partial \delta_C(\boldsymbol{x})$ is the normal cone at $\boldsymbol{x}$. The subgradient inequality theorem has a corollary for an indicator.

Corollary 1 (Indicator Subdifferential) *A subgradient* $\boldsymbol{x}^* \in \partial \delta_C(\boldsymbol{x})$ *if and only if* $\langle \boldsymbol{x}^*, z - \boldsymbol{x}\rangle \leq 0$ *for all* $z \in C$.

This inequality is commonly taken as the definition of the normal cone.

We derive results about the subdifferential at a vector in the effective domain by translating the epigraph to the origin, so $f(\mathbf{0}) = \mathbf{0}$. That the conjugate is nonnegative is basic:

$$f^*(\boldsymbol{x}^*) = \sup_{\boldsymbol{x}} [\langle \boldsymbol{x}^*, \boldsymbol{x}\rangle - f(\boldsymbol{x})] \geq \langle \boldsymbol{x}^*, \mathbf{0}\rangle - f(\mathbf{0}) = 0.$$

The Fenchel equality $f(\boldsymbol{x}) + f^*(\boldsymbol{x}^*) - \langle \boldsymbol{x}^*, \boldsymbol{x}\rangle = 0$ and its subdifferential form $\boldsymbol{x}^* \in \partial f(\boldsymbol{x})$ are equivalent. At $\boldsymbol{x} = \mathbf{0}$, the subdifferential is the $\boldsymbol{x}^*$ such that $f^*(\boldsymbol{x}^*) = 0$, as

$$\begin{aligned} 0 &= f(\boldsymbol{x}) + f^*(\boldsymbol{x}^*) - \langle \boldsymbol{x}^*, \boldsymbol{x}\rangle \\ &= f(\mathbf{0}) + f^*(\boldsymbol{x}^*) - \langle \boldsymbol{x}^*, \mathbf{0}\rangle \\ &= f^*(\boldsymbol{x}^*). \end{aligned}$$

Because the conjugate is nonnegative, the subdifferential is also the contour set of $\boldsymbol{x}^*$ such that $f^*(\boldsymbol{x}^*) \leq 0$ and thus is a closed, convex set. Because

$$f^*(\boldsymbol{x}^*) = \sup_{\boldsymbol{x}} [\langle \boldsymbol{x}^*, \boldsymbol{x}\rangle - f(\boldsymbol{x})],$$

then $f^*(\boldsymbol{x}^*) \leq 0$ if and only if

$$f(\boldsymbol{x}) \geq \langle \boldsymbol{x}^*, \boldsymbol{x}\rangle$$

for all $\boldsymbol{x}$, the subgradient inequality.

A comparative statics result in economic theory says that a change in a variable $\boldsymbol{x}$ changes another variable $\boldsymbol{x}^*$ in a certain way. Many comparative statics results in economic theory boil down to the following inequality.

Theorem 2 (Comparative Statics) *For a convex function f, if $\boldsymbol{x}_1^* \in \partial f(\boldsymbol{x}_1)$ and $\boldsymbol{x}_2^* \in \partial f(\boldsymbol{x}_2)$, then*

$$\langle \boldsymbol{x}_1^* - \boldsymbol{x}_2^*, \boldsymbol{x}_1 - \boldsymbol{x}_2 \rangle \geq 0.$$

The theorem follows from the subgradient inequality:

$$\begin{aligned} f(\boldsymbol{x}_2) &\geq f(\boldsymbol{x}_1) - \langle \boldsymbol{x}_1^*, \boldsymbol{x}_2 - \boldsymbol{x}_1 \rangle \\ f(\boldsymbol{x}_1) &\geq f(\boldsymbol{x}_2) - \langle \boldsymbol{x}_2^*, \boldsymbol{x}_1 - \boldsymbol{x}_2 \rangle. \end{aligned}$$

Adding these inequalities, cancelling, and rearranging obtains the result.

The comparative statics Theorem 2 says that the inner product of the change in $\boldsymbol{x}^*$ and the change in $\boldsymbol{x}$ is greater than or equal to zero. An economist would tend to interpret this inequality as saying that an *increase* in $\boldsymbol{x}$ *raises* $\boldsymbol{x}^*$. Yet the inequality obtains for any Euclidean space. A Euclidean space embodies no concept of order, so one cannot say that one vector is larger than another. Consequently a better interpretation is that a change in $\boldsymbol{x}$ changes $\boldsymbol{x}^*$ in a like direction, rather than in the opposite direction.

In $\operatorname{dom} f$, the subdifferential can be empty only at the relative boundary.

Theorem 3 (Subdifferentiability) *A proper, closed, convex function is subdifferentiable in the relative interior of its effective domain.*

Apply the relative condition in perturbation duality. The "primal"

$$f(\boldsymbol{x}) = \inf f(\boldsymbol{x}) \tag{7.1}$$

(the perturbation is $\boldsymbol{x}$, and the choice variable is empty) has the dual

$$\sup_{\boldsymbol{x}^*} [\langle \boldsymbol{x}^*, \boldsymbol{x} \rangle - f^*(\boldsymbol{x}^*)].$$

The primal satisfies the relative condition if $\boldsymbol{x} \in \operatorname{rint}(\operatorname{dom} f)$. Then there is no duality gap, and the optimum value $f(\boldsymbol{x})$ is attained in the dual. The first-order condition for the dual is $\boldsymbol{x} \in \partial f^*(\boldsymbol{x}^*)$. By subgradient invertibility, the solution set is $\partial f(\boldsymbol{x})$, which is nonempty.

If in (7.1) f is polyhedral, then the polyhedral condition holds throughout $\operatorname{dom} f$.

Theorem 4 (Polyhedral Subdifferentiability) *A proper, polyhedral, convex function is subdifferentiable throughout its effective domain.*

7.2 Subdifferential Calculus

Whereas differential calculus deals with derivatives, subdifferential calculus deals with subdifferentials. In differential calculus, one works with rules for derivatives. Analogously, in subdifferential calculus, one works with rules for subdifferentials. As the subdifferential is a generalization of the derivative, consequently the rules for subdifferential calculus form only a subset of the rules for differential calculus. For example, whereas the chain rule is a key tool of differential calculus, unfortunately this rule does not hold for subdifferential calculus. In general the composition of two convex functions need not be convex, so the chain rule cannot apply in general.

Nonetheless subdifferentials do satisfy certain rules. To a limited extent, the rules of differential calculus do generalize, but with qualifications. For example, in differential calculus, the derivative of a sum of two differentiable functions is the sum of the derivatives. In subdifferential calculus, the subdifferential of the sum of two convex functions is the sum of the subdifferentials (the sum of sets), under general conditions, but not always. Using a perturbation duality argument, we prove this result below.

7.2.1 Subdifferential of a Sum

For proper, closed, convex functions f and g, let us study the subdifferential at a vector where both functions are subdifferentiable.

Condition 1 (Relative Condition for the Subdifferential of a Sum) The relative interiors of the effective domains of f and g have a vector in common.

Condition 2 (Polyhedral Condition for the Subdifferential of a Sum) The effective domains of the polyhedral functions f and g have a vector in common.

Theorem 5 (Subdifferential of a Sum) *If either the relative condition or the polyhedral condition holds, then the subdifferential of the sum is the sum of the subdifferentials:*

$$\partial (f + g) = \partial f + \partial g.$$

We prove this theorem by a perturbation-duality argument. Translate the epigraph of each function so $f(\mathbf{0}) = 0$ and $g(\mathbf{0}) = 0$.

Problem 1 (Sum Primal) The subdifferential $\partial (f + g)(\mathbf{0})$ is the set of $\boldsymbol{x}^*$ such that

$$0 = (f + g)^* (\boldsymbol{x}^*) = \sup_{\boldsymbol{x}} [\langle \boldsymbol{x}^*, \boldsymbol{x}\rangle - f(\boldsymbol{x}) - g(\boldsymbol{x})].$$

Because the choice $\boldsymbol{x} = \boldsymbol{0}$ makes the objective function zero, the supremum is greater than or equal to zero.

We calculate the reciprocal Fenchel dual. Perturb the sum primal as

$$\sup_{x} [\langle \boldsymbol{x}^*, \boldsymbol{x} \rangle - f(\boldsymbol{x}) - g(\boldsymbol{x} - \boldsymbol{y})],$$

in which the perturbation base value $\boldsymbol{y} = \boldsymbol{0}$. Adding the Fenchel inequalities

$$f(\boldsymbol{x}) + f^*(\boldsymbol{x}^* - \boldsymbol{y}^*) - \langle \boldsymbol{x}^* - \boldsymbol{y}^*, \boldsymbol{x} \rangle \geq 0$$
$$g(\boldsymbol{x} - \boldsymbol{y}) + g^*(\boldsymbol{y}^*) - \langle \boldsymbol{y}^*, \boldsymbol{x} - \boldsymbol{y} \rangle \geq 0$$

and rearranging obtains

$$\langle \boldsymbol{y}^*, \boldsymbol{y} \rangle + f^*(\boldsymbol{x}^* - \boldsymbol{y}^*) + g^*(\boldsymbol{y}^*) \\ \geq \langle \boldsymbol{x}^*, \boldsymbol{x} \rangle - f(\boldsymbol{x}) - g(\boldsymbol{x} - \boldsymbol{y}).$$

Problem 2 (Sum Dual) The Fenchel dual is

$$\inf_{y^*} [f^*(\boldsymbol{x}^* - \boldsymbol{y}^*) + g^*(\boldsymbol{y}^*)].$$

That f^* and g^* are both nonnegative verifies that the infimum in the dual is greater than or equal to zero. Reasoning from the sum primal/dual pair, we prove $\partial f(\boldsymbol{0}) + \partial g(\boldsymbol{0}) = \partial (f + g)(\boldsymbol{0})$, by showing inclusion in both directions.[1]

Necessarily $\partial f(\boldsymbol{0}) + \partial g(\boldsymbol{0}) \subset \partial (f + g)(\boldsymbol{0})$, regardless of whether the relative condition or the polyhedral condition holds. Choose $\boldsymbol{y}^* \in \partial g(\boldsymbol{0})$ and select $\boldsymbol{x}^*$ so $\boldsymbol{x}^* - \boldsymbol{y}^* \in \partial f(\boldsymbol{0})$. Thus $f^*(\boldsymbol{x}^* - \boldsymbol{y}^*)$ and $g^*(\boldsymbol{y}^*)$ are both zero. There is no duality gap, and the optimum value is zero. That the optimum value of the primal is zero means $\boldsymbol{x}^* \in \partial (f + g)(\boldsymbol{0})$.

If either the relative condition or the polyhedral condition holds, then the opposite inclusion $\partial (f + g)(\boldsymbol{0}) \subset \partial f(\boldsymbol{0}) + \partial g(\boldsymbol{0})$ follows. Suppose that $\boldsymbol{x}^* \in \partial (f + g)(\boldsymbol{0})$. The optimum value of the primal is zero, so the objective function in the dual is zero for some choice $\boldsymbol{y}^*$. Because f^* and g^* are both nonnegative, each must be zero. Thus $\boldsymbol{x}^* - \boldsymbol{y}^* \in \partial f(\boldsymbol{0})$ and $\boldsymbol{y}^* \in \partial g(\boldsymbol{0})$, so $\boldsymbol{x}^* \in \partial f(\boldsymbol{0}) + \partial g(\boldsymbol{0})$.

7.2.2 Partial Subdifferential

The partial subdifferential is a generalization of the partial derivative. Consider a proper, closed, convex function $f(\boldsymbol{x}, \boldsymbol{y})$ defined on $\mathbf{X} \oplus \mathbf{Y}$. The partial subdifferential $\partial_X f$ is the subdifferential of f with respect to $\boldsymbol{x}$, while holding $\boldsymbol{y}$ constant.

---

[1] Rockafellar demonstrates the inclusion in both directions, but without the simplification obtained via perturbation duality (Rockafellar 1970, Theorem 23.8, p. 223 and Theorem 23.9, p. 225).

The relative and polyhedral conditions are standard.

Condition 3 (Relative Condition) *For the given* $\boldsymbol{y}$ *there is some* $\boldsymbol{x}$ *such that* $(\boldsymbol{x},\boldsymbol{y}) \in \operatorname{rint}(\operatorname{dom} f)$.

Condition 4 (Polyhedral Condition) *For* f *polyhedral, for the given* $\boldsymbol{y}$ *there is some* $\boldsymbol{x}$ *such that* $(\boldsymbol{x},\boldsymbol{y}) \in \operatorname{dom} f$.

Define the projection of $\mathbf{X} \oplus \mathbf{Y}$ onto $\mathbf{X}$ as the mapping $(\boldsymbol{x},\boldsymbol{y}) \mapsto \boldsymbol{x}$.

Theorem 6 (Partial Subdifferential) *If* f *satisfies either the relative condition or the polyhedral condition, then* $\partial_X f(\boldsymbol{x},\boldsymbol{y})$ *is the projection of the subdifferential* $\partial f(\boldsymbol{x},\boldsymbol{y})$ *into* $\mathbf{X}$.

We prove this theorem by a perturbation-duality argument. For simplicity, translate the function so $f(\mathbf{0},\mathbf{0}) = 0$. Just as the subdifferential $\partial f(\mathbf{0})$ is the $(\boldsymbol{x}^*,\boldsymbol{y}^*)$ for which f^* is zero, the analogous property describes the partial subdifferential.

Problem 3 (Partial-Subdifferential Primal) The partial subdifferential $\partial_X f(\mathbf{0})$ is the set of $\boldsymbol{x}^*$ such that

$$0 = \sup_{\boldsymbol{x}} \left[\langle \boldsymbol{x}^*,\boldsymbol{x}\rangle - f(\boldsymbol{x},\mathbf{0})\right].$$

Because the choice $\boldsymbol{x} = \mathbf{0}$ makes the value zero, the supremum is nonnegative.

We calculate the reciprocal dual, in which the perturbation base value $\boldsymbol{y} = \mathbf{0}$. Rearranging the Fenchel inequality

$$f(\boldsymbol{x},\boldsymbol{y}) + f^*(\boldsymbol{x}^*,\boldsymbol{y}^*) - \langle \boldsymbol{x}^*,\boldsymbol{x}\rangle - \langle \boldsymbol{y}^*,-\boldsymbol{y}\rangle \geq 0$$

obtains the reciprocal dual

$$\langle \boldsymbol{y}^*,\boldsymbol{y}\rangle + f^*(\boldsymbol{x}^*,\boldsymbol{y}^*) \geq \langle \boldsymbol{x}^*,\boldsymbol{x}\rangle - f(\boldsymbol{x},-\boldsymbol{y}).$$

Problem 4 (Partial-Subdifferential Dual)

$$\inf_{\boldsymbol{y}^*} f^*(\boldsymbol{x}^*,\boldsymbol{y}^*).$$

Because f^* is nonnegative, the infimum is greater than or equal to zero. If the infimum is zero, then there is no duality gap, as the value of the primal is greater than or equal to zero. If either the relative condition or the polyhedral condition holds, then there is no duality gap; if the optimum value is finite, it is attained in the dual.

That $(\boldsymbol{x}^*,\boldsymbol{y}^*) \in \partial f(\mathbf{0},\mathbf{0})$ implies $\boldsymbol{x}^* \in \partial_X f(\mathbf{0},\mathbf{0})$, regardless of whether the relative or polyhedral condition holds. If $f^*(\boldsymbol{x}^*,\boldsymbol{y}^*) = 0$, the dual attains its minimum value zero at the choice $(\boldsymbol{x}^*,\boldsymbol{y}^*)$. The maximum value of the primal is zero, so $\boldsymbol{x}^* \in \partial_X f(\mathbf{0},\mathbf{0})$.

That $\boldsymbol{x}^* \in \partial_X f(\mathbf{0},\mathbf{0})$ implies $(\boldsymbol{x}^*,\boldsymbol{y}^*) \in \partial f(\mathbf{0},\mathbf{0})$ if there is no duality gap and the solution is attained in the dual. If $\boldsymbol{x}^* \in \partial_X f(\mathbf{0},\mathbf{0})$, the maximum value of the primal is zero. Hence $f^*(\boldsymbol{x}^*,\boldsymbol{y}^*) = 0$ for some dual choice $\boldsymbol{y}^*$, so $(\boldsymbol{x}^*,\boldsymbol{y}^*) \in \partial f(\mathbf{0},\mathbf{0})$.

If the relative condition fails to hold, then this converse can be false. A possibility is that there is no duality gap with optimum value zero, such that the maximum is attained by $\boldsymbol{x} = \mathbf{0}$ in the primal but the infimum in the dual is not attained. Thus $(\boldsymbol{x}^*,\boldsymbol{y}^*) \notin \partial f(\mathbf{0},\mathbf{0})$ for any $\boldsymbol{y}^*$, even though $\boldsymbol{x}^* \in \partial_X f(\mathbf{0},\mathbf{0})$ (the first-order condition for the primal).

At a vector, the subdifferential contains only a single subgradient if and only if the function is differentiable there. Because the relative condition for the partial subdifferential holds, the partial first-order conditions are equivalent to the traditional first-order condition: $\{\mathbf{0}\} = \partial_X f(\boldsymbol{x},\boldsymbol{y})$ and $\{\mathbf{0}\} = \partial_Y f(\boldsymbol{x},\boldsymbol{y})$ if and only if $\{\mathbf{0},\mathbf{0}\} = \partial f(\boldsymbol{x},\boldsymbol{y})$.

7.2.3 Envelope

Consider the following standard perturbed minimization.

Problem 5 (Perturbed Minimization) The value function

$$V(y) = \inf_{\boldsymbol{x}} F(\boldsymbol{x},\boldsymbol{y}), \tag{7.2}$$

in which F is convex in the choice variable $\boldsymbol{x}$ and the perturbation variable $\boldsymbol{y}$.

Let us assume that there is no duality gap.

Condition 5 (No Duality Gap) *For the problem (7.2), there is no duality gap.*

If there is no duality gap, then the subdifferential $\partial V(y)$ of the value function is the solution set of the dual $\sup_{y^*} [\langle y^*, y\rangle - V(y)]$.

Below we prove the following.

Theorem 7 (Envelope Inclusion) *For the problem (7.2), if choice* $\boldsymbol{x}$ *is minimizing for perturbation* $\boldsymbol{y}$*, then*

$$\partial V(y) \subset \partial_Y F(\boldsymbol{x},\boldsymbol{y}).$$

Theorem 8 (Envelope) *For the problem (7.2), if choice* $\boldsymbol{x}$ *is minimizing for perturbation* $\boldsymbol{y}$ *and* F *is differentiable at* $(\boldsymbol{x},\boldsymbol{y})$*, then the dual has a unique solution:*

$$\{y^*\} = \partial V(y) = \partial_Y F(\boldsymbol{x},\boldsymbol{y}).$$

These results generalize the envelope theorem in (Samuelson 1947, p. 34), who assumes that F is differentiable and that the (unique) optimum $\boldsymbol{x}$ is a differentiable function of $\boldsymbol{y}$.

We find the dual of (7.2). Rearranging the Fenchel inequality

$$F(\boldsymbol{x},\boldsymbol{y}) + F^*(\boldsymbol{0},\boldsymbol{y}^*) - \langle(\boldsymbol{0},\boldsymbol{y}^*),(\boldsymbol{x},\boldsymbol{y})\rangle \geq 0 \tag{7.3}$$

obtains the primal and the dual:

$$F(\boldsymbol{x},\boldsymbol{y}) \geq \langle \boldsymbol{y}^*,\boldsymbol{y}\rangle - F^*(\boldsymbol{0},\boldsymbol{y}^*).$$

The dual is

$$\sup_{\boldsymbol{y}^*}\left[\langle \boldsymbol{y}^*,\boldsymbol{y}\rangle - F^*(\boldsymbol{0},\boldsymbol{y}^*)\right].$$

Suppose that $\boldsymbol{x}$ is minimizing in the primal, and $\boldsymbol{y}^*$ is maximizing in the dual. The Fenchel inequality (7.3) holds with equality, and its subdifferential form is

$$(\boldsymbol{0},\boldsymbol{y}^*) \in \partial F(\boldsymbol{x},\boldsymbol{y}).$$

Because the partial subdifferential contains the projection of the subdifferential, therefore

$$\begin{aligned}\boldsymbol{0} &\in \partial_X F(\boldsymbol{x},\boldsymbol{y})\\ \boldsymbol{y}^* &\in \partial_Y F(\boldsymbol{x},\boldsymbol{y}).\end{aligned}$$

The first condition is the first-order condition for the primal. The second condition is the envelope-inclusion theorem: any solution $\boldsymbol{y}^*$ to the dual belongs to the partial subdifferential $\partial_Y F(\boldsymbol{x},\boldsymbol{y})$.

If F is differentiable, then $\{(\boldsymbol{0},\boldsymbol{y}^*)\} = \partial F(\boldsymbol{x},\boldsymbol{y})$ is equivalent to $\{\boldsymbol{0}\} \in \partial_X F(\boldsymbol{x},\boldsymbol{y})$ and $\{\boldsymbol{y}^*\} = \partial_Y F(\boldsymbol{x},\boldsymbol{y})$, and the envelope theorem follows. By independence, the dual has a unique solution, even though the primal perhaps has multiple solutions. (That a convex function is differentiable does not imply that its conjugate is differentiable.)

An example of envelope inclusion is $F(\boldsymbol{x},\boldsymbol{y}) = |\boldsymbol{x}-\boldsymbol{y}|$. The perturbation function is not differentiable at the primal solution $\boldsymbol{x} = \boldsymbol{y}$. The value function $V(\boldsymbol{y}) = 0$ and $\partial V(\boldsymbol{y}) = \{\boldsymbol{0}\}$. At $\boldsymbol{x} = \boldsymbol{y}$, $\partial_Y F(\boldsymbol{x},\boldsymbol{y})$ is the unit ball.

The chapter "Directional Derivative and Subdifferential" presents a more powerful way to discover and prove results of subdifferential calculus. The directional derivative of a function is its derivative in different directions. The conjugate of the directional derivative is the indicator of the subdifferential. A dual of the conjugate yields an alternate formula for the subdifferential.

References

Rockafellar RT (1970) Convex analysis. Princeton, NJ, Princeton University Press
Samuelson PA (1947) Foundations of economic analysis, Vol 80. Harvard University Press, Cambridge, MA

Chapter 8
Lagrangian

Abstract Economists commonly introduce a Lagrangian to analyze an optimization problem with a constraint. Generalizing the traditional *ad hoc* definition employed by economists, we define the Lagrangian as the value of an optimization. Given the perturbation function for a problem, taking the conjugate in the perturbation variable while holding the choice variable fixed obtains the Lagrangian. It is a saddle-value function of the choice variables in the primal and the dual, and its value lies between the values of the primal and dual objective functions. A basic result is the saddle-point theorem that the choice variables form a saddle point of the Lagrangian if and only if they solve the primal and the dual. Let $F(\boldsymbol{x},\boldsymbol{y})$ be a proper, closed, convex, perturbation function. For simplicity of exposition, we analyze the special case of perturbation base values zero. The primal $\inf_{\boldsymbol{x}} F(\boldsymbol{x},\boldsymbol{0})$ has the reciprocal dual $\sup_{\boldsymbol{y}^*}[-F^*(\boldsymbol{0},\boldsymbol{y}^*)]$. By definition, the Lagrangian is

$$L(\boldsymbol{x},\boldsymbol{y}^*) := -\sup_{\boldsymbol{y}}[\langle \boldsymbol{y}^*,\boldsymbol{y}\rangle - F(\boldsymbol{x},\boldsymbol{y})],$$

a saddle-value function—convex in $\boldsymbol{x}$ and concave in $\boldsymbol{y}^*$. One can show that the value of the Lagrangian lies between the value of the primal and the value of the dual:

$$F(\boldsymbol{x},\boldsymbol{0}) \geq L(\boldsymbol{x},\boldsymbol{y}^*) \geq -F^*(\boldsymbol{0},\boldsymbol{y}^*),$$

for any $(\boldsymbol{x},\boldsymbol{y}^*)$. If there is no duality gap and the optimum value is attained in both the primal and the dual, then all three expressions are equal. The value of the Lagrangian is the optimum value of the problem.

Economists commonly introduce a Lagrangian to analyze an optimization problem with a constraint. Generalizing the traditional *ad hoc* definition employed by economists, we define the Lagrangian as the value of an optimization. Given the perturbation function for a problem, taking the conjugate in the perturbation variable while holding the choice variable fixed obtains the Lagrangian. It is a saddle-value function of the choice variables in the primal and the dual, and its value lies between the values of the primal and dual objective functions.

B. C. Dieffenbach, *Conjugate Duality in Economic Analysis*, Contributions to Economics,
https://doi.org/10.1007/978-3-032-21396-9_8

A basic result is the saddle-point theorem that the choice variables form a saddle point of the Lagrangian if and only if they solve the primal and the dual. Two Fenchel equalities link the perturbation function and its conjugate with the Lagrangian. Both hold if there is no duality gap and the optimum value is attained in both the primal and the dual. Together they equivalent to the first-order conditions for a saddle point.

8.1 Saddle Point

Let $F(\boldsymbol{x},\boldsymbol{y})$ be a proper, closed, convex, perturbation function on the Euclidean space $\mathbf{X} \oplus \mathbf{Y}$. For simplicity of exposition, we analyze the special case of perturbation base values $\boldsymbol{x}^* = \mathbf{0}$ and $\boldsymbol{y} = \mathbf{0}$. Rearranging the Fenchel inequality $F(\boldsymbol{x},\mathbf{0}) + F^*(\mathbf{0},\boldsymbol{y}^*) - \langle \mathbf{0},\boldsymbol{x}\rangle - \langle \boldsymbol{y}^*,\mathbf{0}\rangle \geq 0$ obtains the primal and reciprocal dual.

Problem 1 (Zero Perturbation) For the primal

$$\inf_{\boldsymbol{x}} F(\boldsymbol{x},\mathbf{0}),$$

the reciprocal dual is

$$\sup_{\boldsymbol{y}^*} [-F^*(\mathbf{0},\boldsymbol{y}^*)].$$

That the Lagrangian is the value of an optimization makes possible systematic analysis of convex problems.[1]

Definition 1 (Lagrangian for Zero Perturbation)

$$L(\boldsymbol{x},\boldsymbol{y}^*) := -\sup_{\boldsymbol{y}} [\langle \boldsymbol{y}^*,\boldsymbol{y}\rangle - F(\boldsymbol{x},\boldsymbol{y})].$$

The Lagrangian $L(\boldsymbol{x},\boldsymbol{y}^*)$ is a saddle-value function—convex in $\boldsymbol{x}$ and concave in $\boldsymbol{y}^*$, which follows from seeing L as the value function for an optimization with choice $\boldsymbol{y}$ and perturbation $\boldsymbol{x}$. For a given $\boldsymbol{x}$, the conjugate of $\boldsymbol{y} \mapsto F(\boldsymbol{x},\boldsymbol{y})$ is $\boldsymbol{y}^* \mapsto -L(\boldsymbol{x},\boldsymbol{y}^*)$.[2] The conjugate of $\boldsymbol{x} \mapsto L(\boldsymbol{x},\boldsymbol{y}^*)$ is

$$\sup_{\boldsymbol{x}} [\langle \boldsymbol{x}^*,\boldsymbol{x}\rangle - L(\boldsymbol{x},\boldsymbol{y}^*)] = \sup_{\boldsymbol{x}} \left(\langle \boldsymbol{x}^*,\boldsymbol{x}\rangle + \left\{ \sup_{\boldsymbol{y}} [\langle \boldsymbol{y}^*,\boldsymbol{y}\rangle - F(\boldsymbol{x},\boldsymbol{y})] \right\} \right) = F^*(\boldsymbol{x}^*,\boldsymbol{y}^*). \tag{8.1}$$

[1] Even though F is proper, for some $\boldsymbol{x}$ perhaps $\boldsymbol{y} \mapsto F(\boldsymbol{x},\boldsymbol{y})$ is improper, $F(\boldsymbol{x},\boldsymbol{y}) = \infty$ for all $\boldsymbol{y}$. In this case $L(\boldsymbol{x},\boldsymbol{y}^*) = \infty$ also. Consequently, in the arithmetic below, we are careful to avoid the undefined expression $\infty - \infty$.

[2] Whereas $L(\boldsymbol{x},\boldsymbol{y}^*)$ is necessarily closed in $\boldsymbol{y}^*$, there is no guarantee that it is closed in $\boldsymbol{x}$.

Hence

$$F(\boldsymbol{x},\mathbf{0}) \geq L(\boldsymbol{x},\boldsymbol{y}^*) \geq -F^*(\mathbf{0},\boldsymbol{y}^*),$$

for any $(\boldsymbol{x},\boldsymbol{y}^*)$. The value of the Lagrangian lies between the value of the primal and the value of the dual. If there is no duality gap and the optimum value is attained in both the primal and the dual, then all three expressions are equal. The value of the Lagrangian is the optimum value of the problem.

The general relationship is

$$\inf_{\boldsymbol{x}} F(\boldsymbol{x},\mathbf{0}) \geq \inf_{\boldsymbol{x}}\left[\sup_{\boldsymbol{y}^*} L(\boldsymbol{x},\boldsymbol{y}^*)\right] \geq \sup_{\boldsymbol{y}^*}\left[\inf_{\boldsymbol{x}} L(\boldsymbol{x},\boldsymbol{y}^*)\right] = \sup_{\boldsymbol{y}^*}\left[-F^*(\mathbf{0},\boldsymbol{y}^*)\right].$$

That the conjugate of $\boldsymbol{y} \mapsto F(\boldsymbol{x},\boldsymbol{y})$ is $\boldsymbol{y}^* \mapsto -L(\boldsymbol{x},\boldsymbol{y}^*)$ implies the first inequality. The middle inequality holds for any function, independently of convexity/concavity (Rockafellar 1970, Lemma 36.1, p. 379). The final equality follows from (8.1). In the literature, the second term is the minimax; and the third term is the maximin. There is no duality gap if and only if all four terms are equal.

At a saddle point, the Lagrangian is minimized in $\boldsymbol{x}$ and maximized in $\boldsymbol{y}^*$.

Definition 2 (Saddle Point) The pair $\left(\overline{\boldsymbol{x}},\overline{\boldsymbol{y}^*}\right)$ is a saddle point of the Lagrangian if

$$L(\overline{\boldsymbol{x}},\boldsymbol{y}^*) \leq L\left(\overline{\boldsymbol{x}},\overline{\boldsymbol{y}^*}\right) \leq L\left(\boldsymbol{x},\overline{\boldsymbol{y}^*}\right)$$

for all $\boldsymbol{x}$ and all $\boldsymbol{y}^*$.

The two first-order conditions (8.3) and (8.5) are necessary and sufficient for a saddle point. Each first-order condition is equivalent to a Fenchel equality. For a given $\boldsymbol{x}$, the conjugate of the closed, convex function $\boldsymbol{y} \mapsto F(\boldsymbol{x},\boldsymbol{y})$ is $\boldsymbol{y}^* \mapsto -L(\boldsymbol{x},\boldsymbol{y}^*)$. If $F(\boldsymbol{x},\boldsymbol{y})$ is finite for some $\boldsymbol{y}$, then $\boldsymbol{y} \mapsto F(\boldsymbol{x},\boldsymbol{y})$ is proper, so conjugate duality holds. The Fenchel equality

$$F(\boldsymbol{x},\mathbf{0}) - L(\boldsymbol{x},\boldsymbol{y}^*) - \langle \boldsymbol{y}^*,\mathbf{0}\rangle = 0 \tag{8.2}$$

says that the value of the primal equals the value of the Lagrangian. By subdifferential invertibility, equivalently

$$\mathbf{0} \in \partial_Y L(\boldsymbol{x},\boldsymbol{y}^*), \tag{8.3}$$

the first-order condition for $\boldsymbol{y}^*$ to maximize the Lagrangian. For a given $\boldsymbol{y}^*$, the conjugate of the convex function $\boldsymbol{x} \mapsto L(\boldsymbol{x},\boldsymbol{y}^*)$ is $\boldsymbol{x}^* \mapsto F^*(\boldsymbol{x}^*,\boldsymbol{y}^*)$. The Fenchel equality

$$L(\boldsymbol{x},\boldsymbol{y}^*) + F^*(\mathbf{0},\boldsymbol{y}^*) - \langle \mathbf{0},\boldsymbol{x}\rangle = 0 \tag{8.4}$$

says that the value of the Lagrangian equals the value of the dual. By subdifferential invertibility, equivalently

$$\mathbf{0} \in \partial_X L(\boldsymbol{x}, \boldsymbol{y}^*), \tag{8.5}$$

the first-order condition for $\boldsymbol{x}$ to minimize the Lagrangian.

These equivalences imply the following.

Theorem 1 (Saddle-Point Solution)

- *If there is no duality gap and the optimum value is finite, then together primal and dual solutions constitute a saddle point of the Lagrangian, and the saddle value is the optimum value.*
- *If the Lagrangian has a saddle point having a finite saddle value, then there is no duality gap and the optimum value is the saddle value, and the saddle point solves the primal and the dual.*

If there is no duality gap and the optimum value is finite, then the Fenchel equalities (8.2) and (8.4) hold, which implies the first-order conditions (8.3) and (8.5). Conversely, if $(\boldsymbol{x}, \boldsymbol{y}^*)$ is a saddle point of the Lagrangian, then both first-order conditions (8.3) and (8.5) hold. Hence the Fenchel equalities (8.2) and (8.4) hold. Thus $F(\boldsymbol{x}, \mathbf{0}) = -F^*(\mathbf{0}, \boldsymbol{y}^*)$; there is no duality gap, and $\boldsymbol{x}$ solves the primal and $\boldsymbol{y}^*$ solves the dual.

8.2 Karush-Kuhn-Tucker Example

Consider a simple example, a traditional Karush-Kuhn-Tucker problem minimizing a quadratic subject to an inequality constraint:

$$\inf_x \left[\frac{1}{2}(x+1)^2 + \delta_{x \geq 0} \right] = \frac{1}{2},$$

which has solution $x = 0$. The perturbed primal is

$$\inf_x F(x, y) = \inf_x \left[\frac{1}{2}(x+1)^2 + \delta_{x \geq y} \right],$$

in which the base value of the perturbation $y = 0$. The primal satisfies the polyhedral/relative condition, so there is no duality gap. The dual is

$$\sup_{y^*} \left[-F^*(0, y^*) \right] = \sup_{y^*} \left[\frac{1}{2} - \frac{1}{2}(y^* - 1)^2 - \delta_{y^* \geq 0} \right] = \frac{1}{2},$$

with solution $y^* = 1$.

The Lagrangian

$$\begin{aligned} L(x,y^*) &= -\sup_y \left[y^*y - F(x,y)\right] \\ &= -\sup_y \left[y^*y - \frac{1}{2}(x+1)^2 - \delta_{x\geq y}\right] \\ &= \frac{1}{2}(x+1)^2 - \sup_y \left[y^*(x-z) - \delta_{z\geq 0}\right] \ (z := x-y) \\ &= \frac{1}{2}(x+1)^2 - y^*x - \delta_{y^*\geq 0}. \end{aligned}$$

Convex in x and concave in y^*, the Lagrangian reduces to the traditional *ad hoc* definition employed by economists: the quadratic $\frac{1}{2}(x+1)^2$ to minimize minus the Lagrange multiplier y^* multiplying the constraint function x.

Here the Fenchel equality (8.2) is

$$\begin{aligned} 0 &= F(x,0) - L(x,y^*) - y^*0 \\ &= \left[\frac{1}{2}(x+1)^2 + \delta_{x\geq 0}\right] - \left[\frac{1}{2}(x+1)^2 - y^*x - \delta_{y^*\geq 0}\right] \\ &= y^*x + \delta_{x\geq 0} + \delta_{y^*\geq 0}. \end{aligned}$$

This condition reduces to complementary slackness: both x and y^* must be non-negative, and at least one must be zero. Given x, the first-order condition (8.3) is

$$0 \in \partial_Y L(x,y^*) = \begin{cases} \{-x\} \text{ if } y^* > 0 \\ \{-x\} + [0,\infty) = [-x,\infty) \text{ if } y^* = 0 \\ \varnothing \text{ if } y^* < 0 \end{cases}.$$

For there to be a zero subgradient again reduces to complementary slackness.

The Fenchel equality (8.4) is

$$\begin{aligned} 0 &= L(x,y^*) + F^*(0,y^*) - \langle 0,x\rangle \\ &= \left[\frac{1}{2}(x+1)^2 - y^*x\right] + \left[-\frac{1}{2} + \frac{1}{2}(y^*-1)^2\right] \text{ such that } y^* \geq 0 \\ &= \frac{1}{2}(x - y^* + 1)^2 + \delta_{y^*\geq 0}. \end{aligned}$$

For there to be a zero subgradient, $x = 1 - y^*$ and $y^* \geq 0$. Equivalently, given y^*, the first-order condition (8.5)

$$0 \in \partial_X L(x,y^*) = \begin{cases} \{x - y^* + 1\} \text{ if } y^* \geq 0 \\ \varnothing \text{ if } y^* < 0 \end{cases}.$$

8.3 Fenchel Duality

Fenchel duality has simple closed-form expressions for both the Lagrangian and the dual.

Summary 1 (Fenchel Duality) *For convex functions* f and g *and perturbation base value* $\mathbf{y} = \mathbf{0}$, *the Fenchel primal*

$$\inf_{\boldsymbol{x}} [f(\boldsymbol{x}) + g(\boldsymbol{x} + \boldsymbol{y})]$$

has the Fenchel dual

$$\sup_{\boldsymbol{y}^*} [-f^*(-\boldsymbol{y}^*) - g^*(\boldsymbol{y}^*)].$$

The Lagrangian is

$$L(\boldsymbol{x}, \boldsymbol{y}^*) = \begin{cases} \langle \boldsymbol{y}^*, \boldsymbol{x} \rangle + f(\boldsymbol{x}) - g^*(\boldsymbol{y}^*) & \text{if } \boldsymbol{x} \in \operatorname{dom} f \\ \infty \text{ if } \boldsymbol{x} \notin \operatorname{dom} f \end{cases}.$$

Observe that the alternate perturbation

$$\inf_{\boldsymbol{x}} [f(\boldsymbol{x} - \boldsymbol{y}) + g(\boldsymbol{x})]$$

has exactly the same dual (because a change in $\boldsymbol{y}$ affects the optimum value of the primal in exactly the same way) but nonetheless has a different Lagrangian.

For given $\boldsymbol{x} \in \operatorname{dom} f$, the Fenchel equality (8.2) is

$$\begin{aligned} 0 &= F(\boldsymbol{x}, \boldsymbol{0}) - L(\boldsymbol{x}, \boldsymbol{y}^*) - \langle \boldsymbol{y}^*, \boldsymbol{0} \rangle \\ &= [f(\boldsymbol{x}) + g(\boldsymbol{x})] - [\langle \boldsymbol{y}^*, \boldsymbol{x} \rangle + f(\boldsymbol{x}) - g^*(\boldsymbol{y}^*)] \\ &= g(\boldsymbol{x}) + g^*(\boldsymbol{y}^*) - \langle \boldsymbol{y}^*, \boldsymbol{x} \rangle, \end{aligned}$$

a Fenchel equality for g. The equivalent first-order condition (8.3) is

$$0 \in \partial_Y L(\boldsymbol{x}, \boldsymbol{y}^*) = \begin{cases} \{\boldsymbol{x}\} - \partial g^*(\boldsymbol{y}^*) & \text{if } \boldsymbol{x} \in \operatorname{dom} f \\ \varnothing \text{ if } \boldsymbol{x} \notin \operatorname{dom} f \end{cases}.$$

For given $\boldsymbol{y}^* \in \operatorname{dom} g^*$, the Fenchel equality (8.3) is

$$\begin{aligned} 0 &= L(\boldsymbol{x}, \boldsymbol{y}^*) + F^*(\boldsymbol{0}, \boldsymbol{y}^*) - \langle \boldsymbol{0}, \boldsymbol{x} \rangle \\ &= [\langle \boldsymbol{y}^*, \boldsymbol{x} \rangle + f(\boldsymbol{x}) - g^*(\boldsymbol{y}^*)] + [f^*(-\boldsymbol{y}^*) + g^*(\boldsymbol{y}^*)] \\ &= f(\boldsymbol{x}) + f^*(-\boldsymbol{y}^*) - \langle \boldsymbol{y}^*, \boldsymbol{x} \rangle, \end{aligned}$$

a Fenchel equality for f. The equivalent first-order condition (8.5) is

$$0 \in \partial_X L(\boldsymbol{x}, \boldsymbol{y}^*) = \begin{cases} \{\boldsymbol{y}^*\} + \partial f(\boldsymbol{x}) \text{ if } \boldsymbol{y}^* \in \operatorname{dom} g^* \\ \varnothing \text{ if } \boldsymbol{y}^* \notin \operatorname{dom} g^* \end{cases}.$$

8.4 Nonzero Perturbations

Summary 2 (Reciprocal Convex Perturbation Duality) *For a convex function $F(\boldsymbol{x}, \boldsymbol{y})$, the primal*

$$\inf_{\boldsymbol{x}} [\langle \boldsymbol{x}^*, \boldsymbol{x} \rangle + F(\boldsymbol{x}, \boldsymbol{y})]$$

has the reciprocal dual

$$\sup_{\boldsymbol{y}^*} [\langle \boldsymbol{y}^*, \boldsymbol{y} \rangle - F^*(-\boldsymbol{x}^*, \boldsymbol{y}^*)].$$

The Lagrangian is

$$L(\boldsymbol{x}, \boldsymbol{y}^*) := \langle \boldsymbol{x}^*, \boldsymbol{x} \rangle + \langle \boldsymbol{y}^*, \boldsymbol{y} \rangle - \sup_{z} [\langle \boldsymbol{y}^*, z \rangle - F(\boldsymbol{x}, z)].$$

Corresponding to the Fenchel equalities (8.2) and (8.4), set the value of the primal equal to the Lagrangian, and set the value of the Lagrangian equal to the value of the dual. For simplicity we write the Lagrangian $L(\boldsymbol{x}, \boldsymbol{y}^*)$ as a function of the choice variables, even though it is a function of the perturbations as well.

The concave case is analogous.

Summary 3 (Reciprocal Concave Perturbation Duality) *For a concave function $F(\boldsymbol{x}, \boldsymbol{y})$, the primal*

$$\sup_{\boldsymbol{x}} [\langle \boldsymbol{x}^*, \boldsymbol{x} \rangle + F(\boldsymbol{x}, \boldsymbol{y})]$$

has the reciprocal dual

$$\inf_{\boldsymbol{y}^*} [\langle \boldsymbol{y}^*, \boldsymbol{y} \rangle - F_*(-\boldsymbol{x}^*, \boldsymbol{y}^*)].$$

The Lagrangian is

$$L(\boldsymbol{x}, \boldsymbol{y}^*) := \langle \boldsymbol{x}^*, \boldsymbol{x} \rangle + \langle \boldsymbol{y}^*, \boldsymbol{y} \rangle - \inf_{z} [\langle \boldsymbol{y}^*, z \rangle - F(\boldsymbol{x}, z)].$$

This Lagrangian is concave in $\boldsymbol{x}$ and convex in $\boldsymbol{y}^*$.

8.5 Alternative Points-of-View

Overall, the primal/dual approach and the Lagrangian saddle-point approach are equivalent mathematical approaches to optimization, and both contribute to economic understanding. The Fenchel factors obtained by setting the value of the primal equal to the value of the dual have intuitive meaning. Potentially, one can solve both the primal and the dual as separate problems, although in practice this book typically solves both problems simultaneously, via the Fenchel factorization.

Each alternative viewpoint on a convex optimization is potentially valuable. For example, the primal choice variable might be quantity, and one might interpret the dual choice variable as price. To understand the primal and the dual as independent optimizations is important. What is the economic meaning for quantity to solve the primal and for price to solve the dual? What is the meaning of each first-order condition? Perhaps one can interpret the dual as choosing the price to equate demand and supply.

An advantage of the Lagrangian approach is that a saddle-point may reflect an economic interaction between the primal and dual choice variables. For example, the primal may choose the price, and the dual chooses the quantity. There is market equilibrium if and only if the price and quantity form a saddle point. Even though the primal/dual and the Lagrangian approaches are essentially equivalent, together they can provide deeper insight.

The Lagrangian formulation can be more intuitive for a problem in which the dual cannot be expressed in a simple form. A weakness is that solving for a saddle point by direct reasoning can be much more tricky than solving the primal and the dual separately. A systematic solution is always possible via a Fenchel factorization, and this book typically follows this path.

Each of the two first-order conditions for a saddle point is equivalent to a Fenchel equality. As illustrated by the simple Karush-Kuhn-Tucker quadratic example above, the complexity of writing down the subdifferentials of the Lagrangian demonstrates that solving via the first-order conditions can be much more difficult than solving via Fenchel factorization. That the Lagrangian is simultaneously convex and concave in the choice variables sometimes creates a technical problem: one confronts the undefined expression $\infty - \infty$, and dealing with it requires a precise but awkward consideration of the order in which terms are evaluated.

Reference

Rockafellar RT (1970) Convex analysis. Princeton University Press, Princeton, NJ

Chapter 9
Karush-Kuhn-Tucker

Abstract We consider the Karush-Kuhn-Tucker problem

$$\sup_{x} \left[g(\boldsymbol{x}) - \delta_{\text{epi } f}(\boldsymbol{x}, z) \right],$$

in which the perturbation base value $z = 0$. Here g concave, and f is convex. The Lagrangian is

$$L(\boldsymbol{x}, z^*) = g(\boldsymbol{x}) - z^* f(\boldsymbol{x}) - \delta_{\text{dom } f}(\boldsymbol{x}) + \delta_{z^* \geq 0}.$$

For $(\boldsymbol{x}, z^*)$ to be a saddle point of the Lagrangian is necessary and sufficient for $\boldsymbol{x}$ to solve the Karush-Kuhn-Tucker problem and for z^* to solve its dual. The "Lagrange multiplier" is the dual solution set, which may contain zero as well as positive values. Compared to the traditional *ad hoc* Lagrangian analyzed by economists, this expression contains the extra term $\delta_{\text{dom } f}(\boldsymbol{x})$, which comes into play for the exceptional case $z^* = 0$. The dual implies several results concerning the exceptional case. If $\boldsymbol{x}$ solves the Karush-Kuhn-Tucker primal such that $f(\boldsymbol{x}) < 0$, then the unique Lagrange multiplier $z^* = 0$. If the optimum value of the primal is the unconstrained supremum, then there exists a zero Lagrange multiplier, but there may also be positive multipliers. The exceptional case pops up in economic applications. It can be that expenditure minimization does not imply utility maximization: possibly compensated demand does not maximize utility among consumption costing the same or less. This situation is reflected in the zero Lagrange multiplier. We analyze efficiency by maximizing the efficiency function among vectors in a feasible set. In the Karush-Kuhn-Tucker form of the efficiency dual, the exceptional case is inherent: a vector is efficient if and only if the unique Lagrange multiplier is zero.

B. C. Dieffenbach, *Conjugate Duality in Economic Analysis*, Contributions to Economics,
https://doi.org/10.1007/978-3-032-21396-9_9

9.1 Primal

Karush-Kuhn-Tucker optimization maximizes a function subject to an inequality constraint.[1] They analyze the case in which g and f are differentiable, but not necessarily concave and convex. They focus on conditions necessary for a solution $\boldsymbol{x}$. Does there exist a Lagrange multiplier y^* such that $\boldsymbol{x}$ maximizes the Lagrangian? Is there complementary slackness between y^* and $f(\boldsymbol{x})$? Kuhn (1991) evaluates the historical development.

Problem 1 (Karush-Kuhn-Tucker)

$$\sup_{x} \left[g(\boldsymbol{x}) - \delta_{f(x) \leq 0} \right] \tag{9.1}$$

The choice variable $\boldsymbol{x}$ lies in a Euclidean space $\mathbf{X}$. Here g is proper, closed, and concave; and f is proper, closed, and convex. We compare alternative perturbations of the problem.

The traditional formulation introduces a perturbation z in the inequality constraint.

Problem 2 (Karush-Kuhn-Tucker Primal)

$$\sup_{x} \left[g(\boldsymbol{x}) - \delta_{\text{epi } f}(\boldsymbol{x}, z) \right]. \tag{9.2}$$

The base value of the perturbation $z = 0$.

Slater conditions express the relative and polyhedral conditions for no duality gap.

Condition 1 (Relative Condition) *There exists some* $\boldsymbol{x}$ *such that* $\boldsymbol{x} \in \text{rint}(\text{dom } g)$ *and* $(\boldsymbol{x}, 0) \in \text{rint}(\text{epi} f)$.

For convex f, $(\boldsymbol{x}, z) \in \text{rint}[\text{dom}(\text{epi} f)]$ if and only if $\boldsymbol{x} \in \text{rint}(\text{dom} f)$ and $f(\boldsymbol{x}) < z$ (Rockafellar 1970, Lemma 7.3, p. 54). Hence the Slater relative condition is an equivalent statement of the relative condition (Slater 1950, 1959).

Condition 2 (Slater Relative Condition) *There exists some* $\boldsymbol{x} \in \text{rint}(\text{dom } g)$ *and* $\boldsymbol{x} \in \text{rint}(\text{dom} f)$ *such that* $f(\boldsymbol{x}) < 0$.

If f and g are polyhedral, the condition is relaxed.

---

[1] Karush (1939) and Kuhn and Tucker (1951)

Condition 3 (Slater Polyhedral Condition) *There exists some* $\boldsymbol{x} \in \operatorname{dom} g$ *and* $\boldsymbol{x} \in \operatorname{dom} f$ *such that* $f(\boldsymbol{x}) \leq 0$.

Throughout this chapter, we assume that a Slater condition holds. There is no duality gap. If the optimum value is finite, then it is attained by the dual choice variable z^*. The "Lagrange multiplier" is the dual solution set, which may contain zero as well as positive values.

9.2 Lagrangian

Unless the Karush-Kuhn-Tucker primal (9.2) has special structure, it has no simple closed-form dual. Consequently one looks for a saddle point of the Lagrangian:

$$\begin{aligned} L(\boldsymbol{x}, z^*) &= g(\boldsymbol{x}) - \inf_z \left[z^* z + \delta_{\operatorname{epi} f}(\boldsymbol{x}, z)\right] \\ &= g(\boldsymbol{x}) + \begin{cases} -z^* f(\boldsymbol{x}) \text{ if } z^* > 0 \\ -\delta_{\operatorname{dom} f}(\boldsymbol{x}) \text{ if } z^* = 0 \\ \infty \text{ if } z^* < 0. \end{cases} \end{aligned} \tag{9.3}$$

Summary 1 (Karush-Kuhn-Tucker Primal Lagrangian) *The Lagrangian for the Karush-Kuhn-Tucker primal (9.2) is*

$$L(\boldsymbol{x}, z^*) = g(\boldsymbol{x}) - z^* f(\boldsymbol{x}) - \delta_{\operatorname{dom} f}(\boldsymbol{x}) + \delta_{z^* \geq 0}. \tag{9.4}$$

The Lagrangian is concave in $\boldsymbol{x}$ and convex in z^*.[2] We distinguish between the case $z^* > 0$ and the exceptional case $z^* = 0$. Compared to the traditional *ad hoc* Lagrangian analyzed by economists, the Lagrangian (9.4) contains the extra term $\delta_{\operatorname{dom} f}(\boldsymbol{x})$, which comes into play for the exceptional case.

For $(\boldsymbol{x}, z^*)$ to be a saddle point of the Lagrangian is necessary and sufficient for $\boldsymbol{x}$ to solve the Karush-Kuhn-Tucker problem (9.2) and for z^* to solve its dual. For given $\boldsymbol{x}$ belonging to both $\operatorname{dom} g$ and $\operatorname{dom} f$, we calculate the first-order conditions for a saddle point.

---

[2] Examining the derivation of the Lagrangian resolves the ambiguity of the formula (9.4). Unless both $\boldsymbol{x} \in \operatorname{dom} g$ and $\boldsymbol{x} \in \operatorname{dom} f$, then $L = -\infty$. If both $\boldsymbol{x} \in \operatorname{dom} g$ and $\boldsymbol{x} \in \operatorname{dom} f$ and $z^* < 0$, then $L = \infty$. If both $\boldsymbol{x} \in \operatorname{dom} g$ and $\boldsymbol{x} \in \operatorname{dom} f$ and $z^* \geq 0$, then $L = g(\boldsymbol{x}) - z^* f(\boldsymbol{x})$. While the Lagrangian is closed in z^* for given $\boldsymbol{x}$, it is perhaps not closed in $\boldsymbol{x}$ for $z^* = 0$, because $\operatorname{dom} f$ is perhaps not closed.

The first-order condition for z^* to minimize the Lagrangian is

$$\begin{aligned} 0 &\in \partial\left[-z^* f(\boldsymbol{x}) + \delta_{z^*\geq 0}\right] \\ &= \{-f(\boldsymbol{x})\} + \partial\delta_{z^*\geq 0} \\ &= \begin{cases} \{-f(\boldsymbol{x})\} \text{ if } z^* > 0 \\ \{f(\boldsymbol{x})\} - \mathbf{R}_+ \text{ if } z^* = 0 \\ \varnothing \text{ if } z^* < 0. \end{cases} \end{aligned}$$

This first-order condition is equivalent to

$$0 = \delta_{z^*\geq 0} + \delta_{f(x)\leq 0} - z^* f(\boldsymbol{x}) = \delta_{\mathbf{R}_+}(z^*) + \delta^*_{\mathbf{R}_+}[f(\boldsymbol{x})] - z^* f(\boldsymbol{x}). \tag{9.5}$$

As the right-hand expression is a Fenchel factor, it is greater than or equal to zero. There is complementary slackness between z^* and $-f(\boldsymbol{x})$. If the constraint is not binding, $f(\boldsymbol{x}) < 0$, then necessarily $z^* = 0$. And if $z^* > 0$, then the constraint is binding, $f(\boldsymbol{x}) = 0$.

The first-order condition for $\boldsymbol{x}$ to maximize the Lagrangian depends on whether z^* is positive or zero. For $z^* > 0$, the term $\delta_{\text{dom}\, f}(\boldsymbol{x})$ in the Lagrangian is redundant, as $z^* f(\boldsymbol{x})$ is infinite unless $\boldsymbol{x} \in \text{dom}\, f$. Hence the first-order condition is

$$\mathbf{0} \in \partial g(\boldsymbol{x}) - z^* \partial f(\boldsymbol{x}). \tag{9.6}$$

Together this condition and the complementary slackness (9.5) are necessary and sufficient for a saddle point.

For the exceptional case $z^* = 0$,[3]

$$\mathbf{0} \in \partial g(\boldsymbol{x}) - \partial\delta_{\text{cl}(\text{dom}\, f)}(\boldsymbol{x}) \tag{9.7}$$

together with complementary slackness (9.5) are necessary and sufficient for a saddle point. Even though the Lagrange multiplier is zero, the nonnegativity constraint is not irrelevant. The choice $\boldsymbol{x}$ need not be an unconstrained maximum of g, because the requirement $\boldsymbol{x} \in \text{dom}\, f$ remains. If the choice $\boldsymbol{x}$ that maximizes the Lagrangian lies in the interior of dom f, then $\partial\delta_{\text{cl}(\text{dom}\, f)}(\boldsymbol{x}) = \{\mathbf{0}\}$, and $\boldsymbol{x}$ is an unconstrained maximum of g.[4]

---

[3] Because complementary slackness (9.5) requires $\boldsymbol{x} \in \text{dom}\, f$, one can take the closure.

[4] Both the primal and the dual can have multiple solutions. In particular, perhaps the dual is minimized by positive z^*, as well as $z^* = 0$.

9.3 Dual

Adding perturbation to the Karush-Kuhn-Tucker problem yields a closed-form dual.

Problem 3 (Extended Karush-Kuhn-Tucker Primal)

$$\sup_{\boldsymbol{x}} \left[g(\boldsymbol{x}) - \delta_{\text{epi } f}(\boldsymbol{x} - \boldsymbol{y}, z) \right] \tag{9.8}$$

The perturbation base values are $\boldsymbol{y} = \mathbf{0}$ and $z = 0$. The Slater condition implies that there is no duality gap and the optimum value is attained in the dual. Typically this book analyzes a Karush-Kuhn-Tucker via the form (9.8), because the necessary and sufficient conditions for a solution to the primal and the dual are expressed via Fenchel equalities directly in f and g.

We find the dual. Adding the Fenchel inequalities

$$-g(\boldsymbol{x}) - g_*(\boldsymbol{y}^*) + \langle \boldsymbol{y}^*, \boldsymbol{x} \rangle \geq 0 \tag{9.9}$$

$$\delta_{\text{epi } f}(\boldsymbol{x} - \boldsymbol{y}, z) + \delta^*_{\text{epi } f}(\boldsymbol{y}^*, -z^*) - \langle \boldsymbol{y}^*, \boldsymbol{x} - \boldsymbol{y} \rangle - (-z^*) z \geq 0 \tag{9.10}$$

and rearranging obtains

$$\langle \boldsymbol{y}^*, \boldsymbol{y} \rangle + z^* z + \delta^*_{\text{epi } f}(\boldsymbol{y}^*, -z^*) - g_*(\boldsymbol{y}^*) \geq g(\boldsymbol{x}) - \delta_{\text{epi } f}(\boldsymbol{x} - \boldsymbol{y}, z).$$

Problem 4 (Karush-Kuhn-Tucker Dual)

$$\inf_{\boldsymbol{y}^*, z^*} \left[(z^* \star f^*)(\boldsymbol{y}^*) - g_*(\boldsymbol{y}^*) \right]. \tag{9.11}$$

From the Fenchel factors, we derive the same necessary and sufficient conditions found above via the Lagrangian.

For $z^* > 0$, for the Fenchel factors (9.9) and (9.10) to be zero is equivalent to

$$g(\boldsymbol{x}) + g_*(\boldsymbol{y}^*) - \langle \boldsymbol{y}^*, \boldsymbol{x} \rangle = 0 \tag{9.12}$$

$$f(\boldsymbol{x}) + f^*(\boldsymbol{y}^*/z^*) - \langle \boldsymbol{y}^*/z^*, \boldsymbol{x} \rangle = 0 \tag{9.13}$$

$$f(\boldsymbol{x}) = 0. \tag{9.14}$$

By (9.13), $\boldsymbol{y}^*/z^* \in \partial f(\boldsymbol{x})$. By (9.12), $\boldsymbol{y}^* \in g(\boldsymbol{x})$, so (9.6) follows. And (9.14) implies, the complementary slackness (9.5).

For the exceptional case $z^* = 0$, for the Fenchel factors (9.9) and (9.10) to be zero is equivalent to

$$g(\boldsymbol{x}) + g_*(\boldsymbol{y}^*) - \langle \boldsymbol{y}^*, \boldsymbol{x} \rangle = 0 \tag{9.15}$$

$$\delta_{\mathrm{cl(dom}\, f)}(\boldsymbol{x}) + \delta^*_{\mathrm{cl(dom}\, f)}(\boldsymbol{y}^*) - \langle \boldsymbol{y}^*, \boldsymbol{x} \rangle = 0 \tag{9.16}$$

$$f(\boldsymbol{x}) \leq 0. \tag{9.17}$$

By (9.15), $\boldsymbol{y}^* \in \partial g(\boldsymbol{x})$. By (9.16), $\boldsymbol{y}^* \in \partial \delta_{\mathrm{cl(dom}\, f)}(\boldsymbol{x})$, so (9.7) follows. And (9.17) implies the complementary slackness (9.5).

9.4 Exceptional Case

Several results concerning the exceptional case follow quickly from the dual (9.11).

Theorem 1 (Constraint Not Binding) *If $\boldsymbol{x}$ solves the Karush-Kuhn-Tucker primal such that $f(\boldsymbol{x}) < 0$, then the unique Lagrange multiplier $z^* = 0$.*

Above this result was immediate via the complementary slackness (9.5), but it also follows from the dual. A dual solution such that $z^* > 0$ obtains a contradiction: the Fenchel equalities (9.12)–(9.14) imply $f(\boldsymbol{x}) = 0$.

Theorem 2 (Unconstrained Supremum) *If the optimum value of the primal is the unconstrained supremum $\sup_{\boldsymbol{x}} g(\boldsymbol{x}) = -g_*(\boldsymbol{0})$, then there exists a zero Lagrange multiplier.*

Although one can show this result for a saddle point of the Lagrangian, consider a derivation from the dual. A choice $(\boldsymbol{y}^*, z^*)$ obtains the unconstrained supremum if and only if

$$\delta^*_{\mathrm{epi}\, f}(\boldsymbol{y}^*, z^*) - g_*(\boldsymbol{y}^*) = -g_*(\boldsymbol{0}),$$

so the choice $(\boldsymbol{0}, 0)$ solves the dual. The previous theorem shows that the converse is false: a zero Lagrange multiplier does not imply that the optimum value of the primal is the unconstrained supremum.

The exceptional case pops up in economic applications. The effective domain of the concave utility for a "classical consumer" is the nonnegative cone. Consumer demand is the consumption vector maximizing utility, subject to a budget constraint. Compensated demand is the consumption vector minimizing expenditure, subject to the constraint that utility is greater than or equal to a specified value. The Lagrangian contains the extra term, and in a special case the Lagrange multiplier is zero. It can be that expenditure minimization does not imply utility maximization: possibly compensated demand does not maximize utility among consumption costing the same or less. This situation is reflected in the zero Lagrange multiplier.

We analyze efficiency by maximizing the efficiency function among vectors in a feasible set. In the Karush-Kuhn-Tucker form of the efficiency dual, the exceptional case is inherent: a vector is efficient if and only if the unique Lagrange multiplier is zero.

References

Karush W (1939) Minima of functions of several variables with inequalities as side conditions (Master's thesis). M. S. thesis

Kuhn HW (1991) Nonlinear programming: a historical note. In: Lenstra JK, Kan AHGR, Schrijver A (eds) History of mathematical programming: a collection of personal reminiscences. CWI and North-Holland, Amsterdam, pp 82–96

Kuhn HW, Tucker AW (1951) Nonlinear programming. In: Neyman J (ed) Proceeedings of the second berkeley symposium on mathematical statistics and probability. University of California Press, Berkeley, CA, pp 481–492

Rockafellar RT (1970) Convex analysis. Princeton University Press, Princeton, NJ

Slater M (1950) Lagrange multipliers revisited: a contribution to non-linear programming (Tech. Rep.). University of Chicago Mathematics 403

Slater M (1959) Lagrange multipliers revisited (Cowles Foundation Discussion Papers No. 80). Cowles Foundation for Research in Economics, Yale University. http://econpapers.repec.org/RePEc:cwl:cwldpp:80 (Originally 1950)

Part II
Linear Algebra

Chapter 10
Linear Transformation

Abstract In line with the modern mathematical point-of-view, we adopt the coordinate-free point-of-view for linear algebra. A vector in a Euclidean space exists independently of any coordinates used to describe it. A linear transformation from one Euclidean space to another maps one vector to another, independently of any coordinate system. In contrast, in matrix algebra, one works with a basis. A matrix describes how a linear transformation maps one vector to another. In economic theory and econometrics, typically vectors are not seen as coordinate-free. A particular basis is singled out, and one works with coordinates. Despite this tradition, the coordinate-free point-of-view is superior. Not using coordinates reduces the use of subscripts and makes expressions simpler, and theorems are easier to state and to prove. Given a linear transformation $A : \mathbf{X} \to \mathbf{Y}$, then there exists a unique linear transformation (the adjoint) $A^{\top} : \mathbf{Y} \to \mathbf{X}$ that preserves the inner product:

$$\langle y^*, Ax\rangle = \left\langle A^{\top} y^*, x\right\rangle.$$

Let $\mathrm{R}(\cdot)$ and $\mathrm{N}(\cdot)$ denote the range and the null space of a linear transformation. The "fundamental theorem of linear algebra" asserts that the orthogonal complement of the null space of a linear transformation is the range of the adjoint. Given a linear transformation $A : \mathbf{X} \to \mathbf{Y}$, the restricted linear transformation $A : \mathrm{R}\left(A^{\top}\right) \to \mathrm{R}(A)$ is invertible. Extending the inverse of the restricted linear transformation obtains the Moore-Penrose generalized inverse $A^{+} : \mathbf{Y} \to \mathbf{X}$:

- for $\boldsymbol{y} \in \mathrm{R}(A)$, then $A^{+}\boldsymbol{y}$ is the inverse of the restricted linear transformation;
- for $\boldsymbol{y} \in \mathrm{N}\left(A^{\top}\right)$, then $A^{+}\boldsymbol{y} = \mathbf{0}$.

10.1 Coordinate-Free

In line with the modern mathematical point-of-view, we adopt the coordinate-free point-of-view for linear algebra. A vector in a Euclidean space exists independently of any coordinates used to describe it. A linear transformation from one Euclidean space to another maps one vector to another, independently of any coordinate system.

B. C. Dieffenbach, *Conjugate Duality in Economic Analysis*, Contributions to Economics,
https://doi.org/10.1007/978-3-032-21396-9_10

In contrast, in matrix algebra, one works with a basis. Given a basis, any vector can be expressed uniquely as a linear combination of the basis elements. For example, if $\boldsymbol{x} = \sum_i \boldsymbol{y}_i \tilde{x}_i$ for some basis $\boldsymbol{y}_i$, then the $\tilde{x}_i$ are the coordinates of $\boldsymbol{x}$ in terms of this basis. Many linear-algebra textbooks develop all results via a basis. A matrix describes how a linear transformation maps one vector to another. The components of the matrix specify how a basis vector in a Euclidean space is mapped to a linear combination of basis vectors in another Euclidean space.

In economic theory and econometrics, typically vectors are not seen as coordinate-free. A particular basis is singled out, and one works with coordinates. Despite this tradition, the coordinate-free point-of-view is superior. Not using coordinates reduces the use of subscripts and makes expressions simpler, and theorems are easier to state and to prove.

In his "Santa Barbara program" for mathematics education in linear algebra, Steenrod forcefully advocates the coordinate-free formulation (Steenrod 1967). Starting from Steenrod's notes, Fekete's textbook *Real Linear Algebra* presents this approach for a reader with minimal mathematics background (Fekete 1985). Steenrod contends that the coordinate-free approach provides geometric insight missing from matrix algebra:

> [T]he main reason ... to bother with geometry ... is that most problems are presented in geometric form in the first place. Reformulation and solution in analytic terms is merely a second step. To complete the process, there is an indispensable third, namely, the interpretation of the analytic solution in geometric terms. There is another reason, which is psychological. Two views of the same thing reinforce one another. Most of us are able to remember the multitudinous formulas of analysis mainly because we attach to each a geometric picture that keeps us from going astray. Even better than that, the geometric view of a problem helps us to focus on the invariants and to weed out the irrelevant details. A poor choice of coordinates may lead to a horrible mess in the analytical formulation, but with some geometric insight we may be able to choose the "best" coordinate system. (Fekete 1985, p. vii)

Steenrod continues that

> [M]ost textbooks confuse the presentation of theory with the techniques of computation via matrices. One way to see that a book makes this mistake is to see that it has a chapter on determinants and matrices before linear transformations are defined. The linear transformation is an easy conceptual thing to talk about and give examples of without matrices. The matrix is a tool for computation, i. e., it is a set of coordinates subject to prior choice of a basis. The matrix, important as it may be for computation, is of *no* importance in the theoretical or conceptual part of the course, nor in the geometric pictures that come along. Presenting semi-theoretical material on matrices, and teaching the students to work with matrices before introducing linear transformations, is comparable to trying to teach someone to play the piano on a keyboard that isn't attached to any strings. There is no feedback, the student does not see the objective and finds no pleasure in what he is doing. True, historically matrices came first. For a long time a vector space was an $\mathbf{R}^n$ for some n and a linear transformation was a system of linear equations represented by the matrix of coefficients. Thus properties of linear transformations had to be formulated as properties of matrices. In this way an extensive theory of matrices arose. It is a cumbersome theory both in notation and conception. The conceptual point of view, that one could proceed on a different level and work without coordinates, developed during the twenties and thirties. It became clear that the matrix theory tended to obscure the geometric insight. With the new point of view the

picture became easy and lovely, and the theory was disassociated from the mechanism of computation. (Fekete 1985, p. vii)

10.2 Linear Transformation and Adjoint

We analyze linear transformations via the coordinate-free formulation.

Definition 1 (Linear Transformation) A linear transformation from a Euclidean space **X** to a Euclidean space **Y** is a function

$$\begin{aligned} A &: \mathbf{X} \to \mathbf{Y} \\ \boldsymbol{x} &\mapsto \boldsymbol{y} = A\boldsymbol{x} \end{aligned}$$

such that

$$A(\boldsymbol{x}_1 + \boldsymbol{x}_2) = A\boldsymbol{x}_1 + A\boldsymbol{x}_2.$$

Let $\mathrm{N}(A)$ and $\mathrm{R}(A)$ denote the null space and the range.

A linear function is the case $\mathbf{Y} = \mathbf{R}$.

Theorem 1 (Riesz-Fréchet) *A function $f(\boldsymbol{x})$ is linear if and only if $f(\boldsymbol{x}) = \langle \boldsymbol{x}^*, \boldsymbol{x} \rangle$ for a unique $\boldsymbol{x}^*$.*

The theorem follows quickly from conjugate duality. As a proper, closed, positively homogeneous, convex function, a linear function must be the support to some convex set. The support is linear if and only if C is a single point, so $f(\boldsymbol{x}) = \delta^*_{\{x^*\}}(\boldsymbol{x}) = \langle \boldsymbol{x}^*, \boldsymbol{x} \rangle$, for some $\boldsymbol{x}^*$.[1]

The adjoint of a linear transformation in a Euclidean space is basic in applications and has not been appreciated by economists. The adjoint exists independently of any particular choice of bases.

Theorem 2 (Adjoint) *Given a linear transformation $A : \mathbf{X} \to \mathbf{Y}$, then there exists a unique linear transformation (the adjoint)*

$$A^\top : \mathbf{Y} \to \mathbf{X}$$

that preserves the inner product:

$$\langle \boldsymbol{y}^*, A\boldsymbol{x} \rangle = \left\langle A^\top \boldsymbol{y}^*, \boldsymbol{x} \right\rangle \tag{10.1}$$

for all $\boldsymbol{x}$ and $\boldsymbol{y}^*$.

[1] Whereas this theorem is elementary in a Euclidean space, Riesz and Fréchet extend the result to infinite dimensions.

Typically one does not calculate the adjoint directly. Instead one conjectures an expression for the adjoint, and then verifies that the adjoint condition (10.1) holds.

We prove the adjoint theorem using the Riesz-Fréchet theorem. Consider first the following simple adjoint. For some $\boldsymbol{y} \in \mathbf{X}$, the adjoint of the linear transformation

$$\begin{aligned} \boldsymbol{y} : \mathbf{R} &\mapsto \mathbf{X} \\ z &\mapsto \boldsymbol{x} = z\boldsymbol{y} \end{aligned}$$

is

$$\begin{aligned} \boldsymbol{y}^{\top} : \mathbf{X} &\to \mathbf{R} \\ \boldsymbol{x} &\mapsto z = \langle \boldsymbol{y}, \boldsymbol{x} \rangle . \end{aligned}$$

The proof is to verify that the adjoint condition (10.1) holds:

$$\langle \boldsymbol{x}, \boldsymbol{y}z \rangle = \langle \boldsymbol{x}, \boldsymbol{y} \rangle z = \langle \boldsymbol{y}, \boldsymbol{x} \rangle z = \left(\boldsymbol{y}^{\top} \boldsymbol{x} \right) z.$$

Thus

$$\boldsymbol{y}^{\top} \boldsymbol{x} = \langle \boldsymbol{y}, \boldsymbol{x} \rangle ,$$

and the adjoint is unique. Either notation is equivalent, but most often we employ the inner-product notation on the right-hand side.

Extending this result proves the adjoint theorem. In (10.1), $\langle \boldsymbol{y}^*, A\boldsymbol{x} \rangle$ is a linear function of $\boldsymbol{x}$, so $\langle \boldsymbol{y}^*, A\boldsymbol{x} \rangle = \langle \boldsymbol{x}^*, \boldsymbol{x} \rangle$ for a unique $\boldsymbol{x}^*$, by the Riesz-Fréchet theorem. To make the adjoint condition (10.1) hold for any $\boldsymbol{x}$ and $\boldsymbol{y}^*$, define $A^{\top} \boldsymbol{y}^* := \boldsymbol{x}^*$. Since $\langle \boldsymbol{y}^*, A\boldsymbol{x} \rangle$ is a linear function of $\boldsymbol{y}^*$, this formulation indeed defines a linear transformation.

The properties of the adjoint correspond to the properties of the transpose in matrix algebra.

Theorem 3 (Adjoint Properties)

Addition *The adjoint of a sum is the sum of the adjoints:*

$$(A + B)^{\top} = A^{\top} + B^{\top};$$

Composition *The adjoint of a composition is the composition of the adjoints, in reverse order:*

$$(AB)^{\top} = B^{\top} A^{\top};$$

Partitioned Linear Transformation

$$\begin{bmatrix} A & B \end{bmatrix}^{\top} = \begin{bmatrix} A^{\top} \\ B^{\top} \end{bmatrix}.$$

One proves these properties by verifying that the adjoint condition (10.1) holds. For example,

$$\langle \boldsymbol{y}^*, (AB)\boldsymbol{x}\rangle = \langle \boldsymbol{y}^*, A(B\boldsymbol{x})\rangle = \left\langle A^\top \boldsymbol{y}^*, B\boldsymbol{x}\right\rangle = \left\langle B^\top A^\top \boldsymbol{y}^*, \boldsymbol{x}\right\rangle$$

proves the composition property.

The adjoint and the transpose of a matrix representation are different. Consider a linear transformation $A : \mathbf{X} \to \mathbf{Y}$. Let

$$X = \begin{bmatrix} \boldsymbol{x}_1 & \cdots & \boldsymbol{x}_m \end{bmatrix}$$
$$Y = \begin{bmatrix} \boldsymbol{y}_1 & \cdots & \boldsymbol{y}_n \end{bmatrix}$$

denote bases for $\mathbf{X}$ and $\mathbf{Y}$. Each linear transformation is invertible. Let

$$\boldsymbol{x} = X\tilde{\boldsymbol{x}}$$
$$\boldsymbol{y} = Y\tilde{\boldsymbol{y}}$$

denote representations of $\boldsymbol{x}$ and $\boldsymbol{y}$ in terms of the bases. Here $\tilde{\boldsymbol{x}} \in \mathbf{R}^m$ and $\tilde{\boldsymbol{y}} \in \mathbf{R}^n$ are vectors of coefficients.

Then A maps $X\tilde{\boldsymbol{x}}$ to $Y\tilde{\boldsymbol{y}} = AX\tilde{\boldsymbol{x}}$, so the matrix representation of A is

$$\tilde{\boldsymbol{x}} \in \mathbf{R}^m \mapsto \tilde{\boldsymbol{y}} = Y^{-1}AX\tilde{\boldsymbol{x}} \in \mathbf{R}^n$$

The transpose of $Y^{-1}AX$ is $X^\top A^\top \left(Y^{-1}\right)^\top$. Analogously, the matrix representation of $A^\top$ is

$$\tilde{\boldsymbol{y}} \in \mathbf{R}^n \mapsto \tilde{\boldsymbol{x}} = X^{-1}A^\top Y\tilde{\boldsymbol{y}} \in \mathbf{R}^m$$

If the bases are orthonormal,

$$X^{-1} = X^\top$$
$$Y^{-1} = Y^\top,$$

then the matrix representation $X^{-1}A^\top Y$ of $A^\top$ is the transpose $X^\top A^\top \left(Y^{-1}\right)^\top$ of the matrix representation of A; otherwise in general they are not equal.

The quadratic $\langle \boldsymbol{x}, \boldsymbol{x}\rangle$ is the sum of squares of the basis coefficients, if the basis is orthonormal:

$$\langle \boldsymbol{x}, \boldsymbol{x}\rangle = \langle X\tilde{\boldsymbol{x}}, X\tilde{\boldsymbol{x}}\rangle = \left\langle \tilde{\boldsymbol{x}}, X^\top X\tilde{\boldsymbol{x}}\right\rangle = \langle \tilde{\boldsymbol{x}}, \tilde{\boldsymbol{x}}\rangle \text{ if } X \text{ is orthonormal.}$$

Let us define several square linear transformations.[2]

[2] A linear transformation from a Euclidean space to a different Euclidean space having the same dimension is not square.

Definition 2 (Square Linear Transformation) A square linear transformation is a linear transformation from a Euclidean space to itself.

The matrix representation of a square linear transformation is a square matrix. Its matrix representation is $X^{-1}AX$ and its transpose is $X^{\mathsf{T}}A^{\mathsf{T}}\left(X^{-1}\right)^{\mathsf{T}}$.

Definition 3 (Self-Adjoint Linear Transformation) A self-adjoint linear transformation is a square linear transformation equal to its adjoint.

For an orthonormal basis, the matrix representation of a self-adjoint linear transformation is a symmetric matrix. But if the basis is not orthonormal, then in general the matrix representation is not symmetric. The latter situation occurs sometimes in econometrics: a linear transformation is self-adjoint, but its matrix representation is not symmetric.

Definition 4 (Skew Linear Transformation) A skew linear transformation is a square linear transformation equal to the negative of its adjoint.

For an orthonormal basis, the matrix representation of a skew linear transformation equals the negative of its transpose. Each diagonal element is zero. Each off-diagonal element equals the negative of the opposite off-diagonal element.

Since we see vectors as coordinate-free, the matrix representation of a linear transformation is of secondary importance. Apart from simple cases, it may be difficult to write down the matrix representation explicitly. At the same time, perhaps one can describe the adjoint easily, without reference to any basis. In the coordinate-free formulation of linear algebra, the adjoint is a fundamental concept. In contrast, in matrix algebra the transpose of a matrix is not a fundamental concept, but rather is merely a notational convenience.

That the standard practice in econometrics is to express formulas in terms of bases that are not orthonormal accounts for the complexity of many formulas. Because a formula in econometrics commonly involves an adjoint, consequently the expression of the formula is complicated. In contrast, when one expresses the formulas in a coordinate-free form, the formulas are simpler and consequently are far easier to derive and to comprehend.

10.3 Extended Fenchel Duality

The adjoint of a linear transformation is a fundamental component of duality. If the primal optimization problem involves a linear transformation, then the dual optimization problem involves the adjoint. Conversely, if an analytical framework involves the adjoint, then most likely duality is an effective way to present the analysis.

Extended Fenchel duality is a fundamental formulation of perturbation duality. Here f and g are proper, closed, convex functions, but f is defined on $\mathbf{X}$, whereas g is defined on $\mathbf{Y}$.

Problem 1 (Primal)

$$\inf_{x \in \mathbf{X}} \left[\langle x^*, x \rangle + f(x) + g(Ax + y) \right]$$

We calculate the reciprocal dual. Adding the Fenchel inequalities

$$f(x) + f^*\left(-A^\top y^* - x^*\right) - \left\langle -A^\top y^* - x^*, x \right\rangle \geq 0$$
$$g(Ax + y) + g^*(y^*) - \langle y^*, Ax + y \rangle \geq 0$$

and rearranging obtains the reciprocal dual. That the cross terms cancel is required.

Problem 2 (Extended Fenchel Dual)

$$\sup_{y^* \in \mathbf{Y}} \left[\langle y^*, y \rangle - f^*\left(-A^\top y^* - x^*\right) - g^*(y^*) \right].$$

Fenchel duality is the special case in which $\mathbf{Y}$ is a copy of $\mathbf{X}$, and A is the copy map, an orthogonal linear transformation, $A^\top = A^{-1}$.

We state the relative and polyhedral conditions for extended Fenchel duality.

Condition 1 (Relative) *The relative condition for the primal is that there exists some* $x \in \operatorname{rint}(\operatorname{dom} f)$ *such that* $Ax + y \in \operatorname{rint}(\operatorname{dom} g)$.

Condition 2 (Polyhedral) *The polyhedral condition for the primal is that* f *and* g *are polyhedral and that there exists some* $x \in \operatorname{dom} f$ *such that* $Ax + y \in \operatorname{dom} g$.

The polyhedral/relative mixed condition is analogous and applies if either function is polyhedral. If any one of these conditions holds, then there is no duality gap. If the optimum value is finite, it is attained in the dual.

Perturbation duality furnishes a straightforward proof of the fundamental theorem of linear algebra.

Theorem 4 (Fundamental Theorem of Linear Algebra) *(Strang 1993) The orthogonal complement of the null space of a linear transformation is the range of the adjoint:* $\mathbf{X} = \mathrm{N}(A) \oplus \mathrm{R}\left(A^\top\right)$.

The fundamental theorem of linear algebra asserts that $\mathrm{R}\left(A^\top\right)$ is polar to $\mathrm{N}(A)$. The primal is the support

$$\delta^*_{\mathrm{N}(A)} = \sup_x \left(\langle x^*, x \rangle - \delta_{Ax=y} \right),$$

in which the perturbation base values are $\boldsymbol{y} = \mathbf{0}$ and $\boldsymbol{x}^*$. The primal and the dual satisfy the polyhedral condition, so there is no duality gap. Adding the Fenchel inequalities

$$\delta_{\{0\}}(A\boldsymbol{x} - \boldsymbol{y}) + \delta^*_{\{0\}}(\boldsymbol{y}^*) - \langle \boldsymbol{y}^*, A\boldsymbol{x} - \boldsymbol{y}\rangle \geq 0$$
$$\delta_{\{0\}}\left(A^\top \boldsymbol{y}^* - \boldsymbol{x}^*\right) + \delta^*_{\{0\}}(-\boldsymbol{x}) - \left\langle A^\top \boldsymbol{y}^* - \boldsymbol{x}^*, -\boldsymbol{x}\right\rangle \geq 0$$

and rearranging obtains

$$\langle \boldsymbol{y}^*, \boldsymbol{y}\rangle + \delta_{\{0\}}\left(A^\top \boldsymbol{y}^* - \boldsymbol{x}^*\right) + \delta^*_{\{0\}}(\boldsymbol{y}^*) \geq \langle \boldsymbol{x}^*, \boldsymbol{x}\rangle - \delta_{\{0\}}(A\boldsymbol{x} - \boldsymbol{y}) - \delta^*_{\{0\}}(-\boldsymbol{x}).$$

The dual is

$$\inf_{\boldsymbol{y}^*} \delta_{A^\top \boldsymbol{y}^* = \boldsymbol{x}^*} = \delta_{\mathrm{R}(A^\top)}(\boldsymbol{x}^*),$$

as desired.

10.4 Moore-Penrose Generalized Inverse

Given a linear transformation $A : \mathbf{X} \to \mathbf{Y}$, consider its restriction to the subspaces $\mathrm{R}\left(A^\top\right)$ and $\mathrm{R}(A)$.

Theorem 5 (Invertibility of the Restricted Linear Transformation) *The restricted linear transformation*

$$A : \mathrm{R}\left(A^\top\right) \to \mathrm{R}(A) \tag{10.2}$$
$$\boldsymbol{x} \mapsto \boldsymbol{y} = A\boldsymbol{x} \tag{10.3}$$

is invertible.

Via the fundamental theorem of linear algebra, we prove that the restricted linear transformation is onto and one-to-one. If $\boldsymbol{y} \in \mathrm{R}(A)$, then $\boldsymbol{y} = A\boldsymbol{x}$ for some $\boldsymbol{x} = (\boldsymbol{u}, \boldsymbol{v}) \in \mathrm{N}(A) \oplus \mathrm{R}\left(A^\top\right)$. As $A\boldsymbol{x} = A(\boldsymbol{u} + \boldsymbol{v}) = A\boldsymbol{v}$, therefore the restricted linear transformation (10.2) is onto. Furthermore, the restricted linear transformation is one-to-one. If $\boldsymbol{y} = A\boldsymbol{v}$ and $\boldsymbol{y} = A\boldsymbol{w}$ for some $\boldsymbol{v}$ and $\boldsymbol{w}$ in $\mathrm{R}\left(A^\top\right)$, then $A(\boldsymbol{v} - \boldsymbol{w}) = A\boldsymbol{v} - A\boldsymbol{w} = \boldsymbol{y} - \boldsymbol{y} = \mathbf{0}$. Hence $\boldsymbol{v} - \boldsymbol{w} \in \mathrm{N}(A)$, so $\boldsymbol{v} = \boldsymbol{w}$.

Extending the inverse of the restricted linear transformation obtains the Moore-Penrose generalized inverse.

Definition 5 [Moore-Penrose Generalized Inverse] (Desoer and Whalen 1963) The Moore-Penrose generalized inverse is the linear transformation $A^+ : \mathbf{Y} \to \mathbf{X}$ such that

- for $y \in \mathrm{R}(A)$, then $A^{+}y$ is the inverse of the restricted linear transformation (10.2);
- for $y \in \mathrm{N}\left(A^{\mathsf{T}}\right)$, then $A^{+}y = \mathbf{0}$.

By the fundamental theorem of linear algebra, these two conditions define A^{+} uniquely.

The relationship between A and A^{+} is symmetric. A linear transformation is the generalized inverse of its generalized inverse: $\left(A^{+}\right)^{+} = A$. And $AA^{+}A = A$. The decomposition into the null space of a linear transformation and the range of the adjoint is

$$x = \left[\left(\mathsf{I} - A^{+}A\right)x, A^{+}Ax\right] \in \mathrm{N}(A) \oplus \mathrm{R}\left(A^{\mathsf{T}}\right).$$

A corollary is a general formula for the solution to a linear equation.

Corollary 1 (Solutions to a Linear Equation) *If $y \in \mathrm{R}(A)$, the solution set to the linear equation is $Ax = y$ is*

$$\left\{A^{+}y\right\} + \mathrm{N}(A),$$

If $y \notin \mathrm{R}(A)$, there is no solution.

10.5 Trace and Determinant

For a square linear transformation A, the trace $\mathrm{Tr}\,A$ and the determinant $\mathrm{Det}\,A$ are important concepts. They apply to any finite-dimensional vector space, not just to an inner-product space. Commonly a linear-algebra textbook develops these concepts via a basis, but an elegant coordinate-free development is possible (Fekete 1985, ch. 13).

We do not define the determinant in this book, but we do review several properties.

Definition 6 (Characteristic Polynomial) For a square linear transformation A, the characteristic polynomial is $\mathrm{Det}\,(A - \lambda \mathsf{I})$.

Definition 7 (Eigenvalues) The eigenvalues are the roots of the characteristic polynomial.

If $\dim \mathbf{X} = m$, then the characteristic polynomial has degree m:

$$\mathrm{Det}\,(A - \lambda \mathsf{I}) = (-\lambda)^{m} + (\mathrm{Tr}\,A)\,(-\lambda)^{m-1} + \cdots + \mathrm{Det}\,A = (\lambda_1 - \lambda) \cdots (\lambda_m - \lambda).$$

The linear transformation has m eigenvalues λ_i, which might be complex-valued. The sum of the eigenvalues is $\mathrm{Tr}\,A$, and the product of the eigenvalues is $\mathrm{Det}\,A$.

Propostion 1 (Determinant of Composition) *For two square linear transformations A and B on a Euclidean space, the determinant of the composition is the product of the determinants,* $\mathrm{Det}\, AB = (\mathrm{Det}\, A)\,(\mathrm{Det}\, B)$.

The trace is important in this book. The singular-value decomposition presented in the chapter "Decomposition" permits simple coordinate-free derivations of properties of the trace. For example, the trace of a linear transformation equals the trace of the adjoint.

Define the inner product of two linear transformations by the trace.

Definition 8 (Trace Inner Product) For linear transformations A and B from a Euclidean space $\mathbf{X}$ to a Euclidean space $\mathbf{Y}$, the inner product $\langle A, B\rangle := \mathrm{Tr}\, A^{\top} B$.

This definition meets the requirements for an inner product. It is bilinear and symmetric. The inner product $\langle A, A\rangle$ is the sum of the squares of the singular values and is positive for any $A \neq 0$.

Under the trace inner product, each of the following is a Euclidean space.

Definition 9 (Euclidean Spaces of Linear Transformations) For Euclidean spaces $\mathbf{X}$ and $\mathbf{Y}$,

- $\mathbf{T}^{\mathbf{X}\to\mathbf{Y}}$ is the Euclidean space of linear transformations $\mathbf{X} \to \mathbf{Y}$;
- $\mathbf{T}^{\mathbf{X}}$ is the Euclidean space of square linear transformations on $\mathbf{X}$;
- $\mathbf{S}^{\mathbf{X}}$ is the Euclidean space of self-adjoint linear transformations on $\mathbf{X}$;
- $\mathbf{K}^{\mathbf{X}}$ is the Euclidean space of skew linear transformations on $\mathbf{X}$.

If $\mathbf{X} = \mathbf{R}^m$ and $\mathbf{Y} = \mathbf{R}^n$, we use the abbreviated notation $\mathbf{T}^{m\to n}$, $\mathbf{T}^m$, $\mathbf{S}^m$, and $\mathbf{K}^m$. Convex optimization in which the choice variable lies in a Euclidean space of linear transformations is effective. "Semi-definite" programming refers to optimization on $\mathbf{S}^{\mathbf{X}}$, which we apply in econometrics.

References

Desoer CA, Whalen BH (1963) A note on pseudoinverses. J Soc Indust Appl Math 11(2):442–447
Fekete AE (1985) Real linear algebra, Vol 91. Marcel Dekker, New York
Steenrod NE (1967 Sept) The geometric content of freshman and sophomore mathematics courses. In: Report of the committee on undergraduate program in mathematics. Mathematical Association of America, Washington, DC
Strang G (1993) The fundamental theorem of linear algebra. Am Math Month 100(9):848–855

Chapter 11
Decomposition

Abstract The spectral decomposition of a self-adjoint linear transformation expresses it via its real eigenvalues and orthonormal eigenvectors. The singular-value decomposition of a linear transformation (not necessarily square) is a valuable concept not well appreciated by economists. We derive it as a corollary of the spectral decomposition of a self-adjoint linear transformation. In this book an idempotent is a self-adjoint linear transformation equal to its square. There is a one-to-one correspondence between a subspace and an idempotent having rank equal to the dimension of the subspace. The idempotents are the extreme points of the convex hull of the idempotents of a given rank. To find an optimizing subspace, recast the problem as seeking an optimizing idempotent. One finds the singular-value decomposition of a linear transformation from the spectral decomposition of a Jordan-Wielandt partitioned linear transformation. The singular-value decomposition enables the discovery and computation of relationships among linear transformations. A square linear transformation transforms area by the product of the singular values.

11.1 Spectral Decomposition

The following is a fundamental result for a Euclidean space.

Propostion 1 (Real Eigenvalues) *For a Euclidean space* $\mathbf{X}$, *such that* $\dim \mathbf{X} = m$, *a linear transformation* $A \in \mathbf{S}^{\mathbf{X}}$ *has m real eigenvalues* λ_i *and m orthonormal eigenvectors* $\boldsymbol{x}_i \in X$, $A\boldsymbol{x}_i = \lambda_i \boldsymbol{x}_i$.

Write the eigenvectors via a partitioned linear transformation $\mathbf{R}^m \to \mathbf{X}$:

$$A \begin{bmatrix} \boldsymbol{x}_1 & \cdots & \boldsymbol{x}_m \end{bmatrix} = \begin{bmatrix} \lambda_1 \boldsymbol{x}_1 & \cdots & \lambda_m \boldsymbol{x}_m \end{bmatrix}.$$

B. C. Dieffenbach, *Conjugate Duality in Economic Analysis*, Contributions to Economics,
https://doi.org/10.1007/978-3-032-21396-9_11

On the left the adjoint is its inverse, so

$$A = \begin{bmatrix} \lambda_1 \boldsymbol{x}_1 & \cdots & \lambda_m \boldsymbol{x}_m \end{bmatrix} \begin{bmatrix} \boldsymbol{x}_1 & \cdots & \boldsymbol{x}_m \end{bmatrix}^\top = \sum_{i=1}^{m} \lambda_i \boldsymbol{x}_i \boldsymbol{x}_i^\top ,$$

the "spectral decomposition" of A.

Theorem 1 (Spectral Decomposition) *A linear transformation* $A \in \mathbf{S}^{\mathbf{X}}$ *has a spectral decomposition into its real eigenvalues* λ_i *and orthonormal eigenvectors* $\boldsymbol{x}_i \in \mathbf{X}$,

$$A = \sum_{i=1}^{n} \lambda_i \boldsymbol{x}_i \boldsymbol{x}_i^\top . \tag{11.1}$$

If the eigenvalues are distinct, then the spectral decomposition is unique (apart from reversing the sign of $\boldsymbol{x}_i$). If there are repeated roots in the characteristic polynomial, then the orthonormal eigenvectors are unique up to rotation. It is often convenient to work with an *ordered* spectral decomposition, in which $\lambda_1 \geq \lambda_2 \geq \cdots \geq \lambda_n$.

Define a partial ordering of $\mathbf{S}^{\mathbf{X}}$ analogous to the partial ordering of $\mathbf{R}^n$.

Definition 1 (Partial Ordering) For $A \in \mathbf{S}^{\mathbf{X}}$, define the partial ordering:

Nonnegative	$A \geq \mathbf{0}$	$\lambda_i \geq 0$ for all i
Greater than zero	$A > \mathbf{0}$	$\lambda_i \geq 0$ for all i and $\lambda_i > 0$ for some i
Positive	$A \gg \mathbf{0}$	$\lambda_i > 0$ for all i.

One says that $A \geq B$ if $A - B \geq \mathbf{0}$. One can interpret the partial ordering via a quadratic form.

Theorem 2 (Quadratic Form) *For* $A \in \mathbf{S}^{\mathbf{X}}$,

$A \geq \mathbf{0}$ *if and only if* $\langle \boldsymbol{x}, A\boldsymbol{x} \rangle \geq 0$ *for all* $\boldsymbol{x}$;

$A > \mathbf{0}$ *if and only if* $\langle \boldsymbol{x}, A\boldsymbol{x} \rangle \geq 0$ *for all* $\boldsymbol{x}$ *and* $\langle \boldsymbol{x}, A\boldsymbol{x} \rangle > 0$ *for some* $\boldsymbol{x}$;

$A \gg \mathbf{0}$ *if and only if* $\langle \boldsymbol{x}, A\boldsymbol{x} \rangle > 0$ *for all* $\boldsymbol{x} \neq \mathbf{0}$.

11.2 Idempotent

Idempotents are important in convex optimization.

Definition 2 (Idempotent) An idempotent M is a self-adjoint linear transformation such that $M^2 = M$.

(The requirement that the transformation be self-adjoint makes this definition differ from the traditional one.) If a self-adjoint linear transformation has eigenvalue λ, then its square has eigenvalue λ^2.

Theorem 3 (Eigenvalues of an Idempotent) *An idempotent has eigenvalues either one or zero.*

Two idempotents are "orthogonal" if their composition is zero. A "primitive" idempotent is one that cannot be expressed as the sum of orthogonal idempotents. A primitive idempotent has the form $\boldsymbol{x}\boldsymbol{x}^\mathsf{T}$, such that $|\boldsymbol{x}| = 1$. The spectral decomposition (11.1) expresses the linear transformation as a linear combination of orthogonal primitive idempotents.

There is a one-to-one correspondence between a subspace M of dimension m and an idempotent M of rank m. Given a subspace M, define M as the idempotent such that $\mathsf{M}\boldsymbol{x} = \boldsymbol{x}$ for $\boldsymbol{x} \in M$ and $\mathsf{M}\boldsymbol{x} = \boldsymbol{0}$ for $\boldsymbol{x} \in M^\perp$. Conversely, an idempotent M determines a subspace M such that $\mathsf{M}\boldsymbol{x} = \boldsymbol{x}$ for $\boldsymbol{x} \in M$ and $\mathsf{M}\boldsymbol{x} = \boldsymbol{0}$ for $\boldsymbol{x} \in M^\perp$. For an optimization problem involving a constraint $\boldsymbol{x} \in M$, alternatively one can analyze the problem using M, via the constraint $\mathsf{M}\boldsymbol{x} = \boldsymbol{x}$.

For a problem in which one seeks an optimizing subspace M, one can recast the problem as seeking an optimizing idempotent $\mathsf{M} \in \mathbf{S}^\mathbf{X}$. For the convex hull of the idempotents of a given rank, the idempotents are the extreme points. By working with this set, in some cases one can solve for the optimizing subspace as a convex optimization.

11.3 Singular-Value Decomposition

We derive the singular-value decomposition (11.5) for a linear transformation as a corollary of the spectral decomposition (11.1). For a linear transformation $\mathsf{A} \in \mathbf{T}^{\mathbf{X}\to\mathbf{Y}}$, then $\mathsf{A}^\mathsf{T}\mathsf{A} \in \mathbf{S}^\mathbf{X}$. This linear transformation is nonnegative: rank A eigenvalues are positive, and the remaining eigenvalues are zero. Write its ordered spectral decomposition as

$$\mathsf{A}^\mathsf{T}\mathsf{A} = \sum_{i=1}^{\operatorname{rank}\mathsf{A}} \lambda_i \boldsymbol{x}_i \boldsymbol{x}_i^\mathsf{T}, \tag{11.2}$$

in which the eigenvalue $\lambda_i > 0$, and the $\boldsymbol{x}_i$ are orthonormal eigenvectors. Define $\mu_i = \sqrt{\lambda_i} > 0$. For simplicity, omit the $n - \operatorname{rank}\mathsf{A}$ terms for the zero eigenvalues.

The spectral decomposition (11.2) fixes the spectral decomposition (11.3) of the paired linear transformation $\mathsf{A}\mathsf{A}^\mathsf{T} \in \mathbf{S}^\mathbf{Y}$. The vector $\boldsymbol{y}_i = \mathsf{A}\boldsymbol{x}_i/\sqrt{\lambda_i}$ is an eigenvector of $\mathsf{A}\mathsf{A}^\mathsf{T}$ with eigenvalue λ_i:

$$\mathsf{A}\mathsf{A}^\mathsf{T}\boldsymbol{y}_i = \mathsf{A}\mathsf{A}^\mathsf{T}\mathsf{A}\boldsymbol{x}_i/\sqrt{\lambda_i} = \sqrt{\lambda_i}\mathsf{A}\boldsymbol{x}_i = \lambda_i\boldsymbol{y}_i.$$

The $\boldsymbol{y}_i$ are orthonormal, and the spectral decomposition

$$AA^{\mathsf{T}} = \sum_{i=1}^{\operatorname{rank} A} \lambda_i \boldsymbol{y}_i \boldsymbol{y}_i^{\mathsf{T}} \tag{11.3}$$

follows.

The spectral decompositions (11.2) and (11.3) fix the spectral decomposition of the "Jordan-Wielandt" partitioned linear transformation

$$\begin{bmatrix} \boldsymbol{x} \\ \boldsymbol{y} \end{bmatrix} \mapsto \begin{bmatrix} 0 & A^{\mathsf{T}} \\ A & 0 \end{bmatrix} \begin{bmatrix} \boldsymbol{x} \\ \boldsymbol{y} \end{bmatrix},$$

a linear transformation in $\mathbf{S}^{\mathbf{X} \oplus \mathbf{Y}}$.

Theorem 4 (Jordan-Wielandt Spectral Decomposition) *The spectral decomposition*

$$\begin{bmatrix} 0 & A^{\mathsf{T}} \\ A & 0 \end{bmatrix} = \frac{1}{2} \sum_{i=1}^{\operatorname{rank} A} \left(\mu_i \begin{bmatrix} \boldsymbol{x}_i \\ \boldsymbol{y}_i \end{bmatrix} \begin{bmatrix} \boldsymbol{x}_i \\ \boldsymbol{y}_i \end{bmatrix}^{\mathsf{T}} - \mu_i \begin{bmatrix} \boldsymbol{x}_i \\ -\boldsymbol{y}_i \end{bmatrix} \begin{bmatrix} \boldsymbol{x}_i \\ -\boldsymbol{y}_i \end{bmatrix}^{\mathsf{T}} \right), \tag{11.4}$$

omitting the $m + n - 2 \operatorname{rank} A$ *terms for eigenvalues zero.*

If the partitioned vector $\begin{bmatrix} \boldsymbol{x} \\ \boldsymbol{y} \end{bmatrix}$ is an eigenvector with eigenvalue μ, then $A^{\mathsf{T}}\boldsymbol{y} = \mu\boldsymbol{x}$ and $A\boldsymbol{x} = \mu\boldsymbol{y}$. Applying A^{T} to the latter yields $A^{\mathsf{T}}A\boldsymbol{x} = \mu A^{\mathsf{T}}\boldsymbol{y} = \mu^2\boldsymbol{x}$, so $\boldsymbol{x}$ is an eigenvector of $A^{\mathsf{T}}A$ with eigenvalue μ^2. Applying A to the former yields $AA^{\mathsf{T}}\boldsymbol{y} = \mu A\boldsymbol{x} = \mu^2\boldsymbol{y}$, so $\boldsymbol{y}$ is an eigenvector of AA^{T} with eigenvalue μ^2. These relationships also show that the converse holds.

Furthermore, the eigenvectors and eigenvalues come in pairs: $\begin{bmatrix} \boldsymbol{x} \\ -\boldsymbol{y} \end{bmatrix}$ is an eigenvector with eigenvalue $-\mu$. The factor $\frac{1}{2}$ is needed because $\left\| \begin{bmatrix} \boldsymbol{x}_i \\ \boldsymbol{y}_i \end{bmatrix} \right\| = \sqrt{2}$. The Jordan-Wielandt spectral decomposition (11.4) follows.

The lower left of the spectral decomposition (11.4) is the singular-value decomposition.

Theorem 5 (Singular-Value Decomposition) *The singular-value decomposition*

$$A = \sum_{i=1}^{\operatorname{rank} A} \mu_i \boldsymbol{y}_i \boldsymbol{x}_i^{\mathsf{T}}. \tag{11.5}$$

The singular values μ_i *are positive. The* $\boldsymbol{y}_i$ *are orthonormal, and the* $\boldsymbol{x}_i$ *are orthonormal.*

Here

$$\langle y_i, Ax_j \rangle = \begin{cases} \mu_i \text{ if } i = j \\ 0 \text{ if } i \neq j. \end{cases}$$

For a self-adjoint linear transformation, the singular values are the absolute values of the eigenvalues.

The singular-value decomposition enables the discovery and computation of relationships. The adjoint

$$A^\top = \sum_{i=1}^{\operatorname{rank} A} \mu_i x_i y_i^\top, \tag{11.6}$$

and the Moore-Penrose generalized inverse

$$A^+ = \sum_{i=1}^{\operatorname{rank} A} \frac{1}{\mu_i} x_i y_i^\top. \tag{11.7}$$

Using these formulas, one can readily verify other relationships, such as $AA^+A = A$ and $A^+ = \left(A^\top A\right)^+ A^\top$.

A self-adjoint linear transformation is idempotent if and only if it equals its Moore-Penrose generalized inverse. Corresponding to the subspace $M = \mathrm{R}(A)$ is the idempotent $M = AA^+$.

11.4 Trace and Determinant

For a Euclidean space, a coordinate-free definition of the trace is the following.

Definition 3 (Trace of a Square Linear Transformation) The trace Tr is the unique linear function

$$\begin{aligned} \mathrm{Tr} : \mathbf{T}^{\mathbf{X}} &\to \mathbf{R} \\ A &\mapsto \mathrm{Tr}\, A \end{aligned}$$

such that $\mathrm{Tr}\, x^* x^\top := \langle x^*, x \rangle$ for x^* and x in $\mathbf{X}$.

The existence and uniqueness of Tr follows from the singular-value decomposition. One can show that this definition conforms with the traditional definition: the trace is the sum of the diagonal elements of a matrix representation, such that the sum is independent of the basis chosen.

To prove trace relationships via this definition is straightforward. By (11.6), the trace equals the trace of the adjoint:

$$\operatorname{Tr} A = \operatorname{Tr} A^{\top} = \sum_{i=1}^{\operatorname{rank} A} \mu_i \langle \boldsymbol{y}_i, \boldsymbol{x}_i \rangle.$$

Reversing the order of composition leaves the trace unchanged.

Theorem 6 (Trace Reversal) *For linear transformations A and B in $\mathbf{T}^{\mathrm{X}\to\mathrm{Y}}$,*

$$\operatorname{Tr} A^{\top} B = \operatorname{Tr} B^{\top} A.$$

For

$$B = \sum_{j=1}^{\operatorname{rank} B} \mu_j^* \boldsymbol{y}_j^* \boldsymbol{x}_j^{*\top},$$

one evaluates

$$\operatorname{Tr} A^{\top} B = \operatorname{Tr} B^{\top} A = \sum_{ij} \mu_j^* \mu_i \langle \boldsymbol{y}_j^*, \boldsymbol{y}_i \rangle \langle \boldsymbol{x}_j^*, \boldsymbol{x}_i \rangle . \tag{11.8}$$

The trace inner product $\langle A, B \rangle := \operatorname{Tr} A^{\top} B$ makes $\mathbf{T}^{\mathrm{X}\to\mathrm{Y}}$ a Euclidean space. The inner product is bilinear and symmetric. By (11.8), $\langle A, A \rangle$ is the sum of the squared singular values, positive for any nonzero A.

The following is immediate from the derivation of the singular-value decomposition: the determinant of $A^{\top} A$ is the product of the squares of the singular values.

Theorem 7 (Determinant as Product of Singular Values) *For a full-rank, square linear transformation, the absolute value of the determinant is the product of the singular values,*

$$|\operatorname{Det} A| = \prod_{i=1}^{\operatorname{rank} A} \mu_i.$$

The singular-value decomposition furnishes the geometric interpretation of the determinant: A transforms area by the factor $|\operatorname{Det} A|$. In the singular-value representation, $\boldsymbol{x}_i \mapsto \mu_i \boldsymbol{y}_i$. In two dimensions, a square of length one on each side maps to a rectangle of length μ_1 by length μ_2. (This interpretation is simpler than the traditional geometric interpretation, in which a square linear transformation transforms the area of a parallelogram by the factor $|\operatorname{Det} A|$.)

Chapter 12
Linear Equation

Abstract Using perturbation duality, we calculate the general formula for the solution set to a linear equation. We express the solution as the optimum value of a convex optimization. We find the reciprocal dual, and calculate the value function. Its subdifferential is the solution set. To calculate the solution set contrasts the traditional approach, which is not a calculation. Instead, one conjectures the solution and then verifies it via the fundamental theorem of linear algebra. Consider the linear equation

$$Ax = y,$$

in which $A : \mathbf{X} \to \mathbf{Y}$. A convex optimization that models the solution is

$$\inf_x \delta_{Ax=y}.$$

If there is a solution, then the optimum value is zero. If there is no solution, then the infimum is ∞.

Using perturbation duality, we calculate the general formula for the solution set (12.5) to a linear Eq. (12.1). We express the solution as the optimum value of a convex optimization. We find the reciprocal dual, and calculate the value function (12.4). Its subdifferential is the solution set (12.5).

This simple example illustrates how perturbation duality has wide-ranging application, in areas that might seem surprising. To calculate the solution set contrasts the traditional approach, which is not a calculation. Instead, one conjectures the solution and then verifies it via the fundamental theorem of linear algebra. The example also nicely portrays the elements of our methodology.

Consider the linear equation

$$Ax = y, \tag{12.1}$$

B. C. Dieffenbach, *Conjugate Duality in Economic Analysis*, Contributions to Economics,
https://doi.org/10.1007/978-3-032-21396-9_12

in which A is a linear transformation from a Euclidean space $\mathbf{X}$ to a Euclidean space $\mathbf{Y}$. A convex optimization that models the solution is

$$\inf_{x} \delta_{Ax=y}.$$

Its solution set is the solution set to the linear equation. If there is a solution, then the optimum value is zero. If there is no solution, then the infimum is ∞.

We work with some $\boldsymbol{y}$ for which there is a solution $\boldsymbol{x}$.

Condition 1 (Existence of a Solution) *There exists a solution to the linear equation:* $\boldsymbol{y} \in \mathrm{R}(A)$.

Equivalently, $\boldsymbol{y} = AA^{+}\boldsymbol{y}$.

12.1 Primal and Dual

Augment the optimization by a linear term.

Problem 1 (Primal)

$$\inf_{x} \left(\langle \boldsymbol{x}^*, \boldsymbol{x} \rangle + \delta_{Ax=y} \right).$$

Here $\boldsymbol{y}$ is the perturbation for the primal, and $\boldsymbol{x}^*$ is the perturbation for the reciprocal dual. We find the dual and calculate its value function $V(\boldsymbol{x}^*)$, for arbitrary $\boldsymbol{x}^*$. The solution set to the linear equation is the solution set for $\boldsymbol{x}^* = \mathbf{0}$. We calculate it as the subdifferential $\partial V(\boldsymbol{x}^*)$, evaluated at $\boldsymbol{x}^* = \mathbf{0}$.

The primal is polyhedral. By Condition 1, it satisfies the polyhedral condition. There is no duality gap. The optimum value is either finite or ∞. If the optimum value is finite, then this value is attained in both the primal and the dual.

We calculate the reciprocal dual from the Lagrangian. We calculate

$$L(\boldsymbol{x}, \boldsymbol{y}^*) = \langle \boldsymbol{x}^*, \boldsymbol{x} \rangle + \langle \boldsymbol{y}^*, \boldsymbol{y} \rangle - \sup_{z} \left(\langle \boldsymbol{y}^*, z \rangle - \delta_{Ax=z} \right) = \langle \boldsymbol{x}^*, \boldsymbol{x} \rangle + \langle \boldsymbol{y}^*, \boldsymbol{y} \rangle - \langle \boldsymbol{y}^*, A\boldsymbol{x} \rangle.$$

The reciprocal dual is

$$\sup_{y^*} \left[\inf_{x} L(\boldsymbol{x}, \boldsymbol{y}^*) \right] = \sup_{y^*} \left[\langle \boldsymbol{y}^*, \boldsymbol{y} \rangle + \inf_{x} \left(\left\langle \boldsymbol{x}^* - A^{\mathsf{T}}\boldsymbol{y}^*, \boldsymbol{x} \right\rangle \right) \right] = \sup_{y^*} \left(\langle \boldsymbol{y}^*, \boldsymbol{y} \rangle - \delta_{A^{\mathsf{T}} y^* = x^*} \right).$$

Problem 2 (Reciprocal Dual)

$$\sup_{y^*} \left(\langle \boldsymbol{y}^*, \boldsymbol{y} \rangle - \delta_{A^{\mathsf{T}} y^* = x^*} \right).$$

The reciprocal dual has essentially the same form as the primal.[1]

The primal less the dual is necessarily nonnegative:

$$\begin{aligned}0 &\leq \left(\langle \boldsymbol{x}^*, \boldsymbol{x}\rangle - \delta_{A\boldsymbol{x}=\boldsymbol{y}}\right) - \left(\langle \boldsymbol{y}^*, \boldsymbol{y}\rangle - \delta_{A^{\mathsf{T}}\boldsymbol{y}^*=\boldsymbol{x}^*}\right) \\ &= \left[\delta_{\{0\}}\left(A\boldsymbol{x}-\boldsymbol{y}\right) + \delta^*_{\{0\}}\left(-\boldsymbol{y}^*\right) - \langle -\boldsymbol{y}^*, A\boldsymbol{x}-\boldsymbol{y}\rangle\right] \\ &\quad + \left[\delta_{\{0\}}\left(A^{\mathsf{T}}\boldsymbol{y}^* - \boldsymbol{x}^*\right) + \delta^*_{\{0\}}\left(\boldsymbol{x}\right) - \left\langle A^{\mathsf{T}}\boldsymbol{y}^* - \boldsymbol{x}^*, \boldsymbol{x}\right\rangle\right].\end{aligned}$$

Each term in brackets is a Fenchel factor and is nonnegative. If $(\boldsymbol{x}, \boldsymbol{y}^*)$ is a solution to the primal and the dual, then primal equals the dual, so each term must be zero. That the two Fenchel factors be zero is necessary and sufficient for $(\boldsymbol{x}, \boldsymbol{y}^*)$ to solve the primal and the dual:

$$A\boldsymbol{x} = \boldsymbol{y} \tag{12.2}$$

$$A^{\mathsf{T}}\boldsymbol{y}^* = \boldsymbol{x}^*. \tag{12.3}$$

Furthermore, (12.2) is first-order condition for $\boldsymbol{x}$ to minimize the Lagrangian, and (12.3) is the first-order condition for $\boldsymbol{y}^*$ to maximize the Lagrangian, so $(\boldsymbol{x}, \boldsymbol{y}^*)$ is a saddle point.

12.2 Value Function

We calculate the value function $V(\boldsymbol{x}^*)$, the optimum value as a function of $\boldsymbol{x}^*$. If $\boldsymbol{x}^* \notin \mathrm{R}\left(A^{\mathsf{T}}\right)$, then the optimum value is $-\infty$, as there is no solution to (12.3). If $\boldsymbol{x}^* \in \mathrm{R}\left(A^{\mathsf{T}}\right)$, then the optimum value is finite, because there exists a solution to both (12.2) and (12.3). In the dual,

$$\langle \boldsymbol{y}^*, \boldsymbol{y}\rangle = \left\langle \boldsymbol{y}^*, AA^+\boldsymbol{y}\right\rangle = \left\langle A^{\mathsf{T}}\boldsymbol{y}^*, A^+\boldsymbol{y}\right\rangle = \left\langle \boldsymbol{x}^*, A^+\boldsymbol{y}\right\rangle.$$

Theorem 1 (Value Function) *For* $\boldsymbol{y} \in \mathrm{R}(A)$, *the value function*

$$V(\boldsymbol{x}^*) = \left\langle \boldsymbol{x}^*, A^+\boldsymbol{y}\right\rangle - \delta_{\mathrm{R}\left(A^{\mathsf{T}}\right)}(\boldsymbol{x}^*). \tag{12.4}$$

---

[1] The primal/dual pair illustrate how there might be a duality gap: if one can choose $\boldsymbol{y} \notin \mathrm{R}(A)$ and $\boldsymbol{x}^* \notin \mathrm{R}\left(A^{\mathsf{T}}\right)$, then the optimum value of the primal is ∞, whereas the optimum value of the dual is $-\infty$.

As the effective domain of the linear function $\langle \boldsymbol{x}^*, A^+\mathbf{y}\rangle$ is the entire space $\mathbf{X}$, the subdifferential of the sum is the sum of the subdifferentials:

$$\begin{aligned}\partial V(\boldsymbol{x}^*) &= \partial\left[\langle \boldsymbol{x}^*, A^+\mathbf{y}\rangle - \delta_{\mathrm{R}(A^\top)}(\boldsymbol{x}^*)\right]\\ &= \partial\langle \boldsymbol{x}^*, A^+\mathbf{y}\rangle - \partial\delta_{\mathrm{R}(A^\top)}(\boldsymbol{x}^*)\\ &= \begin{cases}\{A^+\mathbf{y}\} - \mathrm{N}(A) \text{ if } \boldsymbol{x}^* \in \mathrm{R}(A^\top)\\ \varnothing \text{ if } \boldsymbol{x}^* \notin \mathrm{R}(A^\top).\end{cases}\end{aligned}$$

At a vector in a subspace, the subdifferential of the indicator is the orthogonal complement of the subspace.

The solution set to the linear equation is the subdifferential $\partial V(\boldsymbol{x}^*)$ evaluated at $\boldsymbol{x}^* = \mathbf{0}$.

Theorem 2 (Solution Set) *For* $\mathbf{y} \in \mathrm{R}(A)$, *the solution set to the linear Eq. (12.1) is*

$$\{A^+\mathbf{y}\} + \mathrm{N}(A). \tag{12.5}$$

For $\mathbf{y} \notin \mathrm{R}(A)$, *there is no solution.*

The chapter "Linear Transformation" derives this same formula from the properties of the generalized inverse.

Part III
Economic Applications

Chapter 13
Benefit and Cost

Abstract For an industry producing a good for consumption, the optimum is to choose the quantity to maximize consumer benefit minus cost. The primal/dual pair below shows that this optimum is attained by a price that achieves market equilibrium. The primal chooses the quantity to maximize benefit minus cost, and the dual chooses the price to minimize the sum of consumers' surplus plus profit. For consumption quantity x, let g denote the dollar benefit to consumers, and let f denote the dollar cost function. Choose x to maximize benefit minus cost:

$$\sup_x [g(x) - f(x)].$$

Consumer demand x maximizes consumers' surplus, benefit minus expenditure:

$$\sup_x [g(x) - x^* x] = -g_*(x^*).$$

Inverting the first-order condition obtains $\{x\} = \partial g_*(x^*)$, demand as a function of price. Because benefit is concave and increasing, demand falls as the price rises. The conjugate of cost is the profit function,

$$f^*(x^*) = \sup_x [x^* x - f(x)],$$

the maximum value of revenue minus cost. Inverting the first-order condition obtains $\{x\} = \partial f^*(x^*)$, the profit-maximizing quantity as a function of price. A higher price raises the supply. The dual chooses the price to minimize the profit plus the consumers' surplus,

$$\inf_{y^*} [f^*(y^*) - g_*(y^*)].$$

A solution (x, y^*) to the primal and the dual is on the demand curve and the supply curve, so market equilibrium maximizes benefit minus cost.

B. C. Dieffenbach, *Conjugate Duality in Economic Analysis*, Contributions to Economics,
https://doi.org/10.1007/978-3-032-21396-9_13

For an industry producing a good for consumption, the optimum is to choose the quantity to maximize consumer benefit minus cost. The primal/dual pair below shows that this optimum is attained by a price that achieves market equilibrium. The primal chooses the quantity to maximize benefit minus cost, and the dual chooses the price to minimize the sum of consumers' surplus plus profit.

Let x denote the consumption quantity. Let $g(x)$ denote the dollar benefit to consumers, an increasing, proper, closed, concave function of x. Let $f(x)$ denote the dollar cost function, an increasing, proper, closed, convex function of the quantity x. The optimum is to choose x to maximize benefit minus cost.

Problem 1 (Benefit and Cost Primal) Choose the quantity x to maximize benefit minus cost,

$$\sup_x [g(x) - f(x)].$$

Assume that g and f are differentiable and furthermore that the conjugates g_* and f^* are also differentiable—for each the subdifferential is a singleton.

13.1 Demand and Supply

Interpret benefit via consumer demand, the demand for the good as a function of its price x^*. An intuitive point-of-view is that the height of the demand curve is the marginal benefit from an additional unit. Assume $g(0) = 0$, so $g(x)$ is the area under the demand curve from 0 to x. Consumer demand x maximizes consumers' surplus, benefit minus expenditure:

$$\sup_x [g(x) - x^*x] = -\inf_x [x^*x - g(x)] = -g_*(x^*).$$

The conjugate $g_*(x^*)$ is the negative of the consumers' surplus. Stating that the cost of an additional unit equals the marginal benefit, the first-order condition $\{x^*\} = \partial g(x)$ says that the demand for price x^* is quantity x. By subdifferential invertibility,

$$\{x\} = \partial g_*(x^*) \tag{13.1}$$

is demand as a function of price. Because benefit is concave and increasing, demand falls as the price rises.

The chapter "Fenchel Factors" interpreted the cost via supply. The height of the supply curve is the marginal cost of an additional unit. Assume $f(0) = 0$, so $f(x)$ is the area under the demand curve from 0 to x. The conjugate of cost is the profit function,

$$f^*(x^*) = \sup_x [x^*x - f(x)],$$

the maximum value of revenue minus cost, a function of price. Here the first-order condition is

$$\{x^*\} = \partial f(x),$$

price equals marginal cost. Inverting the subdifferential obtains supply,

$$\{x\} = \partial f^*(x^*),$$

the profit-maximizing quantity as a function of price. A higher price raises the supply.

13.2 Benefit and Cost Dual

We calculate the Fenchel dual. Perturb the primal as

$$\sup_x [g(x+y) - f(x)],$$

in which the perturbation base value $y = 0$. Adding the Fenchel inequalities

$$f(x) + f^*(y^*) - y^*x \geq 0 \tag{13.2}$$

$$-g(x+y) - g_*(y^*) + y^*(x+y) \geq 0 \tag{13.3}$$

and rearranging obtains

$$y^*y + f^*(y^*) - g_*(y^*) \geq g(x+y) - f(x).$$

Problem 2 (Benefit and Cost Dual) Choose the price y^* to minimize the profit plus the consumers' surplus,

$$\inf_{y^*} [f^*(y^*) - g_*(y^*)].$$

The price is the marginal increase in benefit minus cost if an extra unit were available.

Let us assume that there is no duality gap, and that the optimum value is finite and is attained in both the primal and the dual. For the Fenchel factors (13.2) and (13.3) to be zero is necessary and sufficient for a solution (x, y^*) to the primal and the dual. The first is equivalent to

$$\{x\} = \partial f^* (y^*);$$

(x, y^*) is on the supply curve. The second is equivalent to

$$\{x\} = \partial g_* (y^*);$$

(x, y^*) is on the demand curve. Thus (x, y^*) is a solution if and only if demand equals supply.

Market equilibrium maximizes benefit minus cost. The first-order condition for the dual is

$$\{0\} = \partial g_* (y^*) - \partial f^* (y^*);$$

the price y^* makes demand equal supply.

Figure 13.1 illustrates the primal. Given quantity x, the benefit is the area below the demand curve, whereas the cost is the area below the supply curve. The shaded area is benefit minus cost. The optimum is the quantity at market equilibrium, where demand and supply intersect.

Figure 13.2 illustrates the dual. For price y^*, the darker shaded triangle is the consumers' surplus and the lighter shaded triangle is the profit. The optimum is to minimize the sum of both shaded areas, achieved at the market-equilibrium price, where demand and supply intersect. The figures illustrate how the dual value is necessarily greater than or equal to the primal value.

Fig. 13.1 Supply

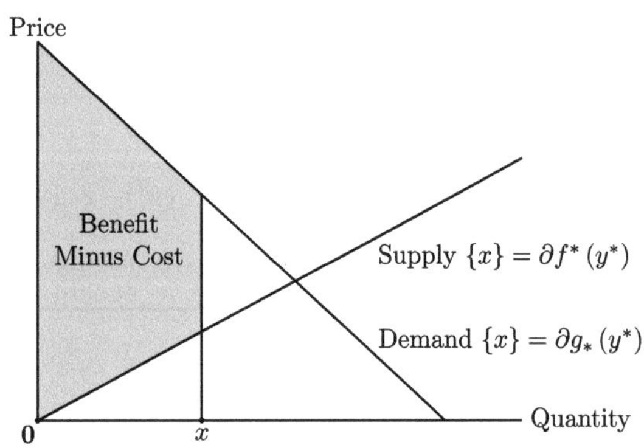

Fig. 13.2 Dual

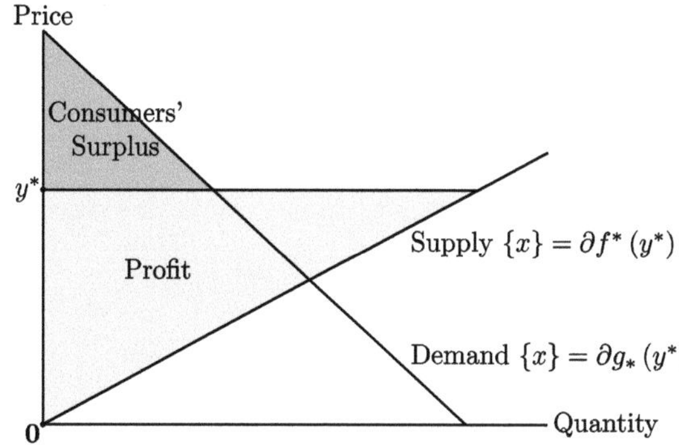

The Lagrangian is

$$\begin{aligned} L(x, y^*) &= -\inf_y \{y^* y - [g(x+y) - f(x)]\} \\ &= -f(x) - \inf_z [y^*(z-x) - g(z)] \quad (z := x+y) \\ &= y^* x - f(x) - g_*(y^*), \end{aligned}$$

expenditure minus cost plus consumers' surplus. The first-order conditions for a saddle point are

$$\{y^*\} = \partial f(x) \tag{13.4}$$
$$\{x\} = \partial g_*(y^*). \tag{13.5}$$

The saddle point is market equilibrium: inverting (13.4) says that (x, y^*) is on the supply curve, and (13.5) says that (x, y^*) is on the demand curve.

Chapter 14
Diet

Abstract Stigler's diet problem is a classic primal/dual application. The problem for the consumer is to minimize the cost of buying food to achieve nutrition targets. Define the linear transformation

$$A := \begin{bmatrix} a_1 & \cdots & a_m \end{bmatrix},$$

in which a_j denotes the nutrition from one unit of food j. Diet $\boldsymbol{x}$ provides nutrition $A\boldsymbol{x}$. The consumer chooses $\boldsymbol{x}$ to minimize the food cost $\langle \boldsymbol{x}^*, \boldsymbol{x} \rangle$ to achieve nutrition target $\boldsymbol{y}$:

$$\inf_{\boldsymbol{x}} \big(\langle \boldsymbol{x}^*, \boldsymbol{x} \rangle + \delta_{A\boldsymbol{x} \succeq \boldsymbol{y}} + \delta_{\boldsymbol{x} \geq \boldsymbol{0}} \big).$$

The reciprocal dual is

$$\sup_{\boldsymbol{y}^*} \big(\langle \boldsymbol{y}^*, \boldsymbol{y} \rangle - \delta_{A^\top \boldsymbol{y}^* \preceq \boldsymbol{x}^*} - \delta_{\boldsymbol{y}^* \geq \boldsymbol{0}} \big).$$

Interpret the dual choice variable $\boldsymbol{y}^*$ as the price (value) of nutrition. Choose $\boldsymbol{y}^*$ to maximize the value $\langle \boldsymbol{y}^*, \boldsymbol{y} \rangle$ of nutrition, subject to food value $A^\top \boldsymbol{y}^*$ less than or equal to food cost $\boldsymbol{x}^*$.

The diet problem (Stigler 1945) is famous in economics, and this polyhedral linear program is a classic primal/dual application (Dorfman et al. 1958, Chaps. 2 and 3). The problem for the consumer is to minimize the cost of buying food to achieve nutrition targets. The consumer buys m foods, to achieve n nutrition requirements. Let $a_j \in \mathbf{R}^n$ denote the nutrition from one unit of food j. Define the linear transformation

$$A := \begin{bmatrix} a_1 & \cdots & a_m \end{bmatrix} \in \mathbf{T}^{m \to n}.$$

Diet $\boldsymbol{x} \in \mathbf{X} = \mathbf{R}^m$ provides nutrition $A\boldsymbol{x}$. The consumer chooses nonnegative $\boldsymbol{x}$ to minimize the food cost $\langle \boldsymbol{x}^*, \boldsymbol{x} \rangle$ to achieve nutrition target $\boldsymbol{y} \in \mathbf{Y} = \mathbf{R}^n$.

B. C. Dieffenbach, *Conjugate Duality in Economic Analysis*, Contributions to Economics,
https://doi.org/10.1007/978-3-032-21396-9_14

Problem 1 (Diet)

$$\inf_{x}\left(\langle x^*, x\rangle + \delta_{Ax\geq y} + \delta_{x\geq 0}\right).$$

The diet is constrained to be in the intersection of $m + n$ halfspaces. Apart from degeneracy the optimum will be an extreme point determined by m halfspaces; m constraints will hold with equality, and the remaining n constraints will not be binding.

The primal perturbation is the nutritional target y, and the dual perturbation is the food price x^*. We calculate the reciprocal dual. Adding the Fenchel inequalities

$$\delta_{X_+}(x) + \delta^*_{X_+}\left(A^\top y^* - x^*\right) - \left\langle A^\top y^* - x^*, x\right\rangle \geq 0 \tag{14.1}$$

$$\delta_{Y_+}(Ax - y) + \delta^*_{Y_+}(-y^*) - \langle -y^*, Ax - y\rangle \geq 0 \tag{14.2}$$

and rearranging obtains

$$\begin{aligned}&\langle x^*, x\rangle + \delta_{Y_+}(Ax - y) + \delta_{X_+}(x)\\ &\geq \langle y^*, y\rangle - \delta^*_{X_+}\left(A^\top y^* - x^*\right) - \delta^*_{Y_+}(-y^*).\end{aligned}$$

The cross term $\langle y^*, Ax\rangle$ cancels the cross term $\left\langle A^\top y^*, x\right\rangle$.

Problem 2 (Reciprocal Dual)

$$\sup_{y^*}\left(\langle y^*, y\rangle - \delta_{A^\top y^* \leq x^*} - \delta_{y^*\geq 0}\right).$$

Again the constraint involves $m+n$ halfspaces, but now n choice variables. Apart from degeneracy the optimum will be an extreme point determined by n halfspaces.

Because the diet problem is polyhedral, there can be a duality gap only if both the primal and dual constraints cannot be satisfied. We asssume that this situation is not the case, and the optimum value is attained in both the primal and the dual.

In the primal, see y as nutrition demanded and Ax as nutrition supplied. See x as input to produce output Ax. Choose nonnegative x to minimize total cost $\langle x^*, x\rangle$, subject to supply greater than or equal to demand.

Interpret the dual choice variable y^* as the price (value) of nutrition. Then $A^\top y^*$ is the value received from foods. One then compares this value to the food cost x^*. Choose nonnegative y^* to maximize the value $\langle y^*, y\rangle$ of nutrition, subject to food value less than or equal to cost.

The Lagrangian provides economic interpretation. The Lagrangian

$$
\begin{aligned}
L(x,y^*) &= \langle x^*,x\rangle + \langle y^*,y\rangle - \sup_z \left[\langle y^*,z\rangle - \delta_{Y_+}(Ax-z) - \delta_{X_+}(x)\right] \\
&= \langle x^*,x\rangle + \langle y^*,y\rangle - \delta_{X_+}(x) - \sup_w \left[\langle y^*,Ax-w\rangle - \delta_{Y_+}(w)\right] \\
(w &:= Ax-z) \\
&= \langle x^*,x\rangle + \langle y^*,y\rangle - \langle y^*,Ax\rangle + \delta_{X_+}(x) - \sup_w \left[\langle -y^*,w\rangle - \delta_{Y_+}(w)\right] \\
&= \langle x^*,x\rangle + \langle y^*,y\rangle - \langle y^*,Ax\rangle + \delta_{X_+}(x) - \delta_{Y_+}(y^*),
\end{aligned}
$$

which is convex in x and concave in y^*. In this Karush-Kuhn-Tucker problem, the "extra term" is absent.

At a saddle point (x,y^*) of the Lagrangian, x minimizes the Lagrangian, so it maximizes

$$\left\langle A^{\mathsf{T}}y^* - x^*, x\right\rangle + \delta_{X_+}(x).$$

The diet maximizes value $A^{\mathsf{T}}y^*$ minus cost x^*, subject to x nonnegative. A finite saddle value requires value less than or equal to cost. If value exceeds cost for a food, let its quantity tend to ∞, to send the Lagrangian toward $-\infty$. If value equals cost, then any nonnegative x_j is minimizing. If value is less than cost, choose $x_j = 0$.

At a saddle point, y^* maximizes

$$\langle y^*, y - Ax\rangle - \delta_{Y_+}(y^*),$$

the value of nutrition demand minus nutrition supply. A finite saddle value requires supply greater than or equal to demand. If supply is greater than demand, choose price zero (a free good). If supply equals demand, then any nonnegative price is maximizing.

For the Fenchel factor (14.1) to be zero is complementary slackness. For the inequality $A^{\mathsf{T}}y^* \preceq x^*$ specifying that value is less than or equal to cost, then the quantity must be zero for any food for which cost exceeds value. If the quantity of a food is positive, then its value must equal cost. The complementary slackness embodies the first-order condition for x to minimize the Lagrangian: its subdifferential expression

$$A^{\mathsf{T}}y^* - x^* \in \partial\delta_{X_+}(x)$$

is the first-order condition.

Analogously, for the Fenchel factor (14.2) to be zero is again complementary slackness. For the inequality $Ax \succeq y$ specifying that the supply of nutrition is greater than or equal to the demand for nutrition, then value must be zero for any nutrition in excess supply. The complementary slackness embodies the first-order condition for $\boldsymbol{y}^*$ to to maximize the Lagrangian: its subdifferential expression

$$A\boldsymbol{x} - \boldsymbol{y} \in \partial\delta_{Y_+}(\boldsymbol{y}^*)$$

is the first-order condition.

Interpret the first-order condition for the dual as market equilibrium—the choice of the nutrition price makes the (inelastic) nutrition demand equal nutrition supply. Let J denote the foods for which value equals cost. Let I denote nutrition having price zero. By the theory of the normal cone to a set, the first-order condition is

$$\boldsymbol{y} \in \sum_{j \in J} a_j \mathbf{R}_+ - \sum_{i \in I} e_i \mathbf{R}_+ \text{ such that } A^\mathsf{T}\boldsymbol{y}^* \leq \boldsymbol{x}^*, \boldsymbol{y}^* \geq \mathbf{0}.$$

(The subdifferential is empty if either constraint does not hold.) Supply is the first sum on the right-hand side, created by food input. A food is not consumed if its value is less than cost, but may be consumed in any nonnegative amount if value equals cost. Demand equals supply for all nutrition having a positive price. The second sum allows supply to exceed demand if the price is zero.

In Stigler's study, $m = 77$ and $n = 9$. Apart from degeneracy, the number of extreme points in the constraint set of either the primal or the dual is

$$\frac{(m+n)!}{m!n!} = \frac{86!}{77!9!} = 459,856,441,980.$$

By clever observation and reasoning, Stigler puts forward an approximate solution. Using the simplex method, (Dantzig and Thapa 1997) solve his problem and show that the cost of his approximate solution is very near the minimum.

References

Dantzig GB, Thapa MN (1997) Linear programming 1: Introduction. Springer, New York
Dorfman R, Samuelson PA, Solow RM (1958) Linear programming and economic analysis. McGraw-Hill, New York
Stigler GJ (1945 May) The cost of subsistence. Journal of Farm Economics 27(2):303–314

Chapter 15
Activity Analysis

Abstract Koopmans studies efficient production in activity analysis. An activity produces output and uses inputs. A production equilibrium is a price/quantity combination such that profit is maximized. The key results are first and second theorems of production equilibrium, analogous to the first and second theorems of welfare economics. The first theorem is that production equilibrium implies that production is efficient. The second theorem is that efficient production can be realized as a production equilibrium; for chosen prices, the production maximizes profit. This chapter obtains his ideas and results via perturbation duality. The primal studies whether activity levels can be changed to increase the net production of goods. The dual chooses goods prices to minimize the value of net output, such that no activity is profitable. Analyzing the primal and the dual obtains the first and second theorems of production equilibrium.

In a classic paper, Koopmans studies efficient production in activity analysis (Koopmans 1951). An activity produces output and uses inputs. A production equilibrium is a price/quantity combination such that profit is maximized. The key results are first and second theorems of production equilibrium, analogous to the first and second theorems of welfare economics. The first theorem is that production equilibrium implies that production is efficient. The second theorem is that efficient production can be realized as a production equilibrium; for chosen prices, the production maximizes profit.

This chapter obtains his ideas and results via perturbation duality. A weak-efficiency primal for production is the initial focus. The dual chooses goods prices to minimize the value of net output, such that there is no opportunity for profit for any activity. One discovers the first and second theorems of production equilibrium by analyzing the primal and the dual. Koopmans studies efficiency, rather than weak efficiency. An efficiency primal for production and its dual lead to his results, comparable to those for weak efficiency.

An activity produces output and uses inputs. There are m activities that produce or use n outputs or inputs. Output is positive, and input is negative. Returns to scale are constant. If activity i is carried out at level $x_i \geq 0$, then net output is $\boldsymbol{a}_i x_i \in \mathbf{R}^n$.

B. C. Dieffenbach, *Conjugate Duality in Economic Analysis*, Contributions to Economics,
https://doi.org/10.1007/978-3-032-21396-9_15

The activity transformation $A = \begin{bmatrix} a_1 & \cdots & a_m \end{bmatrix}$ describes the technology; output $y = Ax$.

Definition 1 (Profit) The profit is the value of net output, $\langle y^*, Ax\rangle$.

Definition 2 (Production Equilibrium) If activity vector $x \geq 0$ maximizes profit for price $y^* \succ 0$, then (x, y^*) is a production equilibrium.

A firm seeks to maximize profit, because higher profit raises the utility of its owners. Maximum profit is

$$\sup_x \left[\langle y^*, Ax\rangle - \delta_{X_+}(x)\right] = \delta^*_{X_+}\left(A^\top y^*\right) = \delta_{A^\top y^* \leq 0}. \tag{15.1}$$

Because returns to scale are constant, the maximum profit is either zero or infinity. If $\langle y^*, a_i\rangle > 0$, then the profit rises toward infinity as the activity level increases. If $\langle y^*, a_i\rangle < 0$, the activity is unprofitable, so its level must be zero. The inequality $A^\top y^* \leq 0$ defines the "no-profit prices." Profit maximization is summarized by the Fenchel equality

$$\delta_{X_+}(x) + \delta^*_{X_+}\left(A^\top y^*\right) - \left\langle A^\top y^*, x\right\rangle = 0, \tag{15.2}$$

complementary slackness. Here (x, y^*) is a production equilibrium if and only if this Fenchel equality holds and also $y^* \succ 0$. The Fenchel equality requires zero profit, $\left\langle A^\top y^*, x\right\rangle = \langle y^*, Ax\rangle = 0$.

Definition 3 (Weakly-Efficient Production) Activity level vector $z \geq 0$ is weakly efficient if there is no $x \geq 0$ such that $Ax \gg Az$.

Definition 4 (Efficient Production) Activity level vector $z \geq 0$ is efficient if there is no $x \geq 0$ such that $Ax > Az$.

Below we solve a weak-efficiency problem and an efficiency problem for activity analysis. That the optimizations are polyhedral simplifies the analysis.

15.1 Weakly Efficient Production

A production equilibrium need not be efficient. Every price need not be positive, as there may be a free good, having price zero. For a free output, profit maximization need not maximize its quantity. For a free input, profit maximization need not minimize (the absolute value) of its quantity. With a free good, typically production is not efficient. Consequently, the natural framework for the analysis of production is weak efficiency.

We study whether activity level $z \geq \mathbf{0}$ is weakly efficient via the following problem.

Problem 1 (Weak-Efficiency Primal)

$$\sup_x \left(\min_i [Ax + y] - \delta_{x \geq \mathbf{0}} \right) = \sup_x \left[-\delta_P^* (-Ax - y) - \delta_{x \geq \mathbf{0}} \right]. \tag{15.3}$$

The base value of the perturbation $y = -Az$. The choice $x = z$ makes the objective function zero, so the optimum value is necessarily greater than or equal to zero. The optimum value is zero if and only if z is weakly efficient. Because the primal is polyhedral and the optimum value is zero or more, there is no duality gap. If the optimum value is finite, then it is attained in both the primal and the dual.

We find the extended Fenchel dual, under the linear transformation $x \mapsto y^* = -Ax$. Adding the Fenchel inequalities

$$\delta_{X_+}(x) + \delta_{X_+}^* \left(A^\top y^* \right) - \left\langle A^\top y^*, x \right\rangle \geq 0 \tag{15.4}$$

$$\delta_P(y^*) + \delta_P^* (-Ax - y) - \langle y^*, -Ax - y \rangle \geq 0 \tag{15.5}$$

and rearranging obtains

$$\langle y^*, y \rangle + \delta_{X_+}^* \left(A^\top y^* \right) + \delta_P(y^*) \geq -\delta_P^* (-Ax - y) - \delta_{X_+}(x).$$

Problem 2 (Weak-Efficiency Dual)

$$\inf_{y^*} \left[\langle y^*, y \rangle + \delta_{A^\top y^* \leq \mathbf{0}} + \delta_P(y^*) \right].$$

Here $\langle y^*, y \rangle = -\langle y^*, Az \rangle = -\left\langle A^\top y^*, z \right\rangle \geq 0$ is the negative of profit. The dual chooses a no-profit price $y^* \in P$ to make profit as large as possible. Because the profit is less than or equal to zero, the optimum value is nonnegative. If z is weakly efficient, then the optimum value is zero; otherwise it is greater than zero.

Analyzing the primal and the dual obtains the following restriction on technology.

Theorem 1 (Existence of Weakly Efficient Production) *There exists a weakly-efficient activity vector if and only if the following two equivalent conditions hold:*

No Output of Everything Without Input *There is no* $x \geq \mathbf{0}$ *such that* $Ax \gg \mathbf{0}$;
No-Profit Price Greater than Zero *There exists some* $y^* > \mathbf{0}$ *such that* $A^\top y^* \leq \mathbf{0}$.

Either condition implies the other. If the second condition does not hold, then the optimum value is ∞, and no z is weakly efficient. In particular $z = \mathbf{0}$ is not weakly-efficient, so the first condition does not hold. If the second condition does hold, then the optimum value is zero for $z = \mathbf{0}$; and the first condition must hold. A weakly-efficient activity vector exists. The conditions hold if and only if the optimum value of the weak efficiency primal and dual is finite.[1]

Theorem 2 (First Theorem of Weakly Efficient Production) *If $(\boldsymbol{x}, \boldsymbol{y}^*)$ is a production equilibrium, then $\boldsymbol{x}$ is weakly-efficient.*

For the Fenchel factors (15.4) and (15.5) to be zero is necessary and sufficient for a solution to the primal and the dual. The first is just the Fenchel equality (15.2) for profit maximization. Let $(z, \boldsymbol{y}^*)$ denote a production equilibrium (normalized so $\boldsymbol{y}^* \in P$). The first Fenchel factor is zero. As $\boldsymbol{y} = -Az$, the second Fenchel factor is also zero. That $\delta_P^* (-Az - \boldsymbol{y}) = 0$ establishes that z is weakly efficient.

The converse also holds.

Theorem 3 (Second Theorem of Weakly Efficient Production) *If $\boldsymbol{x}$ is weakly-efficient, then there exists some $\boldsymbol{y}^*$ such that $(\boldsymbol{x}, \boldsymbol{y}^*)$ is a production equilibrium.*

Suppose that z is weakly efficient, and define $\boldsymbol{y} = -Az$. Let $\boldsymbol{y}^*$ be a solution to the dual. Then $(z, \boldsymbol{y}^*)$ is a production equilibrium. The second Fenchel factor implies $\boldsymbol{y}^* \succ \mathbf{0}$. The first Fenchel factor holds, so profit is maximized, at zero.

To gain additional insight, alternatively one can reason out the first and second theorems of weakly-efficient production via the first-order condition for the weak-efficiency dual. Expand the price constraint in the dual:

$$\inf_{y^*} \left[-\langle y^*, Az \rangle + \delta_{A^\top y^* \leq 0} + \delta_{y^* \geq 0} + \delta_{\langle y^*, 1 \rangle = 1} \right]. \tag{15.6}$$

Let J denote the activities such that profit is zero. Let I denote the goods having price zero. The first-order condition for (15.6) is

$$Az \in \sum_J \boldsymbol{a}_j \mathbf{R}_+ - \sum_I \boldsymbol{e}_i \mathbf{R}_+ + \{\mathbf{1}\} \mathbf{R} \text{ when } A^\top \boldsymbol{y}^* \leq \mathbf{0} \text{ and } \boldsymbol{y}^* \in P. \tag{15.7}$$

(If the constraints are not satisfied, then the subdifferential is empty.)

We show that the first-order condition reduces to

$$Az \in \sum_J \boldsymbol{a}_j \mathbf{R}_+ - \sum_I \boldsymbol{e}_i \mathbf{R}_+ + \{\mathbf{1}\} \langle y^*, Az \rangle \text{ when } A^\top \boldsymbol{y}^* \leq \mathbf{0} \text{ and } \boldsymbol{y}^* \in P. \tag{15.8}$$

---

[1] The equivalence of the two conditions is an example of the efficiency alternative. Later we derive the theorem via an application of weak-efficiency duality.

Applying $\boldsymbol{y}^{*\mathsf{T}}$ to the first-order condition shows

$$\boldsymbol{y}^{*\mathsf{T}} A z \in \sum_J \boldsymbol{y}^{*\mathsf{T}} \boldsymbol{a}_j \mathbf{R}_+ - \sum_I \boldsymbol{y}^{*\mathsf{T}} \boldsymbol{e}_i \mathbf{R}_+ + \mathbf{R} \text{ when } A^{\mathsf{T}} \boldsymbol{y}^* \preceq \mathbf{0} \text{ and } \boldsymbol{y}^* \in P$$

The left-hand side is profit, so the right-hand side must include profit. The two sums are both zero: the activity profit $\boldsymbol{y}^{*\mathsf{T}} \boldsymbol{a}_j = 0$, and the price $\boldsymbol{y}^{*\mathsf{T}} \boldsymbol{e}_i = y_i^* = 0$. That $\boldsymbol{y}^{*\mathsf{T}} \mathbf{1} = 1$ implies the final term in Eq. (15.8).

We prove the first theorem of weakly-efficient production. If $(z, \boldsymbol{y}^*)$ is a production equilibrium, the first-order condition (15.8) holds. Renormalize the price so $\boldsymbol{y}^* \in P$. Profit is zero, so the final term in Eq. (15.8) is zero. The term $\boldsymbol{a}_j \mathbf{R}_+$ expresses how the level of activity j can be any nonnegative value, because the profit of the activity is zero. The term $\boldsymbol{e}_i \mathbf{R}_+$ expresses how a free good can be in excess supply. Because the first-order condition holds and profit is zero, the optimum value of the dual is zero, so z is weakly-efficient.

Conversely, we prove the second theorem of weakly-efficient production. Suppose that z is weakly-efficient and that $\boldsymbol{y}^*$ is a solution to the dual. The optimum value of the dual is zero, so profit is zero. Because the first-order condition holds, z maximizes profit at zero, as explained in the previous paragraph. Consequently $(z, \boldsymbol{y}^*)$ is a production equilibrium.

15.2 Efficient Production

Koopmans chooses to analyze efficiency, rather than weak efficiency. We study whether activity vector $z \succeq \mathbf{0}$ is efficient via the following problem.

Problem 3 (Efficiency Primal)

$$\sup_{\boldsymbol{x}} \left[-\delta^*_{\{1\}+\mathrm{X}_+} (-A\boldsymbol{x} - \boldsymbol{y}) - \delta_{\boldsymbol{x} \geq \mathbf{0}} \right].$$

The base value of the perturbation $\boldsymbol{y} = -Az$. The choice $\boldsymbol{x} = z$ makes the objective function zero, so the optimum value is necessarily greater than or equal to zero. The optimum value is zero if and only if z is efficient. Because the primal is polyhedral and the optimum value is zero or more, there is no duality gap. If the optimum value is finite, then it is attained in both the primal and the dual.

The dual is extended Fenchel duality, under the linear transformation $\boldsymbol{x} \mapsto \boldsymbol{y}^* = -A\boldsymbol{x}$. Adding the Fenchel inequalities

$$\delta_{\mathrm{X}_+}(\boldsymbol{x}) + \delta^*_{\mathrm{X}_+}\left(A^{\mathsf{T}} \boldsymbol{y}^*\right) - \left\langle A^{\mathsf{T}} \boldsymbol{y}^*, \boldsymbol{x} \right\rangle \geq 0 \tag{15.9}$$

$$\delta_{\{1\}+\mathrm{X}_+}(\boldsymbol{y}^*) + \delta^*_{\{1\}+\mathrm{X}_+}(-A\boldsymbol{x} - \boldsymbol{y}) - \langle \boldsymbol{y}^*, -A\boldsymbol{x} - \boldsymbol{y} \rangle \geq 0 \tag{15.10}$$

and rearranging obtains

$$\langle y^*, y\rangle + \delta^*_{X_+}\left(A^\top y^*\right) + \delta_{\{1\}+X_+}(y^*) \geq -\delta^*_{\{1\}+X_+}(-Ax - y) - \delta_{X_+}(x).$$

Problem 4 (Efficiency Dual)

$$\inf_{y^*}\left(\langle y^*, y\rangle + \delta_{A^\top y^* \leq 0} + \delta_{y^* \geq 1}\right).$$

The dual chooses a no-profit price $y^* \geq \mathbf{1}$ to make profit $\langle y^*, Az\rangle = -\langle y^*, y\rangle$ as large as possible. Necessarily the profit is less than or equal to zero, so the infimum is nonnegative. If z is efficient, then the maximum profit is zero and the optimum value is zero; otherwise it is greater than zero. Because the primal is polyhedral and the optimum value of the primal is zero or more, there is no duality gap. If the optimum value is finite, then it is achieved in both the primal and the dual.

Observe that the weak-efficiency and efficiency duals are almost the same, but not quite. The constraint $y^* \in P$ in the former is to achieve $y^* \succ \mathbf{0}$. The constraint $y^* \geq \mathbf{1}$ in the latter is to achieve $y^* \gg \mathbf{0}$. The analyses of efficiency and weak efficiency are very similar.

Analyzing the primal and the dual obtains the following restriction on technology.

Theorem 4 (Existence of Efficient Production) *There exists an efficient activity vector if and only if the following two equivalent conditions hold:*

No Output Without Input *There is no* x *such that* $Ax \succ \mathbf{0}$;
Positive No-Profit Price *There exists some* $y^* \gg \mathbf{0}$ *such that* $A^\top y^* \leq \mathbf{0}$.

If the second condition does not hold, then the optimum value is ∞, and no z is efficient. In particular $z = \mathbf{0}$ is not efficient, so the first condition does not hold. If the second condition does hold, then the optimum value is zero for $z = \mathbf{0}$. As $z = \mathbf{0}$ is efficient, the first condition must hold. An efficient activity vector exists. Either condition implies the other. The conditions hold if and only if the optimum value of the efficiency primal and dual is finite.[2]

For the Fenchel factors (15.9) and (15.10) to be zero is necessary and sufficient for a solution to the primal and the dual. The first is just the Fenchel equality (15.2) for profit maximization. Let (z, y^*) denote a production equilibrium such that the price is positive, normalized so $y^* \gg \mathbf{1}$. The first Fenchel factor is zero. Set $y = -Az$, so the second Fenchel factor is also zero. That $\delta^*_{\{1\}+X_+}(-Az - y) = 0$ establishes that z is efficient.

Theorem 5 (First Theorem of Efficient Production) *If* (x, y^*) *is a production equilibrium such that the price* y^* *is positive, then* x *is efficient.*

[2] The equivalence of the two conditions is an example of the efficiency alternative. Later we derive the theorem via an application of efficiency duality.

Conversely, suppose that z is efficient, and set $\boldsymbol{y} = -Az$. Choose the price $\boldsymbol{y}^*$ as a solution to the dual, which is positive, by the Fenchel factor (15.10). Then $(z, \boldsymbol{y}^*)$ is a production equilibrium. The Fenchel factor (15.9) implies that profit is maximized, at zero.

Theorem 6 (Second Theorem of Efficient Production) *If $\boldsymbol{x}$ is efficient, then there exists some positive price $\boldsymbol{y}^*$ such that $(\boldsymbol{x}, \boldsymbol{y}^*)$ is a production equilibrium.*

Alternatively one can reason out the first and second theorems of efficient production via the first-order condition for the efficiency dual. Let J denote the activities for which profit is zero. Let I denote the goods having price one. The first-order condition for (15.6) is

$$Az \in \{0\} + \sum_J \boldsymbol{a}_j \mathbf{R}_+ - \sum_I \boldsymbol{e}_i \mathbf{R}_+ \text{ when } A^{\mathsf{T}} \boldsymbol{y}^* \leq \mathbf{0} \text{ and } \boldsymbol{y}^* \geq \mathbf{1}.$$

(If the constraints are not satisfied, then the subdifferential is empty.)

We prove the first theorem of efficient production. Consider a production equilibrium $(z, \boldsymbol{y}^*)$ such that the price is positive. Renormalizing the price so $\boldsymbol{y}^* \gg \mathbf{1}$ reduces the first-order condition to

$$Az \in \{0\} + \sum_J \boldsymbol{a}_j \mathbf{R}_+ \text{ when } A^{\mathsf{T}} \boldsymbol{y}^* \leq \mathbf{0}. \tag{15.11}$$

The production equilibrium satisfies this condition. Profit is zero. Any unprofitable activity is not done. Each profitable activity is done at some nonnegative level, which is captured by the sum. Because the first-order condition holds and profit is zero, the optimum value of the dual is zero, so z is efficient.

Conversely, we prove the second theorem of efficient production. Suppose that z is efficient and that $\boldsymbol{y}^*$ is a solution to the dual. The constraint $\boldsymbol{y}^* \geq \mathbf{1}$ is not binding, as multiplying the price by a scalar greater than one leaves the value at zero, so the first-order condition reduces to (15.11). The optimum value of the dual is zero, so profit is maximized at zero, and consequently $(z, \boldsymbol{y}^*)$ is a production equilibrium.

Reference

Koopmans TC (1951) Analysis of production as an efficient combination of activities. In: Koopmans TC (ed) Activity analysis of production and allocation, Vol 13. Wiley, New York, pp 33–97

Chapter 16
Zero-Sum Game

Abstract Nash equilibrium is the cornerstone of game theory. For a two-player, zero-sum game, a Nash equilibrium is a saddle point of the payoff as a function of the strategies of the two players. This function is the Lagrangian of a primal/dual pair. In the primal one player maximizes the minimum payoff across his pure strategies, and in the dual the other player minimizes the maximum payoff across his pure strategies. The game has a Nash equilibrium, and one can calculate the optimum strategies by solving a convex optimization. A mixed strategy is a choice of probabilities for the different possible pure strategies. Let A denote the payoff matrix. Let $\boldsymbol{x} \in P^n$ denote a mixed strategy for player x, and let $\boldsymbol{y}^* \in P^m$ denote a mixed strategy for player y. Then $\langle \boldsymbol{y}^*, A\boldsymbol{x} \rangle$ is the payoff to x and the negative of the payoff to y. A Nash equilibrium $(\boldsymbol{x}, \boldsymbol{y}^*)$ is a saddle point of

$$L(\boldsymbol{x}, \boldsymbol{y}^*) = \langle \boldsymbol{y}^*, A\boldsymbol{x} \rangle - \delta_{P^n}(\boldsymbol{x}) + \delta_{P^m}(\boldsymbol{y}^*).$$

At a Nash equilibrium, $\boldsymbol{x}$ maximizes the payoff given $\boldsymbol{y}^*$, and $\boldsymbol{y}^*$ minimizes the payoff given $\boldsymbol{x}$. Given the strategy of the other player, each player chooses his optimum strategy. Recognizing that the function $L(\boldsymbol{x}, \boldsymbol{y}^*)$ is the Lagrangian of an extended Fenchel duality primal/dual pair, we calculate the primal and the dual from the Lagrangian. Perturbation duality establishes that a Nash equilibrium exists. The objective function in the primal is

$$\min_{\boldsymbol{y}^* \in P^m} L(\boldsymbol{x}, \boldsymbol{y}^*),$$

so the primal is

$$\max_{\boldsymbol{x} \in P^n} \left[\min_i A\boldsymbol{x} \right].$$

Analogously, the dual is

$$\min_{\boldsymbol{y}^* \in P^m} \left[\max_i A^\top \boldsymbol{y}^* \right].$$

B. C. Dieffenbach, *Conjugate Duality in Economic Analysis*, Contributions to Economics,
https://doi.org/10.1007/978-3-032-21396-9_16

Convex optimization and perturbation duality furnish a succinct analysis of a two-player, zero-sum game. There are n possible pure strategies for player x and m possible pure strategies for player y. When x chooses strategy j and y chooses strategy i, then A_{ij} is the payoff to x and the negative of the payoff to y. Let A denote the linear transformation defined by the A_{ij} from the Euclidean Jordan algebra $\mathbf{R}^n$ to the Euclidean Jordan algebra $\mathbf{R}^m$.[1] A mixed strategy is a choice of probabilities for the different possible pure strategies. Let $\boldsymbol{x} \in P^n$ denote a mixed strategy for player x, and let $\boldsymbol{y}^* \in P^m$ denote a mixed strategy for player y. The payoff is then $\langle \boldsymbol{y}^*, A\boldsymbol{x} \rangle$. For simplicity, below we omit the superscripts n and m.

16.1 Nash Equilibrium

Nash equilibrium is the cornerstone for game theory.

Definition 1 (Nash Equilibrium) For the zero-sum game, a Nash equilibrium $(\boldsymbol{x}, \boldsymbol{y}^*)$ is a saddle point of

$$L(\boldsymbol{x}, \boldsymbol{y}^*) = \langle \boldsymbol{y}^*, A\boldsymbol{x} \rangle - \delta_P(\boldsymbol{x}) + \delta_P(\boldsymbol{y}^*). \tag{16.1}$$

At a Nash equilibrium, $\boldsymbol{x}$ maximizes the payoff given $\boldsymbol{y}^*$, and $\boldsymbol{y}^*$ minimizes the payoff given $\boldsymbol{x}$. Given the strategy of the other player, each player chooses his optimum strategy. Recognizing that the function $L(\boldsymbol{x}, \boldsymbol{y}^*)$ is the Lagrangian of an extended Fenchel duality primal/dual pair, we calculate the primal and the dual from the Lagrangian. Perturbation duality establishes that a Nash equilibrium exists.

The objective function in the primal is

$$\begin{aligned} &\min_{y^*} L(\boldsymbol{x}, \boldsymbol{y}^*) \\ &\quad = \min_{y^*} [\langle \boldsymbol{y}^*, A\boldsymbol{x} \rangle - \delta_P(\boldsymbol{x}) + \delta_P(\boldsymbol{y}^*)] \\ &\quad = -\delta_P(\boldsymbol{x}) - \max_{y^*} [\langle \boldsymbol{y}^*, -A\boldsymbol{x} \rangle - \delta_P(\boldsymbol{y}^*)] \\ &\quad = -\delta_P(\boldsymbol{x}) - \delta_P^*(-A\boldsymbol{x}). \end{aligned}$$

Problem 1 (Player x Primal)

$$\max_{x \in P} [-\delta_P^*(-A\boldsymbol{x})] = \max_{x \in P} \left[\min_i A\boldsymbol{x}\right].$$

A Nash-equilibrium mixed strategy $\boldsymbol{x}$ solves the primal. One refers to this problem as the maximin primal, because player x maximizes the minimum of the components of $A\boldsymbol{x}$.

[1] The pure strategies are the primitive idempotents.

The player y dual is analogous. The objective function is

$$\begin{aligned}&\max_x L(\boldsymbol{x},\boldsymbol{y}^*)\\&\quad=\max_x\left[\langle \boldsymbol{y}^*,A\boldsymbol{x}\rangle-\delta_P(\boldsymbol{x})+\delta_P(\boldsymbol{y}^*)\right]\\&\quad=\delta_P(\boldsymbol{y}^*)+\max_x\left[\left\langle A^{\mathsf{T}}\boldsymbol{y}^*,\boldsymbol{x}\right\rangle-\delta_P(\boldsymbol{x})\right]\\&\quad=\delta_P(\boldsymbol{y}^*)+\delta_P^*\left(A^{\mathsf{T}}\boldsymbol{y}^*\right).\end{aligned}$$

Problem 2 (Player y Dual)

$$\min_{y^*\in P}\delta_P^*\left(A^{\mathsf{T}}\boldsymbol{y}^*\right)=\min_{y^*\in P}\left[\max_i A^{\mathsf{T}}\boldsymbol{y}^*\right].$$

A Nash-equilibrium mixed strategy $\boldsymbol{y}^*$ solves the primal. One refers to this problem as the minimax dual, because player y minimizes the maximum of the components of $A^{\mathsf{T}}\boldsymbol{y}^*$.

The expressions above omit the perturbations $\boldsymbol{x}^*$ and $\boldsymbol{y}$, which both have base value zero. Adding the Fenchel inequalities

$$\begin{aligned}\delta_P(\boldsymbol{x})+\delta_P^*\left(-A^{\mathsf{T}}\boldsymbol{y}^*+\boldsymbol{x}^*\right)-\left\langle -A^{\mathsf{T}}\boldsymbol{y}^*+\boldsymbol{x}^*,\boldsymbol{x}\right\rangle\geq 0\\ \delta_P(\boldsymbol{y}^*)+\delta_P^*(A\boldsymbol{x}-\boldsymbol{y})-\langle \boldsymbol{y}^*,A\boldsymbol{x}-\boldsymbol{y}\rangle\geq 0\end{aligned}$$

and rearranging obtains the primal and the reciprocal dual,

$$\langle \boldsymbol{y}^*,\boldsymbol{y}\rangle+\delta_P^*\left(-A^{\mathsf{T}}\boldsymbol{y}^*+\boldsymbol{x}^*\right)+\delta_P(\boldsymbol{y}^*)\geq\langle \boldsymbol{x}^*,\boldsymbol{x}\rangle-\delta_P(A\boldsymbol{x}-\boldsymbol{y})-\delta_P(\boldsymbol{x}).$$

The polyhedral condition holds. There is no duality gap, and the finite optimum is attained in both the primal and the dual.

Theorem 1 (Existence of a Nash Equilibrium) *The zero-sum game has a Nash equilibrium.*

16.2 History

Dimand and Dimand (1996, Chap. 7) and Leonard (2010, Chap. 4) examine at length and in detail the history of zero-sum game theory. The "minimax theorem" is that there is no duality gap: the maximum of the primal equals the minimum of the dual. Borel introduced the concept of a mixed strategy. He proved the minimax theorem for specific, small numbers of pure strategies. For arbitrary numbers of pure strategies, he recognized that the minimum of the dual was greater than or equal to the maximum of the dual, but he was unable to prove that there is no duality gap. By a complicated algebraic argument not involving any general duality theorem (von

Neumann 1928, 1959, translation) proved the minimax theorem (Kjeldsen 2001). Later (Ville 1938) proved it via a convexity argument. Kjeldsen (2001, p. 59) explains how (vonNeumannMorgenstern 1953, pp. 138–141) proved it via a theorem of the alternative for linear inequalities.

Nash extends the analysis of the zero-sum, two-person game to general non-cooperative games having a finite number of players and strategies (Nash 1950, 1951). By invoking a fixed point theorem (and not convex optimization), he proves that the game has a Nash equilibrium.

References

Dimand MA, Dimand RW (1996) The history of game theory, volume I: from the beginnings to 1945. Routledge

Kjeldsen TH (2001 Nov) John von Neumann's conception of the minimax theorem: a journey through different mathematical contexts. Arch Hist Exact Sci 56(1):39–68

Leonard R (2010) Von Neumann, Morgenstern, and the creation of game theory: from chess to social science, 1900–1960. Cambridge University Press, Cambridge

Nash J (1950) Equilibrium points in *n*-person games. Proc Nat Acad Sci 36(1):48–49

Nash J (1951) Non-cooperative games. Annal Math 54(2):286–295

Ville J (1938) Sur la théorie générale des jeux où intervient l'habilité des joueurs. Traité du Calcul des Probabilité et de ses Applications 2(5)

von Neumann J (1928) Zur Theorie der Gesellschaftsspiele. Math Annalen 100:295–320

von Neumann J (1959) On the theory of games of strategy. In: Tucker AW, Luce RD (eds) Contributions to the theory of games IV, Vol 40. Princeton University Press, Princeton, NJ, pp 13–42 (Translation of the original 1928 publication in German)

von Neumann J, Morgenstern O (1953) Theory of games and economic behavior, 2nd edn. Princeton University Press, Princeton, NJ (Originally 1944)

Chapter 17
Assignment

Abstract We analyze the problem of assigning n people to n jobs. The value of person j in job i is X^*_{ij}, and let X^* denote the $n \times n$ matrix of these values. The goal is to maximize the total value. The assignment of people to jobs is a *permutation*. A mathematical model of a permutation is an $n \times n$ matrix X, having components X_{ij}. Each row has a single component one, and each column has a single component one. All other components are zero. Interpret $X_{ij} = 1$ as the assignment of person j to job i. The total value achieved by a permutation X is the linear function $\langle X^*, X \rangle$. The assignment problem was important in the historical development of the theory of linear programming and duality. At first glance, it might seem that the problem is not a convex optimization. Choosing the optimum permutation from among the $n!$ possible permutations is a choice among discrete alternatives, not a convex problem. Nevertheless, it is possible to cast the problem as an interesting convex optimization, having a dual with an insightful economic interpretation. A matrix is doubly stochastic if the sum of each column is one, and the sum of each row is one, and every component is nonnegative. A permutation is doubly stochastic, and any convex combination of permutations is doubly stochastic. Thus any doubly stochastic linear transformation is a convex combination of permutations. That the permutations are the extreme points of the set of doubly stochastic linear transformation is a fundamental theorem. Set the assignment problem as the maximization of $\langle X^*, X \rangle$ among all doubly stochastic linear transformations.

We analyze the problem of assigning n people to n jobs. The value of person j in job i is X^*_{ij}. Let $X^* \in \mathbf{T}^n$ denote the $n \times n$ matrix of these values. The goal is to maximize the total value. Of course the optimum value is finite. The assignment of people to jobs is a *permutation*.

A mathematical model of a permutation is an $n \times n$ matrix $X \in \mathbf{T}^n$, having components X_{ij}. Each row has a single component one, and each column has a single component one. All other components are zero. Interpret $X_{ij} = 1$ as the assignment of person j to job i. The total value achieved by a permutation X is

$$\sum_{ij} X^*_{ij} X_{ij} = \langle X^*, X \rangle,$$

B. C. Dieffenbach, *Conjugate Duality in Economic Analysis*, Contributions to Economics,
https://doi.org/10.1007/978-3-032-21396-9_17

a linear function.

The assignment problem was important in the historical development of the theory of linear programming and duality. At first glance, it might seem that the problem is not a convex optimization. Choosing the optimum permutation from among the $n!$ possible permutations is a choice among discrete alternatives, not a convex problem. Nevertheless it is possible to cast the problem as an interesting convex optimization, having a dual with an insightful economic interpretation.

17.1 Doubly Stochastic Linear Transformations

We invoke a polyhedral representation of the doubly stochastic linear transformations. Here $\mathbf{T}^n$ is the Euclidean Jordan algebra $\mathbf{R}^{n \times n}$, having the standard partial ordering defined by the components X_{ij}.

Definition 1 (Doubly Stochastic Linear Transformation) A linear transformation $X \in \mathbf{T}^n$ is doubly stochastic if

$$\begin{aligned} X\mathbf{1} &= \mathbf{1} \\ X^{\mathsf{T}}\mathbf{1} &= \mathbf{1} \\ X &\geq 0. \end{aligned}$$

The sum of each column is one, and the sum of each row is one. Every component X_{ij} is nonnegative. A permutation is doubly stochastic, and any convex combination of permutations is doubly stochastic.

König (1936) and Birkhoff (1946) prove the following fundamental theorem, commonly known as the Birkhoff theorem. An optimization argument proves it in Borwein and Zhu (2005).

Propostion 1 (Doubly Stochastic Linear Transformations) *The permutations are the extreme points of the set of doubly stochastic linear transformation.*

Thus, any doubly stochastic linear transformation is a convex combination of permutations. The constraints in definition (1) define the convex hull of all permutations.

17.2 Assignment Primal and Dual

Set the assignment problem as the maximization of $\langle X^*, X\rangle$ among all doubly stochastic linear transformations.

Problem 1 (Assignment Primal)

$$\max_X \left(\langle X^*, X\rangle - \delta_{X\mathbf{1}=\boldsymbol{y}} - \delta_{X^\top\mathbf{1}=\boldsymbol{z}} - \delta_{X\geq 0}\right).$$

The permutation base values are $\boldsymbol{y} = \mathbf{1}$ and $\boldsymbol{z} = \mathbf{1}$.

Since the permutations are the extreme points of the constraint set, some permutation will achieve the optimum.

The assignment problem is extended Fenchel duality. The adjoint of the linear transformation $X \mapsto X\mathbf{1}$ is $\boldsymbol{x}^* \mapsto \boldsymbol{x}^*\mathbf{1}^\top$:

$$\langle \boldsymbol{x}^*, X\mathbf{1}\rangle = \boldsymbol{x}^{*\top}X\mathbf{1} = \operatorname{Tr} \mathbf{1}\boldsymbol{x}^{*\top}X = \langle \boldsymbol{x}^*\mathbf{1}^\top, X\rangle.$$

We find the reciprocal dual. Adding the Fenchel inequalities

$$\begin{aligned}
\delta_{\{0\}}(X\mathbf{1} - \boldsymbol{y}) + \delta^*_{\{0\}}(\boldsymbol{y}^*) - \langle \boldsymbol{y}^*, X\mathbf{1} - \boldsymbol{y}\rangle &\geq 0 \\
\delta_{\{0\}}\left(X^\top\mathbf{1} - \boldsymbol{z}\right) + \delta^*_{\{0\}}(\boldsymbol{z}^*) - \left\langle \boldsymbol{z}^*, X^\top\mathbf{1} - \boldsymbol{z}\right\rangle &\geq 0 \\
\delta_{\mathbf{X}_+}(X) + \delta^*_{\mathbf{X}_+}\left(-\boldsymbol{y}^*\mathbf{1}^\top - \mathbf{1}\boldsymbol{z}^{*\top} + X^*\right) - \left\langle -\boldsymbol{y}^*\mathbf{1}^\top - \mathbf{1}\boldsymbol{z}^{*\top} + X^*, X\right\rangle &\geq 0
\end{aligned} \tag{17.1}$$

and rearranging gives

$$\langle \boldsymbol{y}^*, \boldsymbol{y}\rangle + \langle \boldsymbol{z}^*, \boldsymbol{z}\rangle + \delta_{\boldsymbol{y}^*\mathbf{1}^\top + \mathbf{1}\boldsymbol{z}^{*\top} - X^* \geq 0} \geq \left\langle X^*, X\right\rangle - \delta_{X\mathbf{1}=\boldsymbol{y}} - \delta_{X^\top\mathbf{1}=\boldsymbol{z}} - \delta_{X\geq 0}.$$

Because the primal is polyhedral and its optimum value is finite, there is no duality gap; and the optimum value is attained in both the primal and the dual.

Problem 2 (Assignment Dual)

$$\inf_{\boldsymbol{y}^*, \boldsymbol{z}^*} \left(\langle \boldsymbol{y}^*, \mathbf{1}\rangle + \langle \boldsymbol{z}^*, \mathbf{1}\rangle + \delta_{\boldsymbol{y}^*\mathbf{1}^\top + \mathbf{1}\boldsymbol{z}^{*\top} \geq X^*}\right).$$

Interpret the dual choice variables as prices. The price y_i^* is the value of job i, and the price z_j^* is the value of person j. Choose $\boldsymbol{y}^*$ and $\boldsymbol{z}^*$ to minimize the sum of the values of the jobs and the persons. The constraint $y_i^* + z_j^* \geq X_{ij}^*$ requires that the price of job i plus the price of person j be greater than or equal to the value X_{ij}^* of person j in job i. The Fenchel inequality (17.1) is complementary slackness. At an optimum, the final term (the inner product) in the Fenchel factor must be zero. If $y_i^* + z_j^* > X_{ij}^*$, then person j is not assigned to job i ($X_{ij} = 0$): it is more valuable to assign that person to another job, and to assign that job to another person. If the optimum assigns person j to job i ($X_{ij} = 1$), then necessarily $y_i^* + z_j^* = X_{ij}^*$.

The Lagrangian is

$$\begin{aligned} L(X, y^*, z^*) &:= \langle X^*, X \rangle + \langle y^*, \mathbf{1} \rangle + \langle z^*, \mathbf{1} \rangle - \delta_{X \geq 0} + \\ &\quad - \inf_{u,v} \left[\langle y^*, u \rangle + \langle z^*, v \rangle + \delta_{X\mathbf{1}=u} + \delta_{X^\top \mathbf{1}=v} \right] \\ &= \langle X^*, X \rangle + \langle y^*, \mathbf{1} - X\mathbf{1} \rangle + \langle z^*, \mathbf{1} - X^\top \mathbf{1} \rangle - \delta_{X \geq 0} \\ &= \langle X^* - y^* \mathbf{1}^\top - \mathbf{1} z^{*\top}, X \rangle + \langle y^*, \mathbf{1} \rangle + \langle z^*, \mathbf{1} \rangle - \delta_{X \geq 0}. \end{aligned}$$

For given prices y^* and z^*, the profit from assigning person j to job i is $X_{ij}^* - y_i^* - z_j^*$. At the saddle point, choose X_{ij} to maximize profit. If the profit is negative, choose $X_{ij} = 0$. If the profit is positive, let X_{ij} become arbitrarily large, so there is no saddle point. If the profit is zero, then one can assign person j to job i.

The assignment problem underpins the chapter "Spectral Conjugate."

References

Birkhoff G (1946) Tres observaciones sobre el algebra lineal. Universidad Nacionale Tucamán Revista 5:147–151

Borwein JM, Zhu QJ (2005) Techniques of variational analysis, Vol 20. Springer, New York

König D (1936) Theorie der Endlichen und Unendlichen graphen. Akademische Verlagsgesellschaft, Leipzig

Chapter 18
Constant Elasticity of Substitution

Abstract We study constant-returns-to-scale production functions having constant elasticity of substitution $0 \le s \le \infty$ between any two inputs. We calculate the cost function, the minimum cost of producing output one, for given input prices. The elasticity s is the percentage decrease in an input relative to other inputs when its relative price rises by one per cent. Solving the dual of a Karush-Kuhn-Tucker problem obtains the cost function. The cost function resembles the production function: it has the same constant elasticity of substitution functional form, but with the elasticity of substitution inverted. When the elasticity of substitution is high, then a small change in the input price causes a strong substitution of one input for another, but the change in cost is small.

Arrow et al. (1961) derive and analyze the constant elasticity of substitution production function. McFadden (1978) studies at length the production and cost functions.

In Euclidean Jordan algebra $\mathbf{X} = \mathbf{R}^n$, let $\boldsymbol{x}$ denote the input, and $\boldsymbol{x}^* \gg \mathbf{0}$ is the input price. In each case below, the effective domain of the production function contains $\mathbf{X}_{++}$ and is included in $\mathbf{X}_+$.

18.1 Cost-Minimization

By transforming the constant elasticity of substitution production function one can express the minimum cost of producing output one as the optimum value of the following Karush-Kuhn-Tucker problem.

Problem 1 (Cost-Minimization Primal)

$$\inf_{\boldsymbol{x}} \left[\langle \boldsymbol{x}^*, \boldsymbol{x} \rangle + \delta_{\sum_i a_i g(x_i) \ge y} \right]. \tag{18.1}$$

B. C. Dieffenbach, *Conjugate Duality in Economic Analysis*, Contributions to Economics,
https://doi.org/10.1007/978-3-032-21396-9_18

The constraint describes all inputs $\boldsymbol{x}$ that can produce output one or more. Here g is proper, closed, concave. In the constraint, the sum is additively separable in the inputs.

Here $\boldsymbol{a} \in \mathbf{X}$ is the "distribution parameter" (Arrow et al. 1961). For each case below, we specify the following condition.

Condition 1 (Distribution Parameter) *The distribution parameter* $\boldsymbol{a} \gg \mathbf{0}$, *such that* $\langle \mathbf{1}, \boldsymbol{a} \rangle = 1$.

Equivalently, $\boldsymbol{a} \in \operatorname{rint} P$.

The traditional way economists solve a Karush-Kuhn-Tucker problem is to find the primal solution $\boldsymbol{x}$ and the Lagrange multiplier y^* simultaneously, by finding the saddle point of the Lagrangian. In this application this procedure is straightforward but tedious, as the complicated algebra involves repeated substitutions. More directly and simply, we calculate the dual and solve it for y^*.

Define

$$f(\boldsymbol{x}) := \sum_i a_i g(x_i). \tag{18.2}$$

The sum is additively separable in the inputs, so its conjugate is the sum of the conjugates.

$$f_*(\boldsymbol{x}^*) = \sum_i (a_i \star g_*)(x_i).$$

Express the cost-minimization primal as

$$\inf_{\boldsymbol{x}} \left[\langle \boldsymbol{x}^*, \boldsymbol{x} \rangle + \delta_{f(\boldsymbol{x}) \geq y} \right],$$

a Karush-Kuhn-Tucker problem.

Rearranging the Fenchel inequality

$$\delta_{\operatorname{epi} f}(\boldsymbol{x}, y) + \delta^*_{\operatorname{epi} f}(-\boldsymbol{x}^*, y^*) - \langle -\boldsymbol{x}^*, \boldsymbol{x} \rangle - y^* y \geq 0 \tag{18.3}$$

obtains the reciprocal dual:

$$\langle \boldsymbol{x}^*, \boldsymbol{x} \rangle + \delta_{\operatorname{epi} f}(\boldsymbol{x}, y) \geq y^* y - \delta^*_{\operatorname{epi} f}(-\boldsymbol{x}^*, y^*).$$

For a concave function, $(y^* \star f_*)(\boldsymbol{x}^*) = -\delta^*_{\operatorname{epi} f}(-\boldsymbol{x}^*, y^*)$.

Problem 2 (Cost-Minimization Dual)

$$\sup_{y^*} \left[y^* y + (y^* \star f_*)(\boldsymbol{x}^*) \right]. \tag{18.4}$$

In the application below, the relative condition holds for both the primal and the reciprocal dual. There is no duality gap, and the optimum value is attained in both the primal and the dual.

The optimum $y^* > 0$. By the theory of epi-multiplication, for $y^* > 0$ the two conditions

$$f(\boldsymbol{x}) + f_*(\boldsymbol{x}^*/y^*) - \langle \boldsymbol{x}^*/y^*, \boldsymbol{x}\rangle = 0 \tag{18.5}$$

$$y = f(\boldsymbol{x}) \tag{18.6}$$

are equivalent to the Fenchel inequality (18.3) holding with equality.

For $0 < s < \infty$, below we find that the first-order condition for maximizing the dual is

$$y + f_*(\boldsymbol{x}^*/y^*) = 1. \tag{18.7}$$

Substituting this first-order condition into (18.5) and (18.6) obtains $1 = \langle \boldsymbol{x}^*/y^*, \boldsymbol{x}\rangle$, so y^* is the minimum cost. Solving the first-order condition for y^* obtains the cost function.

18.2 Constant-Elasticity Function

The constant-elasticity function is increasing, differentiable, and concave.

Definition 1 (Constant-Elasticity Function) For $0 < s < \infty, s \neq 1$,

$$u^s(x) := \frac{x^{1-s}}{1-s}.$$

For $s < 1$, dom $u = \mathbf{R}_+$; for $s > 1$, dom $u = \mathbf{R}_{++}$.

The defining condition is that the elasticity of the derivative $(u^s)'(x)$ with respect to x is constant: $x(u^s)''/(u^s)' = s$. The conjugate inverts s: $u^s_* = u^{1/s}$.

The constant elasticity of substitution production function is

$$\left[(1-1/s)\sum_i a_i u^{1/s}(x_i)\right]^{1/(1-1/s)}, \tag{18.8}$$

for elasticity s. If $s > 1$, then an isoquant meets each axis; but for $0 < s < 1$, the isoquant does not meet the axes.

We demonstrate that

$$\sum_i a_i u^{1/s}(x_i) \geq 1/(1-1/s) \tag{18.9}$$

is equivalent to the constraint that output is greater than or equal one and holds regardless of whether s is greater or less than one. Taking the natural logarithm of the constraint that output (18.8) is greater than or equal to one gives

$$\frac{1}{(1-1/s)}\ln\left[(1-1/s)\sum_i a_i u^{1/s}(x_i)\right] \geq 0.$$

If $s > 1$, then

$$\ln\left[(1-1/s)\sum_i a_i u^{1/s}(x_i)\right] \geq 0$$

implies

$$(1-1/s)\sum_i a_i u^{1/s}(x_i) \geq 1,$$

so (18.9) holds. If $s < 1$, then

$$\ln\left[(1-1/s)\sum_i a_i u^{1/s}(x_i)\right] \leq 0$$

so

$$(1-1/s)\sum_i a_i u^{1/s}(x_i) \leq 1,$$

and again (18.9) holds.

The conjugate of

$$f(\boldsymbol{x}) := \sum_i a_i u^{1/s}(x_i)$$

is

$$f_*(\boldsymbol{x}^*) := \sum_i (a_i \star u^s)(x_i^*);$$

the conjugate of the sum is the sum of the conjugates. The first-order condition for the dual is

$$\begin{aligned}
\{0\} &= \partial_Y\left[y^*y + y^*f_*(\boldsymbol{x}^*/y^*)\right] \\
&= \{y + f_*(\boldsymbol{x}^*/y^*)\} + y^*\partial_Y\left[f_*(\boldsymbol{x}^*/y^*)\right] \\
&= \{y + f_*(\boldsymbol{x}^*/y^*)\} - \{(1-s)f_*(\boldsymbol{x}^*/y^*)\} \\
&= \{y + sf_*(\boldsymbol{x}^*/y^*)\}.
\end{aligned}$$

The special form u^s makes the two f_* terms combine. Substituting $y = 1/(1 - 1/s)$, one can simplify the first-order condition to

$$f_*(\boldsymbol{x}^*/y^*) = \frac{1}{1-s},$$

and one can verify that (18.7) holds.

Solve the first-order condition for y^* to obtain the cost function.

Theorem 1 (Cost for Constant Elasticity of Substitution) *For the constant elasticity of substitution production function (18.8), the minimum cost of producing output one is*

$$y^* = [(1-s)f_*(\boldsymbol{x}^*)]^{1/(1-s)} = \left[(1-s)\sum_i (a_i \star u^s)(x_i^*)\right]^{1/(1-s)}. \tag{18.10}$$

The cost function (18.10) resembles the production function (18.8): it has the same constant elasticity of substitution functional form, but with the elasticity of substitution inverted (McFadden 1978). When the elasticity of substitution is high, then a small change in the input price causes a strong substitution of one input for another, but the change in cost is small. The epi-multiplication $a_i \star u^s$ in the cost function corresponds to the multiplication $a_i u^{1/s}$ in the production function.

18.3 Cobb–Douglas

The Cobb–Douglas production function

$$\prod_i x_i^{a_i} \tag{18.11}$$

is the case elasticity of substitution $s = 1$. An isoquant does not meet the axes. Taking the limit of the constant elasticity of substitution production function as $s \to 1$ yields the Cobb–Douglas function; one can calculate the limit via l'Hôpital's rule (Arrow et al. 1961). The effective domain is $\mathbf{X}_+$.

We define the concave function $\ln x$.

Definition 2 (Natural Logarithm) With effective domain $\mathbf{R}_{++}$, $\ln x$ is the natural logarithm of x.

The concave conjugate

$$\ln_*(x^*) = 1 + \ln x^*.$$

Taking the logarithm of production greater than or equal to one obtains

$$\sum_i a_i \ln x_i \geq 0.$$

The form (18.2) applies, in which $g(x_i) = \ln x_i$, $f(\boldsymbol{x}) = \sum_i a_i \ln x_i$, and $y = 0$. The conjugate

$$f_*(\boldsymbol{x}^*) = \sum_i (a_i \star \ln_*)(x_i) = \sum_i [a_i \star (1 + \ln)](x_i) = 1 + \sum_i (a_i \star \ln)(x_i).$$

Problem 3 (Cobb–Douglas Cost-Minimization Primal and Dual) For the Cobb–Douglas production function (18.11), the cost-minimization primal (18.1) is

$$\inf_{\boldsymbol{x}} \left[\langle \boldsymbol{x}^*, \boldsymbol{x} \rangle + \delta_{\sum_i a_i \ln x_i \geq y} \right],$$

and the cost-minimization dual (18.4) is

$$\sup_{y^*} \left[\sum_i (a_i y^* \star \ln_*)(x_i^*) \right].$$

The perturbation base value $y = 0$.

The first-order condition for the Cobb–Douglas dual is

$$\begin{aligned} \{0\} &= \partial_Y [y^* f_*(\boldsymbol{x}^*/y^*)] \\ &= \{f_*(\boldsymbol{x}^*/y^*)\} + y^* \partial_Y [f_*(\boldsymbol{x}^*/y^*)] \\ &= \{f_*(\boldsymbol{x}^*/y^*)\} + y^* \partial_Y \left[1 + \sum_i a_i \ln (x_i^*/a_i y^*) \right] \\ &= \{f_*(\boldsymbol{x}^*/y^*) - 1\}. \end{aligned}$$

The first-order condition (18.7) applies, so the optimum y^* is the minimum cost.
Solve the first-order condition for y^* to obtain the cost function.

Theorem 2 (Minimum Cost for Cobb–Douglas) *For the Cobb–Douglas production function, the logarithm of the minimum cost is*

$$\ln y^* = f_*(\boldsymbol{x}^*) = \sum_i (a_i \star \ln_*)(x_i^*).$$

The logarithm of the cost has the same functional form as the logarithm of production, except that epi-multiplication by a_i in the cost function replaces multiplication by a_i in the production function.

18.4 Extremes

The infinite elasticity of substitution production function is

$$\langle \boldsymbol{a}, \boldsymbol{x} \rangle - \delta_{\boldsymbol{x} \geq \boldsymbol{0}}, \tag{18.12}$$

a linear form with a nonnegativity constraint, the limit of the constant elasticity of substitution production function as $s \to \infty$.

The zero elasticity of substitution production function is the fixed-coefficients production function

$$\min_i [x_i/a_i] - \delta_{\boldsymbol{x} \geq \boldsymbol{0}} = \sup_z \left(z - \delta_{\boldsymbol{x} \geq \boldsymbol{a}z} - \delta_{z \geq 0}\right). \tag{18.13}$$

To produce output one requires input $\boldsymbol{a}$, with no opportunity for substitution. The fixed-coefficients production function is the limit of the constant elasticity of substitution production function as $s \to 0$ (Arrow et al. 1961).

For these two extremes, the relationship between the production function and the cost function is analogous to the relationship above. For infinite elasticity of substitution, the cost function has the functional form of a zero elasticity of substitution production function. For zero elasticity of substitution, the cost function has the functional form of an infinite elasticity of substitution production function.

Analyze the infinite elasticity of substitution case via the cost-minimization primal (18.1) and dual (18.4). The minimum cost is

$$\inf_{\boldsymbol{x}} \left(\langle \boldsymbol{x}^*, \boldsymbol{x} \rangle - \delta_{\langle \boldsymbol{a}, \boldsymbol{x} \rangle \geq 1} + \delta_{\boldsymbol{x} \geq \boldsymbol{0}}\right) = \inf_{\boldsymbol{x}} \left(\langle \boldsymbol{x}^*, \boldsymbol{x} \rangle - \delta_{\sum_i a_i g(x_i) \geq 1}\right),$$

in which $g(x_i) = x_i - \delta_{x_i \geq 0}$, the efficiency function, and $y = 1$. The conjugate $g_*(x_i^*) = -\delta_{\{1\}+\mathbf{R}_+}(x_i^*)$, so the conjugate

$$f_*(\boldsymbol{x}^*) = \sum_i (a_i \star g_*)(x_i^*) = -\sum_i \delta_{x_i^*/a_i \geq 1}.$$

For $y^* > 0$, the dual is

$$\sup_{y^*} [y^* + (y^* \star f_*)(\boldsymbol{x}^*/y^*)] = \sup_{y^*} \left[y^* - \sum_i \delta_{x_i^*/a_i \geq y^*}\right] = \min_i [x_i^*/a_i].$$

The functional form of the cost function is the functional form of the fixed-coefficients production function.[1]

[1] It is not the case that the first-order condition is (18.7).

For fixed coefficients, because $\boldsymbol{x}^*$ is positive, the choice $\boldsymbol{x} = \boldsymbol{a}$ is minimizing:

$$\inf_{\boldsymbol{x}} \left[\langle \boldsymbol{x}^*, \boldsymbol{x} \rangle + \delta_{\boldsymbol{x} \geq \boldsymbol{a}} \right] = \langle \boldsymbol{x}^*, \boldsymbol{a} \rangle,$$

the functional form of the infinite elasticity of substitution production function.

References

Arrow KJ, Chenery BH, Minhas BS, Solow RM (1961) Capital-labor substitution and economic efficiency. Rev Econom Stat 43(3):225–250

McFadden D (1978) Cost, revenue, and profit functions. In: Fuss M, McFadden D (eds) Production economics: a dual approach to theory and applications, volume 1, the theory of production, Vol 110. North-Holland, Amsterdam, pp 4–109

Part IV
Cone Representation

Chapter 19
Cone Representation of a Convex Set

Abstract It is possible to replace the analysis of a closed, convex set by an equivalent analysis of a closed, convex cone. That a closed, convex set is the intersection of its containing closed half spaces epi δ_C^* is the key: the cone representation of the set is the polar of the epigraph of the support. We derive various theoretical results about the cone representation via algebraic calculation. For a closed, convex set C,

$$\text{cone } C = \left\{(\boldsymbol{x}, -y) \;\middle|\; \begin{array}{l} \boldsymbol{x} \in yC \text{ if } y > 0 \\ \boldsymbol{x} \in C^\infty \text{ if } y = 0 \end{array}\right\}.$$

The barrier cone is the effective domain of the support, the directions in which the set is bounded. The horizon cone C^∞ is the polar of the barrier cone and describes the set "at infinity," the directions in which the set is unbounded. The concept of the polar of a cone generalizes to the polar of a set. The gauge is the support to the polar.

It is possible to replace the analysis of a closed, convex set by an equivalent analysis of a closed, convex cone. The cone representation "makes it possible, if one so chooses, to deduce many general theorems about convex sets from the corresponding (usually simpler) theorems about convex cones" (Rockafellar 1970, p. 15). Dieudonné (1941) introduces the cone representation. Valentine uses the cone representation to prove Helly's theorem, a geometric result about the intersection of convex sets (Valentine 1964). In the economics literature, Cass explains duality via the cone representation (Cass 1974). That a closed, convex set is the intersection of its containing closed half spaces is the key. The cone representation of the set is the polar of the epigraph of the support.

19.1 Cone Representation

We derive various theoretical results about the cone representation via algebraic calculation. Let $(\boldsymbol{x}^*, y^*)$ refer to the closed half space $\langle \boldsymbol{x}^*, \boldsymbol{x} \rangle \leq y^*$. The cone of closed half spaces containing C is the $(\boldsymbol{x}^*, y^*) \in \text{epi } \delta_C^*$.

B. C. Dieffenbach, *Conjugate Duality in Economic Analysis*, Contributions to Economics,
https://doi.org/10.1007/978-3-032-21396-9_19

Definition 1 (Cone Representation of a Set) The cone representation of a closed, convex set is the polar of the cone of containing closed half spaces:

$$\text{cone}\ C := (\text{epi}\ \delta_C^*)^\circ .$$

Theorem 1 (Cone Representation of a Convex Set) *For a closed, convex set C,*

$$\text{cone}\ C = \left\{ (\boldsymbol{x}, -y) \;\middle|\; \begin{array}{l} \boldsymbol{x} \in yC \text{ if } y > 0 \\ \boldsymbol{x} \in C^\infty \text{ if } y = 0 \end{array} \right\}.$$

We evaluate the polar:

$$\delta_{\text{cone}\ C}(\boldsymbol{x}, -y) = \sup_{\boldsymbol{x}^*, y^*} \left[\langle \boldsymbol{x}^*, \boldsymbol{x} \rangle - y^* y - \delta_{\text{epi}\ \delta_C^*}(\boldsymbol{x}^*, y^*) \right] = (y \star \delta_C)(\boldsymbol{x})$$

For $y > 0$,

$$\delta_{\text{cone}\ C}(\boldsymbol{x}, -y) = y\delta_C(\boldsymbol{x}/y) = \delta_C(\boldsymbol{x}/y).$$

For $y < 0$, $\delta_{\text{cone}\ C}(\boldsymbol{x}, -y) = \infty$.

The horizon cone C^∞ is the exceptional case $y = 0$.[1]

Definition 2 (Horizon Cone) The horizon cone C^∞ is the closed, convex cone having indicator

$$\delta_{C^\infty}(\boldsymbol{x}) := \delta_{\text{cone}\ C}(\boldsymbol{x}, 0) = (0 \star \delta_C)(\boldsymbol{x}) = \delta_C^\infty(\boldsymbol{x}).$$

The horizon cone C^∞ describes C "at infinity": its elements are the limit points of sequences $y\boldsymbol{x}$, as $y \searrow 0$ and $|\boldsymbol{x}| \to \infty$ for $\boldsymbol{x} \in C$. In particular, a closed, convex set C is bounded if and only if $C^\infty = \{\boldsymbol{0}\}$.[2]

The barrier cone is the effective domain of the support, the directions in which the set is bounded.

Definition 3 (Barrier Cone) For a convex set C, the barrier cone is dom δ_C^*, the effective domain of the support.

The barrier cone need not be closed.

Theorem 2 (Polar of the Barrier Cone) *The horizon cone is the polar of the barrier cone.*

---

[1] It is similar to but not identical to the recession cone in Rockafellar (1970) and the horizon cone in Rockafellar and Wets (1998).

[2] Rockafellar (1970, Theorem 8.4, p. 64).

We evaluate

$$
\begin{aligned}
&\delta_{C^\infty}(\boldsymbol{x}) \\
&= \delta_{\text{cone } C}(\boldsymbol{x}, 0) \\
&= \sup_{\boldsymbol{x}^*, y^*} \left[\langle \boldsymbol{x}^*, \boldsymbol{x}\rangle - \delta_{\text{epi } \delta_C^*}(\boldsymbol{x}^*, y^*)\right] \\
&= \sup_{\boldsymbol{x}^*} \left[\langle \boldsymbol{x}^*, \boldsymbol{x}\rangle - \delta_{\text{dom } \delta_C^*}(\boldsymbol{x}^*)\right] \\
&= \delta^*_{\text{dom } \delta_C^*}(\boldsymbol{x}),
\end{aligned}
$$

the polar of the barrier cone.

The cone representation of the empty set is not empty, but rather is the origin. Since $\delta^*_{\varnothing} = -\infty$, epi $\delta^*_{\varnothing} = \mathbf{X} \times \mathbf{R}$. Consequently cone $\varnothing = \{\mathbf{0}, 0\}$. Necessarily $\varnothing^\infty = \{\mathbf{0}\}$. For an entire Euclidean space, the barrier cone is $\{\mathbf{0}\}$; and the horizon cone is the entire space.

The following theorem shows that the horizon cone C^∞ is the directions in which C is unbounded.

Theorem 3 (Horizon-Cone Invariance) *For a closed, convex set C, the sum $C + C^\infty = C$.*

A set C is bounded if and only if $C^\infty = \{\mathbf{0}\}$.

We prove horizon-cone invariance by perturbation duality. The primal has value function $\delta_{C+C^\infty}(\boldsymbol{x})$, whereas the dual has value function $\delta_C(\boldsymbol{x})$. Because the dual satisfies the relative condition, there is no duality gap; and the theorem follows.[3]

Problem 1 (Horizon-Cone Addition Primal)

$$
\delta_{C+C^\infty}(\boldsymbol{x}) = \inf_z [\delta_C(z) + \delta_{C^\infty}(\boldsymbol{x} - z)] = \inf_z [\delta_C(z) + \delta_{\text{cone } C}(\boldsymbol{x} - z, 0)].
$$

We calculate the reciprocal dual. Adding the Fenchel inequalities

$$
\begin{aligned}
\delta_C(z) + \delta_C^*(\boldsymbol{x}^* - z^*) - \langle \boldsymbol{x}^* - z^*, z\rangle &\geq 0 \\
\delta_{\text{cone } C}(\boldsymbol{x} - z, y) + \delta^*_{\text{cone } C}(\boldsymbol{x}^*, y^*) - \langle \boldsymbol{x}^*, \boldsymbol{x} - z\rangle - y^* y &\geq 0
\end{aligned}
$$

and rearranging obtains

$$
\langle z^*, z\rangle + \delta_C(z) + \delta_{\text{cone } C}(\boldsymbol{x} - z, y) \geq \langle \boldsymbol{x}^*, \boldsymbol{x}\rangle + y^* y - \delta_C^*(\boldsymbol{x}^* - z^*) - \delta^*_{\text{cone } C}(\boldsymbol{x}^*, y^*).
$$

The perturbation base values are $\boldsymbol{x}$, $y = 0$, and $z^* = \mathbf{0}$.

---

[3] The primal need not satisfy the relative condition, and indeed the supremum in the dual need not be attained.

Problem 2 (Horizon-Cone Addition Dual)

$$\begin{aligned}
&\sup_{x^*,y^*} \left[\langle x^*, x\rangle - \delta_C^*(x^*) - \delta_{\text{cone } C}^*(x^*, y^*)\right] \\
&= \sup_{x^*,y^*} \left[\langle x^*, x\rangle - \delta_C^*(x^*) - \delta_{\text{epi } \delta_C^*}(x^*, y^*)\right] \\
&= \sup_{x^*} \left[\langle x^*, x\rangle - \delta_C^*(x^*) - \delta_{\text{dom } \delta_C^*}(x^*)\right] \\
&= \sup_{x^*} \left[\langle x^*, x\rangle - \delta_C^*(x^*)\right] \\
&= \delta_C(x).
\end{aligned}$$

Horizon-cone invariance follows.

The horizon cone is independent of translation.

Theorem 4 (Horizon-Cone Translation Independence) *For any* w, $\delta_{(C-\{w\})^\infty}(x) = \delta_{C^\infty}(x)$.

Because

$$\delta_{C-\{w\}}(x) = \delta_C(x + w) \tag{19.1}$$

$$\delta_{\text{cone } C-\{w\}}(x, y) = \delta_{\text{cone } C}(x - yw, y), \tag{19.2}$$

therefore

$$\delta_{(C-\{w\})^\infty}(x) = \delta_{\text{cone } C-\{w\}}(x, 0) = \delta_{\text{cone } C}(x, 0) = \delta_{C^\infty}(x).$$

The middle equality follows from (19.2), by setting $y = 0$.

Let us interpret the Fenchel equality for the cone representation:

$$\delta_{\text{cone } C}(x, -y) + \delta_{\text{cone } C}^*(x^*, y^*) - \langle x^*, x\rangle - y^*(-y) = 0. \tag{19.3}$$

Assume that C is nonempty, closed, and convex. An immediate implication is $\langle x^*, x\rangle = y^* y$. We simply invoke our earlier results for the Fenchel equality for the epigraph, specialized to the support δ_C^*.

Theorem 5 (Standard Case) *For* $y > 0$*, the Fenchel equality (19.3) is equivalent to the Fenchel equality*

$$\delta_C(x/y) + \delta_C^*(x^*) - \langle x^*, x/y\rangle = 0$$

and

$$\delta_C^*(x^*) = y^*.$$

Theorem 6 (Exceptional Case) *For* $y = 0$*, the Fenchel equality (19.3) is equivalent to the Fenchel equality*

$$\delta_{C^\infty}(\boldsymbol{x}) + \delta^*_{C^\infty}(\boldsymbol{x}^*) - \langle \boldsymbol{x}^*, \boldsymbol{x} \rangle = 0$$

and

$$\delta^*_C(\boldsymbol{x}^*) \le y^*.$$

19.2 Cone Generated by a Set

Definition 4 (Cone Generated by a Set) The indicator of the convex cone K generated by a closed, convex set C is the value function

$$\delta_K(\boldsymbol{x}) := \inf_y \delta_{\text{cone } C}(\boldsymbol{x}, -y). \tag{19.4}$$

Defining the linear transformation X as the projection $(\boldsymbol{x}, y) \to \boldsymbol{x}$, then K is the projection of cone C into $\mathbf{X}$: $K = X\text{cone } C$. Figure 19.1 illustrates the relationship. Of course K is not sufficient to determine C.

Breaking the minimization (19.4) into two components obtains a simple formula for K. Excluding $y = 0$, the minimization is

$$\inf_{y>0} \delta_{\text{cone } C}(\boldsymbol{x}, -y) = \mathbf{R}_{++}C.$$

The choice $y = 0$ adds C^∞. Because always $\mathbf{0} \in C^\infty$, the following holds.

Theorem 7 (Cone Generated by a Set) $K = \mathbf{R}_+C \cup C^\infty$.

If C is bounded, then $K = \mathbf{R}_+C$.

In general, the cone K is perhaps not closed, but it is closed if $\mathbf{0} \notin C$.

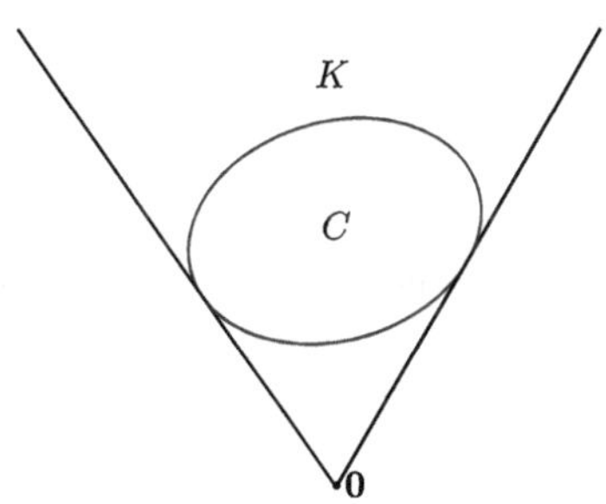

Fig. 19.1 Cone generated by a set

Condition 1 (Slater) $\mathbf{0} \notin C$.

Theorem 8 (Closure of the Cone Generated by a Set) *If the Slater condition holds, then the cone generated by a set is closed.*

We find the reciprocal dual of the definition (19.4). Rearranging the Fenchel inequality

$$\delta_{\text{cone } C}(\boldsymbol{x}, -y) + \delta^*_{\text{cone } C}(\boldsymbol{x}^*, y^*) - \langle \boldsymbol{x}^*, \boldsymbol{x} \rangle - y^*(-y) \geq 0$$

obtains

$$y^* y + \delta_{\text{cone } C}(\boldsymbol{x}, -y) \geq \langle \boldsymbol{x}^*, \boldsymbol{x} \rangle - \delta^*_{\text{cone } C}(\boldsymbol{x}^*, y^*),$$

in which the perturbation base values are $y^* = 0$ and $\boldsymbol{x}$. The dual is a Karush-Kuhn-Tucker problem:

$$\begin{aligned} \sup_{\boldsymbol{x}^*} \left[\langle \boldsymbol{x}^*, \boldsymbol{x} \rangle - \delta^*_{\text{cone } C}(\boldsymbol{x}^*, 0) \right] &= \sup_{\boldsymbol{x}^*} \left[\langle \boldsymbol{x}^*, \boldsymbol{x} \rangle - \delta_{\text{epi } \delta^*_C}(\boldsymbol{x}^*, 0) \right] \\ &= \sup_{\boldsymbol{x}^*} \left[\langle \boldsymbol{x}^*, \boldsymbol{x} \rangle - \delta_{\delta^*_C \leq 0}(\boldsymbol{x}^*) \right], \end{aligned}$$

the support to a closed, convex cone. That $\mathbf{0} \notin C$ means that $\delta^*_C(\boldsymbol{x}^*) < 0$ for some $\boldsymbol{x}^*$. By this Slater condition, there is no duality gap, so the primal value function is also the indicator of a closed, convex cone.

Corollary 1 (Cone Representation) *For a closed, convex set C, the cone generated by $C \times \{-1\}$ is* cone C.

19.3 Polar and Gauge

We generalize the concept of the polar of a cone to the polar of a set. If K is a closed, convex cone, cone $K = K \times -\mathbf{R}_+$ and $(\text{cone } K)^\circ = K^\circ \times \mathbf{R}_+$. Thus

$$\delta_{\text{cone } K^\circ}(\boldsymbol{x}^*, -y^*) = \delta_{(\text{cone } K)^\circ}(\boldsymbol{x}^*, y^*).$$

Definition 5 (Polar of a Set Containing the Origin) For a closed, convex set C containing the origin, its polar C° is the set having the cone representation

$$\delta_{\text{cone } C^\circ}(\boldsymbol{x}^*, -y^*) := \delta_{(\text{cone } C)^\circ}(\boldsymbol{x}^*, y^*). \tag{19.5}$$

The cone representation of the polar is the polar of the cone representation, with a sign reversal. Thus

$$\delta_{\text{cone } C^\circ}(\boldsymbol{x}^*, -y^*) = \delta_{\text{epi } \delta_C^*}(\boldsymbol{x}^*, y^*).$$

Because the support δ_C^* is a nonnegative function if and only if C contains the origin, the stipulation $\mathbf{0} \in C$ is necessary and sufficient for (19.5) to define the cone representation of a set.

The traditional definition of the polar is via the polar inequality.

Theorem 9 (Polar Inequality) *For a closed, convex set C containing the origin, $\boldsymbol{x}^* \in C^\circ$ if and only if $\langle \boldsymbol{x}^*, \boldsymbol{x}\rangle \leq 1$ for all $\boldsymbol{x} \in C$.*

By the definition of the polar,

$$\delta_{C^\circ}(\boldsymbol{x}^*) = \delta_{\text{cone } C^\circ}(\boldsymbol{x}^*, -1) = \delta_{\text{epi } \delta_C^*}(\boldsymbol{x}^*, 1) = \delta_{\delta_C^*(x^*) \leq 1}.$$

By the polar inequality, the polar contains the origin and thus itself has a polar. By the symmetry of the polar inequality theorem, the polar of the polar is the original set.

Theorem 10 (Polar of the Polar) *For a closed, convex set C containing the origin, $C^{\circ\circ} = C$.*

Alternatively, algebraic calculation proves this theorem:

$$\delta_{C^{\circ\circ}}(\boldsymbol{x}) = \delta_{\text{cone } C^{\circ\circ}}(\boldsymbol{x}, -1) = \delta_{(\text{cone } C^\circ)^\circ}(\boldsymbol{x}, 1) = \delta_{\text{cone } C}(\boldsymbol{x}, -1) = \delta_C(\boldsymbol{x}).$$

An interpretation of the polar inequality is to see C° as an "inverse" of C. Under this interpretation, one can understand why the polar turns out to be fundamental in fractional programming, the optimization of one function divided by another.

The gauge function is positively homogeneous.

Definition 6 (Gauge) For a closed, convex set C containing the origin, the gauge $\gamma_C(\boldsymbol{x})$ has epigraph

$$\delta_{\text{epi } \gamma_C}(\boldsymbol{x}, y) := \delta_{\text{cone } C}(\boldsymbol{x}, -y).$$

Because $\mathbf{0} \in C$, this formula defines an epigraph. The gauge

$$\gamma_C(\boldsymbol{x}) = \inf_y [y + \delta_{\text{cone } C}(\boldsymbol{x}, -y)] = \inf_{y \geq 0} \left[y + \delta_{yC}(\boldsymbol{x})\right],$$

the smallest y such that $\boldsymbol{x}$ belongs to yC. The gauge is the support to the polar:

$$\begin{aligned}\gamma_C^*(\boldsymbol{x}^*) &= \sup_{\boldsymbol{x}}\left\{\langle \boldsymbol{x}^*, \boldsymbol{x}\rangle - \inf_y [y + \delta_{\text{cone } C}(\boldsymbol{x}, -y)]\right\} \\ &= \sup_{\boldsymbol{x},y} [\langle \boldsymbol{x}^*, \boldsymbol{x}\rangle - y - \delta_{\text{cone } C}(\boldsymbol{x}, -y)] \\ &= \delta_{\text{cone } C}^*(\boldsymbol{x}^*, 1) \\ &= \delta_{\text{cone } C^\circ}(\boldsymbol{x}^*, -1) \\ &= \delta_{C^\circ}(\boldsymbol{x}^*).\end{aligned}$$

Both the polar and the gauge have economic application.

References

Cass D (1974 Mar) Duality: a symmetric approach from the economist's vantage point. J Econ Theor 7(3):272–295
Dieudonné JA (1941) Sur théorème de Hahn-Banach. Rev Sci 79:642–643
Rockafellar RT (1970) Convex analysis. Princeton University Press, Princeton NJ
Rockafellar RT, Wets RJB (1998) Variational analysis, Vol 317. Springer, Berlin
Valentine FA (1964) Convex sets. McGraw-Hill, New York

Chapter 20
Cone Representation of a Convex Function

Abstract Invoking the cone representation of a set, we define the cone representation of a convex function. We work out its properties via calculation and perturbation duality. The cone representation of a convex function f is the cone representation of its epigraph:

$$\operatorname{cone} f := \operatorname{cone}(\operatorname{epi} f).$$

We calculate

$$\operatorname{cone} f = \left\{ (\boldsymbol{x}, y, -z) \,\middle|\, \begin{array}{l} (\boldsymbol{x}, y) \in z \operatorname{epi} f \text{ if } z > 0 \\ (\boldsymbol{x}, y) \in (\operatorname{epi} f)^{\infty} = \operatorname{epi} f^{\infty} \text{ if } z = 0. \end{array} \right\}$$

The cone representation is the polar of the cone of half spaces containing the epigraph.

The cone representation $\operatorname{cone} f^*$ is almost the polar of $\operatorname{cone} f$, but the final two arguments are in reverse order. Under general conditions, the polar of the cone generated by a contour set is the cone generated by the contour set for the conjugate. We interpret the Fenchel equality for the cone representation, for the standard case, the exceptional case, and the double exceptional case.

20.1 Cone Representation of the Epigraph

Throughout we assume that the function f is proper, closed, and convex.

Definition 1 (Cone Representation of a Convex Function) The cone representation of a function f is the cone representation of its epigraph,

$$\operatorname{cone} f := \operatorname{cone}(\operatorname{epi} f).$$

B. C. Dieffenbach, *Conjugate Duality in Economic Analysis*, Contributions to Economics,
https://doi.org/10.1007/978-3-032-21396-9_20

Summary 1 (Cone Representation of a Convex Function)

$$\operatorname{cone} f = \left\{ (\boldsymbol{x}, y, -z) \;\middle|\; \begin{array}{l} (\boldsymbol{x}, y) \in z \operatorname{epi} f \text{ if } z > 0 \\ (\boldsymbol{x}, y) \in (\operatorname{epi} f)^{\infty} = \operatorname{epi} f^{\infty} \text{ if } z = 0. \end{array} \right\}$$

That $(\operatorname{epi} f)^{\infty} = \operatorname{epi} f^{\infty}$ follows from Eq. (20.2).

The cone representation is the polar of the cone of half spaces containing the epigraph. By the definition,

$$\delta_{\operatorname{cone} f}(\boldsymbol{x}, y, -z) = \delta^{*}_{\operatorname{epi}\, \delta^{*}_{\operatorname{epi} f}}(\boldsymbol{x}, y, -z).$$

The epigraph of the support is the cone of all half spaces (plus zero) containing the epigraph: $(\boldsymbol{x}^{*}, y^{*}, z^{*}) \in \operatorname{epi}\, \delta^{*}_{\operatorname{epi} f}$ if and only the half space $\langle \boldsymbol{x}^{*}, \boldsymbol{x} \rangle + y^{*} y \leq z^{*}$ contains all $(\boldsymbol{x}, y) \in \operatorname{epi} f$. Here $y^{*} = 0$ for a vertical half space, and $y^{*} < 0$ for a nonvertical half space.

20.2 Cone Representation of the Conjugate

The cone representation $\operatorname{cone} f^{*}$ is almost the polar of $\operatorname{cone} f$, but the final two arguments are in reverse order.

Theorem 1 (Cone Representation of the Conjugate) *For a proper, closed, convex function f,*

$$\delta^{*}_{\operatorname{cone} f}(\boldsymbol{x}^{*}, -y^{*}, z^{*}) = \delta_{\operatorname{epi}\,(y^{*} \star f^{*})}(\boldsymbol{x}^{*}, z^{*}) = \delta_{\operatorname{cone} f^{*}}(\boldsymbol{x}^{*}, z^{*}, -y^{*}). \tag{20.1}$$

By the definition of the cone representation,

$$\delta^{*}_{\operatorname{cone} f}(\boldsymbol{x}^{*}, -y^{*}, z^{*}) = \delta_{\operatorname{epi}\, \delta^{*}_{\operatorname{epi} f}}(\boldsymbol{x}^{*}, -y^{*}, z^{*}) = \delta_{\operatorname{epi}\,(y^{*} \star f^{*})}(\boldsymbol{x}^{*}, z^{*}).$$

For $y^{*} = 1$,

$$\delta^{*}_{\operatorname{cone} f}(\boldsymbol{x}^{*}, -1, z^{*}) = \delta_{\operatorname{epi} f^{*}}(\boldsymbol{x}^{*}, z^{*}) = \delta_{\operatorname{cone} f^{*}}(\boldsymbol{x}^{*}, z^{*}, -1)\,.$$

By positive homogeneity, Eq. (20.1) holds for any $y^{*} > 0$. For $y^{*} = 0$,

$$\begin{aligned} \delta_{\operatorname{epi}\,(y^{*} \star f^{*})}(\boldsymbol{x}^{*}, z^{*}) &= \delta_{\operatorname{epi}\,(0 \star f^{*})}(\boldsymbol{x}^{*}, z^{*}) = \delta_{\operatorname{epi} f^{*\infty}}(\boldsymbol{x}^{*}, z^{*}) \\ &= \delta_{\operatorname{cone} f^{*}}(\boldsymbol{x}^{*}, z^{*}, 0). \end{aligned} \tag{20.2}$$

As the chapter "Epigraph" defines the horizon function $f^{\infty} := (0 \star f) = \delta^{*}_{\operatorname{dom} f^{*}}$, the second expression in Eq. (20.2) is the indicator of its epigraph.

20.3 Fenchel Equality for the Cone Representation

We interpret the Fenchel equality

$$\delta_{\text{cone } f}(\boldsymbol{x}, y, -z) + \delta^*_{\text{cone } f}(\boldsymbol{x}^*, -y^*, z^*) - \langle \boldsymbol{x}^*, \boldsymbol{x} \rangle - (-y^*) y - z^* (-z) = 0. \tag{20.3}$$

The implication $\langle \boldsymbol{x}^*, \boldsymbol{x} \rangle = y^* y + z^* z$ is immediate. We analyze the standard case, the exceptional case, and the double exceptional case.

The standard case is $z > 0$ and $y^* > 0$. Dividing (20.3) by z and y^* obtains

$$\begin{aligned}
0 &= \delta_{\text{cone } f}(\boldsymbol{x}, y, -z) + \delta^*_{\text{cone } f}(\boldsymbol{x}^*, -y^*, z^*) - \langle \boldsymbol{x}^*/y^*, \boldsymbol{x}/z \rangle + y/z + z^*/y^* \\
&= \delta_{\text{cone } f}(\boldsymbol{x}, y, -z) + \delta_{\text{cone } f^*}(\boldsymbol{x}^*, z^*, -y^*) - \langle \boldsymbol{x}^*/y^*, \boldsymbol{x}/z \rangle + y/z + z^*/y^* \\
&= \delta_{\text{epi } f}(\boldsymbol{x}/z, y/z) + \delta_{\text{epi } f^*}(\boldsymbol{x}^*/y^*, z^*/y^*) - \langle \boldsymbol{x}^*/y^*, \boldsymbol{x}/z \rangle + y/z + z^*/y^* \\
&= [f(\boldsymbol{x}/z) + f^*(\boldsymbol{x}^*/y^*) - \langle \boldsymbol{x}^*/y^*, \boldsymbol{x}/z \rangle] \\
&\quad + [y/z - f(\boldsymbol{x}/z) + \delta_{\text{epi } f}(\boldsymbol{x}/z, y/z)] \\
&\quad + [z^*/y^* - f^*(\boldsymbol{x}^*/y^*) + \delta_{\text{epi } f^*}(\boldsymbol{x}^*/y^*, z^*/y^*)].
\end{aligned}$$

Each term in brackets is nonnegative, so the sum is zero if and only if each is zero.

Theorem 2 (Standard Case Fenchel Equality) *For $z > 0$ and $y^* > 0$, the Fenchel equality (20.3) is equivalent to the conditions*

$$\begin{aligned}
f(\boldsymbol{x}/z) + f^*(\boldsymbol{x}^*/y^*) - \langle \boldsymbol{x}^*/y^*, \boldsymbol{x}/z \rangle &= 0 \\
f(\boldsymbol{x}/z) &= y/z \\
f^*(\boldsymbol{x}^*/y^*) &= z^*/y^*.
\end{aligned}$$

Next consider the exceptional case $z > 0$ and $y^* = 0$. Dividing (20.3) by z obtains

$$\begin{aligned}
0 &= \delta_{\text{cone } f}(\boldsymbol{x}, y, -z) + \delta^*_{\text{cone } f}(\boldsymbol{x}^*, -0, z^*) - \langle \boldsymbol{x}^*, \boldsymbol{x}/z \rangle + z^* \\
&= \delta_{\text{cone } f}(\boldsymbol{x}, y, -z) + \delta_{\text{cone } f^*}(\boldsymbol{x}^*, z^*, 0) - \langle \boldsymbol{x}^*, \boldsymbol{x}/z \rangle + z^* \\
&= \delta_{\text{epi } f}(\boldsymbol{x}/z, y/z) + \delta_{\text{epi } f^{*\infty}}(\boldsymbol{x}^*, z^*) - \langle \boldsymbol{x}^*, \boldsymbol{x}/z \rangle + z^* \\
&= \delta_{\text{epi } f}(\boldsymbol{x}/z, y/z) + \delta_{\text{dom } f}(\boldsymbol{x}/z) + \delta_{\text{epi } f^{*\infty}}(\boldsymbol{x}^*, z^*) - \langle \boldsymbol{x}^*, \boldsymbol{x}/z \rangle + z^* \\
&\quad + \delta^*_{\text{dom } f}(\boldsymbol{x}^*) - f^{*\infty}(\boldsymbol{x}^*) \\
&= [\delta_{\text{dom } f}(\boldsymbol{x}/z) + \delta^*_{\text{dom } f}(\boldsymbol{x}^*) - \langle \boldsymbol{x}^*, \boldsymbol{x}/z \rangle] \\
&\quad + \delta_{\text{epi } f}(\boldsymbol{x}/z, y/z) + [z^* - f^{*\infty}(\boldsymbol{x}^*) + \delta_{\text{epi } f^{*\infty}}(\boldsymbol{x}^*, z^*)].
\end{aligned}$$

Each of the three terms is nonnegative, so the sum is zero if and only if each is zero.

Theorem 3 (Exceptional Case Fenchel Equality) *For $z > 0$ and $y^* = 0$, the Fenchel equality (20.3) is equivalent to*

$$\delta_{\text{dom } f}(\boldsymbol{x}/z) + \delta^*_{\text{dom } f}(\boldsymbol{x}^*) - \langle \boldsymbol{x}^*, \boldsymbol{x}/z \rangle = 0$$
$$f(\boldsymbol{x}/z) \le y/z$$
$$f^{*\infty}(\boldsymbol{x}^*) = z^*.$$

Here dom f is perhaps not closed; but the equivalence also holds if one replaces dom f by its closure, because the inequality constraint requires $\boldsymbol{x}/z \in \text{dom } f$.

The double exceptional case is $z = 0$ and $y^* = 0$. The Fenchel equality (20.3) is

$$\begin{aligned}
0 &= \delta_{\text{cone } f}(\boldsymbol{x}, y, 0) + \delta^*_{\text{cone } f}(\boldsymbol{x}^*, 0, z^*) - \langle \boldsymbol{x}^*, \boldsymbol{x} \rangle \\
&= \delta_{\text{cone } f}(\boldsymbol{x}, y, 0) + \delta_{\text{cone } f^*}(\boldsymbol{x}^*, z^*, 0) - \langle \boldsymbol{x}^*, \boldsymbol{x} \rangle \\
&= \delta_{\text{epi } f^\infty}(\boldsymbol{x}, y) + \delta_{\text{epi } f^{*\infty}}(\boldsymbol{x}^*, z^*) - \langle \boldsymbol{x}^*, \boldsymbol{x} \rangle \\
&= \delta_{\text{epi } f^\infty}(\boldsymbol{x}, y) + \delta_{\text{dom } f^\infty}(\boldsymbol{x}) + \delta_{\text{epi } f^{*\infty}}(\boldsymbol{x}^*, z^*) + \delta_{\text{ dom } f^{*\infty}}(\boldsymbol{x}^*) - \langle \boldsymbol{x}^*, \boldsymbol{x} \rangle \\
&= \left[\delta_{\text{ dom } f^\infty}(\boldsymbol{x}^*) + \delta_{\text{dom } f^{*\infty}}(\boldsymbol{x}) - \langle \boldsymbol{x}^*, \boldsymbol{x} \rangle\right] + \delta_{\text{epi } f^\infty}(\boldsymbol{x}, y) + \delta_{\text{epi } f^{*\infty}}(\boldsymbol{x}^*, z^*)
\end{aligned}$$

Each of the three terms is nonnegative, so each must be zero.

Theorem 4 (Double Exceptional Case Fenchel Equality) *For the double exceptional case $z = 0$ and $y^* = 0$, the Fenchel equality (20.3) is equivalent to*

$$\delta_{\text{dom } f^\infty}(\boldsymbol{x}) + \delta_{\text{dom } f^{*\infty}}(\boldsymbol{x}^*) - \langle \boldsymbol{x}^*, \boldsymbol{x} \rangle = 0$$
$$f^\infty(\boldsymbol{x}) \le y$$
$$f^{*\infty}(\boldsymbol{x}^*) \le z^*.$$

The effective domain of each horizon function is a cone, but perhaps is not closed. The equivalence also holds if one replaces each effective domain with its closure, because the inequality constraints require that $\boldsymbol{x}$ and $\boldsymbol{x}^*$ each belong to the effective domains.

20.4 Horizon Function

Rearrange the Fenchel factor in Eq. (20.3) to obtain a primal and its reciprocal dual,

$$y^* y + \delta_{\text{cone } f}(\boldsymbol{x}, y, -z) \ge \langle \boldsymbol{x}^*, \boldsymbol{x} \rangle - z^* z - \delta^*_{\text{cone } f}(\boldsymbol{x}^*, -y^*, z^*).$$

The primal has choice variable y and perturbation variables $\boldsymbol{x}$ and z, and the dual has choice variables $\boldsymbol{x}^*$ and z^* and perturbation variable $y^* = 0$. The primal value function

$$V(\boldsymbol{x},z) = \inf_y \delta_{\text{cone } f}(\boldsymbol{x},y,-z) = \inf_y \delta_{\text{epi}(z\star f)}(\boldsymbol{x},y) = \begin{cases} \delta_{\text{dom } f}(\boldsymbol{x}) & \text{if } z > 0 \\ \delta_{\text{dom } f^{\infty}}(\boldsymbol{x}) & \text{if } z = 0 \\ \infty & \text{if } z < 0. \end{cases}$$

If the effective domain of f is not closed, then the value function is not closed. Although there might be a duality gap, that the value function is proper and convex implies cl $V = V^{**}$, the value of the dual.

We obtain two results via the double exceptional case, by evaluating the dual for perturbation base value $z = 0$.

Theorem 5 (Polar Horizon Cones) *The closure of the effective domain of the horizon function is the polar of the effective domain of the horizon function of the conjugate:*

$$\delta_{\text{cl dom } f^{\infty}}(\boldsymbol{x}) = \delta^{*}_{\text{dom } f^{*\infty}}(\boldsymbol{x}). \tag{20.4}$$

The dual

$$\begin{aligned}
&\sup_{\boldsymbol{x}^*,z^*} \left[\langle \boldsymbol{x}^*, \boldsymbol{x}\rangle - \delta^{*}_{\text{cone } f}(\boldsymbol{x}^*, 0, z^*)\right] \\
&= \sup_{\boldsymbol{x}^*,z^*} \left[\langle \boldsymbol{x}^*, \boldsymbol{x}\rangle - \delta_{\text{cone } f^*}(\boldsymbol{x}^*, z^*, 0)\right] \\
&= \sup_{\boldsymbol{x}^*,z^*} \left[\langle \boldsymbol{x}^*, \boldsymbol{x}\rangle - \delta_{\text{epi } f^{*\infty}}(\boldsymbol{x}^*, z^*)\right] \\
&= \sup_{\boldsymbol{x}^*} \left[\langle \boldsymbol{x}^*, \boldsymbol{x}\rangle - \delta_{\text{dom } f^{*\infty}}(\boldsymbol{x}^*)\right] \\
&= \delta^{*}_{\text{dom } f^{*\infty}}(\boldsymbol{x}),
\end{aligned}$$

which proves (20.4).

An alternate evaluation of the dual derives the following.

Theorem 6 (Effective Domain of the Horizon Function) *The closure of the effective domain of the horizon function is the horizon cone of the effective domain,*

$$\text{cl dom } f^{\infty} = (\text{dom } f)^{\infty}.$$

The dual is

$$\begin{aligned}
&\sup_{\boldsymbol{x}^*,z^*} \left[\langle \boldsymbol{x}^*, \boldsymbol{x}\rangle - \delta_{\text{cone } f^*}(\boldsymbol{x}^*, z^*, 0)\right] \\
&\quad = \sup_{\boldsymbol{x}^*,z^*} \left[\langle \boldsymbol{x}^*, \boldsymbol{x}\rangle - \delta_{\text{epi } f^{*\infty}}(\boldsymbol{x}^*, z^*)\right]
\end{aligned}$$

$$= \delta^*_{\text{epi } f^{*\infty}}(\boldsymbol{x}, 0)$$
$$= \delta^*_{\text{epi } \delta^*_{\text{dom } f}}(\boldsymbol{x}, 0)$$
$$= \delta_{\text{cone(dom } f)}(\boldsymbol{x}, 0)$$
$$= \delta_{(\text{dom } f)^\infty}(\boldsymbol{x}).$$

20.5 Polar Cones Generated by Contour Sets

Consider the two contour sets having indicators $\delta_{f(x)\le 0}$ and $\delta_{f^*(x^*)\le 0}$. Under general conditions the cone generated by one contour set is the polar of the cone generated by the other. Figure 20.1 illustrates the theorem.

The cone representation of the contour set having indicator $\delta_{f(x)\le 0}$ has indicator $\delta_{\text{cone } f}(\boldsymbol{x}, 0, z)$. The cone K generated by the contour set has indicator

$$\delta_K(\boldsymbol{x}) = \inf_z \delta_{\text{cone } f}(\boldsymbol{x}, 0, -z).$$

Its polar is

$$\delta_{K^\circ}(\boldsymbol{x}^*) = \sup_{\boldsymbol{x}} [\langle \boldsymbol{x}^*, \boldsymbol{x}\rangle - \delta_K(\boldsymbol{x})] = \sup_{\boldsymbol{x},z} \left[\langle \boldsymbol{x}^*, \boldsymbol{x}\rangle - \delta_{\text{cone } f}(\boldsymbol{x}, 0, -z)\right]. \tag{20.5}$$

We calculate the reciprocal dual. Rearranging the Fenchel factor in Eq. (20.3) obtains

$$y^* y + \delta^*_{\text{cone } f}(\boldsymbol{x}^*, -y^*, z^*) \ge \langle \boldsymbol{x}^*, \boldsymbol{x}\rangle + z^* z - \delta_{\text{cone } f}(\boldsymbol{x}, y, -z).$$

The perturbation base values are $\boldsymbol{x}^*$, $z^* = 0$, and $y = 0$. The reciprocal dual is

$$\inf_{y^*} \delta^*_{\text{cone } f}(\boldsymbol{x}^*, -y^*, 0) = \inf_{y^*} \delta_{\text{cone } f^*}(\boldsymbol{x}^*, 0, -y^*);$$

the value function is the cone generated by the contour set with indicator $\delta_{f^* \le 0}$.

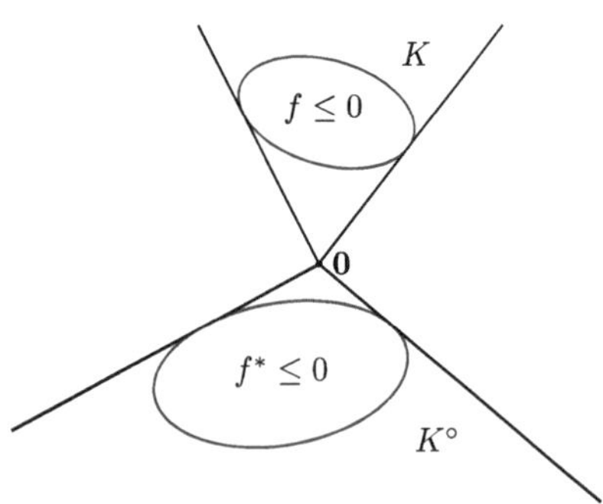

Fig. 20.1 Polar Cones generated by contour sets

The Slater condition is the relative condition for the primal (20.5).

Condition 1 (Slater) $f(\boldsymbol{x}) < 0$ *for some* $\boldsymbol{x}$.

If the Slater condition holds, then there is no duality gap.

Theorem 7 (Polar Cones Generated by Contour Sets) *If the Slater condition holds, then the polar of the cone generated by a contour set is the cone generated by the contour set for the conjugate.*

20.6 Horizon Cone of a Contour Set

We analyze the horizon cone of a contour set.

Theorem 8 (Horizon Cone of a Contour Set) *For a nonempty contour set having indicator* $\delta_{f(\boldsymbol{x})\le y}$*, the indicator of the horizon cone is* $\delta_{f^{\infty}(\boldsymbol{x})\le 0}$.

Thus an increase in y leaves the horizon cone unchanged.

An evaluation proves this result. The cone representation of the contour set satisfies

$$\delta_{\mathrm{cone}(f-y)}(\boldsymbol{x}, v, z) = \delta_{\mathrm{cone}\, f}(\boldsymbol{x}, v - yz, z). \tag{20.6}$$

Define C as the contour set:

$$\delta_C(\boldsymbol{x}) := \delta_{f(\boldsymbol{x})\le y} = \delta_{\mathrm{epi}\, f}(\boldsymbol{x}, y).$$

Then

$$\delta_{C^{\infty}}(\boldsymbol{x}) = \delta_{\mathrm{cone}\ C}(\boldsymbol{x}, 0) = \delta_{\mathrm{cone}(f-y)}(\boldsymbol{x}, 0, 0) = \delta_{\mathrm{cone}\, f}(\boldsymbol{x}, 0, 0),$$

setting $v = 0$ and $z = 0$ in Eq. (20.6).

An evaluation verifies an expression for the horizon cone of a contour set.

Theorem 9 (Horizon Cone of a Contour Set and the Conjugate) *The horizon cone of a contour set is the polar of the cone generated by the effective domain of the conjugate.*

The indicator of the polar is

$$\begin{aligned}
&\sup_{x^*}\left[\langle x^*, x\rangle - \inf_{y^*} \delta_{\text{cone}(\text{dom } f^*)}(x^*, -y^*)\right] \\
&= \sup_{x^*,y^*}\left[\langle x^*, x\rangle - \delta_{\text{cone}(\text{dom } f^*)}(x^*, -y^*)\right] \\
&= \sup_{x^*,y^*}\left[\langle x^*, x\rangle - \delta^*_{\text{epi } \delta^*_{\text{dom } f^*}}(x^*, -y^*)\right] \\
&= \delta_{\text{epi } \delta^*_{\text{dom } f^*}}(x, 0) \\
&= \delta_{\delta^*_{\text{dom } f^*}(x) \le 0} \\
&= \delta_{f\infty(x) \le 0}.
\end{aligned}$$

20.7 Concave Function

There is a parallel theory for a proper, closed, concave function, for which cone f := cone (epi f).

Corollary 1 (Cone Representation of a Concave Function)

$$\text{cone } f = \left\{ (x, y, -z) \;\middle|\; \begin{array}{l} (x, y) \in z \, \text{epi} f \;\; if\, z > 0 \\ (x, y) \in \text{epi} f_\infty \;\; if\, z = 0. \end{array} \right\}$$

Corollary 2 (Cone Representation of the Concave Conjugate)

$$\delta_{\text{cone } f_*}(x^*, y^*, z^*) = \delta^*_{\text{cone } f}(-x^*, -z^*, -y^*).$$

Corollary 3 (Epi-Multiplication)

$$\delta_{\text{epi}(z \star f)}(x, y) = \delta_{\text{cone } f}(x, y, -z).$$

Part V
Local Analysis

Chapter 21
Directional Derivative and Subdifferential

Abstract For a convex function, we define the directional derivative—a positively homogeneous, convex function—the derivative in each direction from a vector. The conjugate of the directional derivative is the indicator of the subdifferential. We study the local behavior of a convex function by translating its epigraph so $f(\mathbf{0}) = 0$ and studying the local behavior at the origin. The conjugate is nonnegative, and the subdifferential is the vectors such that the conjugate is zero. Because the function is convex, one can express the derivative as the value function of a perturbed convex optimization: the derivative in direction x is

$$\inf_{y} (y \star f)(\boldsymbol{x}).$$

The objective function is the epi-multiplication of the function. The perturbation is the direction, and the choice variable is the multiplying factor. One obtains the directional derivative as the choice variable rises to infinity. We obtain Fermat's rule: a vector minimizes a proper, closed, convex function if and only if the directional derivative there is nonnegative.

For a convex function, we define the directional derivative—a positively homogeneous, convex function—the derivative in each direction from a vector. We express the directional derivative as the value function of a perturbed convex optimization. The objective function is the epi-multiplication of the function. The perturbation is the direction, and the choice variable is the multiplying factor. One obtains the directional derivative as the choice variable rises to infinity. The conjugate of the directional derivative is the indicator of the subdifferential.

We study the local behavior of a convex function by translating its epigraph so $f(\mathbf{0}) = 0$ and studying the local behavior at the origin. The conjugate f^* is nonnegative:

$$f^*(\boldsymbol{x}^*) = \sup_{\boldsymbol{x}} [\langle \boldsymbol{x}^*, \boldsymbol{x} \rangle - f(\boldsymbol{x})] \geq \langle \boldsymbol{x}^*, \mathbf{0} \rangle - f(\mathbf{0}) = 0.$$

B. C. Dieffenbach, *Conjugate Duality in Economic Analysis*, Contributions to Economics,
https://doi.org/10.1007/978-3-032-21396-9_21

The subdifferential is the vectors such that the conjugate is zero. The directional derivative at $\mathbf{0}$ is the derivative of the function in a direction.

Definition 1 (Directional Derivative) The derivative of f in direction $\boldsymbol{x}$ is

$$f'(\boldsymbol{x}) := \lim_{\lambda \searrow 0} \frac{f(\lambda \boldsymbol{x}) - f(\mathbf{0})}{\lambda}.$$

Below the notation $f'(\boldsymbol{x})$ always refers to the derivative at $\mathbf{0}$ in direction $\boldsymbol{x}$, not to the derivative of f with respect to $\boldsymbol{x}$.

Because the function is convex, one can express this limit as the value function of a perturbed convex optimization:

$$f'(\boldsymbol{x}) = \lim_{\lambda \searrow 0} \frac{f(\lambda \boldsymbol{x}) - f(\mathbf{0})}{\lambda} = \lim_{y \to \infty} \frac{f(\boldsymbol{x}/y) - 0}{1/y} = \lim_{y \to \infty} y f(\boldsymbol{x}/y) = \inf_y (y \star f)(\boldsymbol{x}).$$

Replacing a limit with a minimization, the final equality holds because $(y \star f)(\boldsymbol{x})$ is nonincreasing in y, as f is convex and $f(\mathbf{0}) = 0$. Typically, the infimum is not attained but is approached as $y \to \infty$.

Theorem 1 (Directional Derivative) *The derivative of f in direction $\boldsymbol{x}$ is*

$$f'(\boldsymbol{x}) = \inf_y (y \star f)(\boldsymbol{x}). \tag{21.1}$$

One can also express the directional derivative via the cone representation:

$$f'(\boldsymbol{x}) = \inf_y (y \star f)(\boldsymbol{x}) = \inf_{y,z} \left[z + \delta_{\mathrm{epi}(y \star f)}(\boldsymbol{x}, z)\right] = \inf_{y,z} \left[z + \delta_{\mathrm{cone}\, f}(\boldsymbol{x}, z, -y)\right].$$

Expressing the directional derivative as the value function of a convex optimization allows one to employ perturbation duality to study the directional derivative and the subdifferential.

The geometry is simple. In Eq. (21.1), because $\mathrm{epi}\,(y \star f) = y\,\mathrm{epi}\, f$ for $y > 0$, the epigraph expands as y increases, containing the epigraph for smaller y. For $\mathbf{X} = \mathbf{R}$, study the local behavior of the quadratic $\vec{f}(x) = \frac{1}{2}\langle x, x\rangle$ at $\vec{x}$ via the translation

$$f(x) = \frac{1}{2}\langle x + \vec{x}, x + \vec{x}\rangle - \frac{1}{2}\langle \vec{x}, \vec{x}\rangle = \frac{1}{2}\langle x, x\rangle + \vec{x}x.$$

The directional derivative

$$f'(x) = \inf_y (y \star f)(x) = \lim_{y \to \infty} \left(\frac{1}{2y}\langle x, x\rangle + \vec{x}x\right) = \vec{x}x,$$

which is linear because f is differentiable, having derivative $\vec{x}$ at 0. For finite $y > 0$, $(y \star f)(x)$ lies above the directional derivative $f'(x)$ by $\frac{1}{2y}\langle x, x\rangle$.

That the directional derivative is the value function of a convex minimization establishes that it is convex. That epi-multiplication is positively homogeneous means that the directional derivative is positively homogeneous. If $\mathbf{0} \in$rbdry(dom f), then possibly it is neither closed nor proper, as a slight change in the direction can make the derivative jump to ∞ or $-\infty$ If $\mathbf{0} \in$ rint (dom f), however, then the directional derivative is both closed and proper (Rockafellar 1970, Theorem 23.4, p. 217). If f is polyhedral, then the directional derivative is closed and proper; and in problem (21.1) the minimum is attained.

Because the directional derivative is positively homogeneous, its conjugate is the indicator of a closed, convex set.

Theorem 2 (Subdifferential as a Conjugate) *The conjugate of the directional derivative is the indicator of the subdifferential,*

$$f'^{*}(\boldsymbol{x}^{*}) = \delta_{\partial f}(\boldsymbol{x}^{*}).$$

The conjugate of the directional derivative is

$$\begin{aligned} f'^{*}(\boldsymbol{x}^{*}) &= \sup_{\boldsymbol{x}} \left[\langle \boldsymbol{x}^{*}, \boldsymbol{x}\rangle - f'(\boldsymbol{x})\right] \\ &= \sup_{\boldsymbol{x}} \left[\langle \boldsymbol{x}^{*}, \boldsymbol{x}\rangle - \inf_{y} (y \star f)(\boldsymbol{x})\right] \\ &= \sup_{\boldsymbol{x}, y} \left[\langle \boldsymbol{x}^{*}, \boldsymbol{x}\rangle - \delta^{*}_{\mathrm{epi}\, f^{*}}(\boldsymbol{x}, -y)\right] \\ &= \delta_{\mathrm{epi}\, f^{*}}(\boldsymbol{x}^{*}, 0) \\ &= \delta_{f^{*}(\boldsymbol{x}^{*}) \le 0} \\ &= \delta_{\partial f}(\boldsymbol{x}^{*}). \end{aligned}$$

The corollary follows by conjugate duality.

Corollary 1 (Conjugate of the Subdifferential) *The directional derivative f' is proper if and only if $\partial f \neq \emptyset$. Then $\delta^{*}_{\partial f} = \mathrm{cl} f'$, and $(\partial f)^{\infty} = (\mathrm{dom}\, f')^{\circ}$.*

An example is the proper, closed, convex function

$$f(x) = \begin{cases} -\sqrt{x} \text{ for } x \ge 0 \\ \infty \text{ for } x < 0. \end{cases}$$

The directional derivative is neither proper nor closed:

$$f'(x) = \begin{cases} -\infty \text{ for } x > 0 \\ 0 \text{ for } x = 0 \\ \infty \text{ for } x < 0. \end{cases}$$

The subdifferential is empty.

The directional derivative first-order condition is equivalent to the standard first-order condition that the subdifferential contains $\mathbf{0}$.

Theorem 3 (Fermat's Rule) *A vector minimizes a proper, closed, convex function if and only if the directional derivative there is nonnegative.*

The result is intuitive, because a vector minimizes a function if and only if the change in the function away from it is nonnegative. Conjugate duality establishes Fermat's rule. The minimization problem

$$\inf_{\boldsymbol{x}} f'(\boldsymbol{x}).$$

has optimum value zero if and only if the directional derivative is nonnegative, and otherwise the infimum is $-\infty$. Since

$$\inf_{\boldsymbol{x}} f'(\boldsymbol{x}) = -f'^{*}(\mathbf{0}) = -\delta_{\partial f(\mathbf{0})}(\mathbf{0}),$$

the infimum is zero if and only if $\mathbf{0} \in \partial f(\mathbf{0})$, the subdifferential condition necessary and sufficient for a minimum.

We state the analogous formulas for a proper, closed, concave function f:

$$f'(\boldsymbol{x}) = \sup_{y} (y \star f)(\boldsymbol{x})$$
$$f'_{*}(\boldsymbol{x}^{*}) = -\delta_{\partial f}(\boldsymbol{x}^{*})$$

The directional derivative is proper if and only if $\partial f \neq \varnothing$. Then

$$\delta^{*}_{\partial f}(-\boldsymbol{x}) = -\mathrm{cl} f'(\boldsymbol{x})$$
$$(\partial f)^{\infty} = -(\mathrm{dom}\, f')^{\circ}.$$

Reference

Rockafellar RT (1970) Convex analysis. Princeton University Press, Princeton, NJ

Chapter 22
Normal Cone

Abstract We analyze the directional derivative and subdifferential for the indicator of a set. After first considering the general case, we specialize to the contour set for a function, the vectors for which the function is less than or equal to a specified value. Translate the epigraph of the function to the origin. The translated contour set has indicator

$$\delta_C(\boldsymbol{x}) = \delta_{f(x)\leq 0} = \delta_{\text{epi } f}(\boldsymbol{x}, 0).$$

Define the projection $X : (\boldsymbol{x}, y) \longmapsto \boldsymbol{x}$. Under general conditions the subdifferential of the indicator of the contour set is the projection of the normal cone of the epigraph,

$$\partial\delta_{f\leq 0} = X\partial\delta_{\text{epi } f}.$$

This subdifferential is also

$$\partial\delta_{f\leq 0} = \mathbf{R}_{+}\partial f \cup \partial\delta_{\text{dom } f},$$

in which the two terms derive from the non-vertical and vertical supporting hyperplanes, respectively.

We analyze the directional derivative and subdifferential for the indicator of a set. We first consider the general case, and we then specialize to the contour set for a function.

22.1 Directional Derivative and Subdifferential

Our method is to study the directional derivative and the subdifferential at a vector in a set by translating the epigraph of the indicator of the set to the origin. For a closed, convex set C, we study the directional derivative δ_C' and the subdifferential $\partial\delta_C$ at $\mathbf{0}$.

A hyperplane $\langle \boldsymbol{x}^*, \boldsymbol{x}\rangle = 0$ supports C at $\mathbf{0}$ if $\delta_C^*(\boldsymbol{x}^*) = 0$. The normal cone is the set of normals $\boldsymbol{x}^*$ of all hyperplanes supporting C at $\mathbf{0}$ (plus $\boldsymbol{x}^* = \mathbf{0}$). Because

B. C. Dieffenbach, *Conjugate Duality in Economic Analysis*, Contributions to Economics,
https://doi.org/10.1007/978-3-032-21396-9_22

$\mathbf{0} \in C$, $\delta_C^*(\boldsymbol{x}^*)$ is nonnegative. The subdifferential is all $\boldsymbol{x}^*$ such that $\delta_C^*(\boldsymbol{x}^*) = 0$, so the subdifferential is the normal cone.

The directional derivative is

$$\delta_C'(\boldsymbol{x}) = \inf_y (y \star \delta_C)(\boldsymbol{x}) = \inf_y \delta^*_{\text{epi } \delta_C^*}(\boldsymbol{x}, -y) = \inf_y \delta_{\text{cone } C}(\boldsymbol{x}, -y),$$

the indicator of the cone generated by C. The directional derivative is either 0 or ∞. The directional derivative is perhaps not closed, as the cone generated by C is perhaps not closed, because $\mathbf{0} \in C$. Going out of the set, the directional derivative is ∞. Yet a direction arbitrarily close may stay in the set, and the directional derivative is zero. Then the directional derivative is not closed.

The indicator of the subdifferential is the conjugate of the directional derivative. The directional derivative is the indicator of its effective domain, $\delta_C'(\boldsymbol{x}) = \delta_{\text{dom } \delta_C'}(\boldsymbol{x})$. Hence the subdifferential is the polar of the effective domain of the directional derivative.

See the situation via the polar cones generated by contour sets. The cone generated by C is dom δ_C', which is also the cone generated by the contour set having indicator $\delta_{\delta_C \le 0}$. Because $\mathbf{0} \in C$, this cone is perhaps not closed. Its polar is the normal cone, the closed cone generated by the contour set having indicator $\delta_{\delta_C^* \le 0}$.

22.2 Contour Set

Given a proper, closed, convex function, consider the contour set of all vectors for which the function is less than or equal to a specified value. We study the directional derivative and the subdifferential at a vector where the function equals this specified value. Translate the epigraph of the function to the origin. The translated contour set has indicator

$$\delta_C(\boldsymbol{x}) = \delta_{f(x) \le 0} = \delta_{\text{epi } f}(\boldsymbol{x}, 0).$$

Because $f(\mathbf{0}) = 0$, the origin lies in the contour set.

We express the directional derivative and the subdifferential in terms of f. The directional derivative is the indicator of the cone generated by the contour set:

$$\delta_C'(\boldsymbol{x}) = \inf_y \delta_{\text{cone } C}(\boldsymbol{x}, -y) = \inf_y \delta_{\text{cone } f}(\boldsymbol{x}, 0, -y) = \inf_y \delta^*_{\text{cone } f^*}(\boldsymbol{x}, -y, 0).$$

The indicator of the subdifferential is the conjugate.

Problem 1 (Contour-Set Subdifferential Primal)

$$\begin{aligned}\delta_{\partial\delta_{f\leq 0}}(\boldsymbol{x}^*) &= \sup_{\boldsymbol{x}}\left[\langle\boldsymbol{x}^*,\boldsymbol{x}\rangle - \delta_C'(\boldsymbol{x})\right]\\ &= \sup_{\boldsymbol{x},y}\left[\langle\boldsymbol{x}^*,\boldsymbol{x}\rangle - \delta_{\mathrm{cone}\,f}(\boldsymbol{x},0,-y)\right]\\ &= \sup_{\boldsymbol{x},y}\left[\langle\boldsymbol{x}^*,\boldsymbol{x}\rangle - \delta^*_{\mathrm{cone}\,f^*}(\boldsymbol{x},-y,0)\right].\end{aligned}$$

The subdifferential is the closed, convex cone polar to the cone generated by C. Calculating the dual obtains an alternate formula for the subdifferential.

We calculate a dual to obtain an alternate formula for the normal cone, involving the subdifferential ∂f. The perturbed primal is

$$\sup_{\boldsymbol{x},y}\left[\langle\boldsymbol{x}^*,\boldsymbol{x}\rangle - y^*y - \delta^*_{\mathrm{cone}\,f^*}(\boldsymbol{x},-y,z)\right],$$

in which the perturbation base values are $\boldsymbol{x}^*$, $y^* = 0$, and $z = 0$. Rearranging the Fenchel inequality

$$\delta_{\mathrm{cone}\,f^*}(\boldsymbol{x}^*,-y^*,-z^*) + \delta^*_{\mathrm{cone}\,f^*}(\boldsymbol{x},-y,z) - \langle\boldsymbol{x}^*,\boldsymbol{x}\rangle - (-y^*)(-y) - (-z^*)z \geq 0$$

obtains the reciprocal dual:

$$z^*z + \delta_{\mathrm{cone}\,f^*}(\boldsymbol{x}^*,-y^*,-z^*) \geq \langle\boldsymbol{x}^*,\boldsymbol{x}\rangle + y^*y - \delta^*_{\mathrm{cone}\,f^*}(\boldsymbol{x},-y,z).$$

Problem 2 (Contour-Set Subdifferential Dual)

$$\inf_{z^*}\delta_{\mathrm{cone}\,f^*}(\boldsymbol{x}^*,-y^*,-z^*)$$

The dual objective function is the indicator of a cone, and the dual value function is the indicator of a cone.

The objective function in the dual

$$\begin{aligned}\delta_{\mathrm{cone}\,f^*}(\boldsymbol{x}^*,0,-z^*) &= \delta^*_{\mathrm{cone}\,f}(\boldsymbol{x}^*,-z^*,0)\\ &= \delta_{\mathrm{epi}\,\delta^*_{\mathrm{epi}\,f}}(\boldsymbol{x}^*,-z^*,0)\\ &= \delta_{\delta^*_{\mathrm{epi}\,f}(\boldsymbol{x}^*,-z^*)\leq 0}\\ &= \delta_{\partial\delta_{\mathrm{epi}\,f}}(\boldsymbol{x}^*,-z^*).\end{aligned}$$

Define the projection $X : (\boldsymbol{x},y) \in \mathbf{X} \oplus \mathbf{R} \longmapsto \boldsymbol{x} \in \mathbf{X}$. The optimum value of the dual is

$$\inf_{z^*}\delta_{\partial\delta_{\mathrm{epi}\,f}}(\boldsymbol{x}^*,-z^*) = \delta_{X\,\partial\delta_{\mathrm{epi}\,f}}(\boldsymbol{x}^*),$$

the indicator of the subdifferential. Here z^* is a solution for perturbation $\boldsymbol{x}^*$ if and only if $\langle \boldsymbol{x}^*, \boldsymbol{x} \rangle + z^* z = 0$ supports $\operatorname{epi} f$ at $(\mathbf{0}, 0)$.

The Slater condition is the relative condition for the contour-set subdifferential primal.

Condition 1 (Slater) $f(\boldsymbol{x}) < 0$ *for some* $\boldsymbol{x}$.

If this condition holds, then there is no duality gap and a finite optimum is attained in the dual.

Theorem 1 (Subdifferential as Projection) *If there is no duality gap, then the subdifferential of the indicator of the contour set is the projection of the normal cone of the epigraph,*

$$\partial \delta_{f \leq 0} = X \partial \delta_{\operatorname{epi} f}.$$

Because the value of the primal is necessarily less than or equal to the value of the dual, the inclusion $\partial \delta_{f \leq 0} \supset X \partial \delta_{\operatorname{epi} f}$ holds, regardless of whether there is a duality gap.

Direct solution of the dual obtains a formula for the subdifferential in terms of ∂f. Given $\boldsymbol{x}^*$, a value $z^* > 0$ is a solution if and only if

$$\begin{aligned} 0 &= \delta_{\operatorname{cone} f^*}(\boldsymbol{x}^*, 0, -z^*) \\ &= \delta_{\operatorname{cone} f^*}(\boldsymbol{x}^*/z^*, 0, -1) \\ &= \delta_{\operatorname{epi} f^*}(\boldsymbol{x}^*/z^*, 0) \\ &= \delta_{\partial f}(\boldsymbol{x}^*/z^*). \end{aligned}$$

The solutions $\boldsymbol{x}^*$ form the cone $\mathbf{R}_{++} \partial f$.

The exceptional case is the solutions $z^* = 0$:

$$0 = \delta_{\operatorname{cone} f^*}(\boldsymbol{x}^*, 0, 0).$$

The hyperplane $\langle \boldsymbol{x}^*, \boldsymbol{x} \rangle + z^* z = 0$ supporting $\operatorname{epi} f$ at $\mathbf{0}$ is vertical. These hyperplanes are all $\boldsymbol{x}^*$ such that $f^{*\infty}(\boldsymbol{x}^*) \leq 0$. Because $\mathbf{0} \in \operatorname{dom} f$, $f^{*\infty}(\boldsymbol{x}^*) = \delta^*_{\operatorname{dom} f}(\boldsymbol{x}^*) \geq 0$, so the hyperplanes are the cone of $\boldsymbol{x}^*$ such that $f^{*\infty}(\boldsymbol{x}^*) = 0$, the normal cone $\partial \delta_{\operatorname{dom} f}$.

Combining both cases obtains the formula.

Theorem 2 (Contour-Set Subdifferential) *If there is no duality gap, then the subdifferential of the indicator of the contour set is*

$$\partial \delta_{f \leq 0} = \mathbf{R}_{+} \partial f \cup \partial \delta_{\operatorname{dom} f}.$$

The two terms derive from the non-vertical and vertical supporting hyperplanes, respectively. Always $\mathbf{0} \in \partial\delta_{\mathrm{dom}\, f}$. If $\partial f = \emptyset$, the first term is empty. If $\partial\delta_{\mathrm{dom}\, f} = \{\mathbf{0}\}$, there is no (nonzero) vertical supporting hyperplane, so $\partial f \neq \emptyset$. If $\mathbf{0} \in \mathrm{int}\,(\mathrm{dom}\, f)$, the second term is $\{\mathbf{0}\}$, and the subdifferential is $\mathbf{R}_+\partial f$.

The exceptional case is pertinent to the analysis of consumer demand.

The chapter "Subdifferential Calculus" employs this same methodology to calculate subdifferentials. One expresses the directional derivative via the cone representation. Its conjugate is the indicator of the subdifferential. A dual for the conjugate then discovers a formula for the subdifferential.

Chapter 23
Subdifferential Calculus

Abstract Subdifferential calculus expresses one subdifferential in terms of another. To derive results of subdifferential calculus directly from the directional derivative is a powerful technique. For example, under general conditions, the subdifferential of a sum is the sum of the subdifferentials. For this example, the primal calculates the conjugate of the directional derivative of the sum of two functions; the value function is the indicator of the subdifferential of the sum. In the dual, the value function is the indicator of the sum of the subdifferentials. Under general conditions there is no duality gap, and the result follows. Using this same technique, we derive various results of subdifferential calculus. The procedure exemplifies how duality leads naturally to the discovery of a result. Calculating the dual discovers an alternate formula for the subdifferential.

23.1 Sum

For two proper, closed, convex functions on a Euclidean space **X**, we show that under general conditions the sum of the subdifferentials is the subdifferential of the sum. Consider a vector where each function is subdifferentiable. Translate each function to the origin.

The directional derivative of the sum is the sum of the directional derivatives:

$$\begin{aligned}(f+g)'(\boldsymbol{x}) &= \lim_{y\to\infty} y(f+g)(\boldsymbol{x}/y)\\ &= \lim_{y\to\infty}[yf(\boldsymbol{x}/y)+yg(\boldsymbol{x}/y)]\\ &= \lim_{y\to\infty} yf(\boldsymbol{x}/y)+\lim_{y\to\infty} yg(\boldsymbol{x}/y)\\ &= f'(\boldsymbol{x})+g'(\boldsymbol{x}).\end{aligned}$$

B. C. Dieffenbach, *Conjugate Duality in Economic Analysis* , Contributions to Economics,
https://doi.org/10.1007/978-3-032-21396-9_23

The third equals sign holds because we assume that neither subdifferential is empty; either directional derivative might be ∞ but cannot be $-\infty$. Also

$$(f+g)'(\boldsymbol{x}) = \lim_{y\to\infty}[yf(\boldsymbol{x}/y) + yg(\boldsymbol{x}/y)] = \inf_y[(y \star f)(\boldsymbol{x}) + (y \star g)(\boldsymbol{x})]$$

expresses the directional derivative of the sum as an optimization.

The indicator of the subdifferential is the conjugate of the directional derivative:

$$\begin{aligned}\delta_{\partial(f+g)}(\boldsymbol{x}^*) &= \sup_{\boldsymbol{x}}\left[\langle \boldsymbol{x}^*, \boldsymbol{x}\rangle - (f+g)'(\boldsymbol{x})\right] \\ &= \sup_{\boldsymbol{x}}[\langle \boldsymbol{x}^*, \boldsymbol{x}\rangle - f'(\boldsymbol{x}) - g'(\boldsymbol{x})] \\ &= \sup_{\boldsymbol{x},y}[\langle \boldsymbol{x}^*, \boldsymbol{x}\rangle - (y \star f)(\boldsymbol{x}) - (y \star g)(\boldsymbol{x})].\end{aligned}$$

Problem 1 (Sum Subdifferential Primal)

$$\delta_{\partial(f+g)}(\boldsymbol{x}^*) = \sup_{\boldsymbol{x},y}\left[\langle \boldsymbol{x}^*, \boldsymbol{x}\rangle + y^*y - \delta^*_{\mathrm{epi}\, f^*}(\boldsymbol{x}-\boldsymbol{u}, -y) - \delta^*_{\mathrm{epi}\, g^*}(\boldsymbol{x}, -y)\right].$$

The base value of the primal perturbation is $\boldsymbol{u} = \boldsymbol{0}$, and the base values of the dual perturbations are $y^* = 0$ and $\boldsymbol{x}^*$.

In this primal and analogously in the primals below, because the supremum is approached as $y \to \infty$, to perturb y has no effect on the optimum value. Thus it is simplifying not to perturb y, and consequently the dual is *not* the reciprocal dual. That there is no duality gap derives from properties of the primal.

We calculate the dual. Adding the Fenchel inequalities

$$\begin{aligned}\delta_{\mathrm{epi}\, f^*}(\boldsymbol{u}^*, 0) + \delta^*_{\mathrm{epi}\, f^*}(\boldsymbol{x}-\boldsymbol{u}, -y) - \langle \boldsymbol{u}^*, \boldsymbol{x}-\boldsymbol{u}\rangle &\ge 0 \\ \delta_{\mathrm{epi}\, g^*}(\boldsymbol{x}^*-\boldsymbol{u}^*, -y^*) + \delta^*_{\mathrm{epi}\, g^*}(\boldsymbol{x}, -y) - \langle \boldsymbol{x}^*-\boldsymbol{u}^*, \boldsymbol{x}\rangle - (-y^*)(-y) &\ge 0\end{aligned}$$

and rearranging obtains

$$\begin{aligned}\langle \boldsymbol{u}^*, \boldsymbol{u}\rangle + \delta_{\mathrm{epi}\, f^*}(\boldsymbol{u}^*, 0) + \delta_{\mathrm{epi}\, g^*}(\boldsymbol{x}^*-\boldsymbol{u}^*, -y^*) \ge \langle \boldsymbol{x}^*, \boldsymbol{x}\rangle \\ + y^*y - \delta^*_{\mathrm{epi}\, f^*}(\boldsymbol{x}-\boldsymbol{u}, -y) - \delta^*_{\mathrm{epi}\, g^*}(\boldsymbol{x}, -y).\end{aligned}$$

The dual is

$$\begin{aligned}&\inf_{\boldsymbol{u}^*}\left[\delta_{\mathrm{epi}\, f^*}(\boldsymbol{u}^*, 0) + \delta_{\mathrm{epi}\, g^*}(\boldsymbol{x}^*-\boldsymbol{u}^*, 0)\right] \\ &= \inf_{\boldsymbol{u}^*}\left[\delta_{f^*(\boldsymbol{u}^*)\le 0} + \delta_{g^*(\boldsymbol{x}^*-\boldsymbol{u}^*)\le 0}\right] \\ &= \inf_{\boldsymbol{u}^*}\left[\delta_{\partial f}(\boldsymbol{u}^*) + \delta_{\partial g}(\boldsymbol{x}^*-\boldsymbol{u}^*)\right] \\ &= \delta_{\partial f+\partial g}(\boldsymbol{x}^*).\end{aligned}$$

The optimum value of the dual is the indicator of the sum of the subdifferentials.

Problem 2 (Sum Subdifferential Dual)

$$\delta_{\partial f+\partial g}(\boldsymbol{x}^*) = \inf_{\boldsymbol{u}^*}\left[\delta_{\partial f}(\boldsymbol{u}^*) + \delta_{\partial g}(\boldsymbol{x}^* - \boldsymbol{u}^*)\right].$$

The relative condition for the sum subdifferential primal is that rint (dom f) and rint (dom g) have a point in common. The polyhedral condition holds if f and g are polyhedral. Under either condition, there is no duality gap, and the optimum value is attained in the dual.

In contrast the sum subdifferential dual fails to satisfy the relative condition. For each function, the conjugate is greater than or equal to zero, and a subgradient is a vector such that the conjugate is zero. This property conforms with how the supremum in the primal is typically not attained but is only approached as the epi-multiplier $y \to \infty$.

We have discovered the following.

Theorem 1 (Subdifferential of a Sum) *If there is no duality gap, then the subdifferential of the sum equals the sum of the subdifferentials,*

$$\partial(f+g) = \partial f + \partial g.$$

That the value of the primal is necessarily greater than or equal to the value of the dual is equivalent to the inclusion $\partial(f+g) \supset \partial f + \partial g$.

This procedure generalizes below to other subdifferentials. The value function of the primal is the indicator of the subdifferential, the conjugate of the directional derivative. The value function of the dual presents an alternate formula. Under general conditions there is no duality gap, so the two formulas for the subdifferential are equal. In the primal the supremum is not attained but is only approached as the epi-multiplier goes to infinity. Hence our standard technique of studying Fenchel equalities necessary and sufficient for simultaneous solutions to the primal and the dual does not apply.

23.2 Composition with Linear Transformation

Consider a linear transformation $A : \mathbf{X} \to \mathbf{Y}$ from one Euclidean space to another, and a proper, closed, convex function f on $\mathbf{Y}$.

Definition 1 (Composition with Linear Transformation) The composition of f with the linear transformation A is

$${}_{A}f(y) := f\left(A^{\mathsf{T}}y\right).$$

Now let f denote the translation to the origin.

Problem 3 (Directional Derivative) The directional derivative

$$({}_A f)'(y) = \inf_z [z \star ({}_A f)](y) = \inf_z \left[(z \star f)\left(A^\top y\right) \right].$$

The indicator of the subdifferential is the conjugate of the directional derivative.

Problem 4 (Composition with Linear Transformation Subdifferential Primal)

$$\delta_{\partial_A f}(y^*) = \sup_{y,z} \left[\langle y^*, y \rangle - \delta^*_{\mathrm{epi}\, f^*}\left(A^\top y, -z\right) \right].$$

Because the objective function is positively homogeneous in the choice variables, the optimum value is either zero or ∞. Perturb the primal as

$$\sup_{y,z} \left[\langle y^*, y \rangle + z^* z - \delta^*_{\mathrm{epi}\, f^*}\left(A^\top y - x, -z\right) \right],$$

in which the perturbation base values are $x = \mathbf{0}$, y^*, and $z^* = 0$. We calculate the dual. Adding the Fenchel inequalities

$$\delta_{\mathrm{epi}\, f^*}(x^*, -z^*) + \delta^*_{\mathrm{epi}\, f^*}\left(A^\top y - x, -z\right) - \left\langle x^*, A^\top y - x \right\rangle - (-z^*)(-z) \geq 0$$
$$\delta_{\{0\}}(Ax^* - y^*) + \delta^*_{\{0\}}(-y) - \langle Ax^* - y^*, -y \rangle \geq 0$$

and rearranging obtains the dual

$$\begin{aligned} \langle x^*, x \rangle + \delta_{\mathrm{epi}\, f^*}(x^*, 0) + \delta_{\{0\}}(Ax^* - y^*) &\geq \langle y^*, y \rangle + z^* z \\ &- \delta^*_{\mathrm{epi}\, f^*}\left(A^\top y - x, -z\right) - \delta^*_{\{0\}}(-y). \end{aligned}$$

Problem 5 (Composition with Linear Transformation Subdifferential Dual)

$$\delta_{A\,\partial f}(y^*) = \inf_{x^*} \left[\delta_{f^*(x^*) \leq 0} + \delta_{Ax^* = y^*} \right] = \inf_{x^*} \left[\delta_{\partial f}(x^*) + \delta_{Ax^* = y^*} \right].$$

By inspection, the value function of the dual is the indicator of $A\,\partial f$. The relative condition for the primal is $A^\top y \in \mathrm{rint}\,(\mathrm{dom}\, f)$ for some y.

Theorem 2 (Composition with Linear Transformation Subdifferential) *If there is no duality gap, then*

$$\partial_A f = A\,\partial f.$$

If there were a duality gap, then $\partial_A f \supset A\,\partial f$.

The partial subdifferential is a special case. Consider a proper, closed, convex function $f(x, y)$ on the Euclidean space $\mathbf{X} \oplus \mathbf{Y}$. Define the projection

$$\begin{aligned} X : \mathbf{X} \oplus \mathbf{Y} &\to \mathbf{X} \\ (x, y) &\mapsto x. \end{aligned}$$

The adjoint is the embedding

$$X^{\mathsf{T}} : \mathbf{X} \to \mathbf{X} \oplus \mathbf{Y}$$
$$\boldsymbol{x} \mapsto (\boldsymbol{x}, \mathbf{0}).$$

Then $\partial_X f$ is the partial subdifferential.

Corollary 1 (Partial Subdifferential) *If there is no duality gap, then the partial subdifferential* $\partial_X f = X \partial f$.

In the previous chapter, the normal cone to a contour set can be written as a partial subdifferential,

$$\partial\delta_{f\leq 0} = X\,\partial\delta_{\mathrm{epi}\,f} = \partial_X \delta_{\mathrm{epi}\,f}.$$

23.3 Chain Rule

Whereas the chain rule is a key tool of differential calculus, unfortunately this rule does not hold for convex calculus. In general the composition of two convex functions need not be convex, so the chain rule cannot apply in general. In contrast to differential calculus, the chain rule holds only in a limited way.

Consider proper, closed, convex functions f and g defined on $\mathbf{R}$ and $\mathbf{X}$, respectively.

Condition 1 (Nondecreasing) f *is nondecreasing.*

Because f is nondecreasing, the following minimization defines the composition of f and g.

Definition 2 (Composition)

$$(f \circ g)(\boldsymbol{x}) := \inf_y \left[f(y) + \delta_{\mathrm{epi}\,g}(\boldsymbol{x}, y)\right] = \inf_{y,z} \left[z + \delta_{\mathrm{epi}\,f}(y, z) + \delta_{\mathrm{epi}\,g}(\boldsymbol{x}, y)\right].$$

The composition is convex, because it is the value function of a convex optimization. Note that $(f \circ g)(\boldsymbol{x}) = \infty$ if $g(\boldsymbol{x}) = \infty$. It follows that

$$\delta_{\mathrm{epi}(f\circ g)}(\boldsymbol{x}, z) = \inf_y \left[\delta_{\mathrm{epi}\,f}(y, z) + \delta_{\mathrm{epi}\,g}(\boldsymbol{x}, y)\right].$$

We calculate the subdifferential of composition via the directional derivative. By definition,

$$[z \bigstar (f \circ g)](\boldsymbol{x}) = \delta^*_{\mathrm{epi}(f\circ g)^*}(\boldsymbol{x}, z). \tag{23.1}$$

For $z > 0$,

$$
\begin{aligned}
[z \star (f \circ g)](\boldsymbol{x}) &= z \inf_{y} \left[f(y) + \delta_{\text{epi } g}(\boldsymbol{x}/z, y)\right] \\
&= \inf_{w} \left[zf(w/z) + \delta_{\text{epi } g}(\boldsymbol{x}/z, w/z)\right] \;\; (w := yz) \\
&= \inf_{w} \left[(z \star f)(w) + \delta_{(z \star g)(\boldsymbol{x}) \leq w}\right] \\
&= \inf_{w} \left[(z \star f)(w) + \delta_{\text{epi}(z \star g)}(\boldsymbol{x}, w)\right].
\end{aligned}
$$

This formula also holds for $z = 0$, because the support (23.1) is closed.
The directional derivative

$$(f \circ g)'(\boldsymbol{x}) = \inf_{z} [z \star (f \circ g)](\boldsymbol{x}) = \inf_{w,z} \left[(z \star f)(w) + \delta_{\text{epi}(z \star g)}(\boldsymbol{x}, w)\right].$$

Here the two epigraphs have been translated so now $f(0) = 0$ and $g(\boldsymbol{0}) = 0$. The indicator of the subdifferential is its conjugate.

Problem 6 (Composition Subdifferential Primal)

$$
\begin{aligned}
\delta_{\partial(f \circ g)}(\boldsymbol{x}^*) &= \sup_{\boldsymbol{x}} \left[\langle \boldsymbol{x}^*, \boldsymbol{x}\rangle - (f \circ g)'(\boldsymbol{x})\right] \\
&= \sup_{\boldsymbol{x}} \left\{\langle \boldsymbol{x}^*, \boldsymbol{x}\rangle - \inf_{w,z} \left[\delta^*_{\text{epi } f^*}(w, -z) + \delta_{\text{cone } g}(\boldsymbol{x}, w, -z)\right]\right\} \\
&= \sup_{\boldsymbol{x},w,z} \left[\langle \boldsymbol{x}^*, \boldsymbol{x}\rangle - \delta^*_{\text{epi } f^*}(w, -z) - \delta_{\text{cone } g}(\boldsymbol{x}, w, -z)\right].
\end{aligned}
$$

We calculate a dual. Adding the Fenchel inequalities

$$\delta^*_{\text{epi } f^*}(w - u, -z) + \delta_{\text{epi } f^*}(u^*, 0) - u^*(w - u) \geq 0$$

$$\delta_{\text{cone } g}(\boldsymbol{x}, w, -z) + \delta^*_{\text{cone } g}(\boldsymbol{x}^*, -u^* + w^*, -z^*) - \langle(\boldsymbol{x}^*, -u^* + w^*, -z^*), (\boldsymbol{x}, w, -z)\rangle \geq 0$$

and rearranging obtains

$$
\begin{aligned}
&u^* u + \delta_{\text{epi } f^*}(u^*, 0) + \delta^*_{\text{cone } g}(\boldsymbol{x}^*, -u^* + w^*, -z^*) \\
&\quad \geq \langle \boldsymbol{x}^*, \boldsymbol{x}\rangle + z^* z + w^* w - \delta^*_{\text{epi } f^*}(w - u, -z) - \delta_{\text{cone } g}(\boldsymbol{x}, w, -z).
\end{aligned}
$$

The perturbation base values are $\boldsymbol{x}^*$, $w^* = 0$, $z^* = 0$, and $u = 0$.

Problem 7 (Composition Subdifferential Dual)

$$\begin{aligned}
&\inf_{u^*}\left[\delta_{\mathrm{epi}\, f^*}(u^*,0)+\delta^*_{\mathrm{cone}\, g}(\boldsymbol{x}^*,-u^*,0)\right]\\
&\quad=\inf_{u^*}\left[\delta_{\mathrm{epi}\, f^*}(u^*,0)+\delta_{\mathrm{epi}(u^*\bigstar g^*)}(\boldsymbol{x}^*,0)\right]\\
&\quad=\inf_{u^*}\left[\delta_{f^*(u^*)\le 0}+\delta_{(u^*\bigstar g^*)(\boldsymbol{x}^*)\le 0}\right]\\
&\quad=\inf_{u^*}\left[\delta_{\partial f}(u^*)+\delta_{\partial\delta_{\mathrm{epi}\, g}}(\boldsymbol{x}^*,-u^*)\right].
\end{aligned}$$

The optimum value is zero if and only if each term is zero. Here

$$\delta_{\partial\delta_{\mathrm{epi}\, g}}(\boldsymbol{x}^*,-u^*)=\begin{cases}\delta_{\partial g}(\boldsymbol{x}^*/u^*) \text{ if } u^*>0\\ \delta_{\partial\delta_{\mathrm{dom}\, g}}(\boldsymbol{x}^*) \text{ if } u^*=0\\ \infty \text{ if } u^*<0.\end{cases}$$

The exceptional case $u^*=0$ is that $\boldsymbol{x}^*$ belongs to the normal cone of dom g.

The chain rule follows immediately from the dual.

Condition 2 (Relative Condition) *The relative condition for the primal is that there exists some* $\boldsymbol{x}\in \mathrm{rint}\,(\mathrm{dom}\, g)$ *such that* $g(\boldsymbol{x})\in \mathrm{rint}\,(\mathrm{dom}\, f)$.

Theorem 3 (Chain Rule) *Assume that f is nondecreasing, and the relative condition holds. If $0\notin\partial f$, then $\partial(f\circ g)=\partial f\,\partial g$. If $0\in\partial f$, then $\partial(f\circ g)=\partial f\,\partial g\cup \partial\delta_{\mathrm{dom}\, g}$.*

If $0\in\partial f$, then $\mathbf{0}\in\partial(f\circ g)$.

23.4 Finite Sum

Using the cone representation, we study the subdifferential of $f=\sum_i f_i$, the sum of n functions. Each function has been translated so $f_i(\mathbf{0})=0$. Assume each $\partial f_i\neq\varnothing$. The directional derivative is the sum of the directional derivatives:

$$f'(\boldsymbol{x})=\lim_{y\to\infty} yf(\boldsymbol{x}/y)=\lim_{y\to\infty}\sum_i\left[yf_i(\boldsymbol{x}/y)\right]=\sum_i\lim_{y\to\infty} yf_i(\boldsymbol{x}/y)=\sum_i f_i'(\boldsymbol{x}).$$

Therefore

$$\begin{aligned} f'(\boldsymbol{x}) &= \sum_i \left[\inf_y (y \star f_i)(\boldsymbol{x}) \right] \\ &= \sum_i \left(\inf_y \left\{ \inf_{z_i} \left[z_i + \delta_{\mathrm{epi}(y \star f_i)}(\boldsymbol{x}, z_i) \right] \right\} \right) \\ &= \sum_i \left(\inf_y \left\{ \inf_{z_i} \left[z_i + \delta_{\mathrm{cone}\, f_i}(\boldsymbol{x}, z_i, -y) \right] \right\} \right) \\ &= \inf_{y,z} \left[\langle \mathbf{1}, z \rangle + \sum_i \delta_{\mathrm{cone}\, f_i}(\boldsymbol{x}, z_i, -y) \right]. \end{aligned}$$

Here $z \in \mathbf{R}^n$.

Problem 8 (Finite-Sum Subdifferential Primal) The indicator of the subdifferential is the conjugate of the directional derivative:

$$\delta_{\partial(\sum_i f_i)}(\boldsymbol{x}^*) = \sup_{\boldsymbol{x},y,z} \left[\langle \boldsymbol{x}^*, \boldsymbol{x} \rangle + y^* y + \langle z^*, z \rangle - \sum_i \delta_{\mathrm{cone}\, f_i}(\boldsymbol{x} - \boldsymbol{u}_i, z_i + w_i, -y - v_i) \right]$$

The perturbation $U = \begin{bmatrix} \boldsymbol{u}_1 & \boldsymbol{u}_2 & \cdots \end{bmatrix} \in \mathbf{T}^{n \to \mathbf{X}}$. The base values of the perturbations are $\boldsymbol{x}^*$, $y^* = 0$, $z^* = -\mathbf{1}$, and $U = 0$.

We find the dual. Adding the Fenchel inequalities

$$\begin{aligned} \delta_{\mathrm{cone}\, f_i}(\boldsymbol{x} - \boldsymbol{u}_i, z_i, -y - v_i) + \delta^*_{\mathrm{cone}\, f_i}(\boldsymbol{u}_i^*, z_i^*, v_i^*) - \langle \boldsymbol{u}_i^*, \boldsymbol{x} - \boldsymbol{u}_i \rangle & \\ - z_i^* z_i - v_i^* (-y - v_i) \geq 0 \text{ for all } i & \\ \delta_{\{0\}}\left(U^* \mathbf{1} - \boldsymbol{x}^* \right) + \delta^*_{\{0\}}(-\boldsymbol{x}) - \left\langle U^* \mathbf{1} - \boldsymbol{x}^*, -\boldsymbol{x} \right\rangle \geq 0 & \\ \delta_{\{0\}}(\langle \boldsymbol{v}^*, \mathbf{1} \rangle + y^*) + \delta^*_{\{0\}}(-y) - (\langle \boldsymbol{v}^*, \mathbf{1} \rangle + y^*)(-y) \geq 0 & \end{aligned}$$

and rearranging obtains

$$\begin{aligned} \left\langle U^*, U \right\rangle + \langle \boldsymbol{v}^*, \boldsymbol{v} \rangle + \sum_i \delta^*_{\mathrm{cone}\, f_i}(\boldsymbol{u}_i^*, z_i^*, v_i^*) + \delta_{\{0\}}\left(U^* \mathbf{1} - \boldsymbol{x}^* \right) + \delta_{\{0\}}(\langle \boldsymbol{v}^*, \mathbf{1} \rangle + y^*) & \\ \geq \langle \boldsymbol{x}^*, \boldsymbol{x} \rangle + y^* y + \langle z^*, z \rangle - \sum_i \delta_{\mathrm{cone}\, f_i}(\boldsymbol{x} - \boldsymbol{u}_i, z_i, -y - v_i). & \end{aligned}$$

The dual is

$$\begin{aligned}
&\inf_{U^*,v^*}\left[\sum_i \delta^*_{\operatorname{cone} f_i}(u_i^*,-1,v_i^*) + \delta_{U^*1=x^*} + \delta_{\langle v^*,1\rangle=0}\right] \\
&= \inf_{U^*,v^*}\left[\sum_i \delta_{\operatorname{cone} f_i^*}(u_i^*,v_i^*,-1) + \delta_{U^*1=x^*} + \delta_{\langle v^*,1\rangle=0}\right] \\
&= \inf_{U^*,v^*}\left[\sum_i \delta_{\operatorname{epi} f_i^*}(u_i^*,v_i^*) + \delta_{U^*1=x^*} + \delta_{\langle v^*,1\rangle=0}\right] \\
&= \inf_{U^*}\left[\sum_i \delta_{\operatorname{epi} f_i^*}(u_i^*,0) + \delta_{U^*1=x^*}\right] \\
&= \inf_{U^*}\left[\sum_i \delta_{\partial f_i}(u_i^*) + \delta_{U^*1=x^*}\right] \\
&= \delta_{\sum_i \partial f_i}(x^*).
\end{aligned}$$

We set $v^* = \mathbf{0}$, because otherwise the objective function is ∞: because f_i^* is nonnegative, $\delta_{\operatorname{epi} f_i^*}(u_i^*, v_i^*) = \infty$ if $v_i^* < 0$.

Problem 9 (Finite-Sum Subdifferential Dual)

$$\delta_{\sum_i \partial f_i}(x^*) = \inf_{U^*}\left[\delta_{U^*1=x^*} + \sum_i \delta_{\partial f_i^*}(u_i^*)\right].$$

By inspection, the value function of the dual is the indicator of the sum of the subdifferentials ∂f_i. The relative condition for the sum subdifferential primal is that the $\operatorname{rint}(\operatorname{dom} f_i)$have a point in common.

Theorem 4 (Finite-Sum Subdifferential) *If there is no duality gap, then the subdifferential of the sum equals the sum of the subdifferentials,*

$$\partial\left(\sum_i f_i\right) = \sum_i \partial f_i.$$

Analogously to the previous examples, a duality gap implies the inclusion $\partial\left(\sum_i f_i\right) \supset \sum_i \partial f_i$, but not equality.

23.5 Maximum

An interesting example is the subdifferential of the maximum of a finite number of functions:

$$f = \max_i [f_i].$$

Each function f_i is proper, closed, and convex; and f is convex. If the value of a function is less than the maximum, then the function is irrelevant to the directional derivative and subdifferential, so consider a vector where each function achieves the maximum. For simplicity, also assume that this vector is in the interior of the effective domain of each function, which ensures that $f_i' > -\infty$, $\partial f_i \neq \varnothing$, and $\operatorname{dom} f_i' = \mathbf{X}$. Translate the epigraph of each function to the origin.

The directional derivative is the maximum of the directional derivatives:

$$f'(\boldsymbol{x}) = \lim_{y\to\infty} y f(\boldsymbol{x}/y) = \lim_{y\to\infty} \max_i \left[y f_i(\boldsymbol{x}/y)\right] = \max_i \left[\lim_{y\to\infty} y f_i(\boldsymbol{x}/y)\right] = \max_i \left[f_i'(\boldsymbol{x})\right].$$

Therefore

$$\begin{aligned} f'(\boldsymbol{x}) &= \max_i \left[\inf_y (y \star f_i)(\boldsymbol{x})\right] \\ &= \max_i \left(\inf_y \left\{\inf_{z_i} \left[z_i + \delta_{\operatorname{epi}(y\star f_i)}(\boldsymbol{x}, z_i)\right]\right\}\right) \\ &= \max_i \left(\inf_y \left\{\inf_{z_i} \left[z_i + \delta_{\operatorname{cone} f_i}(\boldsymbol{x}, z_i, -y)\right]\right\}\right) \\ &= \inf_{y,\boldsymbol{z}} \left[\delta_P^*(\boldsymbol{z}) + \sum_i \delta_{\operatorname{cone} f_i}(\boldsymbol{x}, z_i, -y)\right]. \end{aligned}$$

The optimization proceeds like the finite-sum case, except that the $\delta_P^*(\boldsymbol{z}) = \max_i [z_i]$ replaces the sum $\langle \mathbf{1}, \boldsymbol{z}\rangle$.

Problem 10 (Max Subdifferential Primal) The indicator of the subdifferential is the conjugate of the directional derivative:

$$\delta_{\partial f}(\boldsymbol{x}^*) = \sup_{\boldsymbol{x},y,\boldsymbol{z}} \left[\langle \boldsymbol{x}^*, \boldsymbol{x}\rangle + y^* y + \langle \boldsymbol{z}^*, \boldsymbol{z}\rangle - \delta_P^*(\boldsymbol{z}) - \sum_i \delta_{\operatorname{cone} f_i}(\boldsymbol{x} - \boldsymbol{u}_i, z_i + w_i, -y - v_i)\right]$$

In the primal, the perturbations are $U = \left[\begin{array}{ccc} \boldsymbol{u}_1 & \cdots & \boldsymbol{u}_n \end{array}\right] \in \mathbf{T}^{n\to\mathbf{X}}$, $\boldsymbol{v} \in \mathbf{R}^n$, $\boldsymbol{w} \in \mathbf{R}^n$. The base values are $\boldsymbol{x}^*$, $y^* = 0$, $\boldsymbol{z}^* = \mathbf{0}$, $U = 0$, $\boldsymbol{v} = \mathbf{0}$, and $\boldsymbol{w} = \mathbf{0}$. Because $\mathbf{0} \in \operatorname{int}(\operatorname{dom} f_i)$, the primal satisfies the relative condition.

We find the dual. Adding the Fenchel inequalities

$$\begin{aligned} &\delta_{\operatorname{cone} f_i}(\boldsymbol{x} - \boldsymbol{u}_i, z_i + w_i, -y - v_i) + \delta^*_{\operatorname{cone} f_i}(\boldsymbol{u}_i^*, -w_i^*, v_i^*) \\ &\qquad - \langle \boldsymbol{u}_i^*, \boldsymbol{x} - \boldsymbol{u}_i\rangle - (-w_i^*)(z_i + w_i) - v_i^*(-y - v_i) \geq 0 \text{ for all } i \\ &\delta_{\{0\}}\left(U^*\mathbf{1} - \boldsymbol{x}^*\right) + \delta^*_{\{0\}}(-\boldsymbol{x}) - \left\langle U^*\mathbf{1} - \boldsymbol{x}^*, -\boldsymbol{x}\right\rangle \geq 0 \\ &\delta_P(\boldsymbol{w}^* + \boldsymbol{z}^*) + \delta_P^*(\boldsymbol{z}) - \langle \boldsymbol{w}^* + \boldsymbol{z}^*, \boldsymbol{z}\rangle \geq 0 \\ &\delta_{\{0\}}(\langle \boldsymbol{v}^*, \mathbf{1}\rangle - y^*) + \delta^*_{\{0\}}(y) - \langle \langle \boldsymbol{v}^*, \mathbf{1}\rangle - y^*, y\rangle \geq 0 \end{aligned}$$

and rearranging obtains

$$
\begin{aligned}
&\left\langle U^*, U\right\rangle + \langle \boldsymbol{w}^*, \boldsymbol{w}\rangle + \langle \boldsymbol{v}^*, \boldsymbol{v}\rangle \\
&\quad + \sum_i \delta^*_{\text{cone}\, f_i}\left(\boldsymbol{u}_i^*, -w_i^*, v_i^*\right) + \delta_{\{0\}}\left(U^*\mathbf{1} - \boldsymbol{x}^*\right) + \delta_P\left(\boldsymbol{w}^* + z^*\right) + \delta_{\{0\}}\left(\langle \boldsymbol{v}^*, \mathbf{1}\rangle + y^*\right) \\
&\quad \geq \langle \boldsymbol{x}^*, \boldsymbol{x}\rangle + y^* y + \langle z^*, z\rangle - \sum_i \delta_{\text{cone}\, f_i}\left(\boldsymbol{x}, z_i, -y\right).
\end{aligned}
$$

The dual is

$$
\begin{aligned}
&\inf_{U^*, \boldsymbol{v}^*, \boldsymbol{w}^*} \left[\sum_i \delta^*_{\text{cone}\, f_i}\left(\boldsymbol{u}_i^*, -w_i^*, v_i^*\right) + \delta_{U^*\mathbf{1}=\boldsymbol{x}^*} + \delta_P\left(\boldsymbol{w}^*\right) + \delta_{\langle \boldsymbol{v}^*, \mathbf{1}\rangle = 0}\right] \\
&\quad = \inf_{U^*, \boldsymbol{v}^*, \boldsymbol{w}^*} \left[\sum_i \delta_{\text{cone}\, f_i^*}\left(\boldsymbol{u}_i^*, v_i^*, -w_i^*\right) + \delta_{U^*\mathbf{1}=\boldsymbol{x}^*} + \delta_P\left(\boldsymbol{w}^*\right) + \delta_{\langle \boldsymbol{v}^*, \mathbf{1}\rangle = 0}\right] \\
&\quad = \inf_{U^*, \boldsymbol{w}^*} \left[\sum_i \delta_{\text{cone}\, f_i^*}\left(\boldsymbol{u}_i^*, 0, -w_i^*\right) + \delta_{U^*\mathbf{1}=\boldsymbol{x}^*} + \delta_P\left(\boldsymbol{w}^*\right)\right] \\
&\quad = \inf_{U^*, \boldsymbol{w}^*} \left[\sum_i \delta_{w_i^*\, \partial f_i}\left(\boldsymbol{u}_i^*\right) + \delta_{U^*\mathbf{1}=\boldsymbol{x}^*} + \delta_P\left(\boldsymbol{w}^*\right)\right].
\end{aligned}
$$

The second equality holds because the objective function is ∞ if any $v_i < 0$. The third equality holds even if some $w_i^* = 0$, as $\delta_{\text{cone}\, f_i^*}\left(\boldsymbol{u}_i^*, 0, 0\right) = \delta_{(\text{dom}\, f_i')^\circ}\left(\boldsymbol{u}_i^*\right) = \delta_{\{0\}}\left(\boldsymbol{u}_i^*\right)$.

Problem 11 (Max Subdifferential Dual)

$$
\delta_{\text{co}\, \cup_i \partial f_i}\left(\boldsymbol{x}^*\right) = \inf_{U^*, \boldsymbol{w}^*} \left[\delta_{U^*\mathbf{1}=\boldsymbol{x}^*} + \sum_i \delta_{w_i^*\, \partial f_i}\left(\boldsymbol{u}_i^*\right) + \delta_P\left(\boldsymbol{w}^*\right)\right].
$$

By inspection, the value function of the dual is the indicator of the convex hull of the union of the subdifferentials ∂f_i.

We have discovered the following.

Theorem 5 (Max Subdifferential) *The subdifferential of* $\max_i [f_i]$ *is the convex hull of the union of the subdifferentials,*

$$
\partial f = \text{co}\, \cup_i \partial f_i.
$$

This methodology illustrates how perturbation duality can discover new results. In each case, the alternate formula for the subdifferential arises from a straightforward algebraic calculation, rather than from clever conjecture and proof.

Chapter 24
Polyhedra

Abstract A polyhedron is the intersection of a finite number of closed half spaces, the solutions to linear inequalities. That the polar of a polyhedral cone is polyhedral is fundamental. A polyhedral function is subdifferentiable throughout its effective domain, and the subdifferential is polyhedral. We derive a formula for the subdifferential in terms of the polyhedral representation. The conjugate of a polyhedral function is polyhedral. A primal having a polyhedral perturbation function has a polyhedral value function and a polyhedral dual. The polyhedral condition for no duality gap applies.

A cone is finitely generated if its elements are the nonnegative linear combinations of a finite set of generators. The Minkowski-Weyl theorem is intuitive geometrically, but the proof is subtle: a cone is finitely generated if and only if it is polyhedral. No simple proof by duality is possible. No simple, general formula relates a finitely generated representation and a polyhedral representation. For this reason, sophisticated algorithms are put forward for numerical polyhedral optimization.

That the polar of a polyhedral cone is polyhedral is fundamental. A polyhedral function is subdifferentiable throughout its effective domain, and the subdifferential is polyhedral. The conjugate of a polyhedral function is polyhedral. A primal having a polyhedral perturbation function has a polyhedral value function and a polyhedral dual. The polyhedral condition for no duality gap applies.

A polyhedron is the intersection of a finite number of closed half spaces. A polyhedron is the solutions to linear inequalities

$$A^{\mathsf{T}}\boldsymbol{x} \preceq \boldsymbol{a},$$

in which

$$\begin{aligned} A &= \begin{bmatrix} \boldsymbol{a}_1 & \cdots & \boldsymbol{a}_n \end{bmatrix} \in \mathbf{T}^{n\to\mathbf{X}} \\ \boldsymbol{a} &= \begin{bmatrix} a_1 & \cdots & a_n \end{bmatrix} \in \mathbf{R}^n. \end{aligned}$$

The polyhedron is a cone if $\boldsymbol{a} = \mathbf{0}$.

B. C. Dieffenbach, *Conjugate Duality in Economic Analysis*, Contributions to Economics,
https://doi.org/10.1007/978-3-032-21396-9_24

Theorem 1 (Polyhedral Polar) *The polar of a polyhedral cone is polyhedral.*

This theorem follows from the Minkowski-Weyl proposition stated below.

A cone is finitely generated if its elements are the nonnegative linear combinations of a finite set of generators $\boldsymbol{a}_i$.

Definition 1 (Finitely Generated Cone) A cone is finitely generated if it can be expressed as

$$A\mathbf{R}^n_+,$$

the image under the linear transformation A of the nonnegative cone $\mathbf{R}^n_+$.

The following fundamental proposition of convex analysis is intuitive geometrically, but the proof is subtle. The original references are Minkowski (1911) and Weyl (1935). See Rockafellar (1970, ch. 19), Barvinok (2002), and Borwein and Lewis (2006) for modern treatments.

Propostion 1 (Minkowski-Weyl) *A cone is finitely generated if and only if it is polyhedral.*

No simple proof by duality is possible. For a finitely generated representation $A\mathbf{R}^n_+$ and a polyhedral representation $B^\mathsf{T}\boldsymbol{x} \preceq \mathbf{0}$, $B \in \mathbf{T}^{n\to\mathbf{X}}$, no simple, general formula relates A and B. For this reason, sophisticated algorithms are put forward for numerical polyhedral optimization.

The following together with the Minkowski-Weyl theorem imply the polyhedral polar theorem.

Theorem 2 (Finitely Generated/Polyhedral Polars) *A cone is finitely-generated if and only if its polar is polyhedral: the polar of* $A\mathbf{R}^n_+$ *is the set of solutions to* $A^\mathsf{T}\boldsymbol{x} \preceq \mathbf{0}$.

The polar of $A\mathbf{R}^n_+$ is all $\boldsymbol{x}$ such that $0 \geq \langle A\boldsymbol{y}^*, \boldsymbol{x}\rangle = \left\langle \boldsymbol{y}^*, A^\mathsf{T}\boldsymbol{x}\right\rangle$ for all $\boldsymbol{y}^* \succeq \mathbf{0}$. Thus the polar is all $\boldsymbol{x}$ such that $A^\mathsf{T}\boldsymbol{x} \preceq \mathbf{0}$.

Linear inequalities define a polyhedral function.

Definition 2 (Polyhedral Function) A polyhedral function is a function having a polyhedral epigraph.

The epigraph having indicator $\delta_{f(x)\leq y}$ is polyhedral if it is the solution set to

$$\begin{aligned} A^\mathsf{T}\boldsymbol{x} - \mathbf{1}y &\preceq \boldsymbol{a} \\ B^\mathsf{T}\boldsymbol{x} &\preceq \boldsymbol{b} \end{aligned}$$

(here $B \in \mathbf{T}^{m \to \mathbf{X}}$). The epigraph lies above the nonvertical hyperplanes $A^\top \boldsymbol{x} - \mathbf{1}y = \boldsymbol{a}$ and within the vertical hyperplanes $B^\top \boldsymbol{x} = \boldsymbol{b}$. A polyhedral function is closed and convex.

We calculate the subdifferential at a vector in dom f. The function

$$\begin{aligned} f(\boldsymbol{x}) &= \inf_y \left[y + \delta_{\operatorname{epi} f}(\boldsymbol{x}, y) \right] = \delta_{B^\top \boldsymbol{x} \le \boldsymbol{b}} + \inf_y \left(y + \delta_{A^\top \boldsymbol{x} - \boldsymbol{a} \le \mathbf{1}y} \right) \\ &= \delta_{B^\top \boldsymbol{x} \le \boldsymbol{b}} + \max_i \left[\langle \boldsymbol{a}_i, \boldsymbol{x} \rangle - a_i \right]. \end{aligned}$$

The subdifferential $\partial f(\boldsymbol{x})$ involves only the active constraints in the epigraph. Let an overscore refer to the constraints that hold with equality:

$$\overline{A}^\top \boldsymbol{x} - \overline{\boldsymbol{a}} = \mathbf{1}y \tag{24.1}$$

$$\overline{B}^\top \boldsymbol{x} - \overline{\boldsymbol{b}} = \mathbf{0}. \tag{24.2}$$

($\overline{n}$ constraints in (24.1) and $\overline{m}$ in (24.2). The subdifferential of $\langle \boldsymbol{a}_i, \boldsymbol{x} \rangle - a_i$ is $\{\boldsymbol{a}_i\}$, so the subdifferential of $\max_i (\langle \boldsymbol{a}_i, \boldsymbol{x} \rangle - a_i)$ is all convex combinations $\overline{A}P^{\overline{n}}$. The subdifferential of the indicator of the solutions to $\langle \overline{\boldsymbol{b}}_i, \boldsymbol{x} \rangle \le \overline{b}_i$ is the normal cone $\overline{\boldsymbol{b}}_i \mathbf{R}_+$. That the subdifferential of the sum is the sum of the subdifferentials yields the following.

Theorem 3 (Polyhedral Function Subdifferential) *For* $\boldsymbol{x} \in \operatorname{dom} f$, *the subdifferential* $\partial f(\boldsymbol{x}) = \overline{A}P^{\overline{n}} + \overline{B}\mathbf{R}_+^{\overline{m}}$, *the sum over the active constraints (24.1)–(24.2).*

A polyhedral function is closed and subdifferentiable throughout its effective domain.

The conjugate of a proper polyhedral function is polyhedral. Here cone f is the solutions to

$$\begin{aligned} A^\top \boldsymbol{x} - \mathbf{1}y + \boldsymbol{a}z &\le \mathbf{0} \\ B^\top \boldsymbol{x} + \boldsymbol{b}z &\le \mathbf{0}. \end{aligned}$$

Its polar sets the conjugate

$$\delta_{\operatorname{cone} f^*}(\boldsymbol{x}^*, z^*, y^*) = \delta^*_{\operatorname{cone} f}(\boldsymbol{x}^*, y^*, z^*),$$

so the conjugate is also polyhedral.

The dual of a polyhedral primal is polyhedral. Consider the primal

$$V(\boldsymbol{y}) = \inf_{\boldsymbol{x}} F(\boldsymbol{x}, \boldsymbol{y}),$$

in which F is a proper, polyhedral function, and the perturbation base value $\boldsymbol{y} = \mathbf{0}$. The dual is

$$\sup_{\boldsymbol{y}^*} [-V^*(\boldsymbol{y}^*)] = \sup_{\boldsymbol{y}^*} [-F^*(\mathbf{0}, \boldsymbol{y}^*)].$$

Here

$$F \text{ polyhedral} \Rightarrow F^* \text{ polyhedral} \Rightarrow V^* \text{ polyhedral}.$$

The value function V is polyhedral but may or may not be proper.

If V is not proper, there might be a duality gap, such that the optimum value of the primal is ∞, and the optimum value of the dual is $-\infty$. Because epi V has no nonvertical supporting hyperplanes, the optimum value of the primal is either ∞ or $-\infty$. The conjugate $V^* = \infty$, so the optimum value of the dual is $-\infty$.

The polyhedral condition that there exists some $\boldsymbol{x}$ such that $(\boldsymbol{x}, \boldsymbol{y}) \in \text{dom } F$ implies that V is proper, because the optimum value of the primal is not ∞. That $V = V^{**}$ means that there is no duality gap. If the optimum value is finite, it must attained in the dual, because V is subdifferentiable. In the other direction, the same argument applies to the reciprocal dual.

References

Barvinok A (2002) A course in convexity, Vol 54. American Mathematical Society, Providence, RI

Borwein JM, Lewis AS (2006) Convex analysis and nonlinear optimization: theory and examples, 2nd edn. Springer-Verlag, New York

Minkowski H (1911) Gesammelte abhandlungen II. Leipzig, pp 131–229

Rockafellar RT (1970) Convex analysis. Princeton University Press, Princeton, NJ

Weyl H (1935) Elementare Theorie der konvexen Polyeder. Commentarii Math Helvetici 7:290–306

Part VI
Euclidean Jordan Algebra

Chapter 25
Euclidean Jordan Algebra

Abstract Although the setting for conjugate and perturbation duality is Euclidean space, many important applications involve additional structure. Whereas applied optimization theory commonly adds this extra structure informally, we proceed formally, by working in a "Euclidean Jordan algebra," a Euclidean space having additional algebraic structure. In mathematics, an algebra is a vector space with a bilinear, multiplicative product. A Euclidean Jordan algebra is a finite-dimensional real algebra satisfying certain conditions. The multiplicative product is the Jordan product, written $\boldsymbol{x} \bullet \boldsymbol{y}$. Given a finite-dimensional real vector space, to define the Jordan product specifies the structure of the Euclidean Jordan algebra. The prime example of a Euclidean Jordan algebra is the self-adjoint linear transformations, in which the Jordan product

$$X \bullet Y := \frac{1}{2}(XY + YX).$$

That the product is commutative but not associative is a key property, the reverse of standard matrix multiplication. We present the original axioms of Jordan, von Neumann, and Wigner, which capture very well how a Euclidean Jordan algebra generalizes the real numbers, but not the complex numbers. By definition, $\boldsymbol{x} \succeq \boldsymbol{y}$ means that $\boldsymbol{x} - \boldsymbol{y}$ belongs to the cone of squares—all vectors equal to the square of another. Whereas there is only one type of Euclidean space, in that any two Euclidean spaces of the same dimension are isomorphic, in contrast the classification theorem states that there are five distinct types of Euclidean Jordan algebras.

Although the setting for conjugate and perturbation duality is Euclidean space, many important applications involve additional structure. Whereas applied optimization theory commonly adds this extra structure informally, we proceed formally, by working in a "Euclidean Jordan algebra," a Euclidean space having additional algebraic structure. Euclidean Jordan algebra is a valuable framework for a wide range of convex optimization problems, and we use it extensively.

B. C. Dieffenbach, *Conjugate Duality in Economic Analysis*, Contributions to Economics,
https://doi.org/10.1007/978-3-032-21396-9_25

In mathematics, an algebra is a vector space with a bilinear, multiplicative product. A Euclidean Jordan algebra is a finite-dimensional real algebra satisfying certain conditions. The multiplicative product is the Jordan product, written $\boldsymbol{x} \bullet \boldsymbol{y}$.

For a finite-dimensional real vector space, to define the Jordan product specifies the structure of the Euclidean Jordan algebra. The simplest Euclidean Jordan algebra is the real numbers $\mathbf{R}$, in which the Jordan product is just the usual product of two numbers, $x \bullet y := xy$. The prime example of a Euclidean Jordan algebra is $\mathbf{S}^{\mathbf{X}}$, the self-adjoint linear transformations on a Euclidean space $\mathbf{X}$. The Jordan product of self-adjoint linear transformations X and Y is

$$X \bullet Y := \frac{1}{2}(XY + YX), \tag{25.1}$$

in which XY and YX are the composition of the linear transformations. That the product is commutative but not associative is a key property.

Although it is valuable to see a Euclidean Jordan algebra as a generalization of the real numbers, more specifically it is a generalization of $\mathbf{S}^{\mathbf{X}}$. It is informative to see a result about Euclidean Jordan algebra as a generalization of the same result for $\mathbf{S}^{\mathbf{X}}$. Convex optimization in $\mathbf{S}^{\mathbf{X}}$ is commonly called "semi-definite programming" (Vandenberghe and Boyd 1996; Wolkowicz et al. 2000). Superficially the Euclidean Jordan algebra $\mathbf{S}^{\mathbf{X}}$ is just the Euclidean space $\mathbf{S}^{\mathbf{X}}$ of linear algebra, having the same inner product and the same real eigenvalues and spectral decomposition. Nevertheless there is additional formal structure.

25.1 Jordan, von Neumann, and Wigner

In a remarkable paper, Jordan et al. (1934) provide a far-reaching axiomatic development of Euclidean Jordan algebra.[1] A German mathematical physicist, Jordan was nominated by Einstein for the Nobel prize for his contribution to quantum mechanics, but his membership in the Nazi party precluded any possibility of receiving this honor. Von Neumann was the most famous mathematician of the twentieth century. Wigner was a mathematical physicist, awarded the Nobel prize in 1963 for his work on symmetry groups in physics.

What makes this paper remarkable is not only its thoroughness and comprehensiveness but also its simplicity. In contrast to a typical article in a modern academic journal, it is complete and self-contained. Its derivation of the spectral decomposition is especially straightforward. It even classifies the five possible types of Euclidean Jordan algebra. In his survey, McCrimmon writes:

> I am unable to prove Jordan algebras were known to Archimedes, or that a complete theory has been found in the unpublished papers of Gauss. Their first appearance in recorded history

[1] A Euclidean Jordan algebra is a special case of a Jordan algebra, and the study of the latter by mathematicians has proliferated enormously.

> seems to be in the early 1930's when the theory bursts forth full-grown from the mind, not of Zeus, but of Pascual Jordan, John von Neumann, and Eugene Wigner in their 1934 paper. (McCrimmon 1978, p. 612)

Despite the importance of Euclidean Jordan algebra for applied optimization, the subject has received little attention until recently. For example, the biographies of von Neumann by Bhattacharya (2022) and Macrae (1992) nowhere mention it. Wigner makes no reference to it in his memoir (Wigner and Szanton 1992).

Faraut and Korányi (1994) present an extensive and detailed treatment and make the subject accessible to non-specialists. The long review of their book by Gross is insightful (Gross 1998). The discussion by McCrimmon is wide-ranging and covers the historical development in detail (McCrimmon 2003). Although these authors did not draw attention to its applicability in optimization, others soon did (Faybusovich 1997; Alizadeh and Schmieta 2000; Schmieta and Alizadeh 2001). Read these works for a fuller development.

The Jordan, von Neumann, and Wigner paper is technical and complex, and a full exposition is beyond our scope (Jordan et al. 1934). The discussion below presents the prominent ideas and theorems of the theory and depicts their mathematical development. Each result is interpreted via self-adjoint linear transformations.

Jordan, von Neumann, and Wigner saw a Euclidean Jordan algebra as a generalization of the real numbers. Of course the real numbers can be generalized in various ways, and with enough axioms only the real numbers will satisfy the axioms. Their objective was to develop a theoretical basis for quantum mechanics via abstract algebra (McCrimmon 2003, Part II, Chap. 1). In the Copenhagen interpretation of quantum mechanics, self-adjoint linear transformations represent physical observables, so they sought a generalization of the real numbers that was also a generalization of self-adjoint linear transformations. Ironically, the generalization did not advance quantum mechanics (McCrimmon 2003, Part II, Chap. 1), but it is fundamental to optimization theory.

We present the original axioms of Jordan, von Neumann, and Wigner, which capture very well how a Euclidean Jordan algebra generalizes the real numbers, but not the complex numbers.

Definition 1 (Euclidean Jordan Algebra) A Euclidean Jordan algebra is a finite-dimensional real algebra satisfying the three conditions commutative law (1), power associativity (2), and formal reality (3).

Condition 1 (Commutative Law) *The Jordan product is commutative,*

$$\boldsymbol{x} \bullet \boldsymbol{y} = \boldsymbol{y} \bullet \boldsymbol{x}.$$

Although the Jordan product is not associative, it is power associative: a power of $\boldsymbol{x}$ is independent of the order in which it is evaluated. For example, perhaps $\boldsymbol{x} \bullet \boldsymbol{x}^3 \neq \boldsymbol{x}^2 \bullet \boldsymbol{x}^2$ in an algebra that is not power associative. Power associativity means that both are equal, so the notation $\boldsymbol{x}^4$ is fitting. Power associativity makes the notation $\boldsymbol{x}^n$ unambiguous.

Condition 2 (Power Associativity) *The power x^n is independent of the sequence of evaluation.*

Of course the real numbers satisfy the commutative law and power associativity. Viewed as a two-dimensional real vector space, the complex numbers with their standard product do too.

In contrast the condition formal reality holds for the real numbers, but not for the complex numbers (since $1^2 + i^2 = 0$). Thus the real numbers form a Euclidean Jordan algebra, but not the complex numbers.

Condition 3 (Formal Reality)

$$x^2 + y^2 = 0 \text{ implies } x = 0 \text{ and } y = 0.$$

For $\mathbf{S}^{\mathrm{X}}$, the commutative law is immediate from the definition of the Jordan product. Since $X \bullet X$ is just the composition of X with itself, power associativity holds. Since X^2 is nonnegative and is zero only if $X = \mathbf{0}$, formal reality holds. Consequently $\mathbf{S}^{\mathrm{X}}$ is a Euclidean Jordan algebra.

A Euclidean Jordan algebra is not associative[2]: in general,

$$x \bullet (y \bullet z) \neq (x \bullet y) \bullet z.$$

The Euclidean Jordan algebra $\mathbf{S}^{\mathrm{X}}$ is not associative. That the Jordan product is commutative but not associative is the reverse of standard matrix multiplication, so Euclidean Jordan algebra proofs seem peculiar and tedious to an economist.[3]

The vector $\mathbf{1}$ stands for the unit.

Definition 2 (Unit) A vector $\mathbf{1}$ is a unit if $\mathbf{1} \bullet x = x$ for any x.

Propostion 1 (Existence of Unit) *A Euclidean Jordan algebra has a unique unit (Jordan et al. 1934, Theorem 5, p. 34).*

In $\mathbf{R}$, the unit is the number 1. In $\mathbf{S}^{\mathrm{X}}$, the unit is the identity I, the linear transformation $x \mapsto x$.

---

[2] Faraut and Korányi (1994) present a different but equivalent set of axioms. They first present axioms for a Jordan algebra, a generalization of a Euclidean Jordan algebra. A Jordan algebra satisfies the *Jordan identity* $x \bullet (x^2 \bullet y) = x^2 \bullet (x \bullet y)$, a limited sort of associativity. They then augment the axioms with one additional axiom, to obtain a Euclidean Jordan algebra (Faraut and Korányi 1994, p. 42). The advantage of the Jordan, von Neumann, and Wigner axioms is that the axioms are simple and intuitive, and clearly generalize the real numbers and the self-adjoint linear transformations. The complex numbers do not satisfy the axioms, but the real numbers and the self-adjoint linear transformations do.

[3] To simplify and to organize the proofs, mathematicians have introduced several other operations, defined via the Jordan product. The linear representation of x is the square linear transformation $y \mapsto x \bullet y$. McCrimmon (2003) explains and emphasizes the virtues of the "Jordan quadratic product" and the "Jordan triple product."

25.2 Cone of Squares

In economics, a basic concept is order. One vector of goods may be larger than another vector. Partial order is basic to economic efficiency: one outcome can be better than another in every way, or perhaps one cannot say which outcome is better. Whereas the definition of a Euclidean space involves no concept of order, a Euclidean Jordan algebra embodies a natural partial ordering.

Generalizing the nonnegative real numbers, the cone of squares is all vectors $\boldsymbol{x} \bullet \boldsymbol{x}$, all vectors equal to the square of another. By definition, $\boldsymbol{x} \succeq \boldsymbol{y}$ means that $\boldsymbol{x} - \boldsymbol{y}$ belongs to the cone of squares. In a Euclidean Jordan algebra $\mathbf{X}$, let $\mathbf{X}_+$ denote the cone of squares. For the self-adjoint linear transformations, the cone of squares is the nonnegative transformations. The inequality $X \succeq Y$ means that $X - Y$ is nonnegative.

Formal reality underpins the ordering by the cone of squares. Of course the real numbers are ordered: one real number is greater or less than another. In contrast the complex numbers are not ordered, and one complex number is not greater or less than another.

In a real vector space, one can specify the Jordan product just by defining the square of a vector. In a Euclidean Jordan algebra,

$$\boldsymbol{x} \bullet \boldsymbol{y} = \frac{1}{2}\left[(\boldsymbol{x}+\boldsymbol{y}) \bullet (\boldsymbol{x}+\boldsymbol{y}) - \boldsymbol{x} \bullet \boldsymbol{x} - \boldsymbol{y} \bullet \boldsymbol{y}\right], \tag{25.2}$$

by commutativity. The relationship (25.2) then defines the product of any two vectors.

Define the Euclidean Jordan algebra $-\mathbf{R}$ by the square $x \bullet x = -x^2$. This definition furnishes a formal treatment of a "bad" in economics—more is worse, not better.

Even though the definition of the Jordan product is basic to the definition and understanding of a Euclidean Jordan algebra, the product itself plays only a minimal role in the applications in this book. Instead many applications revolve around the cone of squares. The interest is the partial ordering implied by the Jordan product, a structure introduced to model the partial ordering. The ordering is essential for the analysis of efficiency and weak efficiency in economics.

The ordering can permit an optimization to be expressed as a linear problem. For example, the constraint $A\boldsymbol{x} - \boldsymbol{y} \succeq \boldsymbol{0}$ is a linear inequality. To model the same constraint by setting the problem in $\mathbf{R}^n$ might be a complicated and intractable nonlinear problem. Many applications involve the generalization of complementary slackness to a Euclidean Jordan algebra.

25.3 Direct Sum

Euclidean Jordan algebras $\mathbf{X}$ and $\mathbf{Y}$ determine a natural direct sum Euclidean Jordan algebra $\mathbf{X} \boxplus \mathbf{Y}$ on the product space $(\boldsymbol{x}, \boldsymbol{y}) \in \mathbf{X} \times \mathbf{Y}$.

Definition 3 (Square for Direct Sum)

$$(\boldsymbol{x},\boldsymbol{y}) \bullet (\boldsymbol{x},\boldsymbol{y}) := (\boldsymbol{x} \bullet \boldsymbol{x}, \boldsymbol{y} \bullet \boldsymbol{y}).$$

By Eq. (25.2), the Jordan product is

$$(\boldsymbol{x}^*,\boldsymbol{y}^*) \bullet (\boldsymbol{x},\boldsymbol{y}) := (\boldsymbol{x}^* \bullet \boldsymbol{x}, \boldsymbol{y}^* \bullet \boldsymbol{y}).$$

The unit is $(\mathbf{1},\mathbf{1})$. The cone of squares is $\mathbf{X}_+ \times \mathbf{Y}_+$.

The Euclidean Jordan algebra $\mathbf{R}^n := \mathbf{R} \boxplus \mathbf{R} \boxplus \cdots \boxplus \mathbf{R}$ (n times). An n-tuple $(x_1, x_2, \dots, x_n)$ describes a vector, and the Jordan product

$$(x_1^*, x_2^*, \dots, x_n^*) \bullet (x_1, x_2, \dots, x_n) = (x_1^* x_1, x_2^* x_2, \dots, x_n^* x_n).$$

The unit $\mathbf{1} = (1, 1, \dots, 1)$. The cone of squares is the n-tuples having every component nonnegative. The partial ordering set by the cone of squares is the standard partial ordering for $\mathbf{R}^n$.

25.4 Classification

There is only one type of Euclidean space, in that any two Euclidean spaces of the same dimension are isomorphic. In contrast, two Euclidean Jordan algebras having the same dimension need not be isomorphic. The classification theorem states that there are five distinct types of Euclidean Jordan algebras (Jordan et al. 1934, Part III). The self-adjoint linear transformations $\mathbf{S}^{\mathbf{X}}$ constitute one of the five types.

A Euclidean Jordan algebra is *simple* if it cannot be written as the direct sum of two (non-zero) Euclidean Jordan algebras. The classification theorem states that there are only five types of simple Euclidean Jordan algebras. Any Euclidean Jordan algebra can be expressed uniquely as the direct sum of simple Euclidean Jordan algebras (Jordan et al. 1934, Part III; Faraut and Korányi 1994, Proposition III.4.4, p. 54).

One might question whether the abstract approach of Jordan, von Neumann, and Wigner is justified. There only five distinct types of Euclidean Jordan algebras, and this book employs only three types. Hypothetically, it would be feasible to prove a theorem about Euclidean Jordan algebras by five separate proofs, one for each type. However, the abstract approach unifies the subject, by proving the theorem only once.

The abstract approach has many advantages. That it streamlines the notation is a key benefit. One can apply the basic theorems to prove additional general results applicable to any Euclidean Jordan algebra. The salient features of the analysis stand out better. Abstraction may force one to focus better on why a result holds. In addition, a proof can be simpler, as one does not get bogged down in the details of a specific type. (Nevertheless, effectively we do prove the basic theorems for the three types of Euclidean Jordan algebra used in applications in this book.)

Many simple but important economic applications involve the Euclidean Jordan algebra $\mathbf{R}^n$, such as the analysis of efficiency and weak efficiency. This book applies three of the five distinct types of Euclidean Jordan algebras. Econometric theory applies the Euclidean Jordan algebra $\mathbf{S}^{\mathbf{X}}$. Finance theory involving variance applies the Lorentz Euclidean Jordan algebra $\mathbf{L}^{\mathbf{X}}$. The analysis of the stability of difference and differential equations via eigenvalues applies the Hermitian Euclidean Jordan algebra $\mathbf{H}^{\mathbf{X}}$.

References

Alizadeh F, Schmieta S (2000) Symmetric cones, potential reduction methods. In: Wolkowitz H, Saigal R, Vandenberghe L (eds) Handbook of semidefinite programming: theory, algorithms, and applications. Kluwer Academic, Boston, pp 195–233

Bhattacharya A (2022) The man from the future: the visionary life of John von Neumann. Norton, New York

Faraut J, Korányi A (1994) Analysis on symmetric cones. Clarendon Press, Oxford University Press, Oxford

Faybusovich L (1997) Euclidean Jordan algebras and interior-point algorithms. Positivity 1(4):331–357

Gross KI (1998 Jan) Review of Analysis on Symmetric Cones. Bull Am Math Soc 35(1):77–86

Jordan P, von Neumann J, Wigner E (1934 Jan) On an algebraic generalization of the quantum mechanical formalism. Ann Math 35(1):29–64

Macrae N (1992) John von Neumann. Pantheon Books, New York

McCrimmon K (1978 Jul) Jordan algebras and their applications. Bull Am Math Soc 84(4):612–627

McCrimmon K (2003) A taste of Jordan algebras. Springer, Berlin

Schmieta SH, Alizadeh F (2001) Associative and Jordan algebras, and polynomial-time interior-point algorithms for symmetric cones. Math Oper Res 26(3):543–564

Vandenberghe L, Boyd S (1996 Mar) Semidefinite programming. SIAM Rev 38(1):49–95

Wigner EP, Szanton A (1992) The recollections of Eugene P. Wigner as told to Andrew Szanton. Plenum Press, New York

Wolkowicz H, Saigal R, Vandenberghe L (eds) (2000) Handbook of semidefinite programming: theory, algorithms, and applications. Kluwer Academic, Boston

Chapter 26
Spectral Decomposition

Abstract A Euclidean Jordan algebra admits a spectral decomposition that generalizes the spectral decomposition of the Euclidean space of self-adjoint linear transformations. Each vector has unique real eigenvalues, and the spectral decomposition expresses the vector as the sum of the eigenvalues times idempotents. The cone of squares is the vectors having nonnegative eigenvalues. The trace is the sum of the eigenvalues. The trace inner product makes the algebra into a Euclidean space: the inner product of two vectors is the trace of their Jordan product. The linear representation is the square linear transformation $L_x(y) = \boldsymbol{x} \bullet \boldsymbol{y}$, which is self-adjoint under the trace inner product and is nonnegative if and only if $\boldsymbol{x}$ is nonnegative. The polar of the nonnegative cone is its negative.

A Euclidean Jordan algebra admits a spectral decomposition that generalizes the spectral decomposition of the Euclidean space of self-adjoint linear transformations. Each vector has unique real eigenvalues, and the spectral decomposition expresses the vector as the sum of the eigenvalues times idempotents. The cone of squares is the vectors having nonnegative eigenvalues. The trace is the sum of the eigenvalues. The trace inner product makes the algebra into a Euclidean space: the inner product of two vectors is the trace of their Jordan product. The linear representation is the square linear transformation $L_x(y) = \boldsymbol{x} \bullet \boldsymbol{y}$, which is self-adjoint under the trace inner product and is nonnegative if and only if $\boldsymbol{x}$ is nonnegative. The polar of the nonnegative cone is its negative.

"The formally-real Jordan algebras investigated by Jordan, von Neumann, and Wigner are precisely the finite-dimensional real unital Jordan algebras such that every element has a unique spectral decomposition" (McCrimmon 2003, p. 18). Any vector in a Euclidean Jordan algebra has a spectral decomposition. Conversely, using the spectral decomposition, one can readily verify by direct calculation that the axioms for a Euclidean Jordan algebra hold.

B. C. Dieffenbach, *Conjugate Duality in Economic Analysis*, Contributions to Economics,
https://doi.org/10.1007/978-3-032-21396-9_26

26.1 Idempotents

The idempotents are building blocks of the theory.

Definition 1 (Idempotent) A nonzero vector $\boldsymbol{e}$ is idempotent if it equals its square.

Definition 2 (Orthogonal Idempotents) Two idempotents are orthogonal if their Jordan product is zero.

The sum of two orthogonal idempotents is idempotent. The unit is the largest idempotent. As

$$(\mathbf{1}-\boldsymbol{e}) \bullet (\mathbf{1}-\boldsymbol{e}) = (\mathbf{1}-\boldsymbol{e}),$$

hence $\mathbf{1}-\boldsymbol{e}$ lies in the cone of squares, so $\mathbf{1} \succeq \boldsymbol{e}$.

The primitive idempotents are the smallest idempotents.

Definition 3 (Primitive Idempotent) An idempotent $\boldsymbol{e}$ is primitive if it cannot be written as the sum of two orthogonal idempotents.

In $\mathbf{S}^{\mathbf{X}}$, the primitive idempotents are $\boldsymbol{x}\boldsymbol{x}^{\mathsf{T}}$, $|\boldsymbol{x}| = 1, \boldsymbol{x} \in \mathbf{X}$.

Propostion 1 (Decomposition of an Idempotent) *An idempotent can be decomposed as the sum of orthogonal primitive idempotents,*

$$\boldsymbol{e} = \sum_i \boldsymbol{e}_i$$

(Jordan et al. 1934, Theorem 10, p. 40). The decomposition need not be unique, but the number of terms is uniquely determined (the number is the trace of the idempotent).

26.2 Spectral Decomposition

Jordan, von Neumann, and Wigner derive the spectral decomposition (26.3) for a Euclidean Jordan algebra (Jordan et al. 1934, pp. 35–36). Because the axioms generalize the self-adjoint linear transformations, they essentially repeat a standard linear-algebra derivation of the decomposition. We outline their derivation.

For a given $\boldsymbol{x}$, consider the polynomial

$$\boldsymbol{x}^n + a_1\boldsymbol{x}^{n-1} + \cdots + a_{n-1}\boldsymbol{x} + a_n\mathbf{1}. \tag{26.1}$$

The expression is meaningful, by power associativity and the existence of the unit $\mathbf{1}$.

Definition 4 (Minimal Polynomial) The minimal polynomial is the polynomial of minimum degree equal to zero.

Because $\dim \mathbf{X}$ is finite, for some n and choices $a_1, \ldots, a_n$ the polynomial is zero. The minimal polynomial is unique—otherwise the subtraction of one from another would obtain a polynomial of smaller degree. The factorization

$$\lambda^n + a_1\lambda^{n-1} + \cdots + a_{n-1}\lambda + a_n = \prod_{i=1}^{n} (\lambda - \lambda_i)$$

implies

$$\prod_{i=1}^{n} (\boldsymbol{x} - \lambda_i \mathbf{1}) = \mathbf{0}. \tag{26.2}$$

The roots λ_i are real and distinct. If there were two roots $a \pm b\mathrm{i}$, then

$$[\boldsymbol{x} - (a + b\mathrm{i})\,\mathbf{1}]\,[\boldsymbol{x} - (a - b\mathrm{i})\,\mathbf{1}] = (\boldsymbol{x} - a\mathbf{1})^2 + (b\mathbf{1})^2 = \mathbf{0}.$$

Invoking formal reality yields $\boldsymbol{x} = a\mathbf{1}$ and $b = 0$, a contradiction.

The derivation of the spectral decomposition is constructive, expressing the spectral decomposition via the minimal polynomial.

Definition 5 (Complete System of Orthogonal Idempotents) Idempotents $\boldsymbol{e}_i$ constitute a complete system of orthogonal idempotents if

$$\begin{aligned} \boldsymbol{e}_i^2 &= \boldsymbol{e}_i \\ \boldsymbol{e}_i \bullet \boldsymbol{e}_j &= \mathbf{0} \text{ if } i \neq j \\ \sum_i \boldsymbol{e}_i &= \mathbf{1}. \end{aligned}$$

Theorem 1 (Minimal Polynomial Spectral Decomposition) *There exists a unique spectral decomposition*

$$\boldsymbol{x} = \sum_i \lambda_i \boldsymbol{e}_i, \tag{26.3}$$

in which the λ_i are roots of the minimal polynomial and the

$$\boldsymbol{e}_i = \prod_{j\neq i} (\boldsymbol{x} - \lambda_j \mathbf{1}) \Big/ \prod_{j\neq i} (\lambda_i - \lambda_j)$$

constitute a complete system of orthogonal idempotents.

(Jordan et al. 1934, pp. 35–36; Faraut and Korányi 1994, Proposition III.1.3, p. 45). The proof is a straightforward but complicated evaluation, substituting the minimal polynomial (26.2). (For $n = 1$, then $\boldsymbol{x} = a_1 \mathbf{1}$.)

One can express the spectral decomposition in terms of primitive idempotents.

Definition 6 (Jordan Frame) A Jordan frame is a complete system of orthogonal primitive idempotents.

Every frame comprises the same number of idempotents, the degree of the Euclidean Jordan algebra.

Theorem 2 (Jordan Frame Spectral Decomposition)

$$\boldsymbol{x} = \sum_i \lambda_i \boldsymbol{e}_i, \tag{26.4}$$

in which the λ_i *are eigenvalues and the* $\mathbf{e}_i$ *constitute a Jordan frame. The multiplicities of the* λ_i *are uniquely determined by* $\boldsymbol{x}$.

(Faraut and Korányi 1994, Proposition III.1.3, p. 45). This decomposition follows from the minimal polynomial decomposition, via Proposition 1.

Definition 7 (Eigenvalues) The eigenvalues of $\boldsymbol{x}$ are the coefficients λ_i in the Jordan frame spectral decomposition.

The eigenvalues are uniquely determined, and the number of eigenvalues is the degree of the Euclidean Jordan algebra.

By the spectral decomposition, the cone of squares is the vectors having nonnegative eigenvalues; and the primitive idempotents are its generators.

Definition 8 (Positive Cone) The positive cone $\mathbf{X}_{++}$ is the vectors for which all eigenvalues are positive.

The spectral decomposition furnishes a partial ordering:

Nonnegative	$\boldsymbol{x} \geq \mathbf{0}$	$\boldsymbol{x} \in \mathbf{X}_+$
Greater than zero	$\boldsymbol{x} \succ \mathbf{0}$	$\boldsymbol{x} \in \mathbf{X}_+$ and $\boldsymbol{x} \neq \mathbf{0}$
Positive	$\boldsymbol{x} \gg \mathbf{0}$	$\boldsymbol{x} \in \mathbf{X}_{++}$

The inequality $\boldsymbol{x} \succ \mathbf{0}$ means that each eigenvalue is nonnegative and at least one is positive. An idempotent $\boldsymbol{e} \succ \mathbf{0}$.

The determinant and the trace are functions of the eigenvalues.

Definition 9 (Determinant) The determinant is the product of the eigenvalues,

$$\det \boldsymbol{x} := \prod_i \lambda_i.$$

Definition 10 (Trace) The trace is the sum of the eigenvalues,

$$\operatorname{tr} \boldsymbol{x} := \sum_i \lambda_i.$$

For $X \in \mathbf{S}^{\mathrm{X}}$, there is a conceptual distinction between $\operatorname{tr} X$—the trace as an element of the Euclidean Jordan algebra—and $\operatorname{Tr} X$—its linear-algebra trace as a square linear transformation; but of course $\operatorname{tr} X = \operatorname{Tr} X$.

The product of two idempotents need not be nonnegative, but the trace of the product is nonnegative.

Propostion 2 (Positive Trace) *The trace of the product of two idempotents is positive if their product is nonzero.*

(Jordan et al. 1934, Theorem 21, p. 48). We illustrate this proposition for $\mathbf{S}^{\mathrm{X}}$. The primitive idempotents $\boldsymbol{x}\boldsymbol{x}^\top$ and $\boldsymbol{y}\boldsymbol{y}^\top$ have the Jordan product

$$(\boldsymbol{x}\boldsymbol{x}^\top) \bullet (\boldsymbol{y}\boldsymbol{y}^\top) = \frac{1}{2}\left[(\boldsymbol{x}\boldsymbol{x}^\top)(\boldsymbol{y}\boldsymbol{y}^\top) + (\boldsymbol{y}\boldsymbol{y}^\top)(\boldsymbol{x}\boldsymbol{x}^\top)\right] = \frac{\boldsymbol{x}^\top\boldsymbol{y}}{2}(\boldsymbol{x}\boldsymbol{y}^\top + \boldsymbol{y}\boldsymbol{x}^\top).$$

The two idempotents are orthogonal if and only $\boldsymbol{x}^\top\boldsymbol{y} = 0$. If $\boldsymbol{x}^\top\boldsymbol{y} \neq 0$, the Jordan product fails to be nonnegative unless $\boldsymbol{x} = \pm\boldsymbol{y}$. Yet the trace is

$$\operatorname{tr}\left[(\boldsymbol{x}\boldsymbol{x}^\top) \bullet (\boldsymbol{y}\boldsymbol{y}^\top)\right] = \operatorname{Tr}\left[\frac{\boldsymbol{x}^\top\boldsymbol{y}}{2}(\boldsymbol{x}\boldsymbol{y}^\top + \boldsymbol{y}\boldsymbol{x}^\top)\right] = (\boldsymbol{x}^\top\boldsymbol{y})^2 > 0,$$

in accord with the positive-trace Proposition 2.

26.3 Trace Inner Product

Until this point, it might seem that the adjective "Euclidean" in "Euclidean Jordan algebra" is inappropriate: the Jordan, von Neumann, and Wigner axioms do not say anything about Euclidean space or inner product. Below we justify the definition of the inner product by the trace.

Definition 11 (Trace Inner Product) The inner of product $\boldsymbol{x}$ and $\boldsymbol{y}$ is

$$\langle \boldsymbol{x}, \boldsymbol{y} \rangle := \operatorname{tr}(\boldsymbol{x} \bullet \boldsymbol{y}). \tag{26.5}$$

By the commutative law, this definition is symmetric between the two arguments. By the spectral decomposition, $\langle \boldsymbol{x}, \boldsymbol{x} \rangle > 0$ if $\boldsymbol{x} \neq \mathbf{0}$. Below we explain why the trace is linear, which implies that the trace inner product is bilinear. Thus this definition satisfies the properties required for an inner product. Under the trace inner product, $\operatorname{tr} \boldsymbol{x} = \langle \mathbf{1}, \boldsymbol{x} \rangle$; and the degree of the Euclidean Jordan algebra is $\langle \mathbf{1}, \mathbf{1} \rangle$.

A basic tool to study a Euclidean Jordan algebra is the linear representation.

Definition 12 (Linear Representation) The linear representation L_x of $\boldsymbol{x}$ is the square linear transformation

$$L_x : \mathbf{X} \to \mathbf{X} \boldsymbol{y} \mapsto \boldsymbol{x} \bullet \boldsymbol{y}.$$

Because it is a linear transformation, L_x is associative but not commutative. Of course L_x is linear in $\boldsymbol{x}$.

"Polarization" obtains the possible eigenvalues of the linear representation of an idempotent.

Theorem 3 (Idempotent Linear Representation Eigenvalues) *For an idempotent $\boldsymbol{e}$, the only possible eigenvalues of L_e are* 1, $\frac{1}{2}$, *and* 0.

(Jordan et al. 1934, Theorem 3, p. 33; Faraut and Korányi 1994, Proposition III.1.3, p. 45). By power associativity,

$$(z \bullet z) \bullet (z \bullet z) = [(z \bullet z) \bullet z] \bullet z.$$

Polarization substitutes $z = \lambda \boldsymbol{x} + \mu \boldsymbol{y}$ and expands both expressions. The coefficients of $\lambda \mu^3$ must be equal, which yields

$$4y^2 \bullet (y \bullet \boldsymbol{x}) = y^3 \bullet \boldsymbol{x} + y \bullet \left(y^2 \bullet \boldsymbol{x}\right) + 2y \bullet [y \bullet (y \bullet \boldsymbol{x})],$$

an identity valid for any $\boldsymbol{x}$ and $\boldsymbol{y}$. Replace $\boldsymbol{y}$ by an idempotent $\boldsymbol{e}$ to obtain

$$2\boldsymbol{e} \bullet [\boldsymbol{e} \bullet (\boldsymbol{e} \bullet \boldsymbol{x})] - 3\boldsymbol{e} \bullet (\boldsymbol{e} \bullet \boldsymbol{x}) + \boldsymbol{e} \bullet \boldsymbol{x} = \mathbf{0}.$$

Since $\boldsymbol{x}$ is arbitrary, the linear representation L_e must satisfy

$$2L_e^3 - 3L_e^2 + L_e = 0.$$

An eigenvalue of L_e is therefore a root of

$$0 = 2\lambda^3 - 3\lambda^2 + \lambda = (\lambda - 1)\left(\lambda - \frac{1}{2}\right)(\lambda - 0).$$

Furthermore, the following holds.

Propostion 3 (Idempotent Linear Representation Trace) *In a simple Euclidean Jordan algebra,* Tr L_e *is the same positive number for every primitive idempotent.*

(Jordan et al. 1934, Theorem 22, p. 48). If $\boldsymbol{x}$ has the spectral representation (26.4), then

$$L_x = \sum_i \lambda_i L_{e_i}$$
$$\operatorname{Tr} L_x = \sum_i \lambda_i \operatorname{Tr} L_e.$$

Theorem 4 (Trace of Linear Representation) *In a simple Euclidean Jordan algebra,*

$$\operatorname{tr} \boldsymbol{x} = \operatorname{Tr} L_x / \operatorname{Tr} L_e. \tag{26.6}$$

Thus $\operatorname{tr} \boldsymbol{x}$ is linear, so (26.5) is indeed an inner product.

Even though the Jordan product is not associative, the trace is associative.

Definition 13 (Associator) The associator is

$$\boldsymbol{x} \bullet (\boldsymbol{y} \bullet \boldsymbol{z}) - (\boldsymbol{x} \bullet \boldsymbol{y}) \bullet \boldsymbol{z}.$$

Since typically a Euclidean Jordan algebra is not associative, the associator is not zero. A multi-step application of polarization yields the following.

Propostion 4 (Linear Representation of the Associator) *The linear representation of the associator is*

$$L_y \left(L_x L_z - L_z L_x \right) - \left(L_x L_z - L_z L_x \right) L_y$$

(Jordan et al. 1934, p. 47; Faraut and Korányi 1994, Proposition II.4.1, p. 36). Applying trace reversal quickly obtains that the trace of the linear representation of the associator is zero, which can be restated as the following.

Theorem 5 (Associative Trace) *The trace is associative:*

$$\operatorname{tr} [\boldsymbol{x} \bullet (\boldsymbol{y} \bullet \boldsymbol{z})] = \operatorname{tr} [(\boldsymbol{x} \bullet \boldsymbol{y}) \bullet \boldsymbol{z}]$$

(Faraut and Korányi 1994, Proposition II.4.3, p. 37).

Corollary 1 (Self-Adjoint Linear Representation) *Under the trace inner product, the linear representation is self-adjoint.*

The eigenvalues of L_x and $\boldsymbol{x}$ are all real. The number of eigenvalues of the former is the dimension of **X**, which can be far larger than the number of eigenvalues of the latter, the degree of **X**. In general, the set of all linear representations is a proper subspace of the set of all self-adjoint linear transformations.

To require that the linear representation be self-adjoint effectively sets the inner product uniquely.

Propostion 5 *In a simple Euclidean Jordan algebra, any associative, symmetric bilinear form is proportional to the trace.*

(Faraut and Korányi 1994, Proposition III.4.1, p. 51).

A vector is nonnegative if and only if its linear representation is nonnegative. By this theorem, the partial ordering by the cone of squares is transitive: by linearity, $L_{x+y} = L_x + L_y$, so the sum of nonnegative vectors is nonnegative.

Theorem 6 (Nonnegative Linear Representation) *A vector* $\boldsymbol{x} \succeq \mathbf{0}$ *if and only if* $L_x \succeq 0$.

By Theorem 3, $L_{\boldsymbol{e}} \succ 0$ for an idempotent $\boldsymbol{e}$. If $\boldsymbol{x}$ has the spectral representation (26.4), then

$$L_x = \sum_i \lambda_i L_{e_i}.$$

Hence $L_x \succeq 0$ if each $\lambda_i \geq 0$. If some eigenvalue λ_i is negative, then

$$\langle \boldsymbol{e}_i, L_x \boldsymbol{e}_i \rangle = \langle \boldsymbol{e}_i, \boldsymbol{x} \bullet \boldsymbol{e}_i \rangle = \langle \boldsymbol{e}_i, \lambda_i \boldsymbol{e}_i \rangle = \lambda_i \mathrm{tr}\, \boldsymbol{e}_i < 0,$$

so $L_x \nsucceq 0$.

The polar of the cone of squares is its negative:

$$\begin{aligned}
&\sup_x \left[\langle \boldsymbol{x}^*, \boldsymbol{x} \rangle - \delta_{\boldsymbol{x} \succeq \mathbf{0}} \right] \\
&\quad = \sup_y \langle \boldsymbol{x}^*, \boldsymbol{y} \bullet \boldsymbol{y} \rangle \\
&\quad = \sup_y \langle \boldsymbol{y}, \boldsymbol{x}^* \bullet \boldsymbol{y} \rangle \text{ (associative trace)} \\
&\quad = \sup_y \langle \boldsymbol{y}, L_{x^*} \boldsymbol{y} \rangle \\
&\quad = \delta_{\boldsymbol{x}^* \preceq \mathbf{0}}.
\end{aligned}$$

It follows that the nonnegative cone is closed and convex and has an interior. The positive cone is the interior and is open.

To define the inner product via the trace makes it compatible with the partial ordering of the Euclidean Jordan algebra. Without this compatibility one would not see the Euclidean Jordan algebra as a generalization of the real numbers. In particular, the polar of the nonnegative cone would not be its negative.

26.4 Self-Adjoint Linear Transformations

We illustrate these results for the Euclidean Jordan algebra $\mathbf{S}^{\mathbf{X}}$. Let dim $\mathbf{X} = n$, so dim $\mathbf{S}^{\mathbf{X}} = n(n+1)/2$. Let $X \in \mathbf{S}^{\mathbf{X}}$ have the spectral decomposition

$$X = \sum_{i=1}^{n} \lambda_i \boldsymbol{x}_i \boldsymbol{x}_i^{\mathsf{T}}.$$

The eigenvalue λ_i has eigenvector $\boldsymbol{x}_i$: $X\boldsymbol{x}_i = \lambda_i \boldsymbol{x}_i$, such that $|\boldsymbol{x}_i| = 1$.

The linear representation

$$L_X : \mathbf{S}^{\mathbf{X}} \to \mathbf{S}^{\mathbf{X}}$$
$$Y \mapsto X \bullet Y = \frac{1}{2}(XY + YX)$$

has eigenvectors $\boldsymbol{x}_i\boldsymbol{x}_j^{\mathsf{T}} + \boldsymbol{x}_j\boldsymbol{x}_i^{\mathsf{T}}$, with eigenvalue $\lambda_i + \lambda_j$, for every $i \leq j$:

$$\begin{aligned}
&X \bullet \left(\boldsymbol{x}_i\boldsymbol{x}_j^{\mathsf{T}} + \boldsymbol{x}_j\boldsymbol{x}_i^{\mathsf{T}}\right) \\
&= \frac{1}{2}\left[X\left(\boldsymbol{x}_i\boldsymbol{x}_j^{\mathsf{T}} + \boldsymbol{x}_j\boldsymbol{x}_i^{\mathsf{T}}\right) + \left(\boldsymbol{x}_i\boldsymbol{x}_j^{\mathsf{T}} + \boldsymbol{x}_j\boldsymbol{x}_i^{\mathsf{T}}\right)X\right] \\
&= \frac{1}{2}\left[\left(\lambda_i\boldsymbol{x}_i\boldsymbol{x}_j^{\mathsf{T}} + \lambda_j\boldsymbol{x}_j\boldsymbol{x}_i^{\mathsf{T}}\right) + \left(\lambda_j\boldsymbol{x}_i\boldsymbol{x}_j^{\mathsf{T}} + \lambda_i\boldsymbol{x}_j\boldsymbol{x}_i^{\mathsf{T}}\right)\right] \\
&= \frac{1}{2}\left(\lambda_i + \lambda_j\right)\left(\boldsymbol{x}_i\boldsymbol{x}_j^{\mathsf{T}} + \boldsymbol{x}_j\boldsymbol{x}_i^{\mathsf{T}}\right).
\end{aligned}$$

There are $n(n+1)/2$ orthonormal eigenvectors.

Eigenvalue	Orthonormal Eigenvectors	
$\frac{1}{2}(\lambda_i + \lambda_j)$	$\frac{\sqrt{2}}{2}\left(\boldsymbol{x}_i\boldsymbol{x}_j^{\mathsf{T}} + \boldsymbol{x}_j\boldsymbol{x}_i^{\mathsf{T}}\right)$	$i \neq j$
λ_j	$\boldsymbol{x}_j\boldsymbol{x}_j^{\mathsf{T}}$	j

For a primitive idempotent $\boldsymbol{x}\boldsymbol{x}^{\mathsf{T}}$, the linear representation has one eigenvalue 1, $n-1$ eigenvalues $\frac{1}{2}$, and the remaining $n(n-1)/2$ eigenvalues are zero. Thus Tr $L_{\boldsymbol{x}\boldsymbol{x}^{\mathsf{T}}} = (n+1)/2$. Summing the eigenvalues obtains $\operatorname{Tr} L_X = (n+1)/2 \operatorname{Tr} X$, so the trace ratio (26.6) indeed holds.

A calculation verifies directly that L_X is self-adjoint:

$$\begin{aligned}\langle Z, X \bullet Y\rangle &= \left\langle Z, \frac{1}{2}(XY + YX)\right\rangle \\ &= \mathrm{Tr}\left\{Z\left[\frac{1}{2}(XY + YX)\right]\right\} \\ &= \mathrm{Tr}\left[Z\left(\frac{1}{2}XY\right) + \frac{1}{2}XZY\right] \\ &= \mathrm{Tr}\left[\frac{1}{2}(XZ + ZX)\,Y\right] \\ &= \left\langle \frac{1}{2}(XZ + ZX), Y\right\rangle \\ &= \langle X \bullet Z, Y\rangle.\end{aligned}$$

26.5 Euclidean Jordan Algebra Direct Sum

Consider the Euclidean Jordan algebra $\mathbf{R}^n = \mathbf{R} \boxplus \cdots \boxplus \mathbf{R}$ (n times). A vector is an n-tuple $\boldsymbol{x} = (x_1, \ldots, x_n)$. The Jordan product is multiplication component by component (the Hadamard product). The unit has every component one. The eigenvalues are the components. The trace is the sum of the components. An idempotent has every component is either one or zero. A primitive idempotent has a single component one and other components zero. A nonnegative vector has components greater than or equal to zero. The linear representation is the diagonal matrix of the components.

In many applications, to analyze a Euclidean Jordan algebra $\mathbf{X}$ via the spectral decomposition almost reduces the analysis to a study of $\mathbf{R}^n$, in which n is the degree of $\mathbf{X}$. For example, this reduction makes possible the generalization to a Euclidean Jordan algebra of the Perron theorem for a positive linear transformation. An important example is the spectral conjugate, for a convex function $f(\boldsymbol{x})$ that depends only on the eigenvalues of $\boldsymbol{x}$.

26.6 Isotope

"Isotope" is a recurring theme in McCrimmon (1978, 2003). The term isotope derives from chemistry, in which two isotopes of an element are almost the same, but not quite. Although a Euclidean Jordan algebra has a unique unit $\mathbf{1}$, any vector $\boldsymbol{u}$ in the positive cone can just as well serve as the unit.[1]

[1] The positive cone is a "symmetric cone" (Faraut and Korányi 1994). This terminology expresses the concept that any point in the cone is on an equal footing with any other.

From the original Jordan product $\boldsymbol{x} \bullet \boldsymbol{y}$, define a "twisted" Jordan product $\boldsymbol{x} \bullet_{[u]} \boldsymbol{y}$, to obtain a new, "twisted" Euclidean Jordan algebra, the "$\boldsymbol{u}$-isotope" of the original.

Condition 1 (Positive Cone) $\boldsymbol{u} \gg \mathbf{0}$.

Define the inverse $\boldsymbol{u}^{-1}$ by inverting the eigenvalues in the spectral decomposition of $\boldsymbol{u}$.

Definition 14 (Twisted Jordan Product) In the $\boldsymbol{u}$-isotope the Jordan product is

$$\boldsymbol{x} \bullet_{[u]} \boldsymbol{y} := \boldsymbol{x} \bullet \left(\boldsymbol{u}^{-1} \bullet \boldsymbol{y}\right) + \left(\boldsymbol{x} \bullet \boldsymbol{u}^{-1}\right) \bullet \boldsymbol{y} - (\boldsymbol{x} \bullet \boldsymbol{y}) \bullet \boldsymbol{u}^{-1}.$$

Propostion 6 (Twisted Euclidean Jordan Algebra) *The $\boldsymbol{u}$-isotope is a Euclidean Jordan algebra, in which unit $\mathbf{1}_{[u]} = \boldsymbol{u}$. The positive cone is the same as the original.*

(McCrimmon 2003, pp. 19–20). For self-adjoint linear transformations,

$$X \bullet_{[U]} Y = \frac{1}{2}\left(XU^{-1}Y + YU^{-1}X\right).$$

Here $\mathbf{1}_{[U]} = U$.

Although the underlying real vector space is unchanged, the twisting of the Jordan product causes the idempotents, the eigenvalues, the spectral decomposition, and the trace inner product to change. Nonetheless a critical property is that the positive cone remains the same, and consequently the partial ordering derived from it is the same. An isotope can be invaluable in a Euclidean Jordan algebra application focussing on the positive cone. If the investigation revolves around a positive vector $\boldsymbol{u}$, to reset the analysis in the $\boldsymbol{u}$-isotope will likely be simplifying.

References

Faraut J, Korányi A (1994) Analysis on symmetric cones. Clarendon Press, Oxford University Press, Oxford

Jordan P, von Neumann J, Wigner E (1934 Jan) On an algebraic generalization of the quantum mechanical formalism. Ann Math 35(1):29–64

McCrimmon K (1978 Jul) Jordan algebras and their applications. Bull Am Math Soc 84(4):612–627

McCrimmon K (2003) A taste of Jordan algebras. Springer, Berlin

Chapter 27
Spectral Conjugate

Abstract A spectral function is one expressible as the composition of a symmetric function and the ordered eigenvalues. In a Euclidean Jordan algebra **X** of degree n, the theme is that the calculation of the conjugate and the subdifferential of a spectral function effectively reduces to a calculation in $\mathbf{R}^n$. The trace inner product is less than or equal to the inner product of the ordered eigenvalues:

$$\langle x^*, x\rangle \leq \langle \lambda(x^*), \lambda(x)\rangle.$$

Equality holds if and only if the two vectors have a simultaneous ordered spectral decomposition. A spectral function is one expressible as the composition of a symmetric function and the ordered eigenvalues: $f[\lambda(x)]$. The conjugate $(f \circ \lambda)^* = f^* \circ \lambda$. Spectral conjugacy is widespread in applications.

A spectral function is one expressible as the composition of a symmetric function and the ordered eigenvalues. In a Euclidean Jordan algebra **X** of degree n, the theme is that the calculation of the conjugate and the subdifferential of a spectral function effectively reduces to a calculation in $\mathbf{R}^n$. The presentation below is a simplification of Lewis (1996) and Borwein and Lewis (2006, Sect. 1.2 and Sect. 5.2).

27.1 Simultaneous Ordered Spectral Decomposition

In a Euclidean Jordan algebra of degree n, express $x \in \mathbf{X}$ via a spectral decomposition ordered by the eigenvalues:

$$x = \lambda_1 e_1 + \lambda_2 e_2 + \cdots + \lambda_n e_n = \begin{bmatrix} e_1 & \cdots & e_n \end{bmatrix} \lambda(x) = X\lambda(x) \tag{27.1}$$

B. C. Dieffenbach, *Conjugate Duality in Economic Analysis*, Contributions to Economics,
https://doi.org/10.1007/978-3-032-21396-9_27

The primitive idempotents $\boldsymbol{e}_j$ form a Jordan frame, and $X \in \mathbf{T}^{n \to \mathbf{X}}$ denotes the Jordan frame. Here $X^{\mathsf{T}} X = \mathsf{I}$, and $X^{+} = X^{\mathsf{T}}$. The eigenvalues are ordered $\lambda_1 \geq \lambda_2 \geq \cdots \geq \lambda_n$, and $\boldsymbol{\lambda}(\boldsymbol{x}) \in \mathbf{R}^n$ is the ordered vector of the eigenvalues. The ordered spectral decomposition of a second vector

$$\boldsymbol{x}^* = \lambda_1 \boldsymbol{e}_1^* + \lambda_2 \boldsymbol{e}_2^* + \cdots + \lambda_n \boldsymbol{e}_n^* = \begin{bmatrix} \boldsymbol{e}_1^* & \cdots & \boldsymbol{e}_n^* \end{bmatrix} \boldsymbol{\lambda}(\boldsymbol{x}^*) = X^* \boldsymbol{\lambda}(\boldsymbol{x}^*).$$

Definition 1 (Simultaneous Spectral Decomposition) Two vectors have a simultaneous spectral decomposition if they have a spectral decomposition having the same Jordan frame.

Definition 2 (Simultaneous Ordered Spectral Decomposition) Two vectors have a simultaneous ordered spectral decomposition if they have a simultaneous spectral decomposition having the same ordering.

If there are repeated eigenvalues, then the Jordan frame is not unique, but the ordered vector of eigenvalues is uniquely defined.

27.2 Trace Inequality

The trace inequality is the basis for the spectral conjugate.

Theorem 1 (Trace Inequality) *The trace inner product is less than or equal to the inner product of the ordered eigenvalues.*

$$\langle \boldsymbol{x}^*, \boldsymbol{x} \rangle \leq \langle \boldsymbol{\lambda}(\boldsymbol{x}^*), \boldsymbol{\lambda}(\boldsymbol{x}) \rangle. \tag{27.2}$$

Equality holds if and only if $\boldsymbol{x}^*$ and $\boldsymbol{x}$ have a simultaneous ordered spectral decomposition.

In $\mathbf{R}^n$, the eigenvalues of an n-tuple $\boldsymbol{x}$ are just the values of the components. The n-tuple $\boldsymbol{\lambda}(\boldsymbol{x})$ reorders the components in a nonincreasing sequence. Two n-tuples have a simultaneous ordered spectral decomposition if each is nonincreasing. Hardy et al. (1952, Sect. 368, pp. 261–262) prove the trace inequality for $\mathbf{R}^n$. If one vector is ordered and the other is not, then to switch two components in the latter is nondecreasing.

Propostion 1 (Permutation Inequality) *For any permutation P.*

$$\langle \boldsymbol{\lambda}(\boldsymbol{x}^*), P\boldsymbol{\lambda}(\boldsymbol{x}) \rangle \leq \langle \boldsymbol{\lambda}(\boldsymbol{x}^*), \boldsymbol{\lambda}(\boldsymbol{x}) \rangle$$

The maximum is attained for I and is also attained by other permutations if there are repeated eigenvalues.

Using the spectral decomposition, we invoke the permutation inequality to prove the trace inequality in an arbitrary Euclidean Jordan algebra. Consider the following variant of the assignment problem.

Problem 1 (Assignment)

$$\max_{A \in \mathrm{T}^n} \left\langle \boldsymbol{\lambda}(\boldsymbol{x}^*) \boldsymbol{\lambda}(\boldsymbol{x})^\top, A \right\rangle = \langle \boldsymbol{\lambda}(\boldsymbol{x}^*), A\boldsymbol{\lambda}(\boldsymbol{x}) \rangle$$

such that A is doubly stochastic.

The permutations are the extreme points of the doubly stochastic matrices. By the permutation inequality, the choice $A = \mathrm{I}$ solves the problem, and the optimum value is $\langle \boldsymbol{\lambda}(\boldsymbol{x}^*), \boldsymbol{\lambda}(\boldsymbol{x}) \rangle$.

In the trace

$$\langle \boldsymbol{x}^*, \boldsymbol{x} \rangle = \langle X^* \boldsymbol{\lambda}(\boldsymbol{x}^*), X\boldsymbol{\lambda}(\boldsymbol{x}) \rangle = \left\langle \boldsymbol{\lambda}(\boldsymbol{x}^*), X^{*\top} X \boldsymbol{\lambda}(\boldsymbol{x}) \right\rangle,$$

$X^{*\top}X$ is doubly stochastic. Here

$$X^{*\top}\mathcal{X} = \begin{bmatrix} \langle \boldsymbol{e}_1^*, \boldsymbol{e}_1 \rangle & \langle \boldsymbol{e}_1^*, \boldsymbol{e}_2 \rangle & \\ \langle \boldsymbol{e}_2^*, \boldsymbol{e}_1 \rangle & \langle \boldsymbol{e}_2^*, \boldsymbol{e}_2 \rangle & \\ & & \ddots \end{bmatrix}.$$

By the positive-trace theorem for idempotents, every component in this matrix is nonnegative. Also $X^{*\top}X\mathbf{1} = \mathbf{1}$ and $X^\top X^*\mathbf{1} = \mathbf{1}$. Hence $X^{*\top}X$ is a convex combination of permutations, so the trace inequality (27.2) follows.

Let us next prove the second part of the theorem. For $\boldsymbol{x}$ and $\boldsymbol{x}^*$ to have a simultaneous ordered spectral decomposition means that there exist spectral decompositions such that $X = X^*$, so $X^{*\top}X = \mathrm{I}$ and (27.2) holds with equality. Conversely, suppose that $\langle \boldsymbol{x}^*, \boldsymbol{x} \rangle = \langle \boldsymbol{\lambda}(\boldsymbol{x}^*), \boldsymbol{\lambda}(\boldsymbol{x}) \rangle$. We prove that there exist simultaneous ordered spectral decompositions of $\boldsymbol{x}$ and $\boldsymbol{x}^*$. Let $Y\boldsymbol{\lambda}(\boldsymbol{x}^* + \boldsymbol{x})$ denote an ordered spectral decomposition of $\boldsymbol{x}^* + \boldsymbol{x}$:

$$Y\boldsymbol{\lambda}(\boldsymbol{x}^* + \boldsymbol{x}) = \boldsymbol{x}^* + \boldsymbol{x} = X^*\boldsymbol{\lambda}(\boldsymbol{x}^*) + X\boldsymbol{\lambda}(\boldsymbol{x}).$$

When the trace equality holds, we show that the ordered eigenvalues of the sum equals the sum of the ordered eigenvalues (Equation (27.3) below). Below Equations (27.8) and (27.9) state that $\boldsymbol{x}$ and $\boldsymbol{x}^*$ have simultaneous ordered spectral decompositions, for Jordan frame Y.

Always

$$\langle Y\boldsymbol{\lambda}(\boldsymbol{x}^* + \boldsymbol{x}), Y\boldsymbol{\lambda}(\boldsymbol{x}^* + \boldsymbol{x}) \rangle = \langle \boldsymbol{\lambda}(\boldsymbol{x}^* + \boldsymbol{x}), \boldsymbol{\lambda}(\boldsymbol{x}^* + \boldsymbol{x}) \rangle.$$

By the trace equality,

$$\left\langle X^* \lambda (x^*) + X\lambda (x), X^* \lambda (x^*) + X\lambda (x)\right\rangle = \langle \lambda (x^*) + \lambda (x), \lambda (x^*) + \lambda (x)\rangle.$$

By the trace inequality applied twice,

$$\begin{aligned}\left\langle X^* \lambda (x^*) + X\lambda (x), Y\lambda (x^* + x)\right\rangle &= \left\langle X^* \lambda (x^*), Y\lambda (x^* + x)\right\rangle + \langle X\lambda (x), Y\lambda (x^* + x)\rangle \\ &\leq \langle \lambda (x^*) + \lambda (x), \lambda (x^* + x)\rangle.\end{aligned}$$

Combining yields

$$\langle \lambda (x^*) + \lambda (x) - \lambda (x^* + x), \lambda (x^*) + \lambda (x) - \lambda (x^* + x)\rangle \leq 0,$$

so

$$\lambda (x^* + x) = \lambda (x^*) + \lambda (x) \tag{27.3}$$

when the trace equality holds.

Substituting four instances of the trace inequality

$$\langle Y\lambda (x), X\lambda (x)\rangle \leq \langle \lambda (x), \lambda (x)\rangle \tag{27.4}$$

$$\left\langle Y\lambda (x^*), X^* \lambda (x^*)\right\rangle \leq \langle \lambda (x^*), \lambda (x^*)\rangle \tag{27.5}$$

$$\left\langle Y\lambda (x), X^* \lambda (x^*)\right\rangle \leq \langle \lambda (x), \lambda (x^*)\rangle \tag{27.6}$$

$$\langle Y\lambda (x^*), X\lambda (x)\rangle \leq \langle \lambda (x^*), \lambda (x)\rangle \tag{27.7}$$

obtains

$$\begin{aligned}&\langle \lambda (x^* + x), \lambda (x^* + x)\rangle \\ &= \langle Y\lambda (x^* + x), Y\lambda (x^* + x)\rangle \\ &= \left\langle Y\lambda (x^*) + Y\lambda (x), X^* \lambda (x^*) + X\lambda (x)\right\rangle \\ &= \left\langle Y\lambda (x^*), X^* \lambda (x^*)\right\rangle + \langle Y\lambda (x^*), X\lambda (x)\rangle + \left\langle Y\lambda (x), X^* \lambda (x^*)\right\rangle + \langle Y\lambda (x), X\lambda (x)\rangle \\ &\leq \langle \lambda (x^*), \lambda (x^*)\rangle + \langle \lambda (x^*), \lambda (x)\rangle + \langle \lambda (x), \lambda (x^*)\rangle + \langle \lambda (x), \lambda (x)\rangle, \\ &\qquad \text{by Eq. (27.4)–(27.7)} \\ &= \langle \lambda (x^*) + \lambda (x), \lambda (x^*) + \lambda (x)\rangle.\end{aligned}$$

The final line equals the initial line, so every line is equal. Each of Eqs.(27.4)–(27.7) holds with equality.

As (27.4) and (27.5) are Cauchy–Schwarz inequalities in which both vectors have the same norm, equality requires that both vectors be equal:

$$Y\lambda (x) = X\lambda (x) = x \tag{27.8}$$

$$Y\lambda (x^*) = X^* \lambda (x^*) = x^*. \tag{27.9}$$

Thus x and x^* have a simultaneous ordered spectral decomposition, for Jordan frame Y.

27.3 Spectral Function

Condition 1 (Symmetric Function) *A symmetric function $f(y)$ on $\mathbf{R}^n$ is a function whose value is independent of the order of the components of $\boldsymbol{y}$.*

Any permutation of the components leaves the value unaffected. We assume that f is proper, closed, and convex.

Definition 3 (Spectral Function) A spectral function on a Euclidean Jordan algebra $\mathbf{X}$ is a function expressible as $f[\boldsymbol{\lambda}(\boldsymbol{x})]$, the composition of a symmetric function and the ordered eigenvalues.

Theorem 2 (Spectral Conjugate) *The conjugate of a spectral function*

$$(f \circ \boldsymbol{\lambda})^* = f^* \circ \boldsymbol{\lambda}.$$

The conjugate of a symmetric function is symmetric, so the conjugate of a spectral function is another spectral function. Applying the theorem twice to obtain the conjugate of the conjugate shows that $f \circ \boldsymbol{\lambda}$ is a proper, closed, convex function if f is a proper, closed, convex function.

To prove the spectral-conjugate theorem, reason by the following chain:

$$\begin{aligned}
&(f \circ \boldsymbol{\lambda})^* (\boldsymbol{x}^*) \\
&\quad = \sup_{\boldsymbol{x}} \{\langle \boldsymbol{x}^*, \boldsymbol{x}\rangle - f[\boldsymbol{\lambda}(\boldsymbol{x})]\} \\
&\quad = \sup_{\boldsymbol{x}} \{\langle \boldsymbol{\lambda}(\boldsymbol{x}^*), \boldsymbol{\lambda}(\boldsymbol{x})\rangle - f[\boldsymbol{\lambda}(\boldsymbol{x})]\} \qquad (27.10)\\
&\quad = \sup_{\boldsymbol{y}\in\mathbf{R}^n} [\langle \boldsymbol{\lambda}(\boldsymbol{x}^*), \boldsymbol{y}\rangle - f(\boldsymbol{y})] \ \text{(by the symmetry of } f) \\
&\quad = f^* [\boldsymbol{\lambda}(\boldsymbol{x}^*)].
\end{aligned}$$

The second equality (27.10) follows from the trace inequality: given $\boldsymbol{\lambda}(\boldsymbol{x})$, choosing $\boldsymbol{x}$ so it and $\boldsymbol{x}^*$ have a simultaneous ordered spectral decomposition maximizes $\langle \boldsymbol{x}^*, \boldsymbol{x}\rangle$, at $\langle \boldsymbol{\lambda}(\boldsymbol{x}^*), \boldsymbol{\lambda}(\boldsymbol{x})\rangle$.

This reasoning also interprets the Fenchel equality for a spectral function.

Corollary 1 (Spectral Fenchel Equality) *The Fenchel equality*

$$f \circ \boldsymbol{\lambda}(\boldsymbol{x}) + f^* \circ \boldsymbol{\lambda}(\boldsymbol{x}^*) - \langle \boldsymbol{x}^*, \boldsymbol{x}\rangle = 0$$

is equivalent to

$$f \circ \boldsymbol{\lambda}(\boldsymbol{x}) + f^* \circ \boldsymbol{\lambda}(\boldsymbol{x}^*) - \langle \boldsymbol{\lambda}(\boldsymbol{x}^*), \boldsymbol{\lambda}(\boldsymbol{x})\rangle = 0$$

and $\boldsymbol{x}^*$ and $\boldsymbol{x}$ have a simultaneous ordered spectral decomposition.

Corollary 2 (Spectral Subdifferential) *The following two conditions are equivalent:*

- $\boldsymbol{x}^* \in \partial (f \circ \boldsymbol{\lambda})(\boldsymbol{x})$;
- $\boldsymbol{\lambda}(\boldsymbol{x}^*) \in \partial f[\boldsymbol{\lambda}(\boldsymbol{x})]$, *and* $\boldsymbol{x}^*$ *and* $\boldsymbol{x}$ *have a simultaneous ordered spectral decomposition.*

For a concave function, the situation is similar. The conjugate $(f \circ \boldsymbol{\lambda})_* = f_* \circ \boldsymbol{\lambda}$. In the spectral Fenchel equality the two vectors must have simultaneous spectral decompositions, but the eigenvalues must have opposite order.

27.4 Application

That the polar of the nonnegative cone is its negative is a spectral conjugate. Since the nonnegative cone is the vectors having all nonnegative eigenvalues, the indicator $\delta_{\mathbf{X}_+} = \delta_{\mathbf{R}^n_+} \circ \boldsymbol{\lambda}$. Evaluating

$$\delta_{\mathbf{X}^\circ_+} = \delta^*_{\mathbf{X}_+} = \delta^*_{\mathbf{R}^n_+} \circ \boldsymbol{\lambda} = \delta_{-\mathbf{R}^n_+} \circ \boldsymbol{\lambda} = \delta_{-\mathbf{X}_+}$$

shows that $\mathbf{X}^\circ_+ = -\mathbf{X}_+$. Complementary slackness generalizes to a Euclidean Jordan algebra.

Theorem 3 (Complementary Slackness Fenchel Equality) *The Fenchel equality*

$$\delta_{\mathbf{X}_+}(\boldsymbol{x}) + \delta^*_{\mathbf{X}_+}(\boldsymbol{x}^*) - \langle \boldsymbol{x}^*, \boldsymbol{x} \rangle = 0$$

for complementary slackness is equivalent to the Fenchel equality

$$\delta_{\mathbf{R}^n_+}[\boldsymbol{\lambda}(\boldsymbol{x})] + \delta^*_{\mathbf{R}^n_+}[\boldsymbol{\lambda}(\boldsymbol{x}^*)] - \langle \boldsymbol{\lambda}(\boldsymbol{x}^*), \boldsymbol{\lambda}(\boldsymbol{x}) \rangle = 0$$

and $\boldsymbol{x}$ *and* $\boldsymbol{x}^*$ *have simultaneous ordered spectral decompositions.*

One can also interpret complementary slackness via the Jordan product.

Corollary 3 (Complementary Slackness via Jordan Product) *If both* $\boldsymbol{x}^*$ *and* $\boldsymbol{x}$ *are nonnegative, then* $\langle \boldsymbol{x}^*, \boldsymbol{x} \rangle = 0$ *if and only if* $\boldsymbol{x}^* \bullet \boldsymbol{x} = \mathbf{0}$

(Hertneck 1962).

By spectral conjugacy, the max function on $\mathbf{R}^n$ extends to the maximum-eigenvalue function on a Euclidean Jordan algebra $\mathbf{X}$. Previously we defined the set P on $\mathbf{R}^n$ as the nonnegative n-tuples summing to one. Its extreme points are the primitive idempotents, the n-tuples having a single component one and other components zero. Extend this set to $\mathbf{X}$ by redefining P as the nonnegative vectors having trace one. Its extreme points are the primitive idempotents.

By spectral conjugacy, the maximum-eigenvalue function

$$\max[\boldsymbol{x}^*] := \delta_P^*(\boldsymbol{x}^*) = \max[\boldsymbol{\lambda}(\boldsymbol{x}^*)] = \lambda_1^*.$$

Theorem 4 (max function Fenchel equality) *The Fenchel equality*

$$\delta_P(\boldsymbol{x}) + \delta_P^*(\boldsymbol{x}^*) - \langle \boldsymbol{x}^*, \boldsymbol{x} \rangle = 0$$

is equivalent to

$$\delta_P[\boldsymbol{\lambda}(\boldsymbol{x})] + \delta_P^*[\boldsymbol{\lambda}(\boldsymbol{x}^*)] - \langle \boldsymbol{\lambda}(\boldsymbol{x}^*), \boldsymbol{\lambda}(\boldsymbol{x}) \rangle = 0$$

and $\boldsymbol{x}$ and $\boldsymbol{x}^*$ have simultaneous ordered spectral decompositions.

The maximum is attained for the choice $\boldsymbol{x} = \lambda_1^* \boldsymbol{e}_1^*$.

The sum of the m largest eigenvalues function is analogous. In $\mathbf{R}^n$, an idempotent having trace m is an n-tuple having m components one and other components zero. The convex hull P_m of these idempotents has indicator

$$\delta_{P_m}(\boldsymbol{x}) = \delta_{\langle \mathbf{1}, \boldsymbol{x} \rangle = m} + \delta_{\mathbf{0} \le \boldsymbol{x} \le \mathbf{1}}. \tag{27.11}$$

Its extreme points are these idempotents, so the support $\delta_{P_m}^*(\boldsymbol{x}^*)$ is the sum of the m largest eigenvalues. Extend this set to $\mathbf{X}$ by redefining P_m by Eq. (27.11). Its extreme points are the idempotents having trace m. By spectral conjugacy,

$$\delta_{P_m}^*(\boldsymbol{x}^*) = \delta_{P_m}^*[\boldsymbol{\lambda}(\boldsymbol{x}^*)] = \sum_{i=1}^{m} \lambda_i^*.$$

The support is attained for the choice $\boldsymbol{x} = \sum_{i=1}^m \boldsymbol{e}_1^*$.

Definition 4 (Natural Logarithm of the Determinant) The concave function $(\ln\det)(\boldsymbol{x}) := \sum_i \ln \lambda_i$.

For $\boldsymbol{x} \in \mathbf{R}^n$, the eigenvalues are the components x_i. The concave conjugate

$$\begin{aligned}
(\ln\det)_*(\boldsymbol{x}^*) &= \inf_{\boldsymbol{x}} [\langle \boldsymbol{x}^*, \boldsymbol{x} \rangle - (\ln\det)(\boldsymbol{x})] \\
&= \sum_i \inf_{x_i} (x_i^* x_i - \ln x_i) \\
&= \sum_i \ln_*(x_i^*) \\
&= n + \sum_i \ln x_i^* \\
&= \langle \mathbf{1}, \mathbf{1} \rangle + (\ln\det)(\boldsymbol{x}^*).
\end{aligned}$$

The value is attained for $x_i = 1/x_i^*$.

This formula generalizes to any Euclidean Jordan algebra, via spectral conjugacy. The concave function $(\ln \det)(\boldsymbol{x})$ is the sum of the logarithm of the eigenvalues. The Fenchel equality

$$(\ln \det)(\boldsymbol{x}) + (\ln \det)_*(\boldsymbol{x}^*) - \langle \boldsymbol{x}^*, \boldsymbol{x} \rangle = 0$$

holds if and only if $\boldsymbol{x} \gg \mathbf{0}$ and $\boldsymbol{x}^* = \boldsymbol{x}^{-1}$. The Jordan product $\boldsymbol{x}^* \bullet \boldsymbol{x} = \mathbf{1}$, and the inner product $\langle \boldsymbol{x}^*, \boldsymbol{x} \rangle$ is the degree $\langle \mathbf{1}, \mathbf{1} \rangle$. The subdifferential

$$\partial (\ln \det)(\boldsymbol{x}) = \begin{cases} \{\boldsymbol{x}^{-1}\} & \text{if } \boldsymbol{x} \gg \mathbf{0} \\ \varnothing & \text{otherwise,} \end{cases}$$

so ln det is an increasing function.

Spectral conjugacy is widespread in applications. For example, the trace tr, the maximum-eigenvalue/weak-efficiency function δ_P^*, and the efficiency function $\delta_{\mathbf{X}_+ + \{\mathbf{1}\}}^*$ are all spectral. The trace is the sum of the eigenvalues, the support to $\{\mathbf{1}\}$. The weak-efficiency function is the support to the vectors having all nonnegative eigenvalues, such that the eigenvalues sum to one. The efficiency function is the support to the vectors having all eigenvalues greater than or equal to one.

References

Borwein JM, Lewis AS (2006) Convex analysis and nonlinear optimization: theory and examples, 2nd edn. Springer-Verlag, New York

Hardy GH, Littlewood JE, Pólya G (1952) Inequalities. Cambridge University Press, Cambridge

Hertneck C (1962) Positivitätsbereiche und Jordan-Strukturen. Math Annalen 146:433–455

Lewis AS (1996) Convex analysis on the Hermitian matrices. SIAM J Optim 6:164–177

Chapter 28
Quadratic Function

Abstract We define and analyze the quadratic function, a generalization of the quadratic $\frac{1}{2}\langle \boldsymbol{x}, \boldsymbol{x} \rangle$. The quadratic set having indicator $\delta_Q(\boldsymbol{x}, y) = \delta_{y \geq \frac{1}{2}\boldsymbol{x}^\top \boldsymbol{x}}$ is the epigraph of the quadratic $\frac{1}{2}\langle \boldsymbol{x}, \boldsymbol{x} \rangle$. The quadratic function $\delta_Q^*(\boldsymbol{x}^*, -y^*)$ is the epi-multiplication of the conjugate $\frac{1}{2}\langle \boldsymbol{x}^*, \boldsymbol{x}^* \rangle$ by y^*. We generalize this concept. Let $(A, S) \in \mathbf{T}^{\mathbf{X} \to \mathbf{Y}} \oplus \mathbf{S}^{\mathbf{X}}$. The quadratic set has indicator $\delta_Q(A, S) = \delta_{S \succeq \frac{1}{2}A^\top A}$. The quadratic function is the support $\delta_Q^*(A^*, -S^*)$, which we evaluate by perturbation duality. We interpret the Fenchel factor for the indicator of the quadratic set and its support via equivalent conditions. This book applies the quadratic function to least-squares linear regression and maximum-likelihood linear regression in econometrics.

We define and analyze the quadratic function, a generalization of the quadratic $\frac{1}{2}\langle \boldsymbol{x}, \boldsymbol{x} \rangle$. The quadratic set is a generalization of the epigraph of the quadratic. We define the quadratic function via the support of the quadratic set, and we solve for the quadratic function via perturbation duality. We interpret the Fenchel factor for the indicator of the quadratic set and its support via equivalent conditions.

28.1 Complementary Slackness

We further interpret complementary-slackness for the self-adjoint linear transformations $\mathbf{S}^{\mathbf{X}}$.

Theorem 1 (Complementary Slackness for $\mathbf{S}^{\mathbf{X}}$) *In the Euclidean Jordan algebra $\mathbf{S}^{\mathbf{X}}$, if both X^* and X are nonnegative, then the following three conditions are equivalent:*

- $\langle X^*, X \rangle = 0$;
- $X^* \bullet X = 0$;
- $X^* X = 0$.

B. C. Dieffenbach, *Conjugate Duality in Economic Analysis* , Contributions to Economics,
https://doi.org/10.1007/978-3-032-21396-9_28

"Spectral Conjugacy" presented the equivalence of the first two conditions. Expressed in terms of the orthonormal eigenvectors $\boldsymbol{x}_i$, complementary slackness requires that X and $-X^*$ have simultaneous ordered spectral decompositions

$$X = \lambda_1 \boldsymbol{x}_1 \boldsymbol{x}_1^\top + \cdots + \lambda_j \boldsymbol{x}_j \boldsymbol{x}_j^\top + 0\boldsymbol{x}_{j+1}\boldsymbol{x}_{j+1}^\top + \cdots + 0\boldsymbol{x}_n \boldsymbol{x}_n^\top$$
$$-X^* = 0\boldsymbol{x}_1 \boldsymbol{x}_1^\top + \cdots + 0\boldsymbol{x}_j \boldsymbol{x}_j^\top + \lambda_{j+1}\boldsymbol{x}_{j+1}\boldsymbol{x}_{j+1}^\top + \cdots + \lambda_n \boldsymbol{x}_n \boldsymbol{x}_n^\top,$$

such that $\lambda_i \geq 0$ for $i \leq j$ and $\lambda_i \leq 0$ for $i > j$, for some j. Taking the composition shows that the third condition is equivalent to the first two.

28.2 Schur Complement

By solving a quadratic minimization, we prove the Schur complement theorem, which provides necessary and sufficient conditions for a partitioned, self-adjoint linear transformation to be nonnegative.

Theorem 2 (Schur Complement) *For a partitioned, self-adjoint linear transformation on* $\mathbf{X} \oplus \mathbf{Y}$, *the nonnegativity*

$$\begin{bmatrix} A & B \\ B^\top & C \end{bmatrix} \geq 0 \tag{28.1}$$

is equivalent to the three conditions

$$A \geq 0 \tag{28.2}$$
$$C - B^\top A^+ B \geq 0 \tag{28.3}$$
$$\left(\mathrm{I} - AA^+\right) B = 0 \tag{28.4}$$

Albert (1969).

The expression $C - B^\top A^+ B$ is a "Schur complement" (Haynesworth 1968).

We show that the three conditions are necessary, by analyzing a convex minimization.

Problem 1 (Convex Minimization)

$$\inf_{x} \frac{1}{2} \begin{bmatrix} \boldsymbol{x} \\ \boldsymbol{y} \end{bmatrix} \begin{bmatrix} A & B \\ B^\top & C \end{bmatrix} \begin{bmatrix} \boldsymbol{x} \\ \boldsymbol{y} \end{bmatrix} = \frac{1}{2} \langle \boldsymbol{y}, C\boldsymbol{y} \rangle + \inf_{x} \left(\frac{1}{2} \langle \boldsymbol{x}, A\boldsymbol{x} \rangle + \langle B\boldsymbol{y}, \boldsymbol{x} \rangle \right).$$

Here $\boldsymbol{y}$ is arbitrary. Condition (28.2) is of course necessary for (28.1). We assume this condition, so the minimization is convex.

The minimum value is finite if only if the first-order condition holds:

$$\mathbf{0} = A\boldsymbol{x} + B\boldsymbol{y}$$

There is a solution only if $B\boldsymbol{y} \subset \mathrm{R}(A)$. If this condition fails for any $\boldsymbol{y}$, then the first-order condition does not hold, so the optimum value must be $-\infty$. Hence $\mathrm{R}(B) \subset \mathrm{R}(A)$ is necessary for (28.1). Equivalently, (28.4) holds.

Substituting the first-order condition, the optimum value of the minimization is

$$\begin{aligned}
&\frac{1}{2}\langle y, Cy\rangle + \frac{1}{2}\langle \boldsymbol{x}, A\boldsymbol{x}\rangle + \langle B\boldsymbol{y}, \boldsymbol{x}\rangle \\
&= \frac{1}{2}\langle y, Cy\rangle + \frac{1}{2}\langle \boldsymbol{x}, A\boldsymbol{x}\rangle + \langle -A\boldsymbol{x}, \boldsymbol{x}\rangle \\
&= \frac{1}{2}\langle y, Cy\rangle - \frac{1}{2}\langle \boldsymbol{x}, A\boldsymbol{x}\rangle \\
&= \frac{1}{2}\langle y, Cy\rangle - \frac{1}{2}\left\langle \left(-A^{+}B\boldsymbol{y}\right), A\left(-A^{+}B\boldsymbol{y}\right)\right\rangle \\
&= \frac{1}{2}\left\langle y, \left(C - B^{\mathsf{T}}A^{+}B\right)\boldsymbol{y}\right\rangle,
\end{aligned}$$

so condition (28.3) is necessary. Working backward shows that the three conditions are also sufficient.

28.3 Quadratic Function

The quadratic set having indicator $\delta_Q(\boldsymbol{x}, y) = \delta_{y \geq \frac{1}{2}\boldsymbol{x}^{\mathsf{T}}\boldsymbol{x}}$ is the epigraph of the quadratic $\frac{1}{2}\langle \boldsymbol{x}, \boldsymbol{x}\rangle$. The quadratic function $\delta_Q^*(\boldsymbol{x}^*, -y^*)$ is the epi-multiplication of the conjugate $\frac{1}{2}\langle \boldsymbol{x}^*, \boldsymbol{x}^*\rangle$ by y^*. We generalize this concept.

Definition 1 (Quadratic) Let $(A, S) \in \mathbf{T}^{\mathbf{X}\to\mathbf{Y}} \oplus \mathbf{S}^{\mathbf{X}}$. The quadratic set has indicator $\delta_Q(A, S) = \delta_{S \geq \frac{1}{2}A^{\mathsf{T}}A}$. The quadratic function is the support $\delta_Q^*\left(A^*, -S^*\right)$.

Because

$$\delta_Q(A, S) = \delta_{\mathbf{S}_+^{\mathbf{X}\oplus\mathbf{Y}}}\left(\begin{bmatrix} S & \frac{1}{2}A^{\mathsf{T}} \\ \frac{1}{2}A & \frac{1}{2}I \end{bmatrix}\right),$$

the quadratic set is the solution set of a linear inequality, a closed and convex set.

Theorem 3 (Quadratic Function) *The quadratic function*

$$\delta_Q^*\left(A^*, -S^*\right) = \frac{1}{2}\left\langle A^{*\mathsf{T}}, S^{*+}A^{*\mathsf{T}}\right\rangle + \delta_{A^*(I - S^{*+}S^*) = 0} + \delta_{S^* \geq 0} \tag{28.5}$$

There is no requirement that S^* be positive.

We evaluate the quadratic function by perturbation duality:

$$\delta_Q^*\left(A^*,-S^*\right)=\sup_{A,S}\left(\left\langle A^*,A\right\rangle+\left\langle -S^*,S\right\rangle-\delta_{\begin{bmatrix} S & \frac{1}{2}A^\mathsf{T} \\ \frac{1}{2}A & \frac{1}{2}\mathsf{I}\end{bmatrix}\geq 0}\right).$$

Rearranging the complementary slackness Fenchel inequality

$$\begin{aligned}&\delta_{\mathbb{S}_+^{\mathbb{X}\oplus\mathbb{Y}}}\left(\begin{bmatrix} S & \frac{1}{2}A^\mathsf{T} \\ \frac{1}{2}A & T\end{bmatrix}\right)+\delta^*_{\mathbb{S}_+^{\mathbb{X}\oplus\mathbb{Y}}}\left(\begin{bmatrix} -S^* & A^{*\mathsf{T}} \\ A^* & -T^*\end{bmatrix}\right)\\ &-\left\langle\begin{bmatrix} -S^* & A^{*\mathsf{T}} \\ A^* & -T^*\end{bmatrix},\begin{bmatrix} S & \frac{1}{2}A^\mathsf{T} \\ \frac{1}{2}A & T\end{bmatrix}\right\rangle\geq 0.\end{aligned} \tag{28.6}$$

obtains the primal and the reciprocal dual:

$$\left\langle T^*,T\right\rangle+\delta^*_{\mathbb{S}_+^{\mathbb{X}\oplus\mathbb{Y}}}\left(\begin{bmatrix} -S^* & A^{*\mathsf{T}} \\ A^* & -T^*\end{bmatrix}\right)\geq\left\langle A^*,A\right\rangle+\left\langle -S^*,S\right\rangle-\delta_{\mathbb{S}_+^{\mathbb{X}\oplus\mathbb{Y}}}\left(\begin{bmatrix} S & \frac{1}{2}A^\mathsf{T} \\ \frac{1}{2}A & T\end{bmatrix}\right).$$

The base value of the primal perturbation $T=\frac{1}{2}\mathsf{I}$. The primal satisfies the relative condition, so there is no duality gap.

The dual is

$$\inf_{T^*}\left[\left\langle T^*,\frac{1}{2}\mathsf{I}\right\rangle+\delta_{\mathbb{S}_+^{\mathbb{X}\oplus\mathbb{Y}}}\left(\begin{bmatrix} S^* & -A^{*\mathsf{T}} \\ -A^* & T^*\end{bmatrix}\right)\right].$$

By the Schur complement theorem, necessary and sufficient conditions for this partitioned linear transformation to be nonnegative are

$$\begin{aligned}S^*&\geq 0\\ A^*\left(\mathsf{I}-S^{*+}S^*\right)&=0\\ T^*&\geq A^*S^{*+}A^{*\mathsf{T}}.\end{aligned}$$

The choice $T^*=A^*S^{*+}A^{*\mathsf{T}}$ minimizes the dual, and the formula (28.5) follows.

28.4 Fenchel Equality

The following interprets the Fenchel equality for the quadratic function.

Theorem 4 (Quadratic Factor Interpretation) *The Fenchel equality*

$$\delta_Q(A,S) + \delta_Q^*(A^*,-S^*) - \langle A^*, A\rangle - \langle -S^*, S\rangle = 0$$

holds if and only if

$$A^* = AS^* \tag{28.7}$$

and the complementary slackness Fenchel equality

$$\delta_{\mathbb{S}_+^X}\left(S - \frac{1}{2}A^\top A\right) + \delta_{\mathbb{S}_+^X}^*(-S^*) - \left\langle -S^*, S - \frac{1}{2}A^\top A\right\rangle = 0 \tag{28.8}$$

holds.

If $S^* \gg \mathbf{0}$, then $S = \frac{1}{2}A^\top A$.

A necessary condition for the Fenchel factor (28.6) to be zero is that the composition of the two linear transformations be zero:

$$\begin{bmatrix} -S^* & A^{*\top} \\ A^* & -T^* \end{bmatrix}\begin{bmatrix} S & \frac{1}{2}A^\top \\ \frac{1}{2}A & \frac{1}{2}I \end{bmatrix} = \begin{bmatrix} -S^*S + \frac{1}{2}A^{*\top}A & -\frac{1}{2}S^*A^\top + \frac{1}{2}A^{*\top} \\ A^*S - \frac{1}{2}T^*A & \frac{1}{2}A^*A^\top - \frac{1}{2}T^* \end{bmatrix} = 0.$$

From the lower right,

$$T^* = A^*A^\top = AS^*A^\top = AS^*S^{*+}S^*A^\top = A^*S^{*+}A^{\top *}.$$

From the upper right,

$$A^* = AS^*,$$

condition (28.7). From the upper left,

$$0 = -S^*S + \frac{1}{2}A^{*\top}A = -S^*S + \frac{1}{2}S^*A^\top A = -S^*\left(S - \frac{1}{2}A^\top A\right),$$

and the complementary slackness (28.8) holds.

We show the converse. By (28.7),

$$A^*S^{*+}S^* = AS^*S^{*+}S^* = AS^* = A^*,$$

so the constraints in Eq. (28.5) hold. Substituting

$$\begin{aligned}\delta_Q^*\left(A^*,-S^*\right) &= \frac{1}{2}\operatorname{Tr} A^* S^{*+} A^{*\top} = \frac{1}{2}\operatorname{Tr} A S^* S^{*+} S^* A^\top = \frac{1}{2}\operatorname{Tr} A S^* A^\top \\ &= \frac{1}{2}\operatorname{Tr} S^* A^\top A = \frac{1}{2}\left\langle S^*, A^\top A\right\rangle\end{aligned}$$

gives

$$\begin{aligned}\delta_Q^*\left(A^*,S^*\right) - \langle A^*, A\rangle - \langle -S^*, S\rangle &= \frac{1}{2}\left\langle S^*, A^\top A\right\rangle - \langle AS^*, A\rangle - \langle -S^*, S\rangle \\ &= \left\langle S^*, S - \frac{1}{2}A^\top A\right\rangle = 0,\end{aligned}$$

as desired.

This book applies the quadratic function to least-squares linear regression and maximum-likelihood linear regression in econometrics.

References

Albert A (1969) Conditions for positive and nonnegative definiteness in terms of pseudoinverses. SIAM J Appl Math 17:434–440

Haynesworth EV (1968 Jan) Determination of the inertia of a partitioned Hermitian matrix. Linear Algebra Appl 1(1):73–81

Chapter 29
Lorentz Euclidean Jordan Algebra

Abstract A Lorentz Euclidean Jordan algebra is one of the five possible types of Euclidean Jordan algebra. In a Euclidean space $\mathbf{Z}$, select some vector $\mathbf{1}$ having norm one. Define a Jordan product to make $\mathbf{Z}$ a Euclidean Jordan algebra in which this vector is the unit. Define $\mathbf{X}$ as the orthogonal complement of $\mathbf{R}\{\mathbf{1}\}$, so $\mathbf{Z} = \mathbf{X} \oplus \mathbf{R}\{\mathbf{1}\}$. Express a vector z as a pair $(\boldsymbol{x}, y)$ in the direct sum, such that

$$y = \langle \mathbf{1}, z \rangle$$
$$\boldsymbol{x} = z - y\mathbf{1}.$$

The Lorentz Jordan product is

$$(\boldsymbol{x}^*, y^*) \bullet (\boldsymbol{x}, y) = (y\boldsymbol{x}^* + y^*\boldsymbol{x}, \langle \boldsymbol{x}^*, \boldsymbol{x} \rangle + y^* y).$$

Denote the Lorentz Euclidean Jordan algebra by $\mathbf{L}^{\mathbf{X}}$.

The nonnegative cone is the $(\boldsymbol{x}, y)$ such that $y \geq |\boldsymbol{x}|$, commonly called the "ice-cream" cone. There are an infinite number of primitive idempotents, and the nonnegative cone is not polyhedral.

The linear representation

$$L_{(\boldsymbol{x},y)}(\boldsymbol{x}^*, y^*) = (\boldsymbol{x}^*, y^*) \bullet (\boldsymbol{x}, y)$$

is the arrow-shaped partitioned linear transformation

$$\begin{bmatrix} \boldsymbol{x}^* \\ y^* \end{bmatrix} \mapsto \begin{bmatrix} y\mathsf{I} & \boldsymbol{x} \\ \boldsymbol{x}^\top & y \end{bmatrix} \begin{bmatrix} \boldsymbol{x}^* \\ y^* \end{bmatrix}.$$

Here $L_{(\boldsymbol{x},y)}$ is nonnegative if and only if $(\boldsymbol{x}, y)$ is nonnegative: all eigenvalues are nonnegative if and only if $y \geq |\boldsymbol{x}|$.

This book applies Lorentz Euclidean Jordan algebra to finance theory. In the vector $(\boldsymbol{x}, y)$, $\boldsymbol{x}$ is a portfolio, and the optimum y is standard deviation.

B. C. Dieffenbach, *Conjugate Duality in Economic Analysis*, Contributions to Economics,
https://doi.org/10.1007/978-3-032-21396-9_29

A Lorentz Euclidean Jordan algebra is one of the five possible types of Euclidean Jordan algebra. We define this algebra and develop its properties. The name honors the Dutch physicist Hendrik Lorentz, winner of the 1902 Nobel prize in physics. The Lorentz transformation is the basis for special relativity theory, and the light cone of special relativity theory is the cone of squares of a Lorentz Euclidean Jordan algebra.

29.1 Jordan Product

In a Euclidean space $\mathbf{Z}$, select some vector $\mathbf{1}$ having norm one. We define a Jordan product to make $\mathbf{Z}$ a Euclidean Jordan algebra in which this vector is the unit. Define $\mathbf{X}$ as the orthogonal complement of $\mathbf{R}\{\mathbf{1}\}$, so $\mathbf{Z} = \mathbf{X} \oplus \mathbf{R}\{\mathbf{1}\}$. One can express a vector z as a pair $(\boldsymbol{x}, y)$ in the direct sum, such that

$$\begin{aligned} y &= \langle \mathbf{1}, z \rangle \\ \boldsymbol{x} &= z - y\mathbf{1}. \end{aligned}$$

The unit $\mathbf{1} = (\mathbf{0}, 1)$. Denote the Lorentz Euclidean Jordan algebra by $\mathbf{L}^{\mathbf{X}}$. If $\mathbf{X} = \mathbf{R}^n$, use the notation $\mathbf{L}^n$.

To define the square of a vector sets the Jordan product.

Definition 1 (Lorentz Jordan Product Square) The Lorentz Jordan product square is

$$(\boldsymbol{x}, y) \bullet (\boldsymbol{x}, y) := \left(2y\boldsymbol{x}, \langle \boldsymbol{x}, \boldsymbol{x} \rangle + y^2\right). \tag{29.1}$$

The Lorentz Jordan product

$$\begin{aligned} &(\boldsymbol{x}^*, y^*) \bullet (\boldsymbol{x}, y) \\ &= \frac{1}{2}\{[(\boldsymbol{x}^*, y^*) + (\boldsymbol{x}, y)] \bullet [(\boldsymbol{x}^*, y^*) + (\boldsymbol{x}, y)] - (\boldsymbol{x}^*, y^*) \bullet (\boldsymbol{x}^*, y^*) - (\boldsymbol{x}, y) \bullet (\boldsymbol{x}, y)\} \\ &= \frac{1}{2}\Big\{\Big[2\,(y^* + y)\,(\boldsymbol{x}^* + \boldsymbol{x})\,, \langle \boldsymbol{x}^* + \boldsymbol{x}, \boldsymbol{x}^* + \boldsymbol{x} \rangle + (y^* + y)^2\Big] \\ &- \left(2y^*\boldsymbol{x}^*, \langle \boldsymbol{x}^*, \boldsymbol{x}^* \rangle + y^{*2}\right) - \left(2y\boldsymbol{x}, \langle \boldsymbol{x}, \boldsymbol{x} \rangle + y^2\right)\Big\} \\ &= (y\boldsymbol{x}^* + y^*\boldsymbol{x}, \langle \boldsymbol{x}^*, \boldsymbol{x} \rangle + y^*y). \end{aligned}$$

Theorem 1 (Lorentz Jordan Product) *The Lorentz Jordan product is*

$$(\boldsymbol{x}^*, y^*) \bullet (\boldsymbol{x}, y) = (y\boldsymbol{x}^* + y^*\boldsymbol{x}, \langle \boldsymbol{x}^*, \boldsymbol{x} \rangle + y^*y). \tag{29.2}$$

An idempotent satisfies $(\boldsymbol{x}, y) = (\boldsymbol{x}, y) \bullet (\boldsymbol{x}, y) := \left(2y\boldsymbol{x}, \langle \boldsymbol{x}, \boldsymbol{x} \rangle + y^2\right)$. Of course $\mathbf{1}$ is an idempotent: if $\boldsymbol{x} = \mathbf{0}$, then $y = 1$. The primitive idempotents are the idempotents for $\boldsymbol{x} \neq \mathbf{0}$: the vectors $\frac{1}{2}(\boldsymbol{x}, 1)$ such that $|\boldsymbol{x}| = 1$. Orthogonal primitive idempotents come in pairs: $\frac{1}{2}(\boldsymbol{x}, 1)$ and $\frac{1}{2}(-\boldsymbol{x}, 1) = \mathbf{1} - \frac{1}{2}(\boldsymbol{x}, 1)$, for $|\boldsymbol{x}| = 1$. The algebra has degree two, and a Jordan frame is any such pair.

Theorem 2 (Spectral Decomposition) *The Euclidean Jordan algebra* $\mathbf{L}^{\mathbf{X}}$ *has degree two. For* $\boldsymbol{x} \neq \mathbf{0}$*,*

$$(\boldsymbol{x}, y) = (y + |\boldsymbol{x}|)\left[\frac{1}{2}\left(\frac{\boldsymbol{x}}{|\boldsymbol{x}|}, 1\right)\right] + (y - |\boldsymbol{x}|)\left[\frac{1}{2}\left(-\frac{\boldsymbol{x}}{|\boldsymbol{x}|}, 1\right)\right].$$

The eigenvalues are $y + |\boldsymbol{x}|$ *and* $y - |\boldsymbol{x}|$*.*

For scalar multiples of the unit,

$$(\mathbf{0}, y) = y\left[\frac{1}{2}(\boldsymbol{x}, 1)\right] + y\left[\frac{1}{2}(-\boldsymbol{x}, 1)\right], \text{ for any } |\boldsymbol{x}| = 1.$$

Both eigenvalues are y*.*

Using the spectral decomposition, one can readily verify that the axioms for a Euclidean Jordan algebra hold.

The nonnegative cone is the $(\boldsymbol{x}, y)$ such that $y \geq |\boldsymbol{x}|$, commonly called the "ice-cream" cone. The primitive idempotents generate the nonnegative cone. Orthogonal primitive idempotents lie exactly opposite on its boundary. Of course $\mathbf{L}^0 = \mathbf{R}$. The Euclidean Jordan algebra $\mathbf{L}^1$ is isomorphic to the Euclidean Jordan algebra $\mathbf{R} \boxplus \mathbf{R}$. For $n > 1$, $\mathbf{L}^n$ is simple. There are an infinite number of primitive idempotents, and the nonnegative cone is not polyhedral.

The trace inner product defined by Jordan et al. (1934) on $\mathbf{L}^{\mathbf{X}}$ is twice the native inner product in the Euclidean space $\mathbf{Z}$. As $\operatorname{tr}(\boldsymbol{x}, y) = 2y$, the sum of the eigenvalues, the trace inner product formula follows from (29.2).

Theorem 3 (Trace Inner Product) *The trace inner product in the Euclidean Jordan algebra* $\mathbf{L}^{\mathbf{X}}$ *is*

$$\langle(\boldsymbol{x}^*, y^*), (\boldsymbol{x}, y)\rangle := \operatorname{tr}[(\boldsymbol{x}^*, y^*) \bullet (\boldsymbol{x}, y)] = 2[\langle \boldsymbol{x}^*, \boldsymbol{x}\rangle + y^* y],$$

In applications later in this book, we employ the native inner product, not the trace inner product.

29.2 Linear Representation

We relate the trace inner product to the linear representation. The linear representation

$$L_{(\boldsymbol{x},y)}(\boldsymbol{x}^*, y^*) = (\boldsymbol{x}^*, y^*) \bullet (\boldsymbol{x}, y)$$

is the arrow-shaped partitioned linear transformation

$$\begin{bmatrix} \boldsymbol{x}^* \\ y^* \end{bmatrix} \mapsto \begin{bmatrix} y\mathsf{I} & \boldsymbol{x} \\ \boldsymbol{x}^\top & y \end{bmatrix} \begin{bmatrix} \boldsymbol{x}^* \\ y^* \end{bmatrix}. \tag{29.3}$$

Here

$$\operatorname{Tr} L_{(\boldsymbol{x},y)} = y \dim \mathbf{Z}.$$

For a primitive idempotent $\boldsymbol{e}$,

$$\operatorname{Tr} L_{\boldsymbol{e}} = \frac{1}{2} \dim \mathbf{Z},$$

the same for any primitive idempotent. The definition

$$\operatorname{tr}(\boldsymbol{x}, y) := \operatorname{Tr} L_{(\boldsymbol{x},y)}/\operatorname{Tr} L_{\boldsymbol{e}} = 2y$$

is indeed the sum of the eigenvalues. Under the trace inner product, the linear representation (29.3) is self-adjoint.

We verify that $L_{(\boldsymbol{x},y)}$ is nonnegative if and only if $(\boldsymbol{x}, y)$ is nonnegative: all eigenvalues are nonnegative if and only if $y \geq |\boldsymbol{x}|$. The linear representation has one eigenvalue $y + |\boldsymbol{x}|$, one eigenvalue $y - |\boldsymbol{x}|$, and other eigenvalues y. For a primitive idempotent, one eigenvalue is 1, one eigenvalue is 0, and other eigenvalues $\frac{1}{2}$.

According to the positive-trace theorem for a Euclidean Jordan algebra, if the product of two idempotents is not zero, then the product has a positive trace. We verify this relationship for the Lorentz Euclidean Jordan algebra. Consider primitive idempotents $\frac{1}{2}(\boldsymbol{x}^*, 1)$ and $\frac{1}{2}(\boldsymbol{x}, 1)$, such that $|\boldsymbol{x}^*| = |\boldsymbol{x}| = 1$. Their product $\frac{1}{2}(\boldsymbol{x}^* + \boldsymbol{x}, \langle \boldsymbol{x}^*, \boldsymbol{x}\rangle + 1)$ is zero if and only if $\boldsymbol{x}^* = -\boldsymbol{x}$. Otherwise the trace $\langle \boldsymbol{x}^*, \boldsymbol{x}\rangle + 1$ is positive, by the Cauchy-Schwarz inequality.

29.3 Application

To interpret the meaning of complementary slackness is important for applications. For nonnegative $(\boldsymbol{x}^*, y^*)$ and $(\boldsymbol{x}, y)$, the complementary-slackness theorem says $\langle (\boldsymbol{x}^*, y^*), (\boldsymbol{x}, y)\rangle = 0$ if and only if $(\boldsymbol{x}^*, y^*) \bullet (\boldsymbol{x}, y) = \mathbf{0}$. By spectral conjugacy, $(\boldsymbol{x}, y)$ and $-(\boldsymbol{x}^*, y^*)$ must have a simultaneous ordered spectral decomposition. Following readily from the spectral decomposition, this summary describes the situation for zero, one, and two positive eigenvalues.

Summary 1 (Complementary Slackness)

Zero Positive Eigenvalues If $(\boldsymbol{x}, y) = \mathbf{0}$, then $(\boldsymbol{x}^*, y^*)$ can be any nonnegative vector.
One Positive Eigenvalue If $(\boldsymbol{x}, y) \succ \mathbf{0}$ but not positive, then

$$\begin{aligned} |\boldsymbol{x}| &= y > 0 \\ |\boldsymbol{x}^*| &= y^* \\ \boldsymbol{x}^* &= -y^* \boldsymbol{x}/y. \end{aligned}$$

Geometrically, $(\boldsymbol{x}, y)$ and $(\boldsymbol{x}^*, y^*)$ are opposite boundary elements of the nonnegative cone.
Two Positive Eigenvalues If $(\boldsymbol{x}, y) \gg \mathbf{0}$, then

$$\begin{aligned} y - |\boldsymbol{x}| &> 0 \\ (\boldsymbol{x}^*, y^*) &= \mathbf{0}. \end{aligned}$$

The one-positive-eigenvalue case is key in applications.

The $(\boldsymbol{x}, y)$-decomposition is valuable in applications. Setting an optimization in $\mathbf{L}^{\mathbf{X}}$ may reduce it to a nonpolyhedral linear problem. The norm

$$|\boldsymbol{x}| = \inf_y \left[y + \delta_{\mathbf{L}_+^{\mathbf{X}}} (\boldsymbol{x}, y) \right],$$

a linear minimization. As one can rearrange the inequality $\langle \boldsymbol{x}, \boldsymbol{x} \rangle \leq y$ as $\langle \boldsymbol{x}, \boldsymbol{x} \rangle + \left(\frac{y-1}{2}\right)^2 \leq \left(\frac{y+1}{2}\right)^2$, the quadratic

$$\langle \boldsymbol{x}, \boldsymbol{x} \rangle = \inf_y \left\{ y + \delta_{\mathbf{L}_+^{\mathbf{X} \oplus \mathbf{R}}} \left[\boldsymbol{x}, (y-1)/2, (y+1)/2 \right] \right\}.$$

More generally, setting a quadratic optimization having quadratic constraints in $\mathbf{L}^{\mathbf{X}}$ obtains a nonpolyhedral linear problem.

This book applies Lorentz Euclidean Jordan algebra to finance theory. In the vector $(\boldsymbol{x}, y)$, $\boldsymbol{x}$ is a portfolio, and the optimum y is standard deviation.

References

Alizadeh F, Goldfarb D (2003) Second-order cone programming. Math Prog Ser B 95:3–51

Jordan P, von Neumann J, Wigner E (1934 Jan) On an algebraic generalization of the quantum mechanical formalism. Ann Math 35(1):29–64

Lobo MS, Vandenberghe L, Boyd S, Lebret H (1998 Nov) Applications of second-order cone programming. Linear Algebra Appl 284(1–3):193–228

Chapter 30
Hermitian Euclidean Jordan Algebra

Abstract The complexification of a Euclidean space **X** is its natural extension to a complex vector space having a complex-valued inner product. The complexification expresses a vector as the sum of its real and complex parts,

$$\boldsymbol{x} = \boldsymbol{y} + \boldsymbol{z}\mathrm{i},$$

in which $\boldsymbol{y}$ and $\boldsymbol{z}$ belong to **X**. The complex inner product in the complexification is

$$\langle \boldsymbol{x}^*, \boldsymbol{x} \rangle = \langle \boldsymbol{y}^* + \boldsymbol{z}^*\mathrm{i}, \boldsymbol{y} + \boldsymbol{z}\mathrm{i} \rangle := \left(\boldsymbol{y}^{*\mathsf{T}} - \boldsymbol{z}^{*\mathsf{T}}\mathrm{i} \right) (\boldsymbol{y} + \boldsymbol{z}\mathrm{i}).$$

The concept of the adjoint carries over to the complexification: the square linear transformation $A + B\mathrm{i}$ is self-adjoint if and only if A is self-adjoint and B is skew. A simple Euclidean Jordan algebra, the Hermitian Euclidean Jordan algebra $\mathbf{H}^{\mathbf{X}}$ is the real vector space of self-adjoint linear transformations on the complexification. The Jordan product is

$$X^* \bullet X = \frac{1}{2} \left(X^*X + XX^* \right).$$

Using the spectral decomposition, one can verify that $\mathbf{H}^{\mathbf{X}}$ satisfies the axioms for a Euclidean Jordan algebra.

In addition to the simple Euclidean Jordan algebras $\mathbf{S}^{\mathbf{X}}$ and $\mathbf{L}^{\mathbf{X}}$, a third type of simple Euclidean Jordan algebra is the Hermitian Euclidean Jordan algebra $\mathbf{H}^{\mathbf{X}}$.[1] See $\mathbf{H}^{\mathbf{X}}$ as like $\mathbf{S}^{\mathbf{X}}$, but slightly different. Consequently, the presentation below of $\mathbf{H}^{\mathbf{X}}$ is somewhat cursory.

[1] Charles Hermite, French mathematician, 1822–1901.

B. C. Dieffenbach, *Conjugate Duality in Economic Analysis*, Contributions to Economics,
https://doi.org/10.1007/978-3-032-21396-9_30

30.1 Complexification

The complexification of a Euclidean space $\mathbf{X}$ is its natural extension to a complex vector space having a complex-valued inner product. The complexification expresses a vector as the sum of its real and complex parts,

$$\boldsymbol{x} = \boldsymbol{y} + \boldsymbol{z}\mathrm{i},$$

in which $\boldsymbol{y}$ and $\boldsymbol{z}$ belong to $\mathbf{X}$. Addition of vectors and multiplication by a complex scalar follow the standard rules of complex arithmetic. For a linear transformation A on $\mathbf{X}$, define $A\boldsymbol{x} := A\boldsymbol{y} + A\boldsymbol{z}\mathrm{i}$.

The inner product in $\mathbf{X}$ applied to the real and complex parts sets the inner product in the complexification.

Definition 1 (Complexification Inner Product) The complex inner product in the complexification is

$$\begin{aligned}\langle \boldsymbol{x}^*, \boldsymbol{x}\rangle = \langle \boldsymbol{y}^* + \boldsymbol{z}^*\mathrm{i}, \boldsymbol{y} + \boldsymbol{z}\mathrm{i}\rangle &:= \left(\boldsymbol{y}^{*\top} - \boldsymbol{z}^{*\top}\mathrm{i}\right)(\boldsymbol{y} + \boldsymbol{z}\mathrm{i}) \\ &= (\langle \boldsymbol{y}^*, \boldsymbol{y}\rangle + \langle \boldsymbol{z}^*, \boldsymbol{z}\rangle) + (\langle \boldsymbol{y}^*, \boldsymbol{z}\rangle - \langle \boldsymbol{z}^*, \boldsymbol{y}\rangle)\,\mathrm{i}.\end{aligned}$$

The inner product is not symmetric: $\langle \boldsymbol{x}, \boldsymbol{x}^*\rangle$ is the complex conjugate of $\langle \boldsymbol{x}^*, \boldsymbol{x}\rangle$. For nonzero $\boldsymbol{x}$, the quadratic $\langle \boldsymbol{x}, \boldsymbol{x}\rangle$ is real and positive. For $\mathbf{X} = \mathbf{R}$ the complexification reduces to the complex numbers, having inner product

$$\langle a^* + b^*\mathrm{i}, a + b\mathrm{i}\rangle = (a^* - b^*\mathrm{i})\,(a + b\mathrm{i}).$$

Seen as a complex vector space, the complexification has dimension $\dim \mathbf{X}$.

The concept of the adjoint carries over to the complexification. Define the linear transformation $\boldsymbol{x}$

$$a + b\mathrm{i} \mapsto \boldsymbol{x}\,(a + b\mathrm{i})$$

from the complex numbers to the complexification of $\mathbf{X}$. Its adjoint $\boldsymbol{x}^\top$ is

$$\boldsymbol{x}^* \mapsto \langle \boldsymbol{x}, \boldsymbol{x}^*\rangle = \left(\boldsymbol{y}^\top - \boldsymbol{z}^\top\mathrm{i}\right)\boldsymbol{x}^*.$$

We verify the adjoint condition:

$$\langle \boldsymbol{x}^*, \boldsymbol{x}\,(a + b\mathrm{i})\rangle = \langle \boldsymbol{x}^*, \boldsymbol{x}\rangle\,(a + b\mathrm{i}) = \overline{\langle \boldsymbol{x}, \boldsymbol{x}^*\rangle}\,(a + b\mathrm{i}) = \langle \langle \boldsymbol{x}, \boldsymbol{x}^*\rangle\,, a + b\mathrm{i}\rangle.$$

Consider the adjoint of a square linear transformation.

Theorem 1 (Square Linear Transformation Adjoint) *For A and B in $\mathbf{T}^{\mathbf{X}}$,*

$$\boldsymbol{x} \mapsto (A + B\mathrm{i})\boldsymbol{x}$$

is a square linear transformation on the complexification of **X**, *with adjoint*

$$x^* \mapsto \left(A^\top - B^\top \mathrm{i}\right) x^*.$$

We verify the adjoint condition:

$$\begin{aligned}
\langle x^*, (A + B\mathrm{i})x \rangle &= (y^{*\top} - z^{*\top}\mathrm{i})(A + B\mathrm{i})(y + z\mathrm{i}) \\
&= \left[(y^{*\top}A + z^{*\top}B) + (y^{*\top}B - z^{*\top}A)\,\mathrm{i}\right](y + z\mathrm{i}) \\
&= \left\langle \left(A^\top y^* + B^\top z^*\right) - \left(B^\top y^* - A^\top z^*\right)\mathrm{i}, y + z\mathrm{i} \right\rangle \\
&= \left\langle \left(A^\top - B^\top \mathrm{i}\right)(y^* + z^*\mathrm{i}), y + z\mathrm{i} \right\rangle \\
&= \left\langle \left(A^\top - B^\top \mathrm{i}\right)x^*, x \right\rangle.
\end{aligned}$$

Theorem 2 (Self-Adjoint Linear Transformation) *The square linear transformation* $A + B\mathrm{i}$ *is self-adjoint if and only if* A *is self-adjoint and* B *is skew:*

$$\begin{aligned}
A^\top &= A \\
B^\top &= -B.
\end{aligned}$$

30.2 Hermitian Euclidean Jordan Algebra

Definition 2 (Hermitian Euclidean Jordan Algebra) The Hermitian Euclidean Jordan algebra $\mathbf{H}^{\mathbf{X}}$ is the real vector space of self-adjoint linear transformations on the complexification of **X**, with squared Jordan product

$$X \bullet X = XX.$$

From the squared Jordan product, one obtains the Jordan product

$$X^* \bullet X = \frac{1}{2}\left(X^*X + XX^*\right). \tag{30.1}$$

The Jordan unit is the identity transformation $\mathrm{I} + 0\mathrm{i}$.

The eigenvectors and eigenvalues of $\mathbf{H}^{\mathbf{X}}$ resemble those on $\mathbf{S}^{\mathbf{X}}$.

Propostion 1 (Eigenvectors and Eigenvalues) $X \in \mathbf{H}^{\mathbf{X}}$ *has* $\dim \mathbf{X}$ *orthonormal eigenvectors* x_i, *with real eigenvalues* λ_i. *These eigenvectors constitute a basis for the complexification of* **X**.

Apart from sign, the eigenvectors are unique if there are no repeated eigenvalues. A primitive idempotent has the form $\boldsymbol{x}\boldsymbol{x}^{\top}$, $|\boldsymbol{x}| = 1$. Using the eigenvectors and eigenvalues, the spectral decomposition expresses X as a linear combination of orthogonal primitive idempotents,

$$X = \sum_i \lambda_i \boldsymbol{x}_i \boldsymbol{x}_i^{\top}.$$

The degree of $\mathbf{H}^{\mathbf{X}}$ is dim $\mathbf{X}$. The trace inner product on $\mathbf{H}^{\mathbf{X}}$ is

$$\langle X^*, X \rangle := \operatorname{tr}\left(X^* \bullet X\right) = \operatorname{Tr}\left[\frac{1}{2}\left(X^* X + X X^*\right)\right] = \operatorname{Tr}\left(X^* X\right).$$

Using the spectral decomposition, one can verify that $\mathbf{H}^{\mathbf{X}}$ satisfies the axioms for a Euclidean Jordan algebra.

30.3 Classification

There are five distinct types of simple Euclidean Jordan algebra (Jordan et al. 1934). Any Euclidean Jordan algebra is a unique direct sum of simple Euclidean Jordan algebras. This book applies the three types $\mathbf{L}^{\mathbf{X}}$, $\mathbf{S}^{\mathbf{X}}$, and $\mathbf{H}^{\mathbf{X}}$. See the two remaining types as extensions of $\mathbf{S}^{\mathbf{X}}$ and $\mathbf{H}^{\mathbf{X}}$.

The complex numbers **C** form a two-dimensional real algebra, and see **C** as a doubling of the real numbers **R**. “Cayley-Dickson doubling” extends this idea (Dickson 1923; Baez 2002, Sect. 2.2; Faraut and Korányi 1994, p. 84). Doubling the complex numbers **C** obtains the four-dimensional real algebra of quaternions **Q**. Just as **R** is a subalgebra of **C**, **C** is a subalgebra of **Q**. Just as the complex numbers contain the concept of a complex conjugate, such that the conjugate of the conjugate is the original, the quaternions contain the concept of a quaternion conjugate, such that the conjugate of the conjugate is the original. In the same way, doubling the quaternions obtains the octonians **O**, an eight-dimensional real algebra.

The four algebras **R**, **C**, **Q**, and **O** are famous in the history of mathematics. In the 19th century, mathematicians constructed **Q** and **O** as generalizations of the complex numbers (Baez 2002).[2] Today mathematicians refer to these four real algebras as the “Euclidean Hurwitz algebras” (Faraut and Korányi 1994, Sect. V-1). The table summarizes their properties.

---

[2] The quaternions succinctly describe the geometry of three-dimensional space. Computations with quaternions are useful for spacecraft control and in video-game programming (Kuipers 2002). The Gibbs-Heaviside vector analysis in physics is isomorphic to the quaternion algebra. Baez discusses the octonions at length (Baez 2002), and also see Conway and Smith (2003).

Algebra	Dimension	Commutative?	Associative?
R	1	Yes	Yes
C	2	Yes	Yes
Q	4	No	Yes
O	8	No	No

The Hermitian Euclidean Jordan algebra $\mathbf{H}^n$ is the $n \times n$ Hermitian matrices X with complex components X_{jk}, such that the transpose is the complex conjugate, $X_{kj} = \overline{X_{jk}}$. One obtains the quaternion Euclidean Jordan algebra $\mathbf{Q}^n$ as the $n \times n$ matrices X with quaternion components X_{jk}, such that the transpose is the quaternion conjugate, $X_{kj} = \overline{X_{jk}}$. The definition of the Jordan product is again (30.1), except that one carries out the matrix multiplication via quaternion multiplication. In the same way, one obtains the octonian Euclidean Jordan algebra $\mathbf{O}^3$ as the 3×3 matrices X with octonian components X_{jk}, such that the transpose is the octonian conjugate, $X_{kj} = \overline{X_{jk}}$. The definition of the Jordan product is again (30.1), except that one carries out the matrix multiplication via octonian multiplication. The restriction to 3×3 matrices is needed to satisfy the axioms for a Euclidean Jordan algebra. For any other size of matrix, the Jordan product definition (30.1) is not commutative.

The table summarizes the dimension and the degree (Faraut and Korányi 1994, p. 97):

	Dimension	Degree
$\mathbf{L}^n$	$n+1$	2
$\mathbf{S}^n$	$\frac{1}{2}n(n+1)$	n
$\mathbf{H}^n$	n^2	n
$\mathbf{Q}^n$	$n(2n-1)$	n
$\mathbf{O}^3$	27	3

For $\mathbf{H}^n$, each diagonal element contributes 1 to the dimension and each element above the diagonal contributes 2, so the dimension is $1 \times n + 2 \times \frac{1}{2}n(n-1) = n^2$. For $\mathbf{Q}^n$, each diagonal element contributes 1 to the dimension and each element above the diagonal contributes 4, so the dimension is $1 \times n + 4 \times \frac{1}{2}n(n-1) = n(2n-1)$. For $\mathbf{O}^3$, each diagonal element contributes 1 to the dimension and each element above the diagonal contributes 8, so the dimension is $1 \times 3 + 8 \times 3 = 27$. In this book, there is no application of either $\mathbf{Q}^n$ or $\mathbf{O}^3$.

References

Baez JC (2002 Apr) The octonions. Bull Am Math Soc 39(2):145–206

Conway JH, Smith DA (2003) On quaternions and octonions: their geometry, arithmetic, and symmetry. A. K. Peters, Ltd., Natick, MA

Dickson LE (1923) Algebras and their arithmetics. University of Chicago Press, Chicago

Faraut J, Korányi A (1994) Analysis on symmetric cones. Clarendon Press, Oxford University Press, Oxford
Jordan P, Neumann JV, Wigner E (1934 Jan) On an algebraic generalization of the quantum mechanical formalism. Ann Math 35(1):29–64
Kuipers JB (2002) Quaternions and rotation sequences. Princeton University Press, Princeton

Part VII
Mathematical Programming

Chapter 31
Hahn–Banach

Abstract Using perturbation duality, we prove the Hahn–Banach theorem in Euclidean space. The theorem generalizes basic separation. The primal models a simple condition: for a convex set and an affine set, either the sets have a vector in common (optimum value zero) or they do not (optimum value infinity). For the latter condition, one obtains the Hahn–Banach separation theorem. For the former condition, one obtains the Hahn–Banach extension theorem. That the same primal/dual pair obtains both theorems reveals their intrinsic interconnection. If the two sets have no vector in common, one obtains the Hahn–Banach separation theorem: there exists a hyperplane containing the affine set such the convex set lies entirely on one side of the hyperplane. A value for the dual choice variable defines the hyperplane: select it so the value of the objective function is positive. If the two sets have a vector in common, then the optimum value is zero. For the optimum value of the dual to be zero is the defining condition for the Hahn–Banach extension theorem: there exists a linear function defined on a subspace such that a certain property holds. The theorem asserts that one can extend this linear function to the entire space, such that this same property holds. The perturbation variable for the primal defines the linear function on the subspace. Adding any solution to the primal to the perturbation variable extends the linear function to the entire space.

To obtain the Hahn–Banach theorem in this way is a superb example of how duality enables the discovery of new results. The primal/dual optimization leads naturally to both the separation theorem and the extension theorem. Whether the optimum value is zero or infinity implies the defining condition for each theorem. Then a calculation—solving the primal/dual pair—arrives at both the conclusion of the theorem and a proof of it. That the same primal/dual pair obtains both theorems reveals their intrinsic interconnection.

B. C. Dieffenbach, *Conjugate Duality in Economic Analysis*, Contributions to Economics,
https://doi.org/10.1007/978-3-032-21396-9_31

31.1 Hahn–Banach Primal and Dual

In a Euclidean space, let C denote a proper, closed, convex set. Let M denote a subspace. The affine set $M^\perp + \{y\}$ is the orthogonal complement translated by $\boldsymbol{y}$. The Hahn–Banach optimization models whether the convex set and the affine set have a vector in common:

$$\inf_z \left[\delta_C(z) + \delta_{M^\perp+\{y\}}(z)\right].$$

The optimum value is either zero or infinity: either the two sets have a vector in common, or they do not. The perturbation variable is $\boldsymbol{y}$, and its interesting base values are $\boldsymbol{y} = \boldsymbol{0}$ and $\boldsymbol{y} \in M$.

Transform the problem by the change of variables $\boldsymbol{x} := z - \boldsymbol{y}$.

Problem 1 (Hahn–Banach Primal)

$$\inf_{\boldsymbol{x}} [\delta_C(\boldsymbol{x}+\boldsymbol{y}) + \delta_{M^\perp}(\boldsymbol{x})].$$

Adding the Fenchel inequalities

$$\delta_C(\boldsymbol{x}+\boldsymbol{y}) + \delta_C^*(\boldsymbol{y}^*) - \langle \boldsymbol{y}^*, \boldsymbol{x}+\boldsymbol{y}\rangle \geq 0 \quad (31.1)$$

$$\delta_M(\boldsymbol{y}^* + \boldsymbol{x}^*) + \delta_M^*(-\boldsymbol{x}) - \langle \boldsymbol{y}^*, -\boldsymbol{x}\rangle \geq 0 \quad (31.2)$$

obtains the primal and the dual

$$\delta_C(\boldsymbol{x}+\boldsymbol{y}) + \delta_M^*(-\boldsymbol{x}) \geq \langle \boldsymbol{y}^*, \boldsymbol{y}\rangle - \delta_C^*(\boldsymbol{y}^*) - \delta_M(\boldsymbol{y}^*).$$

Problem 2 (Hahn–Banach Dual)

$$\sup_{\boldsymbol{y}^*} [\langle \boldsymbol{y}^*, \boldsymbol{y}\rangle - \delta_C^*(\boldsymbol{y}^*) - \delta_M(\boldsymbol{y}^*)].$$

31.2 Hahn–Banach Separation

Hahn–Banach separation is the case in which C and $M^\perp$ have no common vector. Set the base value $\boldsymbol{y} = \boldsymbol{0}$. There is no duality gap. Having no common vector means that optimum value of the primal is ∞. Furthermore, the projection of C on M cannot include the origin: expressing any vector in C in the direct sum $M \oplus M^\perp$ necessarily has a nonzero component in M, for otherwise the vector would lie in $M^\perp$.

By basic separation, there exists some $\boldsymbol{y}^* \in M$ such that $\delta_C^*(\boldsymbol{y}^*) < 0$. The objective function in the dual is positively homogeneous in $\boldsymbol{y}^*$, so the optimum value of

Fig. 31.1 No common vector

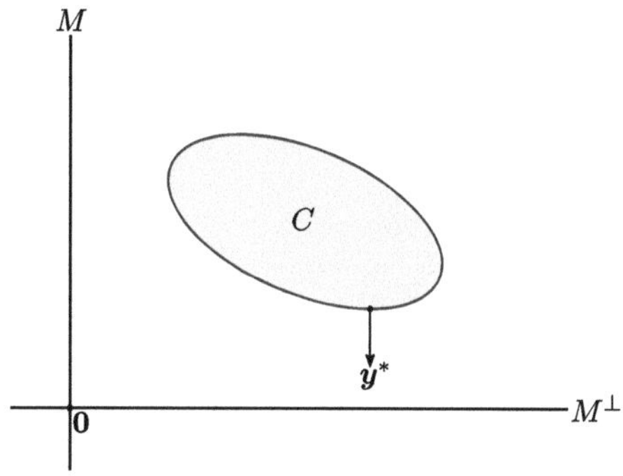

the dual is also ∞. As $\boldsymbol{y}^* \in M$, also $0 = \delta_M(\boldsymbol{y}^*) = \delta^*_{M^\perp}(\boldsymbol{y}^*)$. Thus $\boldsymbol{y}^*$ provides a separating hyperplane.

Theorem 1 (Hahn–Banach Separation) *If C and $M^\perp$ have no common vector, then there exists a hyperplane containing $M^\perp$ such that C lies on one side of the hyperplane: there exists some $\boldsymbol{y}^*$ such that*

$$\langle \boldsymbol{y}^*, \boldsymbol{x}\rangle = 0 \textit{ for all } \boldsymbol{x} \in M^\perp$$
$$\langle \boldsymbol{y}^*, \boldsymbol{x}\rangle < 0 \textit{ for all } \boldsymbol{x} \in C.$$

The case $M^\perp = \{\boldsymbol{0}\}$ is basic separation.

Figure 31.1 illustrates Hahn–Banach separation. At $\boldsymbol{y}^* \in M$, the support $\delta^*_C(\boldsymbol{y}^*) < 0$. Furthermore $\langle \boldsymbol{y}^*, \boldsymbol{x}\rangle = 0$ for any $\boldsymbol{x} \in M^\perp$.

31.3 Hahn–Banach Extension

Hahn–Banach extension is the case in which C and $M^\perp$ have a vector in common. Assume that the base value $\boldsymbol{y} \in M$. The optimum value of the primal and the dual is zero. There is no separating hyperplane.

There is no duality gap. Having a common vector means that the optimum value of the primal is zero. The choice $\boldsymbol{y}^* = \boldsymbol{0}$ is a solution to the dual.

See any arbitrary proper, closed, positively homogeneous function as the support $\delta^*_C(\boldsymbol{y}^*)$ to a nonempty, closed, convex set C, and $\langle \boldsymbol{y}^*, \boldsymbol{y}\rangle$ is a linear function on $\boldsymbol{y}^* \in M$. That the optimum value is zero means that $\delta^*_C(\boldsymbol{y}^*) \geq \langle \boldsymbol{y}^*, \boldsymbol{y}\rangle$ for all $\boldsymbol{y}^* \in M$. The terminology is that the support "dominates" the linear function $\langle \boldsymbol{y}^*, \boldsymbol{y}\rangle$ on M.

Let $\boldsymbol{x} \in M^\perp$ be any solution to the primal. Adding $\boldsymbol{x}$ to $\boldsymbol{y}$ "extends" the linear function to $\langle \boldsymbol{y}^*, \boldsymbol{x} + \boldsymbol{y}\rangle$. Since $\boldsymbol{x} + \boldsymbol{y} \in C$, therefore $\delta^*_C(\boldsymbol{y}^*) \geq \langle \boldsymbol{y}^*, \boldsymbol{x} + \boldsymbol{y}\rangle$ for any $\boldsymbol{y}^*$, so the support dominates the extended linear function everywhere. This result constitutes the Hahn–Banach extension theorem.

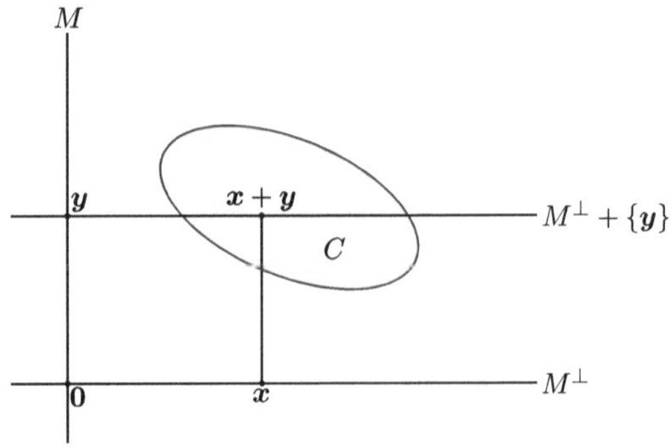

Fig. 31.2 Common vector

Theorem 2 (Hahn–Banach Extension) *For some* $\boldsymbol{y} \in M$*, suppose that the support* $\delta_C^*(\boldsymbol{y}^*)$ *dominates the linear function* $\langle \boldsymbol{y}^*, \boldsymbol{y} \rangle$ *on* M*. Then the linear function can be extended to the entire space as* $\langle \boldsymbol{y}^*, \boldsymbol{y} + \boldsymbol{x} \rangle$, $\boldsymbol{x} \in M^{\perp}$*, such that the support* $\delta_C^*(\boldsymbol{y}^*)$ *dominates the extended linear function everywhere.*

In general the primal has multiple solutions, so $\boldsymbol{x}$ is not unique. Figure 31.2 illustrates the theorem.

We infer that whatever one might prove via the Hahn–Banach theorem can also be proved via convex optimization by duality. Perturbation duality retains its advantage of simplicity and power, by embodying an economic condition into the primal, arriving at the dual by algebraic calculation, and discovering new results. Applications of the Hahn–Banach theorem are typically set in an infinite-dimensional space. Narici and Beckenstein (1997) survey the theorem for various spaces. Zeidler (1985) applies it to many problems.

References

Narici L, Beckenstein E (1997) The Hahn-Banach theorem: the life and times. Topol Appl 77(2):193–211

Zeidler E (1985) Nonlinear functional analysis and its applications, III: variational methods and optimization. Springer-Verlag, New York

Chapter 32
Ekeland Variational Principle

Abstract The standard variational principle is the first-order condition for an optimum. If the optimum is not attained, however, then the first-order condition fails to hold. At an approximate minimum, an approximate first-order condition need not hold. Perhaps the function is not subdifferentiable, or perhaps there is no small subgradient. Given an approximate minimum, the Ekeland variational principle considers a slight variation of the original problem. This variational problem has a minimum that is an appproximate minimum for the original problem. It satisfies an approximate first-order condition for the original problem, and its location is near the approximate minimum. This chapter proves the Ekeland variational principle for a convex optimization, via perturbation duality.

The standard variational principle is the first-order condition for an optimum. If the optimum is not attained, however, then the first-order condition fails to hold. For this case, the Ekeland variational principle is that there exists an approximate minimum that satisfies an approximate first-order condition.

This chapter proves the Ekeland variational principle for a convex optimization, via perturbation duality. Consider the minimization

$$-f^*(0) = \inf_{x} f(x), \tag{32.1}$$

for a proper, closed, convex function f. The minimum is attained at y if and only if the first-order condition $\mathbf{0} \in \partial f(y)$ holds.

Suppose, however, that the minimum is not attained. Suppose that y is an approximate minimum:

$$f(y) = \varepsilon - f^*(0),$$

in which $\varepsilon > 0$ is "small." Even though y is an approximate minimum, an approximate first-order condition need not hold at y. Possibly f is not subdifferentiable at y. If f is subdifferentiable at y, there is no assurance that some subgradient $y^* \in \partial f(y)$ is "small" (that $|y^*|$ is small).

B. C. Dieffenbach, *Conjugate Duality in Economic Analysis*, Contributions to Economics,
https://doi.org/10.1007/978-3-032-21396-9_32

We establish the Ekeland variational principle (Theorem 1) by analyzing the following variational problem and its dual.

Problem 1 (Variational Problem)

$$\inf_{x}[f(x) + \lambda|x - y|].$$

If λ is small, then this problem is a slight variation of the original problem. We show the following. For any $\lambda > 0$, the minimum is attained. If z is a minimum of the variational problem, then z is an approximate minimum near y that satisfies the approximate first-order condition.

In this analysis, one normally sees ε and λ as small: the vector y is an approximate optimum, and the variation of the problem is small. However in fact the mathematical development below applies for any positive values, large or small.

From the chapter "Support," recall the subdifferential of the norm,

$$\partial|w^*| = \partial\delta^*_{\mathbf{B}}(w^*) = \begin{cases} \{w^*/|w^*|\} \text{ if } w^* \neq \mathbf{0} \\ \mathbf{B} \text{ if } w^* = \mathbf{0}. \end{cases}$$

For $w^* \neq \mathbf{0}$, the norm of the subgradient is one.

We analyze the primal

Problem 2 (Variational Primal)

$$\inf_{x}[f(x) + \lambda|x - y|] = \inf_{x}\left[f(x) + \delta^*_{\lambda\mathbf{B}}(x - y)\right],$$

in which the perturbation is y. Since the extra term is nonnegative, the minimum here is greater than or equal to the minimum of the original problem. We show that the primal and the reciprocal dual both satisfy the relative condition, so for both the minimum is attained, and there is no duality gap. Adding the Fenchel inequalities

$$f(x) + f^*(y^* - x^*) - \langle y^* - x^*, x\rangle \geq 0 \tag{32.2}$$

$$\delta_{\lambda\mathbf{B}}(y^*) + \delta^*_{\lambda\mathbf{B}}(y - x) - \langle y^*, y - x\rangle \geq 0 \tag{32.3}$$

and rearranging gives

$$\langle x^*, x\rangle + f(x) + \delta^*_{\lambda\mathbf{B}}(y - x) \geq \langle y^*, y\rangle - f^*(y^* - x^*) - \delta_{\lambda\mathbf{B}}(y^*).$$

The perturbation base values are $x^* = \mathbf{0}$ and y.

Problem 3 (Variational Dual)

$$\sup_{y^*}[\langle y^*, y\rangle - f^*(y^* - x^*) - \delta_{\lambda\mathbf{B}}(y^*)].$$

The reciprocal dual satisfies the relative condition. Since $f^*(\mathbf{0})$ is finite, $\mathbf{0} \in \operatorname{dom} f^*$. Since $\mathbf{0} \in \operatorname{int}(\operatorname{dom} \delta_{\lambda \mathbf{B}})$, it follows that the relative condition holds. The primal has a finite optimum, greater than or equal to $-f^*(\mathbf{0})$, and the optimum is attained. The primal also satisfies the relative condition, so the optimum is also attained in the dual.

Any solution z to the variational problem satisfies the three conditions of the theorem.

Theorem 1 (Ekeland Variational Principle) *Let* $\mathbf{y}$ *be an approximate minimum in (32.1),* $f(y) = \varepsilon - f^*(0), \varepsilon > 0$*. For any* $\lambda > 0$*, there exists some* z *having the following three properties:*

Approximate Minimum $f(z) \leq f(y)$;
Approximate First-Order Condition *there exists a subgradient* $z^* \in \partial f(z)$ *such that* $|z^*| \leq \lambda$;
Proximity $|z - y| \leq \varepsilon/\lambda$.

If there exists $\mathbf{y}^* \in \partial f(y)$ such that $|y^*| \leq \lambda$, then $\boldsymbol{x} = \mathbf{y}$ solves the primal and $\mathbf{y}^*$ solves the dual, as these choices satisfy the Fenchel equalities. Thus $z = \mathbf{y}$ satisfies the three properties in the theorem.

If there is no $\mathbf{y}^* \in \partial f(y)$ such that $|y^*| \leq \lambda$, then a solution z to the primal must differ from $\mathbf{y}$. Necessarily $f(z) < f(y)$. This case prevails whenever λ is sufficiently small, because $\mathbf{0} \notin \partial f(y)$. We show that z satisfies the three conditions in the theorem.

We first show proximity. When $z \neq \mathbf{y}$, then a simple but key fact is that z cannot be farther than ε/λ from $\mathbf{y}$. Because z is minimizing,

$$f(z) + \lambda |z - y| \leq f(y).$$

If

$$\lambda |z - y| > \varepsilon$$

were true, then

$$f(z) + \lambda |z - y| > -f^*(0) + \varepsilon = f(y),$$

a contradiction.

That the Fenchel factors (32.2)–(32.3) are both zero implies that z satisfies the approximate first-order condition. When $z \neq \mathbf{y}$, necessary and sufficient conditions for solutions to the primal and the dual are

$$\begin{gathered} y^* \in \partial f(z) \\ y^* \in \partial \delta^*_{\lambda \mathbf{B}}(y - z) = \{\lambda (y - z) / |y - z|\}. \end{gathered}$$

The latter implies that the solutions z and y^* are unique, and $|y^*| = \lambda$. The approximate first-order condition holds.

In the applications in this book, the focus is on the approximate first-order condition. An existence condition, it implies that there exists a subgradient satisfying an interesting property. Proximity plays a lesser role, because the matter of whether z is near y is commonly unimportant.

A key application is to the analysis of efficiency. The efficiency primal examines whether an allocation is efficient, and the efficiency dual chooses an efficiency price vector. The effective domain of the efficiency function is the nonnegative cone, not the entire space. In the efficiency primal, efficiency means that the optimum is at the origin, and the primal fails to satisfy the relative condition. The Ekeland variational principle explains the case in which the optimum in the dual is not attained.

Ekeland proves the variational principle for any proper, closed function, regardless of whether it is convex (Ekeland 1974, 1979). Furthermore, it holds in more general mathematical spaces. It is especially useful in infinite dimensions, because non-attainment of the optimum is more common in applications in infinite dimensions than in finite dimensions. See Borwein and Zhu (2005, Chap. 2) and (Güler 2010, Chap. 3) for proofs and applications. The use here of the variational Problem (1) follows (Hiriart-Urruty 1983), specialized here to the convex case.

References

Borwein JM, Zhu Q (2005) Techniques of variational analysis, Vol 20. Springer

Ekeland I (1974) On the variational principle. J Math Anal Appl 47:324–353

Ekeland I (1979) Nonconvex minimization problems. Bull Am Math Soc 1:443–474

Güler O (2010) Foundations of optimization. Springer

Hiriart-Urruty JB (1983 Mar) A short proof of the variational principle for approximate solutions of a minimization problem. Am Math Month 90(3):206–207

Chapter 33
Homogeneous Fractional Programming

Abstract We analyze the maximization of the ratio of a concave positively homogeneous function to a convex positively homogeneous function:

$$\sup_{x \neq 0} \frac{g(\boldsymbol{x})}{f(\boldsymbol{x})} = \sup_{x \neq 0} \frac{-\delta_C^*(-\boldsymbol{x})}{\delta_D^*(\boldsymbol{x})},$$

in which C and D are nonempty, closed, convex sets, neither equal to the entire space. We define a "standard homogeneous fractional program," the situation in many economic applications. Because in general the ratio is not a concave function of $\boldsymbol{x}$, reduce it to a Karush-Kuhn-Tucker maximization:

$$\sup_{x \neq 0} \left[\frac{g(\boldsymbol{x})}{f(\boldsymbol{x})}\right] = \sup_{x} \left[g(\boldsymbol{x}) - \delta_{f(x) \leq z}\right],$$

in which the perturbation base value $z = 1$. Under general conditions, the zero-maximum condition

$$0 = \max_{x \neq 0} \left[g(\boldsymbol{x}) - z^* f(\boldsymbol{x})\right] \text{ for some } z^* > 0.$$

is attained by a choice $\boldsymbol{x}$ if and only if $\boldsymbol{x}$ solves the homogeneous fractional program and z^* is the optimum value. Alternatively, to solve primal and dual problems can be insightful.

Working in a Euclidean space, we analyze the maximization of the ratio of a concave positively homogeneous function to a convex positively homogeneous function.

Problem 1 (Homogeneous Fractional Program)

$$\sup_{x \neq 0} \frac{g(\boldsymbol{x})}{f(\boldsymbol{x})} = \sup_{x \neq 0} \frac{-\delta_C^*(-\boldsymbol{x})}{\delta_D^*(\boldsymbol{x})}, \tag{33.1}$$

B. C. Dieffenbach, *Conjugate Duality in Economic Analysis*, Contributions to Economics,
https://doi.org/10.1007/978-3-032-21396-9_33

in which C and D are nonempty, closed, convex sets, neither equal to the entire space.

If $\boldsymbol{x}$ maximizes the ratio, then so does any positive scalar multiple. Require $\boldsymbol{x} \neq \mathbf{0}$, since $f(\mathbf{0}) = 0$.

We impose conditions on the denominator and the numerator.

Condition 1 (Positive Denominator) *$f > 0$ for $\boldsymbol{x} \neq \mathbf{0}$.*

This condition rules out division by zero. It holds if and only if $\mathbf{0} \in \operatorname{int} D$.

Condition 2 (Positive Optimum Value) *$g(\boldsymbol{x}) > 0$ for some $\boldsymbol{x} \in \operatorname{dom} f$.*

That $\mathbf{0} \notin C$ is necessary but not sufficient for the positive-optimum-value condition. Together these two conditions imply that the optimum value of the fractional program is positive.

A standard homogeneous fractional program is the situation in many economic applications.

Definition 1 (Standard Homogeneous Fractional Program) The homogeneous fractional program (33.1) is standard if the positive-denominator condition and the positive-optimum-value condition both hold.

For a standard homogeneous fractional program, the optimum value is positive, where both $g > 0$ and $f > 0$. That g is proper rules out $g = \infty$, and the optimum value cannot be ∞. Neither dom g nor dom f need be the entire space. A solution $\boldsymbol{x}$ must belong to both effective domains, because otherwise the ratio is not positive.

Because in general the ratio (33.1) is not a concave function of $\boldsymbol{x}$, it is traditional in economics to reduce it to a Karush-Kuhn-Tucker maximization:

$$\sup_{\boldsymbol{x} \neq \mathbf{0}} \left[\frac{g(\boldsymbol{x})}{f(\boldsymbol{x})} \right] = \sup_{\boldsymbol{x}} \left[g(\boldsymbol{x}) - \delta_{f(\boldsymbol{x}) \leq z} \right]. \tag{33.2}$$

In the Karush-Kuhn-Tucker problem on the right, the perturbation base value $z = 1$. See the constraint as a normalization of $\boldsymbol{x}$. If $\boldsymbol{x}$ solves the problem on the right, then any positive scalar multiple also solves the fractional program on the left.

By the positive-optimum-value condition, the Slater condition holds, so there is no duality gap and in the dual the optimum value is attained. Furthermore, the constraint set is bounded. Because g is proper, in the primal the optimum value is attained. The constraint is binding, which rules out the choice $\boldsymbol{x} = \mathbf{0}$.

The Lagrangian furnishes necessary and sufficient conditions for the fractional program. The Lagrangian is

$$L(\boldsymbol{x}, z) = g(\boldsymbol{x}) - z^* [f(\boldsymbol{x}) - z] - \delta_{\operatorname{dom} f}(\boldsymbol{x}) + \delta_{z^* \geq 0}.$$

The solutions to the primal and the dual are the saddle points of the Lagrangian. Given z^*, the Lagrangian is positively homogeneous in $\boldsymbol{x}$, so a necessary condition for a saddle point is

$$0 = \max_{\boldsymbol{x}} [g(\boldsymbol{x}) - z^* f(\boldsymbol{x})]. \tag{33.3}$$

The maximum here cannot be negative, because choosing $\boldsymbol{x} = \boldsymbol{0}$ makes the value zero. It cannot be positive, because taking a positive scalar multiple of $\boldsymbol{x}$ and letting the multiple rise to infinity would drive the Lagrangian to infinity. By the positive-optimum-value condition, at a saddle point $z^* > 0$, because the choice $z^* = 0$ would be inconsistent with (33.3). That the maximum is zero means $z^* = g(\boldsymbol{x}) / f(\boldsymbol{x})$, the optimum value of the fractional program.

We obtain necessary and sufficient conditions for the fractional program.

Theorem 1 (Zero Maximum) *The zero-maximum condition*

$$0 = \max_{\boldsymbol{x} \neq \boldsymbol{0}} [g(\boldsymbol{x}) - z^* f(\boldsymbol{x})] \textit{ for some } z^* > 0. \tag{33.4}$$

is attained by a choice $\boldsymbol{x}$ *if and only if* $\boldsymbol{x}$ *solves the homogeneous fractional program (33.1) and* z^* *is the optimum value.*

Alternatively, analyze the concave reduction by extended Fenchel duality. The constraint is

$$\delta_{\delta^*_D(\boldsymbol{x}) \leq 1} = \delta_{\mathrm{epi}\ \delta^*_D}(\boldsymbol{x}, 1) = \delta^*_{\mathrm{cone}\ D}(\boldsymbol{x}, 1) = \delta_{\mathrm{cone}\ D^\circ}(\boldsymbol{x}, -1) = \delta_{D^\circ}(\boldsymbol{x}).$$

We calculate the reciprocal dual of the problem

$$\sup_{\boldsymbol{x}} [-\delta^*_C(-\boldsymbol{x}) - \delta^*_{\mathrm{cone}\ D}(\boldsymbol{x}, 1)].$$

Adding the Fenchel inequalities

$$\delta_{\mathrm{cone}\ D}(\boldsymbol{y}^*, -z^*) + \delta^*_{\mathrm{cone}\ D}(\boldsymbol{x} - \boldsymbol{y}, z) - \langle \boldsymbol{y}^*, \boldsymbol{x} - \boldsymbol{y} \rangle - (-z^*) z \geq 0 \tag{33.5}$$

$$\delta_C(\boldsymbol{y}^* - \boldsymbol{x}^*) + \delta^*_C(-\boldsymbol{x}) - \langle \boldsymbol{y}^* - \boldsymbol{x}^*, -\boldsymbol{x} \rangle \geq 0 \tag{33.6}$$

and rearranging obtains

$$\langle \boldsymbol{y}^*, \boldsymbol{y} \rangle + z^* z + \delta_{\mathrm{cone}\ D}(\boldsymbol{y}^*, -z^*) + \delta_C(\boldsymbol{y}^* - \boldsymbol{x}^*) \geq \langle \boldsymbol{x}^*, \boldsymbol{x} \rangle - \delta^*_C(-\boldsymbol{x}) - \delta^*_{\mathrm{cone}\ D}(\boldsymbol{x} - \boldsymbol{y}, z).$$

The perturbation base values are $\boldsymbol{y} = \boldsymbol{0}$, $z = 1$, and $\boldsymbol{x}^* = \boldsymbol{0}$. The extended Fenchel dual is

$$\inf_{\boldsymbol{y}^*, z^*} [z^* + \delta_{\mathrm{cone}\ D}(\boldsymbol{y}^*, -z^*) + \delta_C(\boldsymbol{y}^*)]. \tag{33.7}$$

The optimum value z^* is positive. The dual seeks the smallest z^* such that $\boldsymbol{y}^*/z^* \in D$ and $\boldsymbol{y}^* \in C$.

For the Fenchel factor (33.5) to be zero is equivalent to

$$\begin{aligned} \delta_D\left(\boldsymbol{y}^*/z^*\right) + \delta_D^*\left(\boldsymbol{x}\right) - \langle \boldsymbol{y}^*/z^*, \boldsymbol{x}\rangle &= 0 \\ \delta_D^*\left(\boldsymbol{x}\right) &= 1. \end{aligned}$$

Thus

$$\begin{aligned} \boldsymbol{x} &\in \partial\delta_D\left(\boldsymbol{y}^*/z^*\right) \\ \langle \boldsymbol{y}^*, \boldsymbol{x}\rangle &= z^*. \end{aligned}$$

For the Fenchel factor (33.6) to be zero means

$$-\boldsymbol{x} \in \delta_C\left(\boldsymbol{y}^*\right).$$

Figure 33.1 illustrates the dual. Here $\mathbf{0} \in C$, and $\mathbf{0} \notin D$. The solution to the dual is $(\boldsymbol{y}^*, z^*)$. In the figure, the optimum value $z^* > 1$. The solution to the primal is $\boldsymbol{x}$, and the normal $\boldsymbol{x}$ to C and the normal $-\boldsymbol{x}$ to D lie on the ray from the origin passing through $\boldsymbol{y}^*$.

The primal is also

$$\sup_{\boldsymbol{x}} \left[-\delta_C^*\left(-\boldsymbol{x}\right) - \delta_{D^\circ}\left(\boldsymbol{x}\right)\right].$$

Its Fenchel dual is

$$\inf_{\boldsymbol{y}^*} \left[\gamma_D\left(\boldsymbol{y}^*\right) + \delta_C\left(\boldsymbol{y}^*\right)\right], \tag{33.8}$$

which one can also obtain by minimizing the dual (33.7) in z^*.

A valuable application of the Fenchel duality is to reason backwards from either the primal or the dual to discover the corresponding homogeneous fractional program. We have seen that the optimum value of the primal is finite and is attained. By the positive-denominator condition, the relative condition holds for the

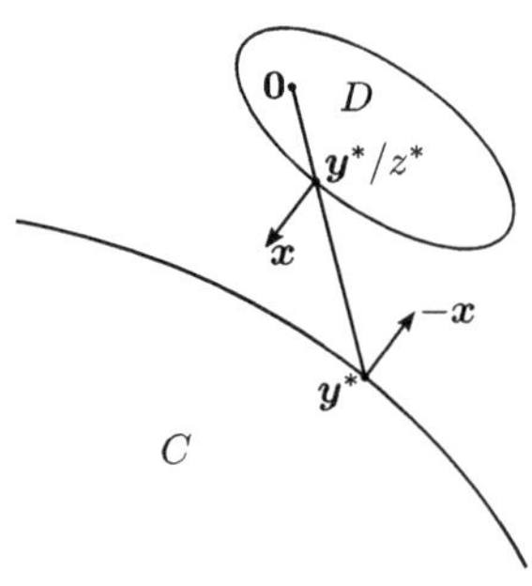

Fig. 33.1 Homogeneous fractional programming dual

Fenchel primal, so there is no duality gap and the optimum value is attained in the Fenchel dual. Consequently below we express each fractional program as max rather than sup.

We express the gauge as the optimum value of a fractional program. Consider a bounded, closed, convex set, such that $\mathbf{0} \in \operatorname{int} D$. Its polar is also a bounded, closed, convex set, and $\mathbf{0} \in \operatorname{int} D^{\circ}$. The gauge is the support to its polar,

$$\gamma_D(\boldsymbol{x}^*) = \sup_x [\langle \boldsymbol{x}^*, \boldsymbol{x} \rangle - \delta_{D^\circ}(\boldsymbol{x})].$$

For $\boldsymbol{x}^* \neq \mathbf{0}$ this problem is the Fenchel primal of the standard homogeneous fractional program,

$$\max_{x \neq 0} \frac{-\delta^*_{\{x^*\}}(-\boldsymbol{x})}{\delta^*_D(\boldsymbol{x})},$$

which obtains the following.

Theorem 2 (Gauge as a Homogeneous Fractional Program) *If* $\mathbf{0} \in \operatorname{int} D$, *a bounded, closed, convex set, then*

$$\gamma_D(\boldsymbol{x}^*) = \max_{x \neq 0} \frac{\langle \boldsymbol{x}^*, \boldsymbol{x} \rangle}{\gamma_{D^\circ}(\boldsymbol{x})}$$

(Rockafellar 1970). If the set is the unit ball B, then this theorem is the Cauchy-Schwarz inequality.

The minimum distance from the origin to the set not containing the origin is the optimum value of a fractional program. Consider a proper, closed, convex set C not containing the origin.

Problem 2 (Minimum Distance)

$$\inf_{y^*} [|y^*| + \delta_C(y^*)]$$

The unit ball is its own polar, $B^\circ = B$. Here $|y^*| = \delta^*_B(y^*) = \gamma_{B^\circ}(y^*)$. The minimum-distance problem has the form of the Fenchel dual of the standard homogeneous fractional program

$$\max_{x \neq 0} \frac{-\delta^*_C(\boldsymbol{x})}{|\boldsymbol{x}|}.$$

Reference

Rockafellar RT (1970) Convex analysis. Princeton University Press, Princeton, NJ

Chapter 34
Fractional Programming

Abstract We generalize the analysis of a homogeneous fractional program to the case in which the denominator and numerator are not positively homogeneous:

$$\sup_{x} \frac{g(\boldsymbol{x})}{f(\boldsymbol{x})},$$

for a concave function g and a convex function f. In general the ratio g/f is not a concave function, but for a "standard fractional program," one can set up an equivalent concave maximization. We focus on the concave and nondecreasing value function of the following Karush-Kuhn-Tucker problem:

$$V(z) := \sup_{x} \left[g(\boldsymbol{x}) - \delta_{f(x) \leq z} \right].$$

We show

$$\sup_{x} \frac{g(\boldsymbol{x})}{f(\boldsymbol{x})} = \sup_{z>0} \frac{V(z)}{z}.$$

The optimum value is the unique z^* such that $V_*(z^*) = 0$. Alternatively, even if the numerator and the denominator in the fractional program are not positively homogeneous, nevertheless one can reduce the analysis to the positively homogeneous case, via epi multiplication. The zero-supremum condition

$$0 = \sup_{x} [g(\boldsymbol{x}) - z^* f(\boldsymbol{x})] \text{ for some } z^* > 0.$$

holds if and only if z^* is the optimum value of the fractional program.

B. C. Dieffenbach, *Conjugate Duality in Economic Analysis*, Contributions to Economics,
https://doi.org/10.1007/978-3-032-21396-9_34

Problem 1 (Fractional Program)

$$\sup_x \frac{g(x)}{f(x)}. \tag{34.1}$$

for a proper, closed, concave function g and a proper, closed, convex function f.

We first derive necessary and sufficient conditions for an optimum in terms of the value function of a Karush-Kuhn-Tucker problem. The conjugate of the value function is zero at optimum value of the ratio. We then explain how the fractional program (34.1) can alternatively be analyzed via a homogeneous fractional problem involving epi-multiplication.

34.1 Standard Fractional Program

In general the ratio (34.1) is not a concave function of $\boldsymbol{x}$, but under additional conditions one can set up an equivalent concave maximization.

Condition 1 (Positive Denominator) $f > 0$.

Condition 2 (Positive Horizon Function) $f^{\infty} > 0$ *for* $\boldsymbol{x} \neq \boldsymbol{0}$.

The effective domain of each function need not be the entire space. From the chapter "Cone Representation of a Convex Function," recall that the positive-horizon-function condition means that, for any z, the contour set having indicator $\delta_{f(x)\le z}$ is bounded. The positive-denominator condition implies $\inf_x f(\boldsymbol{x}) = -f^*(\boldsymbol{0}) \geq 0$. When the positive-horizon-function condition holds as well, then $\inf_x f(\boldsymbol{x}) > 0$.

Condition 3 (Positive Optimum Value) $g(\boldsymbol{x}) > 0$ *for some* $\boldsymbol{x} \in \operatorname{dom} f$.

That g is proper rules out $g = \infty$, so the optimum value cannot be ∞.

Below we analyze a standard fractional program.

Definition 1 (Standard Fractional Program) The fractional program (34.1) is standard if the positive-denominator condition, the positive-horizon function condition, and the positive-optimum-value condition all hold.

For a standard fractional program, the optimum value is finite and positive.

34.2 Zero Conjugate

We focus on the value function of the following Karush-Kuhn-Tucker problem.[1]

Problem 2 (Constrained Maximization)

$$V(z) := \sup_{x} \left[g(x) - \delta_{f(x)\leq z} \right]. \tag{34.2}$$

The value function is concave, proper, and closed and is nondecreasing. If z is less than $\min_x f(x) > 0$, then $V(z) = -\infty$. Where the optimum value is finite, it is attained. If z is large enough, then $V(z) > 0$.

To express the fractional program via the value function reduces the analysis to optimization on $\mathbf{R}$.

Theorem 1 (Value-Ratio Maximization) *For a standard fractional program,*

$$\sup_{x} \frac{g(x)}{f(x)} = \sup_{z>0} \frac{V(z)}{z}. \tag{34.3}$$

(That g is concave and f is convex is not used in the proof.) We have

$$\begin{aligned} &\sup_{z>0} \frac{g(x) - \delta_{f(x)\leq z}}{z} \\ &\quad = \begin{cases} \frac{g(x)}{f(x)} \text{ if } g(x) > 0 \text{ (choose } z = f(x)\text{)} \\ 0 \text{ if } g(x) \leq 0 \text{ (choose } z \to \infty\text{)}. \end{cases} \end{aligned} \tag{34.4}$$

Then

$$\begin{aligned} &\sup_{z>0} \frac{V(z)}{z} \\ &\quad = \sup_{z>0} \left\{ \frac{1}{z} \sup_{x} \left[g(x) - \delta_{f(x)\leq z} \right] \right\} \\ &\quad = \sup_{x} \left\{ \sup_{z>0} \left[\frac{g(x) - \delta_{f(x)\leq z}}{z} \right] \right\} \\ &\quad = \sup_{x} \frac{g(x)}{f(x)}, \end{aligned}$$

The final equals sign invokes (34.4) and the positive-optimum-value condition.

[1] The Slater condition perhaps does not hold. For example, it might be that $V(z)$ is attained only at some x on the relative boundary of dom g.

Invert the choice variable:

$$\sup_{z>0} \frac{V(z)}{z} = \sup_{y>0} \frac{V(1/y)}{1/y} = \sup_{y>0} (y \star V)(1) = \sup_{y} \left[-\delta^*_{\text{epi } V_*}(-1, y) \right],$$

by the closure of the support. The exceptional case $y = 0$ corresponds to the case $z \to \infty$, for which the supremum of the fractional program is not attained.

We calculate the reciprocal dual. Rearranging the Fenchel inequality

$$\delta_{\text{epi } V_*}(z^*, 0) + \delta^*_{\text{epi } V_*}(-1, y) - z^*(-1) - 0y \geq 0$$

obtains

$$z^* + \delta_{\text{epi } V_*}(z^*, 0) \geq -\delta^*_{\text{epi } V_*}(-1, y).$$

The dual finds the smallest z^* such that $V_*(z^*) \geq 0$:

$$\inf_{z^*} \left[z^* + \delta_{V_*(z^*) \geq 0} \right].$$

Both the primal and the dual satisfy the relative condition. There is no duality gap, and attained in both the primal and the dual. Because V_* is increasing, there is a unique z^* such that the conjugate is zero. (In the primal, the solution y need not be unique.)

Theorem 2 (Zero Conjugate) *In a standard fractional program, the optimum value is the unique z^* such that $V_*(z^*) = 0$.*

Figure 34.1 illustrates the dual. The value function $V(z)$ in the Karush-Kuhn-Tucker maximization (34.2) is concave and nondecreasing. Its subdifferential is the Lagrange multiplier for the constraint. The optimum z maximizes $V(z)/z$. The optimum value of the fractional program (34.1) is z^*, at which the conjugate $V_*(z^*) = 0$.

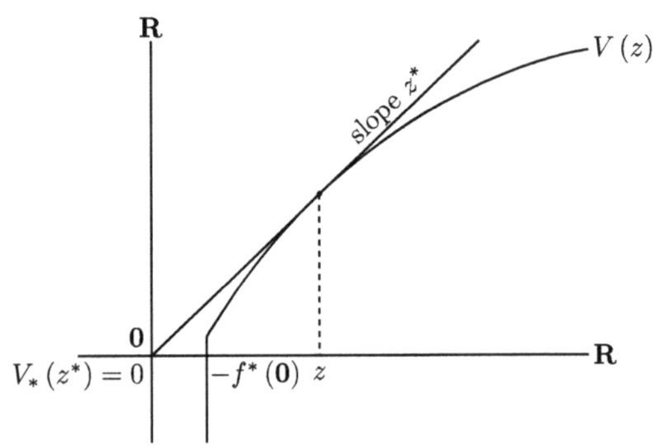

Fig. 34.1 Zero conjugate

34.3 Homogeneous Fractional Program

A drawback of zero-conjugate Theorem 2 is that it does not reduce the problem to the solution of a concave maximization having the optimum ratio as the answer. Even if the numerator and the denominator in the fractional program (34.1) are not positively homogeneous, nevertheless one can reduce the analysis to the positively homogeneous case, via epi multiplication (Schaible 1981).

The following three problems have the same optimum value:

$$\sup_{x} \frac{g(\boldsymbol{x})}{f(\boldsymbol{x})} = \sup_{\boldsymbol{u},v>0} \frac{vg(\boldsymbol{u}/v)}{vf(\boldsymbol{u}/v)} = \sup_{(\boldsymbol{u},v)\neq 0} \frac{(v \star g)(\boldsymbol{u})}{(v \star f)(\boldsymbol{u})}. \tag{34.5}$$

See the second problem as essentially a restatement of the first. If $\boldsymbol{u}$ and v solve the second problem, then $\boldsymbol{x} = \boldsymbol{u}/v$ solves the first. Conversely, if $\boldsymbol{x} = \boldsymbol{u}/v$ solves the first problem for some $v > 0$, then $\boldsymbol{u}$ and v solve the second problem.

The third problem is a standard homogeneous fractional program. The positive-denominator and positive-horizon-function conditions for the fractional program (34.1) are equivalent to a positive denominator for $(\boldsymbol{u}, v) \neq 0$. The extension to $v = 0$ in the third problem follows from the closure of the support to the epigraph. For a choice $(\boldsymbol{u}, 0)$ to solve this problem means that the supremum in the second problem is not attained, but is only approached as the choice $v \searrow 0$.

For the homogeneous fractional program (34.5), the zero-maximum necessary and sufficient condition for an optimum is

$$0 = \max_{(\boldsymbol{u},v)\neq 0} [(v \star g)(\boldsymbol{u}) - z^* (v \star f)(\boldsymbol{u})] \text{ for some } z^* > 0. \tag{34.6}$$

Theorem 3 (Zero Maximum) *The zero-maximum condition (34.6) holds if and only if z^* is the optimum value of the fractional program (34.1).*

One can restate the zero-maximum condition (34.6) as the zero-supremum condition

$$0 = \sup_{x} [g(\boldsymbol{x}) - z^* f(\boldsymbol{x})] \text{ for some } z^* > 0. \tag{34.7}$$

If the optimum in Eq. (34.6) is attained for $v > 0$, one can choose $v = 1$, and in Eq. (34.7) the optimum is attained. If the optimum in Eq. (34.6) is attained for $v = 0$, then in Eq. (34.7) the supremum is approached but not attained.

Corollary 1 (Zero Supremum) *The zero-supremum condition (34.7) holds if and only if z^* is the optimum value of the fractional program (34.1).*

Jagannathan (1966) and Dinkelbach (1967) present fractional programming via (34.7).[2]

---

[2] That $V(z)$ is nondecreasing in z underpins for their analysis, and they make no use of the concavity of g or the convexity of f. They assume that the choice variables lie in a compact, convex set, so the supremum is necessarily attained.

The follwing linear fractional program illustrates the epi-multiplication formulation (Charnes and Cooper 1962).

Problem 3 (Charnes-Cooper Linear Fractional Program)

$$\sup_{x}\left[\frac{\langle c^*, x\rangle + c^*}{\langle d^*, x\rangle + d^* + \delta_{Ax\leq b}}\right] \tag{34.8}$$

Condition 4 (Positive Denominator) If $Ax \preceq b$, then $\langle d^*, x\rangle + d^* > 0$:

$$\langle d^*, x\rangle + d^* + \delta_{Ax\leq b} > 0.$$

Here

$$(v \star g)(u) = \langle c^*, u\rangle + c^* v - \delta_{v\geq 0}$$
$$(v \star f)(u) = \langle d^*, u\rangle + d^* v + \delta_{Au\leq bv} + \delta_{v\geq 0}.$$

The equivalent concave reduction is

$$\sup_{(u,v)}\left[\langle c^*, u\rangle + c^* v - \delta_{\langle d^*, u\rangle + d^* v\leq 1} - \delta_{Au\leq bv} - \delta_{v\geq 0}\right],$$

a polyhedral maximization. For a solution (u, v), necessarily $v > 0$, and the choice $x = u/v$ solves the fractional program (34.8).

References

Charnes A, Cooper WW (1962) Programming with linear fractional functionals. Naval Res Logist Quart 9:181–186

Dinkelbach W (1967 Mar) On nonlinear fractional programming. Manage Sci 13:492–498

Jagannathan R (1966 Mar) On some properties of programming problems in parametric form pertaining to fractional programming. Manage Sci 12(7):609–615

Schaible S (1981) A survey of fractional programming. In: Schaible S, Ziemba WT (eds) Generalized concavity in optimization and economics. Academic Press, pp 417–440

Chapter 35
The Alternative

Abstract A theorem of the alternative asserts that either of two conditions holds, but not both simultaneously. Theorems of the alternative have played a major role in the development of mathematical economics. See a theorem of the alternative as an existence theorem: the failure of one condition to hold implies the other condition. This chapter develops the concept of the alternative via perturbation duality. The primal/dual pair has no duality gap. The optimum value is zero if the first alternative holds and the second does not, and is infinity if the second alternative holds and the first does not. In the historic development of the alternative, the mathematician posits the alternative out of a hat, without explanation or motivation, and then proves the result. In contrast, the duality approach arrives at the alternative by calculating the dual. In this chapter the starting point is a cone alternative. Express one alternative as a closed, convex set. One derives the second alternative from the polar of the cone representation of this set. An example is the efficiency alternative, which reveals the close linkage between efficiency and weak efficiency. Setting the analysis in a general Euclidean Jordan algebra advances the theory. The classical examples of the alternative involve linear inequalities for which the solution set is polyhedral. Even though this latter property does not hold for a general Euclidean Jordan algebra, the classical theorems nevertheless remain valid.

A theorem of the alternative asserts that either of two conditions holds, but not both simultaneously. Theorems of the alternative have played a major role in the development of mathematical economics. See a theorem of the alternative as an existence theorem: the failure of one condition to hold implies the other condition.

The Farkas alternative is especially well known to economists. Consider a linear transformation $\mathsf{A} \in \mathbf{T}^{\mathbf{X}\to\mathbf{Y}}$, in which $\mathbf{X}$ is a Euclidean space and $\mathbf{Y}$ is a Euclidean Jordan algebra.

Theorem 1 (Farkas Alternative) *Exactly one of the following has a solution:*

1. $\mathsf{A}^{\mathsf{T}}\boldsymbol{y}^* = \boldsymbol{x}^*, \boldsymbol{y}^* \geq \mathbf{0}$;
2. $\mathsf{A}\boldsymbol{x} \geq \mathbf{0}, \langle \boldsymbol{x}^*, \boldsymbol{x} \rangle < 0$

B. C. Dieffenbach, *Conjugate Duality in Economic Analysis*, Contributions to Economics,
https://doi.org/10.1007/978-3-032-21396-9_35

(Farkas 1902). In the history of mathematics, mathematicians proved various theorems of the alternative, like the Farkas alternative. In many cases, one theorem of the alternative proves another.

This chapter develops the concept of the alternative via perturbation duality. The primal/dual pair has no duality gap. The optimum value is zero if the first alternative holds and the second does not, and is infinity if the second alternative holds and the first does not. In the historic development of the alternative, the mathematician posits the alternative out of a hat, without explanation or motivation, and then proves the result. In contrast, the duality approach arrives at the alternative by calculating the dual.

In this chapter the starting point is a cone alternative. Express one alternative as a closed, convex set. One derives the second alternative from the polar of the cone representation of this set.

The classical examples of the alternative involve linear inequalities. Setting the analysis in a general Euclidean Jordan algebra advances the theory. In the Euclidean Jordan algebra $\mathbf{R}^n$—the traditional setting for the alternative—the solution set for a linear inequality is polyhedral. Even though this property does not hold for a general Euclidean Jordan algebra, the classical theorems nevertheless remain valid.

The efficiency alternative involves a closed, convex cone and its polar, in a Euclidean Jordan algebra. This alternative is valuable for the analysis of both efficiency and weak efficiency and reveals the close linkage between them.

35.1 Cone Alternative

The empty-set primal/dual pair motivates the cone alternative (2). For a closed, convex set C in a Euclidean space, the following problem establishes whether C is empty: the optimum value is zero if $C \neq \varnothing$ but is ∞ if $C = \varnothing$.

Problem 1 (Empty-Set Primal)

$$\inf_{\boldsymbol{x}} \delta_{\text{cone } C}(\boldsymbol{x}, y),$$

in which the perturbation y has base value -1.

We calculate a dual. Rearranging the Fenchel inequality

$$\delta_{\text{cone } C}(\boldsymbol{x}, -1) + \delta^*_{\text{cone } C}(\mathbf{0}, y^*) - \langle \mathbf{0}, \boldsymbol{x} \rangle - y^*(-1) \geq 0$$

obtains the dual.

Problem 2 (Empty-Set Dual)

$$\sup_{y^*} \left[-y^* - \delta_{(\text{cone } C)^\circ}(\mathbf{0}, y^*) \right].$$

There is no duality gap. If $C \neq \emptyset$, the optimum value is zero. Any $\boldsymbol{x} \in C$ makes the primal zero. Setting $y^* = 0$ solves the dual. If $C = \emptyset$, the optimum value is ∞. The primal is ∞ for any $\boldsymbol{x}$. Since any half space contains the empty set, $(\text{cone } C)^\circ = \mathbf{X} \times \mathbf{R}$, the entire space. Letting $y^* \to -\infty$ sends the value of the dual to infinity.

The empty-set primal/dual pair reduces to the following theorem of the alternative.

Theorem 2 (Cone Alternative) *For a closed, convex set C, exactly one of the following holds:*

1. $(\boldsymbol{x}, -1) \in \text{cone } C$ *for some* $\boldsymbol{x}$;
2. $(\mathbf{0}, -1) \in (\text{cone } C)^\circ$.

Independently of the primal/dual pair, it is easy to verify this theorem directly. Alternative (1) holds if C is not empty. Alternative (2) holds if C is empty, as $(\text{cone } C)^\circ$ is the entire space. Both alternatives cannot hold simultaneously, since the inner product $\langle (\boldsymbol{x}, -1), (\mathbf{0}, -1) \rangle = 1$: two vectors can belong to a cone and its polar only if their inner product is less than or equal to zero.

35.2 Linear-Transformation Alternative

We apply the cone alternative to the cone generated by a linear transformation.

Definition 1 (Cone Generated by a Linear Transformation) $A\mathbf{X}_+$ is the cone generated by A.

In general this cone is not polyhedral, because the nonnegative cone is not finitely generated. We calculate the polar:

$$
\begin{aligned}
&\sup_{y^*} \left[\langle y^*, y \rangle - \delta_{A\mathbf{X}_+}(y) \right] \\
&\quad = \sup_{y} \left[\langle y^*, y \rangle - \inf_{\boldsymbol{x}} \left(\delta_{y=A\boldsymbol{x}} + \delta_{\boldsymbol{x} \geq \mathbf{0}} \right) \right] \\
&\quad = \sup_{\boldsymbol{x}, y} \left(\langle y^*, y \rangle - \delta_{y=A\boldsymbol{x}} - \delta_{\boldsymbol{x} \geq \mathbf{0}} \right) \\
&\quad = \sup_{\boldsymbol{x}} \left(\langle y^*, A\boldsymbol{x} \rangle - \delta_{\boldsymbol{x} \geq \mathbf{0}} \right) \\
&\quad = \sup_{\boldsymbol{x}} \left(\left\langle A^\top \boldsymbol{y}^*, \boldsymbol{x} \right\rangle - \delta_{\boldsymbol{x} \geq \mathbf{0}} \right) \\
&\quad = \delta_{A^\top \boldsymbol{y}^* \leq \mathbf{0}}.
\end{aligned}
$$

The polar cone of $A\mathbf{X}_+$ is the solution set to $A^\top \boldsymbol{y}^* \leq \mathbf{0}$.

Below we use this polar relationship to derive theorems of the alternative. In each case, we write a cone representation of alternative (1) as a linear inequality. The polar cone is a generation and defines alternative (2). We invoke alternative (2) in the cone alternative Theorem 2 and interpret the meaning.

In applications, the inequality $\boldsymbol{y}^* \geq \mathbf{1}$ replaces $\boldsymbol{y}^* \gg \mathbf{0}$, and together $\mathbf{1}^\mathsf{T}\boldsymbol{y}^* \geq 1$ and $\boldsymbol{y}^* \geq \mathbf{0}$ replace $\boldsymbol{y}^* > \mathbf{0}$.

Theorem 3 (Linear Inequality Alternative) *Exactly one of the following has a solution:*

1. $A^\mathsf{T}\boldsymbol{y}^* \preceq \boldsymbol{x}^*$;
2. $A\boldsymbol{x} = \mathbf{0}, \boldsymbol{x} \geq \mathbf{0}, \langle \boldsymbol{x}^*, \boldsymbol{x}\rangle < 0$.

Define C as the solutions $\boldsymbol{y}^*$ to the linear inequality

$$A^\mathsf{T}\boldsymbol{y}^* \preceq \boldsymbol{x}^*.$$

The first alternative says $(\boldsymbol{y}^*, -1) \in \operatorname{cone} C$, and the second alternative says $(\mathbf{0}, -1) \in (\operatorname{cone} C)^\circ$. The cone representation cone C consists of the solutions $(\boldsymbol{y}^*, z^*)$ to the linear inequalities

$$\begin{aligned} A^\mathsf{T}\boldsymbol{y}^* + \boldsymbol{x}^* z^* &\preceq \mathbf{0} \\ z^* &\leq 0. \end{aligned}$$

Its polar is

$$\begin{bmatrix} A \\ \boldsymbol{x}^{*\mathsf{T}} \end{bmatrix} \mathbf{X}_+ + \begin{bmatrix} \mathbf{0} \\ 1 \end{bmatrix} \mathbf{R}_+,$$

so cone alternative (2) is

$$\begin{bmatrix} \mathbf{0} \\ -1 \end{bmatrix} \in \begin{bmatrix} A \\ \boldsymbol{x}^{*\mathsf{T}} \end{bmatrix} \mathbf{X}_+ + \begin{bmatrix} \mathbf{0} \\ 1 \end{bmatrix} \mathbf{R}_+.$$

Equivalently, there exists $\boldsymbol{x} \geq \mathbf{0}$ and a scalar $z \geq 0$ such that

$$\begin{aligned} \mathbf{0} &= A\boldsymbol{x} \\ -1 &= \langle \boldsymbol{x}^*, \boldsymbol{x}\rangle + z. \end{aligned}$$

Eliminate z to reduce these conditions to

$$\begin{aligned} \mathbf{0} &= A\boldsymbol{x} \\ -1 &\geq \langle \boldsymbol{x}^*, \boldsymbol{x}\rangle, \end{aligned}$$

and alternative (2) follows.

Theorem 4 (Fredholm Alternative) *Exactly one of the following has a solution:*

1. $A^{\mathsf{T}}\boldsymbol{y}^* = \boldsymbol{x}^*$;
2. $A\boldsymbol{x} = \boldsymbol{0}, \langle \boldsymbol{x}^*, \boldsymbol{x}\rangle \neq 0$

(Fredholm 1903). Because the Fredholm alternative involves no partial ordering, it applies for Euclidean spaces.

The cone representation of alternative (1) is

$$\begin{aligned} A^{\mathsf{T}}\boldsymbol{y}^* + \boldsymbol{x}^* z^* &\preceq \boldsymbol{0} \\ -A^{\mathsf{T}}\boldsymbol{y}^* - \boldsymbol{x}^* z^* &\preceq \boldsymbol{0} \\ z^* &\leq 0, \end{aligned}$$

so the generation of the polar is

$$\begin{bmatrix} A \\ \boldsymbol{x}^{*\mathsf{T}} \end{bmatrix} \mathbf{X}_+ + \begin{bmatrix} -A \\ -\boldsymbol{x}^{*\mathsf{T}} \end{bmatrix} \mathbf{X}_+ + \begin{bmatrix} \boldsymbol{0} \\ 1 \end{bmatrix} \mathbf{R}_+.$$

Thus the cone alternative (2) says

$$\begin{bmatrix} \boldsymbol{0} \\ -1 \end{bmatrix} \in \begin{bmatrix} A \\ \boldsymbol{x}^{*\mathsf{T}} \end{bmatrix} \mathbf{X} + \begin{bmatrix} \boldsymbol{0} \\ 1 \end{bmatrix} \mathbf{R}_+.$$

Equivalently, there exist $\boldsymbol{x}$ and $z \geq 0$ such that

$$\begin{aligned} A\boldsymbol{x} &= \boldsymbol{0} \\ -\langle \boldsymbol{x}^*, \boldsymbol{x}\rangle &= 1 + z \geq 1, \end{aligned}$$

which is equivalent to Fredholm alternative (2).

In the chapter "Stochastic Discount Factor," the Fredholm alternative proves the equivalence of the law of one price—that two assets having the same payoff must have the same price—and the existence of a stochastic discount factor.

Theorem 5 (Stiemke Alternative) *Exactly one of the following has a solution:*

1. $A^{\mathsf{T}}\boldsymbol{y}^* = \boldsymbol{0}, \boldsymbol{y}^* \gg \boldsymbol{0}$;
2. $A\boldsymbol{x} \prec \boldsymbol{0}$

(Stiemke 1915).

Because the solution set for alternative (1) is not closed, replace it with the solution set to $A^{\mathsf{T}}\boldsymbol{y}^* = \boldsymbol{0}; \boldsymbol{y}^* \succeq \boldsymbol{1}$, which is closed. Alternative (1) has a solution if and only if the replacement has a solution. The cone representation for the replacement is

$$\begin{aligned} A^{\mathsf{T}}\boldsymbol{y}^* &\leq \mathbf{0} \\ -A^{\mathsf{T}}\boldsymbol{y}^* &\leq \mathbf{0} \\ -\boldsymbol{y}^* - \mathbf{1}z^* &\leq \mathbf{0} \\ z^* &\leq 0. \end{aligned}$$

The generation of its polar is

$$\begin{bmatrix} A \\ \mathbf{0} \end{bmatrix}\mathbf{X}_+ + \begin{bmatrix} -A \\ \mathbf{0} \end{bmatrix}\mathbf{X}_+ + \begin{bmatrix} -I \\ -\mathbf{1}^{\mathsf{T}} \end{bmatrix}\mathbf{Y}_+ + \begin{bmatrix} \mathbf{0} \\ 1 \end{bmatrix}\mathbf{R}_+.$$

By the cone-alternative theorem,

$$\begin{bmatrix} \mathbf{0} \\ -1 \end{bmatrix} \in \begin{bmatrix} A \\ \mathbf{0} \end{bmatrix}\mathbf{X} + \begin{bmatrix} -I \\ -\mathbf{1}^{\mathsf{T}} \end{bmatrix}\mathbf{Y}_+ + \begin{bmatrix} \mathbf{0} \\ 1 \end{bmatrix}\mathbf{R}_+.$$

Equivalently, there exist $\boldsymbol{x}, \boldsymbol{y} \geq \mathbf{0}$ and $z \geq 0$ such that

$$\begin{aligned} A\boldsymbol{x} &= \boldsymbol{y} \\ -1 &= -\mathbf{1}^{\mathsf{T}}\boldsymbol{y} + z. \end{aligned}$$

The second implies $\mathbf{1}^{\mathsf{T}}\boldsymbol{y} \geq 1$, so $\boldsymbol{y} > \mathbf{0}$ and hence $A\boldsymbol{x} > \mathbf{0}$.

The proof is similar for a generalized Stiemke alternative that adds a second linear transformation $B \in \mathbf{T}^{\mathbf{X}\to\mathbf{Z}}$.

Theorem 6 (Generalized Stiemke Alternative) *Exactly one of the following has a solution:*

1. $A^{\mathsf{T}}\boldsymbol{y}^* + B^{\mathsf{T}}\boldsymbol{z}^* = \mathbf{0}; \boldsymbol{y}^* \gg \mathbf{0}$;
2. $A\boldsymbol{x} > \mathbf{0}; B\boldsymbol{x} = \mathbf{0}$.

In the chapter "Stochastic Discount Factor," the Stiemke alternative proves the fundamental theorem of finance, that there is no arbitrage opportunity if and only if there exists a positive stochastic discount factor. The generalized Stiemke alternative establishes necessary and sufficient conditions for the existence of a cost/mean/variance efficient portfolio.

Kjeldsen (2001, p. 59) explains how (vonNeumannMorgenstern 1953, pp. 138–141) employ the following alternative to prove the minimax theorem for a zero-sum, two-person game.

Theorem 7 (von Neumann-Morgenstern Alternative) *Exactly one of the following has a solution:*

1. $A^{\mathsf{T}}\boldsymbol{y}^* \gg \mathbf{0}, \boldsymbol{y}^* \gg \mathbf{0}$;
2. $A\boldsymbol{x} \preceq \mathbf{0}, \boldsymbol{x} \succ \mathbf{0}$.

The chapter "Stability" uses this alternative to develop the theory of stability for difference and differential equations. Alternative (1) holds for a stable equation, whereas alternative (2) holds for an equation that is not stable.

Because the solution set for alternative (1) is not closed, replace it with the closed solution set to

$$\begin{aligned} -A^{\mathsf{T}}\boldsymbol{y}^* &\preceq -\mathbf{1} \\ -I\boldsymbol{y}^* &\preceq \mathbf{1}. \end{aligned}$$

Alternative (1) has a solution if and only if the replacement has a solution. The cone representation is the solutions to

$$\begin{aligned} -A^{\mathsf{T}}\boldsymbol{y}^* - \mathbf{1}z^* &\preceq \mathbf{0} \\ -I\boldsymbol{y}^* - \mathbf{1}z^* &\preceq \mathbf{0} \\ z^* &\leq 0, \end{aligned}$$

so cone alternative (2) says

$$\begin{bmatrix} \mathbf{0} \\ -1 \end{bmatrix} \in \begin{bmatrix} -A \\ -\mathbf{1}^{\mathsf{T}} \end{bmatrix} \mathbf{X}_+ + \begin{bmatrix} -I \\ -\mathbf{1}^{\mathsf{T}} \end{bmatrix} \mathbf{Y}_+ + \begin{bmatrix} \mathbf{0} \\ 1 \end{bmatrix} \mathbf{R}_+.$$

Equivalently, there exist $\boldsymbol{x} \succeq \mathbf{0}$, $\boldsymbol{y} \succeq \mathbf{0}$, and $z \geq 0$ such that

$$\begin{aligned} -\boldsymbol{y} &= A\boldsymbol{x} \\ \mathbf{1}^{\mathsf{T}}\boldsymbol{x} + \mathbf{1}^{\mathsf{T}}\boldsymbol{y} &= 1 + z. \end{aligned}$$

Eliminating $\boldsymbol{y}$ and z reduces these conditions to

$$\begin{aligned} \mathbf{1}^{\mathsf{T}}\boldsymbol{x} - \mathbf{1}^{\mathsf{T}}A\boldsymbol{x} &\geq 1 \\ \boldsymbol{x} &\succeq \mathbf{0} \\ A\boldsymbol{x} &\preceq \mathbf{0}. \end{aligned}$$

Necessarily $\boldsymbol{x} \succ \mathbf{0}$, because otherwise $A\boldsymbol{x} = \mathbf{0}$, and the first inequality would be violated. The stability alternative (2) follows.

Other types of theorem of the alternative are possible. For the Euclidean Jordan algebra $\mathbf{R}^n$, that the solution set for a linear inequality is polyhedral leads to alternatives not valid in a general Euclidean Jordan algebra.

35.3 Efficiency Alternative

Theorem 8 (Efficiency Alternative) *In a Euclidean Jordan algebra, for a closed, convex cone K, exactly one of the following has a solution:*

1. *Some $\boldsymbol{x} \gg \mathbf{0}$ belongs to K;*
2. *Some $z^* \succ \mathbf{0}$ belongs to K°.*

We arrive at the efficiency alternative via the cone-alternative theorem. Alternative (1) is equivalent to $\boldsymbol{x} \in K$ for some $\boldsymbol{x} \succeq \mathbf{1}$. The cone representation of this set has indicator

$$\delta_{\text{cone } C}(\boldsymbol{x}, y) = \delta_K(\boldsymbol{x}) + \delta_{\boldsymbol{x}+y\mathbf{1}\succeq\mathbf{0}} + \delta_{y\leq 0}.$$

If alternative (1) is false,

$$\begin{aligned} 0 &= \delta_{(\text{cone } C)^\circ}(\mathbf{0}, -1) \\ &= \sup_{\boldsymbol{x},y}\left[-y - \delta_K(\boldsymbol{x}) - \delta_{\boldsymbol{x}+y\mathbf{1}\succeq\mathbf{0}} - \delta_{y\leq 0}\right] \\ &= \sup_{\boldsymbol{x}}\left[\min[\boldsymbol{x}] - \delta_K(\boldsymbol{x})\right], \end{aligned} \tag{35.1}$$

(In (2), the objective function is $-\infty$ if $\boldsymbol{x}$ is not nonnegative.) This efficiency-alternative primal conforms with alternative (1): because the objective function is positively homogeneous, the optimum value is ∞ if there is some $\boldsymbol{x} \gg \mathbf{0}$ in K, and the optimum value is 0 if this alternative is false.

The reciprocal dual is Fenchel duality. Adding the Fenchel inequalities

$$\begin{aligned} \delta_K(\boldsymbol{x}) + \delta_K^*(z^* + \boldsymbol{x}^*) - \langle z^* + \boldsymbol{x}^*, \boldsymbol{x}\rangle &\geq 0 \\ \delta_P(z^*) + \delta_P^*(-\boldsymbol{x} - z) - \langle z^*, -\boldsymbol{x} - z\rangle &\geq 0 \end{aligned}$$

and rearranging obtains

$$\langle z^*, z\rangle + \delta_K^*(z^* + \boldsymbol{x}^*) + \delta_P(z^*) \geq \langle \boldsymbol{x}^*, \boldsymbol{x}\rangle - \delta_P^*(-\boldsymbol{x} - z) - \delta_K(\boldsymbol{x}).$$

The perturbation base values are $\boldsymbol{x}^* = \mathbf{0}$ and $z = \mathbf{0}$.

Problem 3 (Efficiency-Alternative Dual)

$$\inf_{z^*}\left[\delta_{K^\circ}(z^*) + \delta_P(z^*)\right].$$

The primal satisfies the polyhedral/relative condition, so there is no duality gap. For the optimum value to be zero is efficiency alternative (2).

References

Farkas J (1902) Über die Theorie der einfachen Ungleichungen. J für die reine und angewandte Mathematik 124:1–27

Fredholm EI (1903) Sur une classe d'equations fonctionnelles. Acta Math 27:365–390

Kjeldsen TH (2001 Nov) John von Neumann's conception of the minimax theorem: a journey through different mathematical contexts. Arch Hist Exact Sci 56(1):39–68

Stiemke E (1915 June) Über positive Lösungen homogener linearer Gleichungen. Math Annal 76:340–342

von Neumann J, Morgenstern O (1953) Theory of games and economic behavior, 2nd edn. Princeton University Press, Princeton, NJ (Originally 1944)

Part VIII
Efficiency

Chapter 36
Efficiency Duality

Abstract We seek conditions for the feasible set to contain an efficient vector or a weakly efficient vector. For a closed, convex cone K in a Euclidean Jordan algebra, the "efficiency alternative" states that exactly one of the following has a solution:

1. Some $\boldsymbol{x} \gg \mathbf{0}$ belongs to K;
2. Some $\boldsymbol{x}^* \succ \mathbf{0}$ belongs to K°.

The efficiency alternative leads to both efficiency duality and weak-efficiency duality. Consider a nonempty, closed, convex feasible set in a Euclidean Jordan algebra. "Efficiency duality" establishes that three properties are equivalent.

Efficiency The feasible set contains an efficient vector;
Horizon Cone The horizon cone contains no vector greater than zero;
Barrier Cone The barrier cone contains a positive vector.

"Weak-efficiency duality" establishes that three properties are equivalent.

Weak Efficiency The feasible set contains a weakly efficient vector;
Horizon Cone The horizon cone contains no positive vector;
Barrier Cone The barrier cone contains a vector greater than zero.

Invoke the Ekeland variational principle to prove these statements for the barrier cone. The chapter applies these relationships to activity analysis.

We seek conditions for the feasible set to contain an efficient vector or a weakly-efficient vector. The conditions involve the barrier cone and its polar, the horizon cone.

Condition 1 (Feasible Set) *In a Euclidean Jordan algebra* $\mathbf{X}$, *the feasible set* C *is nonempty, closed, and convex.*

Definition 1 (Efficiency) A vector $\boldsymbol{x} \in C$ is efficient if there is no $\boldsymbol{y} \in C$ such that $\boldsymbol{y} \succ \boldsymbol{x}$.

B. C. Dieffenbach, *Conjugate Duality in Economic Analysis*, Contributions to Economics,
https://doi.org/10.1007/978-3-032-21396-9_36

Definition 2 (Weak Efficiency) A vector $\boldsymbol{x} \in C$ is weakly efficient if there is no $\boldsymbol{y} \in C$ such that $\boldsymbol{y} \gg \boldsymbol{x}$.

Below we invoke the efficiency alternative, proved previously in the chapter "The Alternative."

Theorem 1 (Efficiency Alternative) *In a Euclidean Jordan algebra, for a closed, convex cone K, exactly one of the following has a solution:*

1. *Some* $\boldsymbol{x} \gg \boldsymbol{0}$ *belongs to* K;
2. *Some* $\boldsymbol{x}^* > \boldsymbol{0}$ *belongs to* K°.

36.1 Efficiency Duality

Efficiency duality establishes three equivalent conditions.

Theorem 2 (Efficiency Duality) *In a Euclidean Jordan algebra, for a nonempty, closed, convex feasible set, the following three properties are equivalent:*

Efficiency *The feasible set contains an efficient vector;*
Horizon Cone *The horizon cone contains no vector greater than zero;*
Barrier Cone *The barrier cone contains a positive vector.*

By the definition of efficiency, the efficiency property implies the horizon-cone property.

Apply the efficiency alternative, defining K° as the horizon cone and K as the closure of the barrier cone. The horizon-cone property and the barrier-cone property are equivalent. (If the closure of the barrier cone contains a positive vector, then the barrier cone must also.) Using the Ekeland variational principle, below we prove that the barrier-cone property implies the efficiency property.

We invoke the traditional sufficient condition for efficiency.

Definition 3 (Positively Supported) A vector $\boldsymbol{x} \in C$ is positively supported if some subgradient $\boldsymbol{x}^* \in \partial\delta_C(\boldsymbol{x})$ is positive.

Equivalently, by invertibility, $\boldsymbol{x} \in \partial\delta_C^*(\boldsymbol{x}^*)$ for some $\boldsymbol{x}^* \gg \boldsymbol{0}$.

Theorem 3 (Positive Support and Efficiency) *If a vector is positively supported, then it is efficient.*

By translation, reduce the analysis to the vector $\boldsymbol{x} = \boldsymbol{0}$. That $\boldsymbol{0}$ is positively supported means $\boldsymbol{x}^* \in \partial\delta_C(\boldsymbol{0})$ for some positive $\boldsymbol{x}^*$, which is just the first-order condition for the support

$$\delta_C^*(\boldsymbol{x}^*) = \sup_{\boldsymbol{x}} [\langle \boldsymbol{x}^*, \boldsymbol{x} \rangle - \delta_C(\boldsymbol{x})] \tag{36.1}$$

to be attained at $\boldsymbol{x} = \boldsymbol{0}$; thus $\delta_C^*(\boldsymbol{x}^*) = 0$. It follows that $\boldsymbol{0}$ is efficient: if $\boldsymbol{x} > \boldsymbol{0}$ for some $\boldsymbol{x} \in C$, then $\delta_C^*(\boldsymbol{x}^*) \geq \langle \boldsymbol{x}^*, \boldsymbol{x} \rangle > 0$, a contradiction. (The converse is false: a vector can be efficient but not be positively supported.) One can also interpret the sufficient condition via value maximization. Interpret $\boldsymbol{x}^* \gg \boldsymbol{0}$ as a price. If $\boldsymbol{0}$ maximizes the value $\langle \boldsymbol{x}^*, \boldsymbol{x} \rangle$ among all feasible $\boldsymbol{x}$, then it is efficient.

For the barrier cone to contain a positive vector $\boldsymbol{x}^*$ implies that there exists an efficient vector. If the support $\delta_C^*(\boldsymbol{x}^*)$ is attained by a vector, then this vector is efficient, by Theorem 3. If the support is not attained, we show via the convex Ekeland variational principle that an efficient vector exists, because there is a positive vector arbitrarily close to $\boldsymbol{x}^*$ for which the support is attained.

Supposing that support $\delta_C^*(\boldsymbol{x}^*)$ is not attained for some $\boldsymbol{x}^* \gg \boldsymbol{0}$, consider the variational problem

$$\sup_{\boldsymbol{x}} [\langle \boldsymbol{x}^*, \boldsymbol{x} \rangle - \delta_C(\boldsymbol{x}) - \lambda |\boldsymbol{x} - \boldsymbol{y}|]. \tag{36.2}$$

The final term is the variation. Here $\lambda > 0$, and the perturbation $\boldsymbol{y} \in C$ is an approximate maximum. Define $\varepsilon := \delta_C^*(\boldsymbol{x}^*) - \langle \boldsymbol{x}^*, \boldsymbol{y} \rangle > 0$. The Fenchel dual is

$$\inf_{\boldsymbol{y}^*} \left[\langle \boldsymbol{y}^*, \boldsymbol{y} \rangle + \delta_C^*(\boldsymbol{x}^* - \boldsymbol{y}^*) + \delta_{|\boldsymbol{y}^*| \leq \lambda} \right].$$

The Ekeland variational principle states the following. There is no duality gap between the primal and the dual. The optimum value is less than $\delta_C^*(\boldsymbol{x}^*)$ and greater than or equal to $\langle \boldsymbol{x}^*, \boldsymbol{y} \rangle$ and is attained in both the primal and the dual. If z is a solution to the variational problem, then it is an approximate maximum of the support (36.1); and it satisfies an approximate first-order condition for it.

Specifically, $z \in C$ and $\delta_C^*(\boldsymbol{x}^*) > \langle \boldsymbol{x}^*, z \rangle \geq \langle \boldsymbol{x}^*, \boldsymbol{y} \rangle$, and greater than if $z \neq \boldsymbol{y}$. Let z^* denote any solution to the dual. The approximate first-order condition is $z^* \in \{\boldsymbol{x}^*\} - \partial \delta_C(z)$, such that $|z^*| \leq \lambda$.

We show that there exists an efficient vector. The approximate first-order condition rearranges as $\boldsymbol{x}^* - z^* \in \partial \delta_C(z)$. Taking λ small makes $|z^*|$ small, so choose λ small enough so that $\boldsymbol{x}^* - z^* \gg \boldsymbol{0}$. Hence z is positively supported by $\boldsymbol{x}^* - z^*$ and by Theorem 3 is efficient.

Setting an upper limit to the distance between z and $\boldsymbol{y}$, proximity says $|z - \boldsymbol{y}| \leq \varepsilon/\lambda$ but plays no role in this reasoning.

36.2 Weak-Efficiency Duality

Weak-efficiency duality is analogous to efficiency duality. We now require that the feasible set satisfy free disposal: for any feasible vector, any vector less is also feasible.

Condition 2 (Free Disposal) *The horizon cone C^∞ contains $-\mathbf{X}_+$.*

The horizon cone C^∞ is the closed, convex cone of directions in which C is unbounded. The barrier cone is dom δ_C^*, and its polar is the horizon cone, so under free disposal the barrier cone is contained in $\mathbf{X}_+$. Unless $C = \mathbf{X}$, the barrier cone is not $\{\mathbf{0}\}$. It must contain a vector greater than zero, but perhaps no positive vector.

Weak-efficiency duality establishes three equivalent conditions.

Theorem 4 (Weak-Efficiency Duality) *In a Euclidean Jordan algebra, for a nonempty, closed, convex feasible set satisfying free disposal, the following three properties are equivalent:*

Weak Efficiency *The feasible set contains a weakly efficient vector;*
Horizon Cone *The horizon cone contains no positive vector;*
Barrier Cone *The barrier cone contains a vector greater than zero.*

According to the theorem, the feasible set contains a weakly efficient vector as long as it is not the entire space.

By the definition of weak efficiency, the weak-efficiency property implies the horizon-cone property.

Apply the efficiency alternative, defining K° as the horizon cone and K as the closure of the barrier cone. Then the horizon-cone property and the barrier-cone property are equivalent. (If the closure of the barrier cone contains a vector greater than zero, then the barrier cone must also.)

Using the Ekeland variational principle, below we prove that the barrier-cone property implies the weak-efficiency property.

We invoke the traditional sufficient condition for weak efficiency.[1]

Definition 4 (Supported by a Vector Greater Than Zero) A vector $\boldsymbol{x}$ is supported by a vector greater than zero if the subdifferential $\partial\delta_C(\boldsymbol{x})$ contains a subgradient $\boldsymbol{x}^* \succ \mathbf{0}$.

Equivalently, by invertibility, $\boldsymbol{x} \in \partial\delta_C^*(\boldsymbol{x}^*)$ for some $\boldsymbol{x}^* \succ \mathbf{0}$.

Theorem 5 (Support and Weak Efficiency) *If a vector is supported by vector greater than zero, then it is weakly efficient.*

---

[1] In the chapter "Weak Efficiency," a weak efficiency primal/dual pair shows that the converse also holds: if a vector is weakly efficient, then it is supported by a vector greater than zero.

That $\boldsymbol{x}$ is supported by $\boldsymbol{x}^*$ is the first-order condition for the support (36.1) to be attained at $\boldsymbol{x}$; thus $\delta_C^*(\boldsymbol{x}^*) = \langle \boldsymbol{x}^*, \boldsymbol{x} \rangle$. If $\boldsymbol{x}^* \succ \mathbf{0}$, it follows that $\boldsymbol{x}$ is weakly efficient: if some $z \gg \boldsymbol{x}$ belongs to C, then $\delta_C^*(\boldsymbol{x}^*) \geq \langle \boldsymbol{x}^*, z \rangle > \langle \boldsymbol{x}^*, \boldsymbol{x} \rangle$, a contradiction. One can also interpret this sufficient condition for weak efficiency via value maximization. Interpret $\boldsymbol{x}^* \succ \mathbf{0}$ as a price. If $\boldsymbol{x}$ maximizes the value $\langle \boldsymbol{x}^*, \boldsymbol{x} \rangle$ among all feasible $\boldsymbol{x}$, then it is weakly efficient.

We show that a weakly efficient vector exists if the barrier cone contains a vector greater than zero. By free disposal, the barrier cone is contained in $\mathbf{X}_+$. If the barrier cone contains a positive vector, then an efficient (and weakly efficient) vector exists, by efficiency duality. So let us assume that the barrier cone contains a vector greater than zero, but no positive vector.

Again consider the Ekeland variational problem (36.2) for the support, for some $\boldsymbol{x}^* \succ \mathbf{0}$. Choose λ "small": $|\boldsymbol{x}^*| > \lambda > 0$. There is no duality gap between the variational problem and its dual, for perturbation $\boldsymbol{y}$. The finite optimum value is attained in both problems.

Whereas $\boldsymbol{y}$ is perhaps not weakly efficient, any solution z to the variational problem is weakly efficient. Let z^* denote any solution to the dual. The approximate first-order condition rearranges as $\boldsymbol{x}^* - z^* \in \partial \delta_C(z)$. Necessarily $z^* \neq \boldsymbol{x}^*$, because $|\boldsymbol{x}^*| > \lambda \geq |z^*|$. As $\boldsymbol{x}^* - z^*$ lies in the barrier cone, it must be greater than zero (but not positive). By Theorem 5, z is weakly efficient.

Proximity plays no role in this reasoning.

36.3 Activity Analysis

In the chapter "Activity Analysis," the feasible set is a polyhedral cone. The cone K of net output $\boldsymbol{y}$ is the cone generated by the linear transformation A:

$$\delta_K(\boldsymbol{y}) = \inf_{\boldsymbol{x}} \left(\delta_{\boldsymbol{y}=A\boldsymbol{x}} + \delta_{\boldsymbol{x}\geq 0} \right).$$

Here A defines the activities, and $\boldsymbol{x}$ is the activity vector.

The horizon cone is the feasible set, and its polar is the barrier cone:

$$\begin{aligned}
\delta_K^*(\boldsymbol{y}^*) &= \sup_{\boldsymbol{y}} \left[\langle \boldsymbol{y}^*, \boldsymbol{y} \rangle - \delta_K(\boldsymbol{y}) \right] \\
&= \sup_{\boldsymbol{y}} \left[\langle \boldsymbol{y}^*, \boldsymbol{y} \rangle - \inf_{\boldsymbol{x}} \left(\delta_{\boldsymbol{y}=A\boldsymbol{x}} + \delta_{\boldsymbol{x}\geq 0} \right) \right] \\
&= \sup_{\boldsymbol{x},\boldsymbol{y}} \left(\langle \boldsymbol{y}^*, \boldsymbol{y} \rangle - \delta_{\boldsymbol{y}=A\boldsymbol{x}} - \delta_{\boldsymbol{x}\geq 0} \right) \\
&= \sup_{\boldsymbol{x}} \left(\langle \boldsymbol{y}^*, A\boldsymbol{x} \rangle - \delta_{\boldsymbol{x}\geq 0} \right) \\
&= \sup_{\boldsymbol{x}} \left(\left\langle A^\top \boldsymbol{y}^*, \boldsymbol{x} \right\rangle - \delta_{\boldsymbol{x}\geq 0} \right) \\
&= \delta_{A^\top \boldsymbol{y}^* \leq 0},
\end{aligned}$$

the cone of prices such that no activity has a positive profit.

The following is the efficiency-duality Theorem 2 applied to activity analysis.

Theorem 6 (Efficiency Duality for Activity Analysis) *For activity analysis, the following three properties are equivalent:*

Efficiency *For some* $\mathbf{y} = A\mathbf{x}$, $\mathbf{x} \geq \mathbf{0}$, *production is efficient;*
Horizon Cone *There exists no* $\mathbf{x} \geq \mathbf{0}$ *such that net output* $A\mathbf{x} > \mathbf{0}$*;*
Barrier Cone *For some price* $\mathbf{y}^* \gg \mathbf{0}$, *activity profit* $A^{\mathsf{T}}\mathbf{y}^* \leq \mathbf{0}$.

The following is the weak-efficiency duality Theorem 4 applied to activity analysis.[2]

Theorem 7 (Weak-Efficiency Duality for Activity Analysis) *For activity analysis, the following three properties are equivalent:*

Weak Efficiency *For some* $\mathbf{y} = A\mathbf{x}$, $\mathbf{x} \geq \mathbf{0}$, *production is weakly efficient;*
Horizon Cone *There exists no* $\mathbf{x} \geq \mathbf{0}$ *such that net output* $A\mathbf{x} \gg \mathbf{0}$*;*
Barrier Cone *For some price* $\mathbf{y}^* > \mathbf{0}$, *activity profit* $A^{\mathsf{T}}\mathbf{y}^* \leq \mathbf{0}$.

[2] Include a free-disposal activity for each input and output.

Chapter 37
Weak Efficiency

Abstract Whether a particular vector is weakly efficient is an important application of perturbation duality: maximize the weak-efficiency function subject to constraints. We present a weak-efficiency primal and a weak-efficiency dual. To calculate the dual and the Lagrangian and to find the primal and dual solutions provide insight into meaning of weak efficiency. A value maximization theorem applies: a vector is weakly efficient if and only if for some price greater than zero it maximizes value in the feasible set. Alternatively, a vector is weakly efficient if and only if a particular Karush-Kuhn-Tucker problem has a zero Lagrange multiplier.

Whether a particular vector is weakly efficient is an important application of perturbation duality. Maximize the weak-efficiency function subject to constraints. To calculate the dual and the Lagrangian and to find the primal and dual solutions provide insight into meaning of weak efficiency. Below we apply this methodology to the matter of whether a vector in a set is weakly efficient.

Let C denote the feasible set in a Euclidean Jordan algebra $\mathbf{X}$.

Definition 1 (Weak Efficiency) A vector $\boldsymbol{x} \in C$ is weakly efficient if there is no $\boldsymbol{w} \in C$ such that $\boldsymbol{w} \gg \boldsymbol{x}$.

Condition 1 (Feasible Set) *The feasible set is nonempty, closed, and convex and satisfies free disposal.*

Condition 2 (Existence) *The feasible set contains a weakly efficient vector.*

According to the weak-efficiency duality theorem, the feasible set contains a weakly efficient vector as long as it is not the entire space.

B. C. Dieffenbach, *Conjugate Duality in Economic Analysis*, Contributions to Economics,
https://doi.org/10.1007/978-3-032-21396-9_37

37.1 Weak-Efficiency Primal and Dual

By translation, reduce the analysis to whether the vector $\mathbf{0}$ is weakly efficient.

Problem 1 (Weak-Efficiency Primal)

$$\sup_{x \in C} (\min \lfloor x \rfloor).$$

Here $\min \lfloor x \rfloor := -\delta_P^* (-x)$ is the smallest eigenvalue of x. By the existence condition, the optimum value is finite.

Whether the optimum value is zero, positive, or negative tests whether $\mathbf{0}$ is weakly efficient:

- the optimum value is 0 if and only if $\mathbf{0}$ is weakly-efficient;
- the optimum value is positive if and only if $\mathbf{0}$ is feasible but not weakly efficient;
- the optimum value is negative if and only if $\mathbf{0}$ is infeasible.

If the optimum value is zero, then C contains some vector that is nonnegative but no positive vector. By free disposal, $\mathbf{0} \in C$ and is weakly efficient. If the optimum value is positive, then $\mathbf{0}$ is not weakly efficient, because there exists some $x \gg \mathbf{0}$ in C. By free disposal, $\mathbf{0} \in C$. If the optimum value is negative, then $\mathbf{0} \notin C$. If $\mathbf{0}$ is weakly efficient but not efficient, then there are multiple solutions to the weak-efficiency primal: $\mathbf{0}$ is not the only solution to the weak-efficiency primal, for there are other solutions that are greater.

We perturb the primal to obtain a dual.

Problem 2 (Perturbed Weak-Efficiency Primal)

$$\sup_{x} [\langle x^*, x \rangle - \delta_P^* (-x - y) - \delta_C (x)]. \tag{37.1}$$

The perturbation base values are $y = \mathbf{0}$ and $x^* = \mathbf{0}$.

Here $-y$ is efficient if and only if the optimum is zero. The minus sign on the perturbation makes the price y^* in the dual greater than zero.

We calculate the reciprocal dual. Adding the Fenchel inequalities

$$\delta_C (x) + \delta_C^* (y^* + x^*) - \langle y^* + x^*, x \rangle \geq 0$$
$$\delta_P (y^*) + \delta_P^* (-x - y) - \langle y^*, -x - y \rangle \geq 0$$

and rearranging gives the primal and the reciprocal dual:

$$\langle y^*, y \rangle + \delta_C^* (y^* + x^*) + \delta_P (y^*) \geq \langle x^*, x \rangle - \delta_P^* (-x - y) - \delta_C (x).$$

Interpret the dual choice-variable y^* as a price.

Problem 3 (Weak-Efficiency Dual)

$$\inf_{y^*}[\delta_C^*(y^*) + \delta_P(y^*)] \tag{37.2}$$

The constraint $y^* \in P$ captures the requirement that y^* be greater than zero.

Because the effective domain of the weak-efficiency function is the entire space, the primal satisfies the relative condition. There is no duality gap, and the optimum value is attained in the dual.

For the relative condition to hold in the reciprocal dual means that the barrier cone contains a positive price vector. By efficiency duality, this condition holds if and only if there exists an efficient vector. On the other hand, if there exists a weakly-efficient vector but no efficient vector, then the relative condition does not hold.

37.2 Exceptional Case

Nonetheless the optimum value in the primal is attained. Express the weak-efficiency dual as a Karush-Kuhn-Tucker problem:

$$\begin{aligned}\inf_{y^*}[\delta_C^*(y^*) - \delta_P(y^*)] &= \inf_{y^*}\left[\delta_C^*(y^*) + \delta_{\langle \mathbf{1}, y^*\rangle = z^*} + \delta_{y^* \geq \mathbf{0}}\right] \\ &= \inf_{y^*}\left[\delta_C^*(y^*) + \delta_{\mathbf{1}^\top y^* = z^*}\right].\end{aligned}$$

By free disposal, the barrier cone of C is contained in $\mathbf{X}_+$, so the requirement that y^* be nonnegative is redundant. The perturbation base value $z^* = 1$.

We calculate its dual. Adding the Fenchel inequalities

$$\begin{gathered}\delta_C(\mathbf{1}z) + \delta_C^*(y^*) - \langle y^*, \mathbf{1}z\rangle \geq 0 \\ \delta_{\{0\}}(\langle \mathbf{1}, y^*\rangle - z^*) + \delta_{\{0\}}^*(-z) - (\langle \mathbf{1}, y^*\rangle - z^*)(-z) \geq 0\end{gathered}$$

and rearranging obtains

$$\delta_C^*(y^*) + \delta_{\{0\}}(\langle \mathbf{1}, y^*\rangle - z^*) \geq z^* z - \delta_C(\mathbf{1}z).$$

The dual is

$$\sup_z [z - \delta_C(\mathbf{1}z)].$$

The Karush-Kuhn-Tucker problem satisfies the polyhedral/relative condition, so there is no duality gap and the optimum value is attained. See that $\boldsymbol{x} = \mathbf{1}z$ solves the weak-efficiency primal (37.1).[1]

See weak-efficiency as the exceptional case of a zero Lagrange multiplier.

[1] The primal may have additional solutions: $\mathbf{1}z$ plus a vector greater than zero but not positive.

Theorem 1 (Zero Lagrange Multiplier) *The vector* $\boldsymbol{y} = \boldsymbol{0}$ *is weakly-efficient if and only if the Karush-Kuhn-Tucker problem has Lagrange multiplier* $z = 0$.

When the vector is weakly-efficient, the Lagrange multiplier is zero, because changing the perturbation z^* to any positive value leaves the optimum value unchanged at zero.

37.3 Value Maximization

That the Lagrangian for Eq. (37.1) has a saddle point proves the value-maximization theorem.

Theorem 2 (Value Maximization) *A vector is weakly-efficient if and only if for some price greater than zero it maximizes value in the feasible set.*

The Lagrangian is

$$\begin{aligned} L(\boldsymbol{x},\boldsymbol{y}^*) &= \sup_{\boldsymbol{y}} [-\langle \boldsymbol{y}^*,\boldsymbol{y}\rangle - \delta_P^*(-\boldsymbol{x}-\boldsymbol{y}) - \delta_C(\boldsymbol{x})] \\ &= -\delta_C(\boldsymbol{x}) + \sup_{\boldsymbol{y}} [-\langle \boldsymbol{y}^*, -\boldsymbol{x}-\boldsymbol{w}\rangle - \delta_P^*(\boldsymbol{w})] \quad (\boldsymbol{w} := -\boldsymbol{x}-\boldsymbol{y}) \\ &= \langle \boldsymbol{y}^*,\boldsymbol{x}\rangle - \delta_C(\boldsymbol{x}) + \delta_P(\boldsymbol{y}^*). \end{aligned}$$

Here $\boldsymbol{0}$ is weakly-efficient if and only if the Lagrangian has a saddle point $(\boldsymbol{0},\boldsymbol{y}^*)$ with saddle-value zero.

The value-maximization Theorem 2 follows immediately. If $\boldsymbol{0}$ is weakly-efficient, the saddle-point property says that $\boldsymbol{x} = \boldsymbol{0}$ must be value-maximizing for price $\boldsymbol{y}^*$. Conversely, suppose that $\boldsymbol{x} = \boldsymbol{0}$ is value-maximizing for some price $\boldsymbol{y}^* \in P$. Then $(\boldsymbol{0},\boldsymbol{y}^*)$ constitutes a saddle point of the Lagrangian having saddle-value zero, because for $\boldsymbol{x} = \boldsymbol{0}$ the Lagrangian is zero for any $\boldsymbol{y}^* \in P$. Consequently $\boldsymbol{0}$ is weakly-efficient.

Chapter 38
Efficiency

Abstract Paralleling the chapter "Weak Efficiency," an efficiency primal and an efficiency dual obtain a necessary and sufficient condition for efficiency: a vector is efficient if and only if the optimum value is zero. The efficiency primal chooses a vector to maximize the efficiency function on the feasible set. The efficiency dual looks for a positive price to minimize the support to the feasible set. Alternatively, the efficiency dual is equivalent to a Karush-Kuhn-Tucker efficiency dual. The exceptional case applies: a vector is efficient if and only if the Lagrange multiplier is zero. The primal/dual pair leads to a direct proof of the fundamental characterization of vector efficiency, by invoking the Ekeland variational principle.

Paralleling the chapter "Weak Efficiency," an efficiency primal and an efficiency dual obtain a necessary and sufficient condition for efficiency: a vector is efficient if and only if the optimum value is zero. The efficiency primal chooses a vector to maximize the efficiency function on the feasible set. The efficiency dual looks for a positive price to minimize the support to the feasible set. Alternatively, the efficiency dual is equivalent to a Karush-Kuhn-Tucker efficiency dual. The exceptional case applies: a vector is efficient if and only if the Lagrange multiplier is zero. The primal/dual pair leads to a direct proof of the fundamental characterization of vector efficiency (Arrow et al. 1953).[1]

Let C denote the feasible set in a Euclidean Jordan algebra $\mathbf{X}$.

Definition 1 (Efficiency) A vector $\boldsymbol{x} \in C$ is efficient if there is no $\boldsymbol{w} \in C$ such that $\boldsymbol{w} \succ \boldsymbol{x}$.

A vector is efficient if one cannot choose a larger vector.

Condition 1 (Feasible Set) *The feasible set is nonempty, closed, and convex and satisfies free disposal.*

[1] Stadler surveys the efficiency literature (Stadler 1979).

B. C. Dieffenbach, *Conjugate Duality in Economic Analysis*, Contributions to Economics,
https://doi.org/10.1007/978-3-032-21396-9_38

Condition 2 (Existence) *The feasible set contains an efficient vector.*

By efficiency duality, that the barrier cone contains a positive vector is equivalent.

38.1 Efficiency Primal and Dual

By translation, reduce the analysis to whether the vector $\mathbf{0}$ is efficient. The efficiency primal tests whether $\mathbf{0}$ is efficient.

Problem 1 (Efficiency Primal)

$$\sup_{x \in C} \left[\langle \mathbf{1}, x \rangle + \delta_{x \geq 0} \right] = \sup_{x \in C} \left[-\delta^*_{\mathrm{X}_+ + \{1\}}(-x) \right]. \tag{38.1}$$

Maximize the sum of the eigenvalues of $x \in C$, subject to the constraint that x is nonnegative.

Whether the optimum value is zero, positive, or negative tests whether $\mathbf{0}$ is efficient:

- the optimum value is 0 if and only if $\mathbf{0}$ is efficient;
- the optimum value is positive if and only if $\mathbf{0}$ is feasible but not efficient;
- the optimum value is $-\infty$ if and only if $\mathbf{0}$ is infeasible.

If $\mathbf{0}$ is efficient, then there is no $x \succ \mathbf{0}$ in C, so the optimum value must be zero. Then $\mathbf{0}$ is the unique solution to the efficiency primal. If $\mathbf{0} \in C$ but is not efficient, then there is some $x \succ \mathbf{0}$ in C, so the optimum value must be positive. If $\mathbf{0} \notin C$, then no greater vector belongs to C, by free disposal, so the optimum value must be $-\infty$.

Perturb the primal to test whether the vector $-y$ is efficient.

Problem 2 (Perturbed Efficiency Primal) For base value y,

$$\sup_{x} \left[\langle x^*, x \rangle - \delta^*_{\mathrm{X}_+ + \{1\}}(-x - y) - \delta_C(x) \right]. \tag{38.2}$$

The perturbation base values $y = \mathbf{0}$ and $x^* = \mathbf{0}$.

Here $-y$ is efficient if and only if the optimum is zero. Interpret the dual choice variable y^* as a price. Placing a minus sign on the perturbation makes y^* positive.

We calculate the reciprocal dual. Adding the Fenchel inequalities

$$\delta_C(x) + \delta^*_C(y^* + x^*) - \langle y^* + x^*, x \rangle \geq 0$$
$$\delta_{\mathrm{X}_+ + \{1\}}(y^*) + \delta^*_{\mathrm{X}_+ + \{1\}}(-x - y) - \langle y^*, -x - y \rangle \geq 0$$

and rearranging obtains the primal and the reciprocal dual:

$$\langle y^*, y \rangle + \delta^*_C(y^* + x^*) + \delta_{\mathrm{X}_+ + \{1\}}(y^*) \geq \langle x^*, x \rangle - \delta^*_{\mathrm{X}_+ + \{1\}}(-x - y) - \delta_C(x).$$

Problem 3 (Efficiency Dual)

$$\inf_{y^*} \delta_C^*(y^*) + \delta_{y^* \geq 1}.$$

Effectively the dual chooses the price to minimize the support, subject to the constraint that the price be positive.

That there exists an efficient vector means that the barrier cone contains a positive vector, which is the relative condition for the efficiency dual. There is no duality gap. The optimum value is finite and is attained in the primal. (If there were no efficient vector, by efficiency duality the optimum value in both the efficiency primal and the efficiency dual would be ∞.)

On the other hand, if $\mathbf{0}$ is efficient, then the efficiency primal does not satisfy the relative condition. The condition requires that there exists some $\boldsymbol{x}$ in the relative interior of the effective domain of the objective function. If $\mathbf{0}$ is efficient, the effective domain of the efficiency function is $\mathbf{X}_+$, and the relative interior is $\mathbf{X}_{++}$. For some vector in C to be positive would contradict the efficiency of $\mathbf{0}$. Thus the optimum zero value in the dual is perhaps only approached as an infimum.

The traditional analysis of efficiency puts forward a sufficient condition and a necessary condition for efficiency, in terms of value maximization for an efficiency-price vector $\boldsymbol{y}^*$. A vector in C is efficient if it maximizes the value $\langle \boldsymbol{y}^*, \boldsymbol{x} \rangle$ among all $\boldsymbol{x}$ in C, for some $\boldsymbol{y}^* \gg \mathbf{0}$. This condition is sufficient but not necessary. A necessary condition for efficiency is to maximize the value for some $\boldsymbol{y}^* > \mathbf{0}$, which we have seen is both necessary and sufficient for weak efficiency. Equivalently, express these conditions via the support to the feasible set. That a vector is supported by some $\boldsymbol{y}^* \gg \mathbf{0}$ is sufficient for efficiency. That a vector is supported by some $\boldsymbol{y}^* > \mathbf{0}$ is necessary for efficiency.

The efficiency primal and dual advance this traditional formulation, by obtaining a necessary and sufficient condition for efficiency: a vector is efficient if and only if the optimum value is zero. If $\boldsymbol{y}^*$ solves the dual and the optimum value is zero, then so does any scalar multiple $\lambda \boldsymbol{y}^*$ for any $\lambda \geq 1$, because the value of the dual objective function is unchanged, at zero. The optimum $\boldsymbol{y}^*$ is not unique. The first-order condition reduces to

$$\mathbf{0} \in \partial \delta_C^*(y^*), \tag{38.3}$$

because $\partial \delta_{y^* \geq 1} = \{\mathbf{0}\}$. Inverting the first-order condition obtains $\boldsymbol{y}^* \in \partial \delta_C(\mathbf{0})$, the sufficient condition for efficiency that $\mathbf{0}$ be positively supported. If the optimum value zero is approached but not attained, then the vector is efficient, but the traditional sufficient condition does not hold.

Figure 38.1 illustrates the efficiency primal and dual. The curved line is the upper boundary of C. The origin $\mathbf{0}$ is efficient. In the primal, the choice $\boldsymbol{x} = \mathbf{0}$ makes the optimum value zero. This choice is the optimum, because there is no $\boldsymbol{x} \in C$ greater than zero. The vector $\boldsymbol{y}^* \gg \mathbf{0}$ is normal to C at $\mathbf{0}$, and the support $\delta_C^*(y^*) = 0$. For any vector not proportional to $\boldsymbol{y}^*$, the support is positive. Hence any positive multiple of $\boldsymbol{y}^*$ greater than or equal to $\mathbf{1}$ is minimizing and solves the dual.

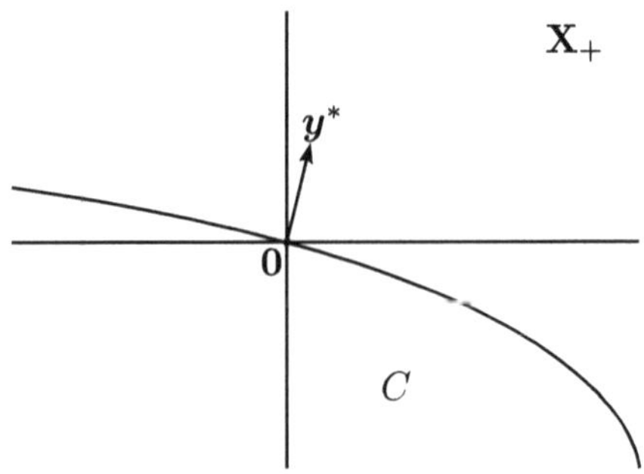

Fig. 38.1 Efficient vector

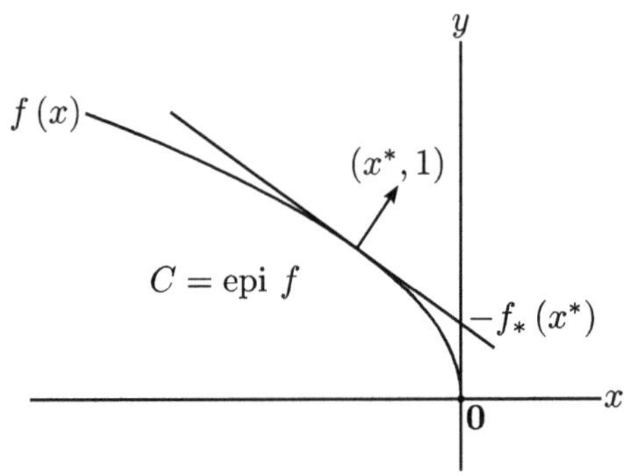

Fig. 38.2 Efficient vector not positively supported

Figure 38.2 presents a graphical example in which **0** is efficient, such that in the dual the optimum zero value is only approached as an infimum. Let **X** be the Euclidean Jordan algebra $\mathbf{R}^2$. The set C is the epigraph of a concave function f on **R**, for which $x \mapsto y = f(x)$. Here $f(0) = 0$. Free disposal requires that f be nonincreasing.

The origin **0** is efficient but is not supported by a positive vector. The graph of f is vertical at $x = 0$; f is not subdifferentiable there. Because epi f contains **0**, the support $\delta^*_{\text{epi } f}$ is nonnegative. The support

$$\delta^*_{\text{epi } f}(x^*, 1) = -f_*(x^*),$$

the vertical intercept of the tangent. As $x^* \to \infty$, the support converges to zero; the infimum of the efficiency dual is zero.

38.2 Exceptional Case

See the proof that a vector is efficient as an exceptional case, by restating the efficiency dual as a Karush-Kuhn-Tucker efficiency dual:

$$\inf_{y^*}\left[\delta_C^*(y^*) + \delta_{y^*\geq x^*}\right].$$

The perturbation base value $\boldsymbol{x}^* = \mathbf{1}$. Because there exists an efficient vector, the Slater condition holds. We find the reciprocal dual. Because we are merely reversing which term in a Fenchel duality problem is perturbed, the reciprocal dual must be the same as the efficiency primal. Adding and rearranging the Fenchel inequalities

$$\delta_C(z-y) + \delta_C^*(y^*) - \langle y^*, z-y\rangle \geq 0$$
$$\delta_{X_+}(y^*-z^*) + \delta_{X_+}^*(-z) - \langle y^*-z^*, -z\rangle \geq 0$$

(the second Fenchel factor is complementary slackness between z and $\boldsymbol{y}^*-\mathbf{1}$) obtains

$$\langle y^*, y\rangle + \delta_C^*(y^*) + \delta_{X_+}(y^*-z^*) \geq \langle z^*, z\rangle - \delta_C(z-y) - \delta_{X_+}^*(-z).$$

Testing whether $-\boldsymbol{y} = \mathbf{0}$ is efficient, the reciprocal dual is

$$\sup_z\left[\langle \mathbf{1}, z\rangle + \delta_{X_+}(z) - \delta_C(z-y)\right],$$

which is the efficiency primal (38.1). The solutions z are the Lagrange multipliers for the constraint $\boldsymbol{y}^* \geq \mathbf{1}$. Thus $-\boldsymbol{y} = \mathbf{0}$ is efficient if and only if the unique solution is $z = \mathbf{0}$.

Efficiency is the exceptional case of a zero Lagrange multiplier.

Theorem 1 (Zero Lagrange Multiplier) *The vector $\boldsymbol{y} = \mathbf{0}$ is efficient if and only if the Karush-Kuhn-Tucker efficiency Lagrange multiplier $z = \mathbf{0}$.*

The result holds regardless of whether the optimum value in the efficiency dual is attained. That the Lagrange multiplier is zero if $\mathbf{0}$ is efficient is intuitive: if $\boldsymbol{y}^*$ solves the dual, then any scalar multiple $\lambda \boldsymbol{y}^*$ for $\lambda > 1$ also solves the dual.

By assumption $\mathbf{0} \in C$, so $\delta_C^*(y^*) \geq 0$. If $\boldsymbol{y} = \mathbf{0}$ is efficient, the optimum value in the efficiency problem is the unconstrained minimum $\inf_{y^*} \delta_C^*(y^*) = 0$.

38.3 Arrow-Barankin-Blackwell

The Arrow-Barankin-Blackwell theorem is the fundamental characterization of vector efficiency (Arrow et al. 1953). The first part of the theorem is the standard sufficient condition for efficiency. Its converse is false, but the second part of the theorem applies.[2]

Theorem 2 (Arrow-Barankin-Blackwell) *In a nonempty, closed, convex set in a Euclidean Jordan algebra,*

- *if a vector is positively supported, then it is efficient;*
- *if a vector is efficient but is not positively supported, then there is another efficient vector arbitrarily close that is positively supported.*

We prove the second part of the theorem by applying the Ekeland variational principle to the efficiency dual. Suppose that $\mathbf{0}$ is efficient, but the optimum value zero in the efficiency dual is not attained. Choose an approximate solution $\boldsymbol{x}^*$ to the efficiency dual.[3] The variational problem for the efficiency dual is

$$\inf_{y^*}\left[\delta_C^*(y^*) + \delta_{y^*\geq 1} + \lambda\, \lfloor y^* - \boldsymbol{x}^* \rfloor\right].$$

The minimum in the variational problem is attained for some $\boldsymbol{y}^*$, a "better" approximate solution. Let $\boldsymbol{x}$ denote a solution to the dual of the variational problem, for perturbation $\boldsymbol{x}^*$. That $\boldsymbol{y}^*$ is positive and supports $\boldsymbol{x}$ implies that $\boldsymbol{x}$ is efficient.

The proof invokes the Ekeland variational principle. Because the infimum of the efficiency dual is zero, for any $\varepsilon > 0$ one can choose $\boldsymbol{x}^* \gg \mathbf{21}$ such that $\delta_C^*(\boldsymbol{x}^*) = \varepsilon$ (every eigenvalue of $\boldsymbol{x}^*$ is greater than two). Set $\lambda = \varepsilon$.

The Ekeland variational principle states the following. Let $\boldsymbol{y}^*$ denote the solution to the variational problem. The approximate minimum condition says $\delta_C^*(y^*) \leq \delta_C^*(\boldsymbol{x}^*) = \varepsilon$. By proximity, $\lfloor y^* - \boldsymbol{x}^* \rfloor \leq \varepsilon/\lambda = 1$, so $\boldsymbol{y}^* \gg \mathbf{1}$. As the constraint $\boldsymbol{y}^* \geq \mathbf{1}$ in the variational problem is not binding, the subdifferential $\partial\delta_{X_+ + \{1\}}(y^*) = \{\mathbf{0}\}$. Consequently the approximate first-order condition reduces to

$$\boldsymbol{x} \in \partial\delta_C^*(y^*),$$

in which $\lfloor \boldsymbol{x} \rfloor \leq \lambda$. Here $\boldsymbol{x}$ is a solution to the dual of the variational problem, for perturbation $\boldsymbol{x}^*$. Inversion gives

$$y^* \in \partial\delta_C(\boldsymbol{x}).$$

Thus $\boldsymbol{x}$ is positively supported by $\boldsymbol{y}^*$, so it is efficient. Take the limit as $\lambda = \varepsilon$ approaches zero. Because $\lfloor \boldsymbol{x} \rfloor \leq \lambda$, the sequence of $\boldsymbol{x}$ converges to zero. Thus there exists an efficient vector arbitrarily close to $\mathbf{0}$.

---

[2] Arrow, Barankin, and Blackwell also demonstrate the following, by an example. If the Euclidean Jordan algebra has dimension three or more, then a limit vector of a sequence of efficient vectors can be inefficient. (In two dimensions, a limit vector is efficient.)

[3] Although $\boldsymbol{x}^*$ is positive, perhaps $\partial\delta_C^*(\boldsymbol{x}^*)$ is empty. Equivalently, perhaps there is no $\boldsymbol{x}$ such that $\boldsymbol{x}^* \in \partial\delta_C(\boldsymbol{x})$.

The polyhedral case avoids the complications analyzed above. The efficiency function is polyhedral if **X** is the Euclidean Jordan algebra $\mathbf{R}^n$, but not otherwise. If C is also polyhedral, then the efficiency primal and dual are polyhedral. The primal satisfies the polyhedral condition when **0** is efficient. The optimum value zero is attained in the dual. Hence **0** is efficient if and only if it is positively supported.

References

Arrow KJ, Barankin EW, Blackwell D (1953) Admissible points of convex sets. In: Kuhn HW, Tucker AW (eds) Contributions to the theory of games, Vol II. Princeton University Press, Princeton, NJ, pp 87–91

Stadler W (1979) A survey of multicriteria optimization, or the vector maximum problem, part I: 1776–1960. J Optim Theor Appl 29:1–52

Part IX
Production

Chapter 39
Domestic Product and Income

Abstract Let $\boldsymbol{x}$ denote a vector of output and input quantity, and $\boldsymbol{x}^*$ is a price. The profit is $\langle \boldsymbol{x}^*, \boldsymbol{x} \rangle$. A price/quantity pair $(\boldsymbol{x}^*, \boldsymbol{x})$ is a production equilibrium if the quantity $\boldsymbol{x}$ maximizes the profit $\langle \boldsymbol{x}^*, \boldsymbol{x} \rangle$ for the price $\boldsymbol{x}^*$. Returns to scale are constant: the technology set K is a cone. The polar K° is all prices for which the maximum profit is zero. The Fenchel equality

$$\delta_K(\boldsymbol{x}) + \delta_{K^\circ}(\boldsymbol{x}^*) - \langle \boldsymbol{x}^*, \boldsymbol{x} \rangle = 0$$

is necessary and sufficient for the price/quantity pair $(\boldsymbol{x}^*, \boldsymbol{x})$ to be a production equilibrium. One can model production equilibrium via a domestic-product and income duality. Decompose the quantity and the price into separate components for output and input:

$$\begin{aligned} \boldsymbol{x} &= (\boldsymbol{y}, -\boldsymbol{z}) \\ \boldsymbol{x}^* &= (\boldsymbol{y}^*, \boldsymbol{z}^*). \end{aligned}$$

The primal maximizes the domestic product: choose output to maximize its value $\langle \boldsymbol{y}^*, \boldsymbol{y} \rangle$, such that this output can be produced with input $\boldsymbol{z}$.

$$\sup_{\boldsymbol{y}} \left[\langle \boldsymbol{y}^*, \boldsymbol{y} \rangle - \delta_K(\boldsymbol{y}, -\boldsymbol{z}) \right].$$

The dual minimizes the domestic income: choose the input price $\boldsymbol{z}^*$ to minimize the input value $\langle \boldsymbol{z}^*, \boldsymbol{z} \rangle$, among all prices for which the maximum profit is zero.

$$\inf_{\boldsymbol{z}^*} \left[\langle \boldsymbol{z}^*, \boldsymbol{z} \rangle + \delta_{K^\circ}(\boldsymbol{y}^*, \boldsymbol{z}^*) \right].$$

If $(\boldsymbol{y}^*, \boldsymbol{z}^*, \boldsymbol{y}, \boldsymbol{z})$ is a production equilibrium, then it solves the domestic-product primal and the domestic-income dual, and there is no duality gap. If $(\boldsymbol{y}^*, \boldsymbol{z}^*, \boldsymbol{y}, \boldsymbol{z})$ solves the primal and the dual and there is no duality gap, then it is a production equilibrium.

B. C. Dieffenbach, *Conjugate Duality in Economic Analysis*, Contributions to Economics,
https://doi.org/10.1007/978-3-032-21396-9_39

39.1 Production Equilibrium

Consider profit maximization in the Euclidean Jordan algebra $\mathbf{R}^n$. Let $\boldsymbol{x}$ denote a vector of output and input quantity, and $\boldsymbol{x}^*$ is a price. A component x_i is positive for an output and negative for an input. The profit is $\langle \boldsymbol{x}^*, \boldsymbol{x} \rangle$. Returns to scale are constant: $\boldsymbol{x} \in K$, the technology set, a proper, closed, convex cone. There is free disposal: $-\mathbf{X}_+ \subset K$. The horizon cone is K, which is the polar of the barrier cone K°. Free disposal is equivalent to $K^\circ \subset \mathbf{X}_+$. Assume that there exists a weakly efficient $\boldsymbol{x}$ (K is not the entire space).

Definition 1 (Production Equilibrium) A price/quantity pair $(\boldsymbol{x}^*, \boldsymbol{x})$ is a production equilibrium if the quantity $\boldsymbol{x}$ maximizes the profit $\langle \boldsymbol{x}^*, \boldsymbol{x} \rangle$ for the price $\boldsymbol{x}^*$.

Profit maximization amounts to the evaluation of a conjugate:

$$\delta_{K^\circ}(\boldsymbol{x}^*) = \delta_K^*(\boldsymbol{x}^*) = \sup_{\boldsymbol{x}} \left[\langle \boldsymbol{x}^*, \boldsymbol{x} \rangle - \delta_K(\boldsymbol{x}) \right]. \tag{39.1}$$

The support to K is profit as a function of price. Under constant returns to scale, the maximum profit is either zero or infinity. The polar K° is all prices for which the maximum profit is zero, which we refer to as the cone of zero-profit prices. Necessarily the price is nonnegative. The profit maximization (39.1) implies the following.

Theorem 1 (Production Equilbrium) *The Fenchel equality*

$$\delta_K(\boldsymbol{x}) + \delta_{K^\circ}(\boldsymbol{x}^*) - \langle \boldsymbol{x}^*, \boldsymbol{x} \rangle = 0 \tag{39.2}$$

is necessary and sufficient for the price/quantity pair $(\boldsymbol{x}^*, \boldsymbol{x})$ *to be a production equilibrium.*

39.2 Domestic-Product and Income Duality

In the domestic-product and income accounts, profit is defined as the value of output minus the cost of input. Domestic product is the value of output, and domestic income is the cost of input plus profit. By the definition of profit, domestic product equals domestic income, an accounting identity.

One can model production equilibrium via a domestic-product and income duality. Decompose the quantity and the price into separate components for output and input.

Definition 2 (Output/Input Decomposition)

$$\boldsymbol{x} = (y, -z)$$
$$\boldsymbol{x}^* = (y^*, z^*).$$

The first component is output, and the second is input. The motive for the minus sign is to make z nonnegative.

Using the output/input decomposition, consider the Fenchel inequality

$$\delta_K(y, -z) + \delta_K^*(y^*, z^*) - \langle y^*, y\rangle - \langle z^*, -z\rangle \geq 0. \tag{39.3}$$

The Fenchel factor is zero if and only if (y^*, z^*, y, z) is a production equilibrium.

Obtain a domestic-income and product duality by taking y and z^* as choice variables and y^* and z as perturbations. Rearrange the Fenchel inequality (39.3) as

$$\langle z^*, z\rangle + \delta_K^*(y^*, z^*) \geq \langle y^*, y\rangle - \delta_K(y, -z)$$

to obtain a primal and its reciprocal dual.

Problem 1 (Domestic-Product Primal)

$$\sup_y [\langle y^*, y\rangle - \delta_K(y, -z)].$$

Problem 2 (Domestic-Income Dual)

$$\inf_{z^*} [\langle z^*, z\rangle + \delta_{K^\circ}(y^*, z^*)].$$

The primal maximizes the domestic product: choose output to maximize its value $\langle y^*, y\rangle$, such that this output can be produced with input z. The dual minimizes the domestic income: choose the input price z^* to minimize the input value $\langle z^*, z\rangle$, among all prices for which the maximum profit is zero.

No duality gap means that domestic product equals domestic income. The value function $V(y^*, z)$ expresses the domestic product and income as a function of the output price and the input quantity.[1]

Theorem 2 (Domestic Income and Product) *If (y^*, z^*, y, z) is a production equilibrium, then it solves the domestic-product primal and the domestic-income dual, and there is no duality gap. If (y^*, z^*, y, z) solves the primal and the dual and there is no duality gap, then it is a production equilibrium.*

[1] Cass (1976) refers to the value function as the Hamiltonian.

Variable interchange obtains a second domestic-income and product primal/dual pair. Rearrange the decomposition Fenchel inequality (39.3) as

$$\langle z^*, z\rangle + \delta_K(y, -z) \geq \langle y^*, y\rangle - \delta_K^*(y^*, z^*).$$

Take y^* and z as the choice variables and y and z^* as the perturbation variables to obtain the following.

Problem 3 (Variable-Interchange Primal and Dual)

$$\sup_{y^*} [\langle y^*, y\rangle - \delta_{K^\circ}(y^*, z^*)] \tag{39.4}$$

$$\inf_{z} [\langle z^*, z\rangle + \delta_K(y, -z)]. \tag{39.5}$$

This primal maximizes the domestic product, such that profit is zero: choose the output price y^* to maximize the value of the output y among all zero maximum-profit prices. The dual minimizes the input cost: choose the input z to minimize its value such that output y can be produced. No duality gap means that domestic product equals domestic income.

One obtains the following analogue of the theorem above.

Theorem 3 (Domestic Income and Product) *If (y^*, z^*, y, z) is a production equilibrium, then it solves the domestic-product primal (39.4) and the domestic-income dual (39.5), and there is no duality gap. If (y^*, z^*, y, z) solves the primal (39.4) and dual (39.5) and there is no duality gap, then it is a production equilibrium.*

Samuelson (1953–1954) presents the domestic-income and product duality. The Heckscher-Ohlin international trade model takes the world price of output and the domestic supply of input as given, and one solves for the output quantity to maximize domestic product and the input price to minimize domestic income. Cass and Shell (1976) model economic growth via domestic-product and income duality.

References

Cass D (1976) The Hamiltonian representation of static competitive or efficient allocation. In: Brown M, Sato K, Zarembka P (eds) Essays in modern capital theory. North-Holland, Amsterdam, pp 159–178

Cass D, Shell K (1976 Feb) Introduction to Hamiltonian dynamics in economics. J Econ Theor 12(1):1–10

Samuelson PA (1953–1954) Prices of factors and goods in general equilibrium. Rev Econ Stud 21(1):1–20

Chapter 40
Shephard Duality

Abstract Shephard's book *Cost and Production Functions* explains and proves the equivalence between the production function and the cost function: either determines the other. Furthermore, the relationship is symmetric: the relationship of cost to product is the same as the relationship of product to cost. This chapter uses conjugate and perturbation duality to develop Shephard's ideas. We study the "classical technology with constant returns to scale." The unit-isocost set is the set of input prices such that the cost is greater than or equal to one. Its lower boundary—where the cost is one—is the factor-price frontier, the input prices compatible with supply equilibrium with positive production. The unit-isoquant set is the set of all input quantities for which production is one or more. This chapter develops the relationship between the production function, the cost function, the unit-isoquant set, and the unit-isocost set. The relationship of the cost function to the unit-isoquant set is the same as the relationship of the production function to the unit-isocost set. Shephard introduces the distance-function representation of production and cost. The Shephard equality holds if and only if prices and quantities constitute a production equilibrium.

Shephard's book *Cost and Production Functions* explains and proves the equivalence between the production function and the cost function: either determines the other (Shephard 1981). Furthermore, the relationship is symmetric: the relationship of cost to product is the same as the relationship of product to cost. In his forward to its reprinting three decades later, Jorgenson declares that this "remarkable book" "ranks as one of the most original contributions to economic theory of all time"(Shephard 1981, pp. iii–iv). "Shephard's central idea is that technologies can be determined from two alternative and completely equivalent points of view: the production function and ... the cost function." "Shephard's book has now attained the status of a true classic in the literature of economic theory, ... still unsurpassed in lucidity and intuitive appeal."

B. C. Dieffenbach, *Conjugate Duality in Economic Analysis*, Contributions to Economics,
https://doi.org/10.1007/978-3-032-21396-9_40

This chapter uses conjugate and perturbation duality to develop Shephard's ideas. Shephard (1970) extends his earlier work (Shephard 1981). Uzawa shows the equivalence of the production function and the cost function in a mathematical proof more understandable to economists (Uzawa 1964). McFadden develops a comprehensive treatment of production and cost (McFadden 1978). However none of these authors achieve the simplification possible via conjugate and perturbation duality.

40.1 Production and Cost

Consider the production of a single output by input $\boldsymbol{x} \in \mathbf{X}$, the Euclidean Jordan algebra $\mathbf{R}^n$. The inner product $\langle \boldsymbol{x}^*, \boldsymbol{x} \rangle$ is the cost of input $\boldsymbol{x}$ for input price $\boldsymbol{x}^*$. The production function $f(\boldsymbol{x})$ is the maximum output for input $\boldsymbol{x}$. The following condition defines the "classical technology with constant returns to scale" via the production function.

Condition 1 (Production) For the classical technology with constant returns to scale, the production function f satisfies the following:

- f is a proper, closed, concave function;
- $\operatorname{dom} f = \mathbf{X}_+$;
- f is positively homogeneous;
- $f > 0$ for $\boldsymbol{x} \gg \mathbf{0}$.

Consequently $f(\boldsymbol{x}) = -\delta_C^*(-\boldsymbol{x})$ for a unique nonempty, closed, convex set C. The input must be nonnegative, and production is then nonnegative.

For output price one and input price $\boldsymbol{x}^*$, the maximum profit is the negative of the conjugate:

$$\sup_{x} [f(\boldsymbol{x}) - \langle \boldsymbol{x}^*, \boldsymbol{x} \rangle] = -\inf_{x} [\langle \boldsymbol{x}^*, \boldsymbol{x} \rangle - f(\boldsymbol{x})] = -f_*(\boldsymbol{x}^*) = \delta_C(\boldsymbol{x}^*).$$

The profit is zero if the cost is greater than one, as the firm produces zero. The profit is infinite if the cost is less than one, as the firm produces as much as possible. Thus the "unit-isocost set" C is the set of input prices such that the cost is greater than or equal to one. Its lower boundary—where the cost is one—is the factor-price frontier, the input prices compatible with supply equilibrium with positive production.

Alternatively, one can express the relationship of f and C as a relationship between polar cones:

$$\delta_{\operatorname{epi} f}(\boldsymbol{x}, y) = \delta_{\operatorname{epi}(-\delta_C^*)}(-\boldsymbol{x}, y) = \delta_{\operatorname{epi} \delta_C^*}(-\boldsymbol{x}, -y) = \delta_{\operatorname{cone} C}^*(-\boldsymbol{x}, -y).$$

Theorem 1 (Polar Relationship)

$$\operatorname{cone} C = -(\operatorname{epi} f)^\circ. \tag{40.1}$$

The relationship of the cost function to the unit-isoquant set is the same as the relationship of the production function to the unit-isocost set.

Definition 1 (Cost Function) The cost function $g(\boldsymbol{x}^*)$ is the minimum cost of producing one unit of output.

Definition 2 (Unit-Isoquant Set) The unit-isoquant set D is the set of all input quantities for which production is one or more:

$$\delta_D(\boldsymbol{x}) := \delta_{f(\boldsymbol{x})\geq 1}.$$

Problem 1 (Cost Minimization) The cost function

$$g(\boldsymbol{x}^*) := \inf_{\boldsymbol{x}} \left[\langle \boldsymbol{x}^*, \boldsymbol{x}\rangle + \delta_{f(\boldsymbol{x})\geq 1}\right].$$

By this definition, $g(\boldsymbol{x}^*) = -\delta_D^*(-\boldsymbol{x}^*)$, and its conjugate is $g_*(\boldsymbol{x}) = -\delta_D(\boldsymbol{x})$. Also,

$$\text{cone } D = -(\text{epi } g)^\circ. \tag{40.2}$$

A simple formula connects C and D:

$$\delta_D(\boldsymbol{x}) = \delta_{\text{epi } f}(\boldsymbol{x}, 1) = \delta^*_{\text{cone } C}(-\boldsymbol{x}, -1).$$

The analogous relationship holds with f and g and with C and D reversed. By conjugate duality, f, g, C, and D are equivalent: any one determines the others. Furthermore, the relationship between production and cost is symmetric.

Figure 40.1 shows the symmetrical relationship between D and C, how D determines C and how C determines D, in the same way. The parallelogram formed by $\boldsymbol{x}$ and $\boldsymbol{x}^*$ illustrates this symmetry. The input cost $\langle \boldsymbol{x}^*, \boldsymbol{x}\rangle = 1$. For input price $\boldsymbol{x}^*$, input $\boldsymbol{x}$ minimizes the cost of producing output one, with cost one. In the figure, $-\boldsymbol{x}^*$ is a normal to D at $\boldsymbol{x}$. The figure shows how D determines C: $\boldsymbol{x}^*$ lies on the isocost frontier (the factor-price frontier), and $-\boldsymbol{x}$ is a normal to C at $\boldsymbol{x}^*$.

Apart from a sign reversal, epi f and cone D are *almost* the same:

$$\delta_{\text{cone } D}(\boldsymbol{x}, -y) = \delta_{f(\boldsymbol{x})\geq y} + \delta_{y\geq 0}. \tag{40.3}$$

This relationship derives from the following:

- For $y > 0$, $\delta_{\text{epi } f}(\boldsymbol{x}, y) = \delta_{\text{cone } D}(\boldsymbol{x}, -y)$;
- $\delta_{\text{epi } f}(\boldsymbol{x}, 0) = \delta_{\text{cone } D}(\boldsymbol{x}, 0) = \delta_{D^\infty}(\boldsymbol{x}) = \delta_{\boldsymbol{x}\geq \boldsymbol{0}}$;
- for $y < 0$, $\delta_{\text{epi } f}(\boldsymbol{x}, y) = \delta_{\boldsymbol{x}\geq \boldsymbol{0}}$ versus $\delta_{\text{cone } D}(\boldsymbol{x}, -y) = \infty$.

The analogous relationship holds between cone C and epi g:

$$\delta_{\text{cone } C}(\boldsymbol{x}^*, -y^*) = \delta_{g(\boldsymbol{x}^*)\geq y^*} + \delta_{y^*\geq 0}. \tag{40.4}$$

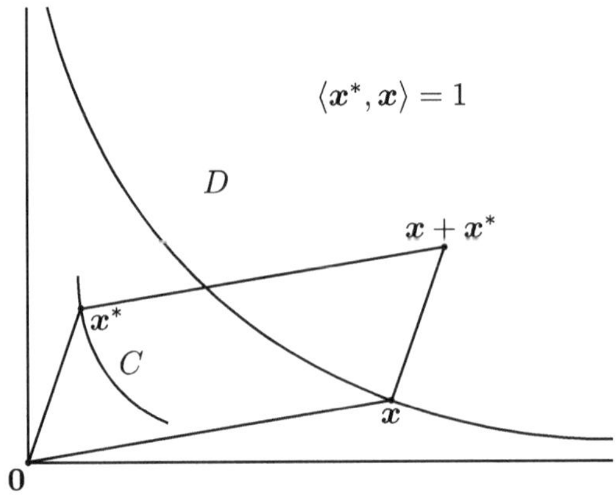

Fig. 40.1 Symmetry of production and cost

View Hanoch's exhaustive analysis as an application of Shephard duality (Hanoch 1978). Given the unit-isoquant and unit-isocost sets for a technology, obtain a reverse technology by reversing the meaning of the two sets. For the two-input constant elasticity of substitution case, the reverse technology is also constant elasticity of substitution, but with the elasticity inverted.

40.2 Four Equivalent Conditions

The production condition is equivalent to the following three conditions; any one implies the others.

Condition 2 (Unit-Isoquant) *For the classical technology with constant returns to scale, the unit-isoquant set D satisfies the following:*

- D is a nonempty, closed, convex set;
- $\mathbf{0} \notin D$;
- $D^\infty = \mathbf{X}_+$.

The production condition implies the unit-isoquant condition. Because $f(\mathbf{0}) = 0$, therefore $\mathbf{0} \notin D$. Because production increases on any ray from $\mathbf{0}$ in $\mathbf{X}_+$, therefore $D^\infty = \mathbf{X}_+$.

The cost condition is analogous to the production condition.

Condition 3 (Cost) *For the classical technology with constant returns to scale, the cost function g satisfies the following:*

- *g is a proper, closed, concave function;*

- dom $g = \mathbf{X}_+$;
- *g is positively homogeneous;*
- $g > 0$ for $\boldsymbol{x} \gg \mathbf{0}$.

The unit-isoquant condition implies the cost-function condition. Because D^∞ is the polar of dom δ_D^*, $D^\infty = \mathbf{X}_+$ implies dom $g = \mathbf{X}_+$. Together $\mathbf{0} \notin D$ and $D^\infty = \mathbf{X}_+$ imply $D \subset \mathbf{X}_+$. Furthermore, $\boldsymbol{x} > \mathbf{0}$ for every $\boldsymbol{x} \in D$. For any $\boldsymbol{x}^* \gg \mathbf{0}$, therefore $\langle \boldsymbol{x}^*, \boldsymbol{x} \rangle > 0$ for all $\boldsymbol{x} \in D$. Because D is closed, $g(\boldsymbol{x}^*) = -\delta_D^*(-\boldsymbol{x}^*) > 0$.

The unit-isocost condition for C is analogous to the condition for D and is implied by the cost condition.

Condition 4 (Unit-Isocost) *For the classical technology with constant returns to scale, the unit-isocost set C satisfies the following:*

- *C is a nonempty, closed, convex set;*
- $\mathbf{0} \notin C$;
- $C^\infty = \mathbf{X}_+$.

How the unit-isocost condition implies the production condition is analogous to how the unit-isoquant condition implies the cost condition.

40.3 Shephard Equality

Shephard introduces the distance-function representation of production and cost.

Theorem 2 (Distance-Function Representation of Production) *For the classical technology,*

$$f(\boldsymbol{x}) = \sup_y \left[y - \delta_{\text{cone } D}(\boldsymbol{x}, -y) \right].$$

Because f is nonnegative on its effective domain, this relationship follows immediately from Eq. (40.3). Figure 40.2 illustrates the distance function. Given a nonnegative $\boldsymbol{x}$, consider the ray from $\mathbf{0}$ through $\boldsymbol{x}$. Find the largest y such that $\boldsymbol{x}/y$ lies in D; one looks for the "distance" y along the ray from the origin to the unit isoquant. Production $f(\boldsymbol{x}) = y$.

The production primal multiplies the distance-function representation of production by the output price y^*.

Problem 2 (Production Primal)

$$\sup_y \left[y^* y - \delta_{\text{cone } D}(\boldsymbol{x}, -y) \right]. \tag{40.5}$$

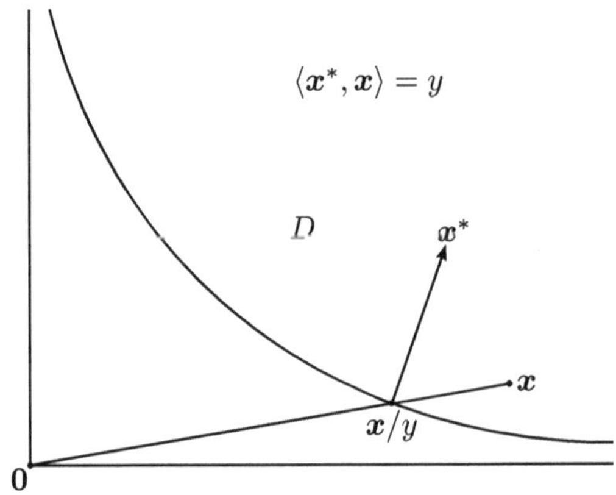

Fig. 40.2 Distance function

The interesting case is primal perturbation for which $f(\boldsymbol{x}) > 0$ and dual perturbation $y^* > 0$, and below we work with this case. The optimum value is the value of production $y^* f(\boldsymbol{x}) > 0$.

We calculate the reciprocal dual via the Lagrangian:

$$L(\boldsymbol{x}^*, y) = \langle \boldsymbol{x}^*, \boldsymbol{x} \rangle + y^* y - \inf_{\boldsymbol{x}} [\langle \boldsymbol{x}^*, \boldsymbol{x} \rangle + \delta_{\text{cone } D}(\boldsymbol{x}, -y)] = \langle \boldsymbol{x}^*, \boldsymbol{x} \rangle + y [y^* - g(\boldsymbol{x}^*)].$$

The infimum is the minimum cost of producing output y.

The dual objective function is

$$\begin{aligned} \sup_y L(\boldsymbol{x}^*, y) &= \sup_y \{\langle \boldsymbol{x}^*, \boldsymbol{x} \rangle + y [y^* - g(\boldsymbol{x}^*)]\} \\ &= \langle \boldsymbol{x}^*, \boldsymbol{x} \rangle + \delta_{\text{epi } g}(\boldsymbol{x}^*, y^*) \\ &= \langle \boldsymbol{x}^*, \boldsymbol{x} \rangle + \delta_{\text{cone } C}(\boldsymbol{x}^*, -y^*), \text{ by Eq. (40.4).} \end{aligned}$$

Problem 3 (Production Dual) The reciprocal dual of the production primal is

$$\inf_{\boldsymbol{x}^*} [\langle \boldsymbol{x}^*, \boldsymbol{x} \rangle + \delta_{\text{cone } C}(\boldsymbol{x}^*, -y^*)]. \tag{40.6}$$

The dual chooses the input price $\boldsymbol{x}^*$ to minimize the cost of input $\boldsymbol{x}$, subject to the condition that the minimum cost be greater than or equal to the output price. Because $y^* > 0$, the production dual satisfies the relative condition, so there is no duality gap. The optimum value of the primal and the dual is positive. If $\boldsymbol{x} \gg \boldsymbol{0}$, the production primal satisfies the relative condition.

The first-order conditions are simple. As $y^* > 0$,

$$\inf_{\boldsymbol{x}^*} [\langle \boldsymbol{x}^*, \boldsymbol{x} \rangle + \delta_{\text{cone } C}(\boldsymbol{x}^*, -y^*)] = \inf_{\boldsymbol{x}^*} [\langle \boldsymbol{x}^*, \boldsymbol{x} \rangle + \delta_C(\boldsymbol{x}^*/y^*)],$$

the first-order condition for the production dual is

$$-\boldsymbol{x} \in \partial\delta_C(\boldsymbol{x}^*/y^*).$$

The first-order conditions for a saddle point of the Lagrangian are

$$g(\boldsymbol{x}^*) = y^*$$
$$\boldsymbol{x}/y \in \partial g(\boldsymbol{x}^*)$$

The subdifferential form is "Shephard's lemma": the subdifferential of the cost function provides the cost-minimizing inputs $\boldsymbol{x}$ to produce y for input price $\boldsymbol{x}^*$.

Setting the production primal equal to the production dual obtains the Shephard equality

$$\delta_{\text{cone } D}(\boldsymbol{x}, -y) + \delta_{\text{cone } C}(\boldsymbol{x}^*, -y^*) - (y^*y - \langle \boldsymbol{x}^*, \boldsymbol{x}\rangle) = 0. \tag{40.7}$$

The expression on the left is necessarily nonnegative. Profit is output value minus input cost, $y^*y - \langle \boldsymbol{x}^*, \boldsymbol{x}\rangle$. In primal/dual problems applying Shephard duality, cone C appears in one problem, and cone D appears in the other.

The Shephard equality holds if and only if $(\boldsymbol{x}^*, y^*, \boldsymbol{x}, y)$ is a production equilibrium.[1] The first term is the indicator of the pairs $(\boldsymbol{x}, y)$ such that $f(\boldsymbol{x}) \geq y$, for $y \geq 0$. The second term is the indicator of the pairs $(\boldsymbol{x}^*, y^*)$ such that $g(\boldsymbol{x}^*) \geq y^*$, for $y^* \geq 0$. Under these conditions, in the third term

$$y^*y - \langle \boldsymbol{x}^*, \boldsymbol{x}\rangle \leq g(\boldsymbol{x}^*)f(\boldsymbol{x}) - \langle \boldsymbol{x}^*, \boldsymbol{x}\rangle \leq 0. \tag{40.8}$$

For input $\boldsymbol{x}$ and input price $\boldsymbol{x}^*$, the cost is $\langle \boldsymbol{x}^*, \boldsymbol{x}\rangle$. If this input produces output $f(\boldsymbol{x})$, then the minimum cost of producing this output would be $g(\boldsymbol{x}^*)f(\boldsymbol{x})$, which must be less than or equal to the actual cost $\langle \boldsymbol{x}^*, \boldsymbol{x}\rangle$. The Shephard equality requires that profit be zero. For positive y and y^*, zero profit requires $f(\boldsymbol{x}) = y$ and $g(\boldsymbol{x}^*) = y^*$. And $g(\boldsymbol{x}^*) = y^* = \langle \boldsymbol{x}^*, \boldsymbol{x}\rangle / y$, so $\boldsymbol{x}$ minimizes the cost of producing output y, for input price $\boldsymbol{x}^*$.

Analogous relationships hold for cost.

Theorem 3 (Distance-Function Representation of Cost) *For the classical technology,*

$$g(\boldsymbol{x}^*) = \sup_{y^*} [y^* - \delta_{\text{cone } C}(\boldsymbol{x}^*, -y^*)].$$

Problem 4 (Cost Primal)

$$\sup_{y^*} [y^*y - \delta_{\text{cone } C}(\boldsymbol{x}^*, -y^*)].$$

[1] By Eq. (40.3) and (40.4), the Shephard equality is not quite a Fenchel equality,

The Lagrangian is

$$L(\boldsymbol{x}, y^*) = \langle \boldsymbol{x}^*, \boldsymbol{x} \rangle + y^* [y - f(\boldsymbol{x})].$$

For $y^* > 0$, the first-order conditions for a saddle point are

$$\begin{aligned} \boldsymbol{x}^*/y^* &\in \partial f(\boldsymbol{x}) \\ f(\boldsymbol{x}) &= y. \end{aligned}$$

Problem 5 (Cost Dual)

$$\inf_{\boldsymbol{x}} [\langle \boldsymbol{x}^*, \boldsymbol{x} \rangle + \delta_{\text{cone } D}(\boldsymbol{x}, -y)].$$

The interesting case is primal perturbation for which $g(\boldsymbol{x}^*) > 0$ and dual perturbation $y > 0$. The optimum value is $y\, g(\boldsymbol{x}^*)$, the minimum cost of producing output y for input price $\boldsymbol{x}^*$. The first-order condition for the cost dual is

$$-\boldsymbol{x}^* \in \partial \delta_D(\boldsymbol{x}/y).$$

Observe that the cost primal and dual constitute an example of variable interchange: the choice and perturbation variables are reversed. Setting the primal equal to the dual obtains the same Shephard equality (40.7).

References

Hanoch G (1978) Symmetric duality and polar production functions. In: Fuss M, McFadden D (eds) Production economics: a dual approach to theory and applications, volume 1, the theory of production, Vol 110. North-Holland, Amsterdam, pp 111–131

McFadden D (1978) Cost, revenue, and profit functions. In: Fuss M, McFadden D (eds) Production economics: a dual approach to theory and applications, volume 1, the theory of production, Vol 110. North-Holland, Amsterdam, pp 4–109

Shephard RW (1970) Theory of cost and production functions. Princeton University Press, Princeton, NJ

Shephard RW (1981) Cost and production functions, Vol 194. Springer-Verlag, Berlin (originally 1953)

Uzawa H (1964) Duality principles in the theory of cost and production. Int Econ Rev 5(2):216–220

Chapter 41
Heckscher-Ohlin Via Shephard Duality

Abstract Heckscher-Ohlin international-trade theory develops the theory of comparative advantage, by focussing on how factor endowments determine the direction of trade. A country having a high ratio of capital to labor endowment exports capital-intensive goods and imports labor-intensive goods. The natural modelling framework for the theory is a domestic-product primal and a domestic-income dual, in which the world price of outputs and the factor endowments are given, and one solves for the output quantities and the factor prices. Model production by Shephard's distance function. Rather than deriving the dual by adding and rearranging Fenchel inequalities, the Lagrangian sets the dual, by invoking the Shephard relationship between the unit isoquant and the unit isocost sets. The optimum value of the primal is domestic product, and the optimum value of the dual is domestic income. We analyze the special case of two sectors and two inputs, capital and labor. Fundamental arguments of Heckscher-Ohlin theory follow immediately. Compare two countries with identical technology but slightly different inputs. One country has a higher ratio of capital to labor than the other. This country will produce relatively more of the capital-intensive good and relatively less of the labor-intensive good, compared to the other country. Furthermore "factor-price equalization" applies: There is no economic incentive for capital or labor to flow between countries. The Stolper-Samuelson theorem asserts that imposing a tariff on the labor-intensive good raises the real wage. One models this idea by raising the output price of the labor-intensive good. For each sector, the capital/labor ratio rises. In each sector the marginal product of labor rises, so the real wage rises, measured in terms of either good. That there is no duality gap means that domestic product equals domestic income.

Heckscher-Ohlin international-trade theory develops the theory of comparative advantage, by focussing on how factor endowments determine the direction of trade. A country having a high ratio of capital to labor endowment exports capital-intensive goods and imports labor-intensive goods. The natural modelling framework for the theory is a domestic-product primal and a domestic-income dual, in which the world

B. C. Dieffenbach, *Conjugate Duality in Economic Analysis*, Contributions to Economics,
https://doi.org/10.1007/978-3-032-21396-9_41

price of outputs and the factor endowments are given, and one solves for the output quantities and the factor prices. We model production by Shephard duality, and its graphical presentation conforms to the textbook presentation of Heckscher-Ohlin theory.

The Heckscher-Ohlin theory is a fundamental component of international-trade theory (Heckscher 1919, 1991; Ohlin 1933; Ruffin 2011; Deardorff 1994). Krugman praises its formalization by Stolper and Samuelson (1941) as a revolutionary change in economic methodology (Krugman 1994). Economics became a "model-oriented field, with an analytical core based on the concepts of maximization and equilibrium" (p. 275). "Stolper-Samuelson used general equilibrium to demonstrate, in an irrefutable way, that informal intuition can be badly misleading" (p. 277). The rigorous and insightful application of duality by economists has been instrumental in the further development of the theory.

41.1 Domestic-Product Primal and Domestic-Income Dual

There are n goods and m inputs. In each sector, the inputs produce the output, via a classical technology with constant returns to scale. Model production by Shephard's distance function: the value of output in sector i is

$$\sup_{z_i} \left[z_i^* z_i - \delta_{\text{cone } D_i}(\boldsymbol{x}_i, -z_i) \right].$$

Input $\boldsymbol{x}_i$ produces output z_i, which has price z_i^*. The cone representation of the unit-isoquant set D_i defines the feasible input and output. Let z and z^* denote the vectors of output and output prices, in the Euclidean Jordan algebra $\mathbf{R}^n$.

An input vector lies in the Euclidean Jordan algebra $\mathbf{R}^m$. The resource constraint that total input is less than or equal to input supply is

$$X\mathbf{1} \preceq \boldsymbol{y}.$$

Here

$$X = \left[\begin{array}{ccc} \boldsymbol{x}_1 & \boldsymbol{x}_2 & \cdots \end{array} \right] \in \mathbf{T}^{n \to m}.$$

is the partitioned linear transformation of input vectors, and $\boldsymbol{y}$ is input supply.

The domestic-product primal chooses inputs X and output z to maximize domestic product, subject to the resource constraint.

Problem 1 (Domestic-Product Primal)

$$\sup_{X,z} \left[\langle z^*, z \rangle - \sum_i \delta_{\text{cone } D_i}(\boldsymbol{x}_i, -z_i) - \delta_{\boldsymbol{y} \succeq X\mathbf{1}} \right].$$

Assume that the primal perturbation $\boldsymbol{y} \gg \mathbf{0}$, and the dual perturbation $z^* \gg \mathbf{0}$.

In contrast to our usual methodology of deriving the dual by adding and rearranging Fenchel inequalities, here we find the dual from the Lagrangian, by invoking the Shephard relationship between the unit isoquant and the unit isocost sets. The Lagrangian for this Karush-Kuhn-Tucker problem is standard:

$$L(X, z, \boldsymbol{y}^*) = \langle z^*, z \rangle - \sum_i \delta_{\text{cone } D_i}(\boldsymbol{x}_i, -z_i) + \langle \boldsymbol{y}^*, \boldsymbol{y} - X\mathbf{1} \rangle. \tag{41.1}$$

Interpret the Lagrange multiplier $\boldsymbol{y}^*$ as the input-price vector.

The reciprocal dual is

$$\inf_{\boldsymbol{y}^*}\left[\sup_{X,z} L(X, z, \boldsymbol{y}^*)\right] = \inf_{\boldsymbol{y}^*}\left\{\langle \boldsymbol{y}^*, \boldsymbol{y} \rangle + \sum_i \sup_{\boldsymbol{x}_i, z_i}\left[z_i^* z_i - \langle \boldsymbol{y}^*, \boldsymbol{x}_i \rangle - \delta_{\text{cone } D_i}(\boldsymbol{x}_i, -z_i)\right]\right\}$$

The inner maximization is profit maximization: choose production $(\boldsymbol{x}_i, z_i)$ to maximize profit $z_i^* z_i - \langle \boldsymbol{y}^*, \boldsymbol{x}_i \rangle$. By constant returns to scale, the maximum profit is either zero or infinity. The maximum profit is zero if and only if $(\boldsymbol{y}^*, -z_i^*) \in$ cone C_i, which sets the reciprocal dual.

Problem 2 (Domestic-Income Dual)

$$\inf_{\boldsymbol{y}^*}\left[\langle \boldsymbol{y}^*, \boldsymbol{y} \rangle + \sum_i \delta_{\text{cone } C_i}(\boldsymbol{y}^*, -z_i^*)\right].$$

Choose the nonnegative input price $\boldsymbol{y}^*$ to minimize the input value $\langle \boldsymbol{y}^*, \boldsymbol{y} \rangle$ subject to the constraints that prices $(\boldsymbol{y}^*, z_i^*)$ allow no positive profit. It is implicit in the primal and the dual that all quantities $(\boldsymbol{x}_i, z_i)$ and prices $(\boldsymbol{y}^*, z_i^*)$ are nonnegative.

The dual amounts to the dual of Woodland, in which the constraint specifies that in each sector the minimum cost is greater than or equal to the output price (Woodland 1977). Others also analyze this dual (Mussa 1979; Dixit and Norman 1980, Chap. 2).

That $\boldsymbol{y} \gg \mathbf{0}$ implies that the primal satisfies the polyhedral/relative condition. That $z^* \gg \mathbf{0}$ implies that the dual satisfies the relative condition. There is no duality gap. The optimum value is finite and is attained in both the primal and the dual. The optimum value of the primal is domestic product $\langle z^*, z \rangle$, and the optimum value of the dual is domestic income $\langle \boldsymbol{y}^*, \boldsymbol{y} \rangle$. That there is no duality gap means that domestic product equals domestic income.

Because there is no duality gap, for any solution to the primal and the dual the subtraction of the primal from the dual must be zero. That $\boldsymbol{y}^* \geq \mathbf{0}$ is implicit in the cone C_i constraint in the dual, so one is free to augment the dual objective function with the constraint $\delta_{\boldsymbol{y}^* \geq \mathbf{0}}$. The subtraction factors into the sum of nonnegative terms:

$$\begin{aligned} 0 = &\sum_i \left[\delta_{\text{cone } D_i}(\boldsymbol{x}_i, -z_i) + \delta_{\text{cone } C_i}(\boldsymbol{y}^*, -z_i^*) - (z_i^* z_i - \langle \boldsymbol{y}^*, \boldsymbol{x}_i \rangle)\right] \\ &+ \left(\delta_{\boldsymbol{y} \geq X\mathbf{1}} + \delta_{\boldsymbol{y}^* \geq \mathbf{0}} + \langle \boldsymbol{y}^*, \boldsymbol{y} - X\mathbf{1} \rangle\right). \end{aligned}$$

On the first line, the expression in brackets is nonnegative, by the Shephard inequality. The second line is also nonnegative, by complementary slackness. That each expression be zero is therefore necessary and sufficient for a solution to the primal and the dual.

For a domestic-product primal and a domestic-income dual, it is a general proposition that there is production equilibrium if and only if the prices and quantities solve the primal and the dual. Nevertheless let us show this relationship via the necessary and sufficient conditions.

Production equilibrium requires profit maximization with zero profit, so for each sector the Shephard equality holds. That an input price must be zero if there is excess supply implies the complementary slackness. The necessary and sufficient conditions hold, so a production equilibrium solves the primal and the dual.

Conversely, suppose that the prices and quantities solve the primal and the dual. That $\boldsymbol{z}^* \gg \mathbf{0}$ implies $\boldsymbol{y}^* > \mathbf{0}$. That the Shephard equality holds for each sector implies profit maximization with zero profit. The complementary slackness implies that an input price is zero if there is excess input supply. Hence there is production equilibrium.

41.2 Two Sectors and Two Inputs

Let us analyze the special case of two sectors and two inputs, capital and labor. In our framework, a production equilibrium might potentially exhibit specialization, in which only one good is produced. Furthermore, there might be excess supply of one input. However below we suppose that both goods are produced, $\boldsymbol{z} \gg \mathbf{0}$; and both inputs are fully utilized, with $\boldsymbol{y}^* \gg \mathbf{0}$.

The figures below graph the solution to the primal and the dual. Precise drawing makes the figures mutually consistent. Each sector has Cobb-Douglas production, so in production equilibrium the factor income shares are constant. In sector 1, the labor share is $1/3$ and the capital share is $2/3$. In sector 2, the shares are both $1/2$. Production in sector 1 is relatively labor-intensive, whereas sector 2 is capital-intensive.

Graphed in input-price space, Fig. 41.1 depicts the domestic-income dual. The isocost set $z_i^* C_i$ is all input prices $\boldsymbol{y}^*$ such that the minimum cost is greater than or equal to the output price z_i^*. The dual solution $\boldsymbol{y}^*$ is constrained to lie in the intersection of these sets.

In the figure, the dual solution is $\boldsymbol{y}^*$, where $\langle \boldsymbol{y}^*, \boldsymbol{y} \rangle$ is lowest. The downward-sloping, constant-value line has normal $\boldsymbol{y}$; the value of input $\boldsymbol{y}$ is constant along it. The solution is at the intersection of the lower boundaries of the isocost sets. For each good, the output price is the minimum cost of production.

That $\boldsymbol{y}^*$ is positive requires full utilization of input, $\boldsymbol{y} = \boldsymbol{x}_1 + \boldsymbol{x}_2$. The Shephard equality for sector i requires

$$-\boldsymbol{x}_i \in \partial \delta_{z_i^* C_i}(\boldsymbol{y}^*); \tag{41.2}$$

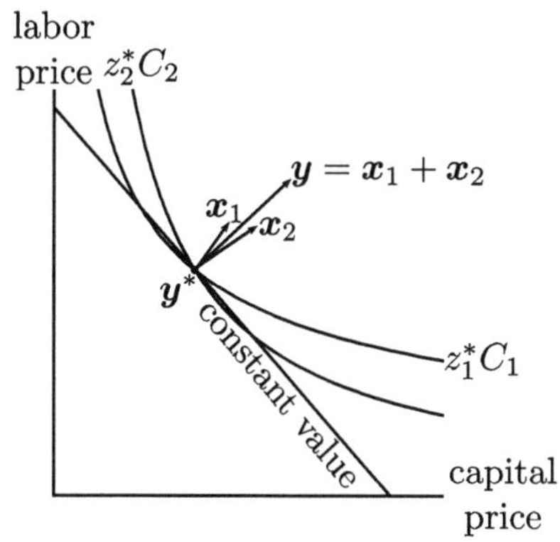

Fig. 41.1 Woodland diagram

otherwise cost is not minimized.[1] In the figure, this condition sets the capital/labor ratio in each sector. The choices $\boldsymbol{x}_i$ are normalized so all input is utilized. Each output z_i is positive.

Figure 41.1 applies as long as $\boldsymbol{y}$ lies within the "diversification" cone generated by the $\boldsymbol{x}_i$. Holding the world prices fixed, a small change in $\boldsymbol{y}$ leaves the capital/labor ratio unchanged in both sectors. The only change is renormalization of the $\boldsymbol{x}_i$, so all input is utilized. For example, if $\boldsymbol{y}$ changes so labor rises relative to capital, then output in the labor-intensive sector rises relative to output in the capital-intensive sector.

Fundamental arguments of Heckscher-Ohlin theory follow immediately. Compare two countries with identical technology but slightly different inputs. One country has a higher ratio of capital to labor than the other. This country will produce relatively more of the capital-intensive good and relatively less of the labor-intensive good, compared to the other country, as argued by Heckscher and Ohlin. Furthermore "factor-price equalization" applies: $\boldsymbol{y}^*$ is the same in both countries (Fig. 41.2) There is no economic incentive for capital or labor to flow between countries. (On the other hand, if $\boldsymbol{y}$ shifts outside the diversification cone, then factor-price equalization no longer applies.)

The Stolper-Samuelson theorem asserts that imposing a tariff on the labor-intensive good raises the real wage (Stolper and Samuelson 1941). One models this idea by raising the output price z_1^*. Here z_1^* rises 5% to $\widetilde{z_1^*}$. The isocost set $\widetilde{z_i^*} C_i$ has shifted 5% away from the origin. At the new dual solution $\widetilde{\boldsymbol{y}^*}$, domestic income (and therefore domestic product) is higher. The price of labor is higher, but the price of capital is lower. Now most of the input is used to produce good 1. For each sector, the capital/labor ratio rises: $\boldsymbol{x}_i$ is steeper than $\widetilde{\boldsymbol{x}_i}$. In each sector the marginal product of labor rises, so the real wage rises, measured in terms of either good.

[1] In the chapter "Shephard Duality," the first-order condition for the cost dual is Eq. (41.2).

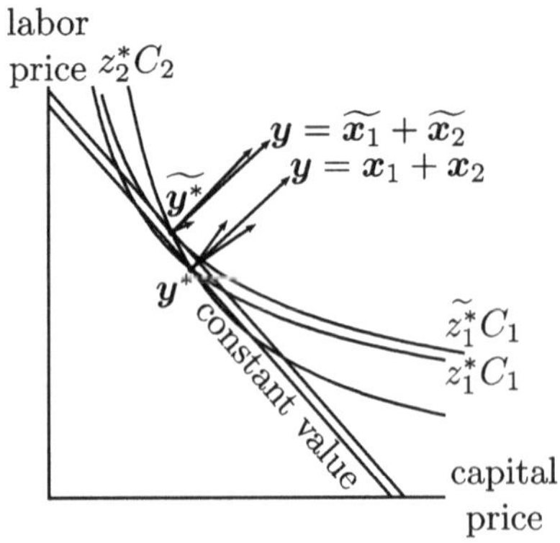

Fig. 41.2 Comparative statics

41.3 Lerner Diagram

To present Heckscher-Ohlin ideas, textbooks focus on the Lerner diagram (Fig. 41.3), plotted in input-quantity space (Lerner 1952). The Shephard equality for sector i requires

$$-\boldsymbol{y}^* \in \partial\delta_{D_i}(\boldsymbol{x}_i/z_i), \tag{41.3}$$

otherwise cost is not minimized.[2] Lerner introduces the dollar-isoquant set D_i/z_i^* of input that can produce one dollar of output or more. Given $\boldsymbol{y}^*$, consider the cost-one hyperplane of all inputs costing one dollar. Restate (41.3) as

$$-\boldsymbol{y}^* \in \partial\delta_{D_i/z_i^*}(\boldsymbol{x}_i/z_i^* z_i).$$

Together with the zero-profit requirement $z_i^* z_i = \langle \boldsymbol{y}^*, \boldsymbol{x}_i \rangle$, this condition says that the cost-one hyperplane supports D_i/z_i^* at $\boldsymbol{x}_i/z_i^* z_i$, the input per dollar of revenue.

For a production equilibrium in which both goods are produced, the cost-one hyperplane must be the same for both sectors: it must be a supporting hyperplane to both dollar-isoquant sets. In Fig. 41.3, the two tangent points are $\boldsymbol{x}_i/z_i^* z_i$, input per dollar of revenue. Each tangent point is cost-minimizing, with cost one. This solution requires that input $\boldsymbol{y}$ lie in the diversification cone generated by the two points. The input price $\boldsymbol{y}^*$ is the normal such that cost is one. Full utilization of input leads to the solution values X, $\boldsymbol{z}$, and $\boldsymbol{z}^*$.

Factor-price equalization applies: holding the world prices fixed, a small change in $\boldsymbol{y}$ leaves $\boldsymbol{y}^*$ unaffected. The Lerner diagram is unchanged, and the tangent points remain the same. The capital/labor ratio in each sector is unaffected. Output and input change in proportion, so that all input is utilized.

[2] In the chapter "Shephard Duality," the first-order condition for the production dual is Eq. (41.3).

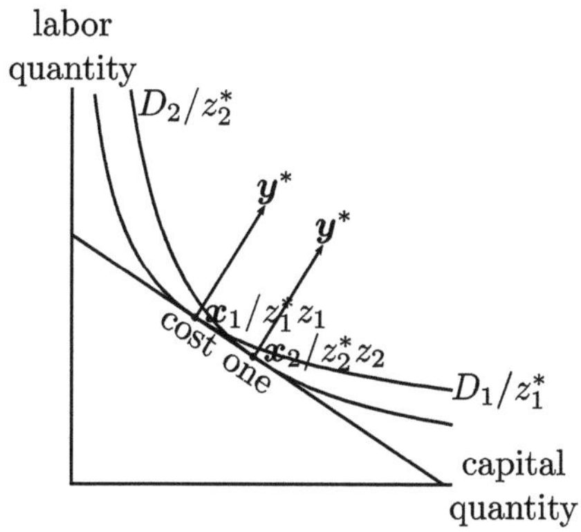

Fig. 41.3 Lerner diagram

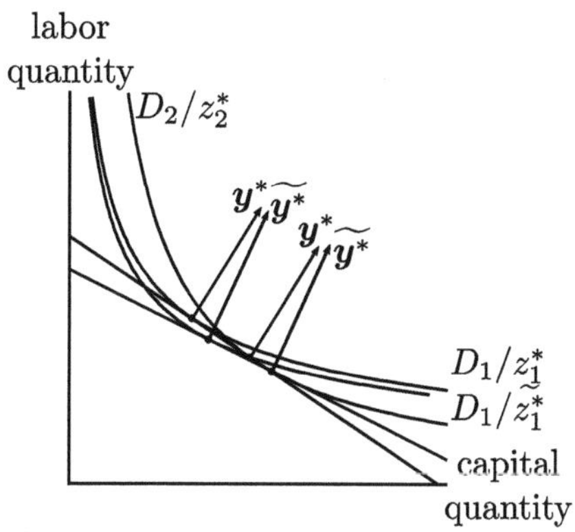

Fig. 41.4 Comparative statics

One also obtains the Stolper-Samuelson theorem. Figure 41.4 graphs the comparative statics for a 5% rise in z_1^* to $\widetilde{z_1^*}$. The dollar isoquant set $D_1/\widetilde{z_1^*}$ has shifted 5% toward the origin. The cost-one tangent hyperplane is flatter, so the price of labor rises relative to the price of capital. The two tangent points shift down and right, so the capital/labor ratio rises in both sectors. The marginal product of labor increases in both sectors, so the real wage rises in terms of either output.

References

Deardorff AV (1994) Overview of the Stolper-Samuelson theorem. In: Deardorff AV, Stern RM (eds) The Stolper-Samuelson theorem: a golden jubilee. University of Michigan Press, Ann Arbor, MI, pp 7–34

Dixit A, Norman V (1980) Theory of international trade: a dual, general equilibrium approach. James Nisbet & Co. Ltd and Cambridge University Press, Welwyn, Hertfordshire

Heckscher EF (1919) The effect of foreign trade on the distribution of income. Ekonomisk Tidskrift 21:497–512
Heckscher EF (1991). The effect of foreign trade on the distribution of income. MIT Press, Cambridge, MA
Krugman P (1994 Nov–Dec) The myth of Asia's miracle. Foreign Aff 73(6):62–78
Lerner AP (1952 Feb) Factor prices and international trade. Econ New Ser 19(73):1–15
Mussa M (1979 Nov) The two-sector model in terms of its dual: a geometric exposition. J Int Econ 9(4):513–526
Ohlin BG (1933) Interregional and international trade. Harvard University Press, Cambridge, MA
Ruffin RJ (2011) The development of international trade theory. In: Bernhofen D et al (eds) Palgrave handbook of international trade. Palgrave Macmillan, New York, pp 15–38
Stolper WF, Samuelson PA (1941 Nov) Protection and real wages. Rev Econ Stud 9(1):58–73
Woodland AD (1977) A dual approach to equilibrium in the production sector in international trade theory. Canad J Econom 10(1):50–68

Chapter 42
Factor-Intensive Production Function

Abstract A pervasive concept in economic thinking is factor intensity, the ratio of capital to labor in production. The neoclassical growth model of Solow concerns the determination of the long-run capital/labor ratio and how it sets the wage and the interest rate and consumption per worker (Solow, R. M. (1956, February). A contribution to the thory of economic growth. Quarterly Journal of Economics, 70 (1), 65–94). The Heckscher-Ohlin approach to international trade posits a key role for the capital/labor ratio and how it differs between sectors in the economy. Assuming constant returns to scale, Solow models production by the factor-intensive production function: output per worker is a nondecreasing, concave function of the capital/labor ratio. Epi-multiplication of the factor-intensive production function recovers the original production function. The conjugate of the factor-intensive production function is the factor-price frontier, the combinations of capital cost and labor cost such that the minimum cost of production is one. The conjugate evaluated at the capital cost is the negative of the labor cost. In combination the factor-intensive production function and the factor-price frontier are valuable for economic applications. Below an application of this relationship is the golden rule of saving, the saving rate that maximizes long-run per capita consumption in the neoclassical one-sector growth model (Phelps, E. (1961, September). The golden rule of accumulation: A fable for growthmen. American Economic Review, 51 (4), 638–643). Choosing the optimum saving rate reduces to the evaluation of the conjugate of the factor-intensive production function. Per capita consumption is the negative of the conjugate evaluated at the rate of population growth, and the steady-state capital/labor ratio is a subgradient. For a competitive economy in market equilibrium, that the conjugate is the factor-price frontier links the golden rule of saving to the real interest rate and the real wage. The golden rule of saving is achieved if the real interest rate equals the rate of population growth and the real wage equals per capita consumption. Saving equals profit.

B. C. Dieffenbach, *Conjugate Duality in Economic Analysis*, Contributions to Economics,
https://doi.org/10.1007/978-3-032-21396-9_42

42.1 Factor-Intensive Production Function

Supposing that capital and labor produce output under constant returns to scale, Solow models production by the factor-intensive production function: output per capita is a nondecreasing, concave function $f(x)$ of the capital/labor ratio x. Let y denote the capital input and z denote the labor input. For $z > 0$, the capital/labor ratio (the capital intensity) $x = y/z$ (Solow 1956).

Let $h(y, z)$ denote a classical production function having constant returns to scale.

Definition 1 (Factor-Intensive Production Function) The factor-intensive production function $f(y) := h(y, 1)$.

Epi-multiplication recovers the production function.

Theorem 1 (Epi-Multiplication Production)

$$h(y, z) = (z \star f)(y) = -\delta^*_{\text{epi } f_*}(-y, z).$$

For $z > 0$,

$$(z \star f)(y) = zf(y/z) = zh(y/z, 1) = h(y, z),$$

by constant returns to scale. The result also holds for $z = 0$, by the closure of the support, so $h(y, 0) = f_\infty(y)$. For $z < 0$, both $(z \star f)(y)$ and $h(y, z)$ are $-\infty$.

It is straightforward to show that the following conditions are necessary and sufficient for $(y \star f)(x)$ to be a classical production function having constant returns to scale.

Condition 1 (Factor-Intensive Production Function)

- *f is proper, closed, and concave*;
- $\operatorname{dom} f = \mathbf{R}_+$;
- $f > 0$ *for* $x > 0$;
- *f is nondecreasing.*

Perhaps $f(0) > 0$.

For the Cobb-Douglas production function

$$(z \star f)(y) = \frac{1}{1-a} y^{1-a} z^a, y \geq 0, z \geq 0, 0 < a < 1, \tag{42.1}$$

the factor-intensive form is

$$f(x) = \frac{1}{1-a} x^{1-a}, \operatorname{dom} f = \mathbf{R}_+,$$

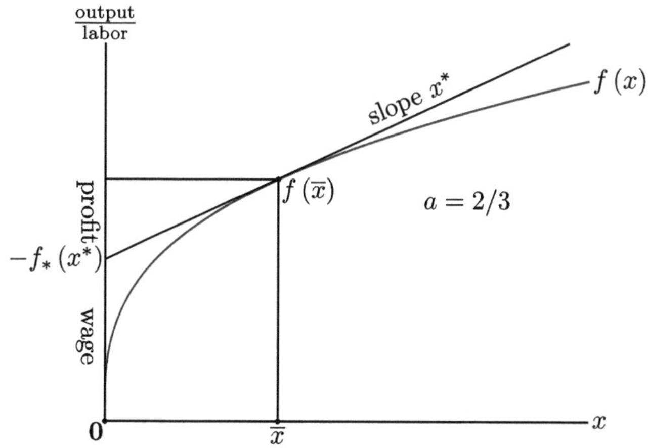

Fig. 42.1 Cobb-Douglas factor-intensive production function

the constant elasticity function with elasticity a. Its conjugate

$$f_*(x^*) = \frac{1}{1-\frac{1}{a}}(x^*)^{1-\frac{1}{a}}, \text{dom} f = \mathbf{R}_{++},$$

the constant elasticity function with elasticity $1/a$. Figure 42.1 graphs the Cobb-Douglas factor-intensive production function for $a = 2/3$.

Consider the fixed-coefficient production function in which one unit of output requires $a > 0$ units of capital and $b > 0$ units of labor,

$$(z \star f)(y) = \min\left[\frac{1}{a}y, \frac{1}{b}z\right] - \delta_{y\geq 0} - \delta_{z\geq 0}.$$

The factor-intensive form is

$$f(x) = \min\left[\frac{1}{a}x, \frac{1}{b}\right] - \delta_{x\geq 0}.$$

Figure 42.2 graphs the fixed-coefficients factor-intensive production function. The kink is $x = a/b$ (where (y, z) is a positive multiple of (a, b)). For a lower capital/labor ratio, there is surplus labor and capital is the constraint on output. For a higher ratio, there is surplus capital and labor is the constraint on output.

42.2 Factor-Price Frontier

We calculate the factor-price frontier. Expressed in terms of output, let y^* and z^* denote the factor prices of capital and labor, respectively. The cost of production is

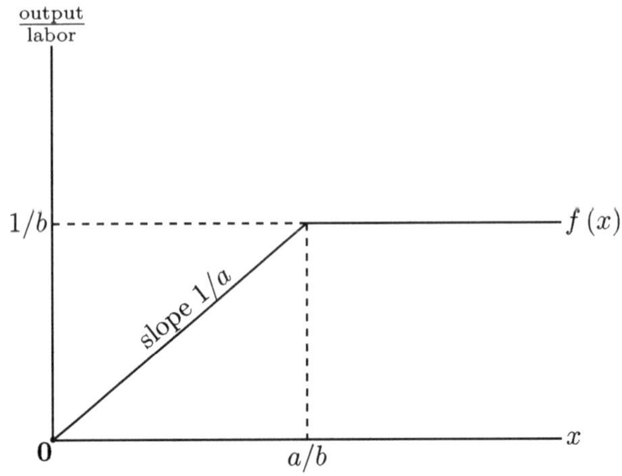

Fig. 42.2 Fixed-coefficients factor-intensive production function

the profit y^*y plus wages z^*z. The economic profit is the output minus the cost. Under constant returns to scale, the maximum economic profit is either zero or infinity. In Shephard duality, the unit-cost set C is the factor prices (y^*, z^*) such that maximum profit is zero:

$$\begin{aligned}\delta_C(y^*, z^*) &= \sup_{y,z}\left[(z \bigstar f)(y) - y^*y - z^*z\right] \\ &= \sup_{y,z}\left[-y^*y - z^*z - \delta^*_{\text{epi } f_*}(-y, z)\right] \\ &= \delta_{\text{epi } f_*}(y^*, -z^*)\,.\end{aligned} \tag{42.2}$$

The factor-price frontier is the locus $z^* = -f_*(y^*)$; the conjugate f_* is a nondecreasing function. Positive output with zero profit requires that the factor prices lie on the frontier. Figure 42.3 graphs the factor-price frontier for the Cobb-Douglas production function for $a = 2/3$.

We calculate the subdifferential of the production function in terms of the factor-intensive production function. For the profit maximization (42.2), the equivalent Fenchel equality is

$$\delta_{\text{epi } f_*}(y^*, -z^*) + \delta^*_{\text{epi } f_*}(-y, z) + y^*y + z^*z = 0.$$

For positive z, equivalently

$$f(y/z) + f_*(y^*) - y^*y/z = 0 \tag{42.3}$$

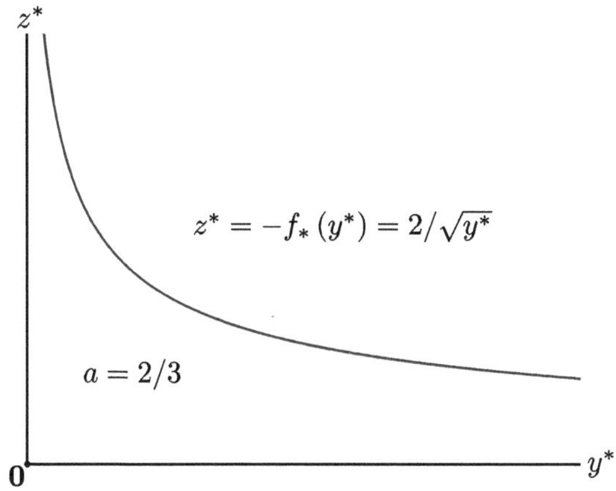

Fig. 42.3 Cobb-Douglas factor-price frontier

$$-f_*(y^*) = z^*. \tag{42.4}$$

The subdifferential $\partial(z \star f)(y)$ of the production function is all solutions (y^*, z^*) to these two equations.

Express the subdifferential in terms of the subdifferential of the factor-intensive production function. The Fenchel equality (42.3) says

$$f(x) + f_*(y^*) - y^*x = 0. \tag{42.5}$$

See this Fenchel equality (42.5) as a zero economic-profit condition: per capita, output $f(x)$ equals the wage $z^* = -f_*(y^*)$ plus the capital cost y^*x. By Eqs. (42.4) and (42.5),

$$z^* = -f_*(y^*) = f(x) - y^*x, \text{ such that } y^* \in \partial f(x).$$

Theorem 2 (Subdifferential of the Factor-Intensive Production Function) *For* $z > 0$, *the subdifferential*

$$(y^*, z^*) \in \partial(z \star f)(y) = \{[0, f(x)]\} + \begin{bmatrix} 1 & -x \end{bmatrix} \partial f(x). \tag{42.6}$$

The first-order condition for the profit maximization (42.2) is the subgradient condition $(y^*, z^*) \in \partial(z \star f)(y)$, equivalent to the marginal condition $y^* \in \partial f(x)$ together with the zero economic-profit condition $z^* = f(x) - y^*x$.

The subdifferential $\partial f_*(y^*)$ is all capital/labor ratios x that are cost-minimizing for cost of capital y^* and wage $z^* = -f_*(y^*)$, such that economic profit is zero. In Fig. 42.3, the slope of the frontier is the negative of the optimum x. Moving down the frontier reduces the factor price of labor and raises the factor price of capital, so the optimum x falls.

In Fig. 42.3, the capital/labor ratio $\overline{x}$ is profit-maximizing with economic-profit zero if the slope x^* at $f(\overline{x})$ is the cost of capital and the intercept $-f_*(x^*)$ is the cost of labor. The vertical distance from $-f_*(x^*)$ to $f(\overline{x})$ is $x^*\overline{x}$, the slope times the capital/labor ratio. Hence the zero economic-profit condition is the Fenchel equality $f(x)+f_*(x^*)-x^*\overline{x}=0$.

For the Cobb-Douglas production function, the subdifferential

$$\begin{aligned}\partial(z\star f)(y) &= \{[0,f(x)]\}+\begin{bmatrix}1 & -x\end{bmatrix}\partial f(x)\\ &= \left\{\left(0,\frac{1}{1-a}x^{1-a}\right)\right\}+\begin{bmatrix}1 & -x\end{bmatrix}\{x^{-a}\}\\ &= \left\{\left(x^{-a},\frac{a}{1-a}x^{-a}\right)\right\},\end{aligned}$$

in agreement with the differentiation of the production function (42.1).

For the fixed-coefficients production function, the factor-price frontier is the conjugate

$$f_*(x^*)=\begin{cases}0 & \text{if } x^*\geq 1/a\\ (ax^*-1)/b & \text{if } 1/a\geq x^*\geq 0\\ -\infty & \text{if } x^*<0.\end{cases}$$

For y^* in the middle range $1/a > y^* > 0$, the subdifferential $\{a/b\}$ is the cost-minimizing capital/labor ratio for factor prices y^* and $z^*=-f_*(y^*)$, for which economic profit is zero.

At the kink $x=a/b$ (where (y,z) is a positive multiple of (a,b)), the subdifferential

$$\begin{aligned}\partial(z\star f)(y) &= \{(0,f(x))\}+\begin{bmatrix}1 & -x\end{bmatrix}\partial f(x)\\ &= \{(0,1/b)\}+\begin{bmatrix}1 & -a/b\end{bmatrix}\{(0,1/a)\},\end{aligned}$$

the line segment from $(0,1/b)$ to $(1/a,0)$ (Fig. 42.2).

The sloped line is the factor-price combinations such that production at the kink is cost-minimizing, with zero economic profit.

42.3 Golden Rule of Saving

The one-sector technology is the special case in which there is only one produced good. All goods are the same—output, capital, and consumption are the same good. In a competitive economy in market equilibrium, then y^* is the real interest rate and z^* is the real wage.

Solow's neoclassical growth model analyzes the long-run implication of an unchanging one-sector technology, saving at a constant fraction $1>s>0$ of domestic income, and constant population growth $n^* > 0$. All resources are fully employed, and there is no government. Domestic income is consumption plus saving. Excluding extreme cases, he argues that the economy converges to long-run steady-state growth, in which aggregates (capital, consumption, investment, saving) all grow at rate n^*. Per capita magnitudes are constant. In a competitive economy in market equilibrium, the real interest rate and the real wage are constant.

Capital widening refers to capital accumulation to equip newborn people with capital. Capital deepening refers to capital accumulation that raises the capital/labor ratio. In long-run steady-state growth, all saving is for capital widening; no saving remains for capital deepening. The per capita condition for long-run steady-state growth is

$$sf(x) = n^*x. \tag{42.7}$$

The left-hand side is per capita saving. The right-hand side is per capita capital widening. The capital widening is proportional to n^*, as twice as many newborns calls for twice as much saving. The capital widening is also proportional to x, as doubling the capital/labor ratio calls for twice as much saving. There is steady-state growth where capital deepening is zero.

Phelps analyzes the "golden rule of saving": find the saving fraction s that maximizes per capita consumption in the long-run steady-state (Phelps 1961). The name reflects the maxim of Jesus, "do unto others as you would have them do unto you" (Matthew 7:12). The current generation saves at the high golden-rule rate, which benefits the future generation. (One need not see the golden rule of saving as the optimum: consumption in the long run is not all that matters.)

In steady-state growth, per capita consumption is $(1-s)f(x)$, output less saving. Substituting the steady-state growth condition (42.7) eliminates s, and steady-state per capita consumption is $f(x) - n^*x$. Solving for the golden rule of saving reduces to the evaluation of the conjugate of the factor-intensive production function:

$$\sup_x [f(x) - n^*x] = -\inf_x [n^*x - f(x)] = -f_*(n^*). \tag{42.8}$$

At the golden rule of saving, per capita consumption is $-f_*(n^*)$. The steady-state capital/labor ratio is any subgradient $x \in \partial f_*(n^*)$, and per capita saving is $n^*x = f(x) + f_*(n^*)$, output less consumption.

In a competitive economy in market equilibrium, the real interest rate and the real wage lie on the factor-price frontier. The first-order condition for Eq. (42.8) is $n^* \in \partial f(x)$, so the golden rule of saving is achieved if the real interest rate is n^*. The real wage $-f_*(n^*)$ is per capita consumption. Per capita, saving is profit n^*x, domestic income less the wage.

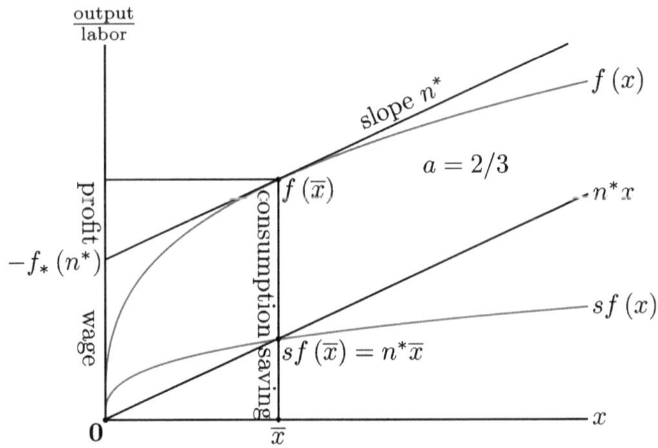

Fig. 42.4 Golden rule of saving

The Cobb-Douglas production function illustrates the golden rule of saving. The golden-rule per capita consumption is

$$-f_*(n^*) = -\frac{1}{1-\frac{1}{a}}(n^*)^{1-\frac{1}{a}}.$$

Taking the subdifferential obtains

$$x = (n^*)^{-1/a}.$$

Substituting into (42.7) obtains

$$\frac{s}{1-a}\left[(n^*)^{-1/a}\right]^{1-a} = n^*(n^*)^{-1/a},$$

which requires that $s = 1 - a$, the profit share of domestic income.

Figure 42.4 portrays the golden rule of saving for the Cobb-Douglas production function. The optimum capital/labor ratio is $\{\overline{x}\} = \partial f_*(n^*) = \left\{(n^*)^{-3/2}\right\}$. The tangent to $f(x)$ has slope n^*, so in a market economy the real interest rate is n^* and the real wage is $-f_*(n^*) = 2/\sqrt{n^*}$. Because the capital widening ray n^*x is parallel to the tangent, per capita consumption equals the real wage, and saving equals profit. The optimum saving rate $s = 1/3$.

Phelps (1961) finds the golden rule of saving for fixed-coefficients production. Assume $n^*a < 1$. Production is at the kink of the production function, $x = a/b$. Per capita consumption is $-f_*(n^*) = -(1 - n^*a)/b > 0$. Taking the subdifferential verifies $\{x\} = \{a/b\} = \partial f_*(n^*)$. Per capita output is $1/b$, and one calculates $s = n^*a$. The golden rule of saving is achieved in a competitive market equilibrium if the real interest rate $y^* = n^*$ and the real wage z^* is per capita consumption. (The golden rule of saving can also be achieved with a real interest rate that is slightly higher or lower.)

References

Phelps E (1961 Sept) The golden rule of accumulation: a fable for growthmen. Am Econom Rev 51(4):638–643

Solow RM (1956 Feb) A contribution to the thory of economic growth. Quart J Econom 70(1):65–94

Chapter 43
Heckscher-Ohlin Via Factor-Intensive Production Function

Abstract We develop further the Heckscher-Ohlin international-trade theory, by analyzing the two-sector, two-input model of production. The domestic-product primal maximizes the output value, subject to the technology and input constraints. Because the theory focuses on factor endowments and the capital/labor ratio in production, it is natural to set up the primal via factor-intensive production functions. The domestic-income dual then involves the factor-price frontier, the conjugate of the factor-intensive production function. The dual chooses input prices to minimize the value of the resource endowments, subject to the constraint that the prices lie on or above the factor-price frontiers. The necessary and sufficient conditions for solution to the primal and the dual provide the basis for a traditional comparative-statics analysis. The theorems of the Heckscher-Ohlin theory—factor-price equalization, Rybczynski, Heckscher-Ohlin direction of trade, and Stolper-Samuelson—revolve around factor prices and factor intensity. Differentiating the necessary and sufficient conditions obtains simple formulas that prove these theorems. The envelope condition explains key results.

We analyze production equilibrium in a two-sector model. In sector i, capital input x_i and labor input y_i produce z_i of the good. Define input and output vectors $\boldsymbol{x}, \boldsymbol{y}$, and $\boldsymbol{z}$. Returns to scale are constant. Let f_i denote the concave intensive production function, and the epi-multiplication $(y_i \star f_i)(x_i)$ is the production function.

The resource endowments of capital and labor are $x > 0$ and $y > 0$, respectively. Let $\boldsymbol{z}^* = (z_1^*, z_2^*) \gg \boldsymbol{0}$ denote the world prices for the outputs. Below x and y are primal perturbations, and $\boldsymbol{z}^*$ is the dual perturbation.

43.1 Domestic-Product Primal

The domestic-product primal maximizes the value of output, subject to resource endowment constraints.

B. C. Dieffenbach, *Conjugate Duality in Economic Analysis*, Contributions to Economics,
https://doi.org/10.1007/978-3-032-21396-9_43

Problem 1 (Domestic-Product Primal)

$$\sup_{x,y}\left[z_1^*(y_1 \star f_1)(x_1) + z_2^*(y_2 \star f_2)(x_2)\right]$$

such that

$$x_1 + x_2 \leq x$$
$$y_1 + y_2 \leq y.$$

This problem is just like the domestic-product primal set up previously via Shephard duality. Both problems maximize the value of the domestic product. Both have the same choice variables and the same perturbations. Yet the formulation via factor-intensive production rather than via unit-isoquant and unit-isocost sets makes the presentation quite different.

To calculate the dual, restate the primal via cone f_i. The indicator of the production cone of feasible inputs and output (x_i, y_i, z_i) is the indicator of the epigraph of the production function:

$$\delta_{(y_i \star f_i)(x_i) \geq z_i} = \delta_{\mathrm{epi}(y_i \star f_i)}(x_i, z_i) = \delta_{\mathrm{cone}\, f_i}(x_i, z_i, -y_i). \tag{43.1}$$

Below we describe production by the cone representation.

For prices (x_i^*, y_i^*, z_i^*), the maximum profit is either zero or ∞, by constant returns to scale. Maximum profit is a conjugate (apart from sign):

$$\sup_{x_i, y_i, z_i}\left[z_i^* z_i - x_i^* x_i - y_i^* y_i - \delta_{\mathrm{cone}\, f_i}(x_i, z_i, -y_i)\right] = \delta^*_{\mathrm{cone}\, f_i}(-x_i^*, z_i^*, y_i^*). \tag{43.2}$$

The indicator of the price cone such that no choice of inputs and output gives a positive profit is therefore

$$\begin{aligned}\delta^*_{\mathrm{cone}\, f_i}(-x_i^*, z_i^*, y_i^*) = \delta_{\mathrm{cone}\, f_{i*}}(x_i^*, -y_i^*, -z_i^*) &= \delta_{\mathrm{epi}(z_i^* \star f_{i*})}(x_i^*, -y_i^*) \\ &= \delta_{(z_i^* \star f_{i*})(x_i^*) \geq -y_i^*}.\end{aligned} \tag{43.3}$$

The factor-price frontier is $(z_i^* \star f_{i*})(x_i^*) = -y_i^*$, the prices for which maximum profit is zero. For $z_i^* > 0$, at a point on the frontier, the subdifferential $\partial f_{i*}(x_i^*/z_i^*)$ is the set of profit-maximizing capital/labor ratios x_i/y_i. The domestic-product primal involves the two production functions, and we interpret the domestic-income dual below in terms of the two factor-price frontiers.

We interpret the Fenchel inequality for the cone representation:

$$\begin{aligned}0 &\leq \delta_{\mathrm{cone}\, f_i}(x_i, z_i, -y_i) + \delta^*_{\mathrm{cone}\, f_i}(-x^*, z_i^*, y^*) - \langle(-x^*, z_i^*, y^*), (x_i, z_i, -y_i)\rangle \\ &= \delta_{(y_i \star f_i)(x_i) \geq z_i} + \delta_{(z_i^* \star f_{i*})(x^*) \geq -y^*} - (z_i^* z_i - x^* x_i - y^* y_i)\end{aligned}$$

Each expression contains three terms. For the first term to be zero means that the inputs and the output lie in the production cone. The production is feasible. For the second term to be zero means that the prices lie in the price cone. The maximum profit is either zero or negative. The input prices lie on or above the factor-price frontier for the output price. The third term is the profit. The Fenchel inequality that the sum of the three terms is greater than or equal to zero means that profit is less than or equal to zero. For the Fenchel equality to hold with equality, production is feasible, there is no profit opportunity, and profit is zero.

The domestic-product primal maximizes the output value, subject to the technology and input constraints. We restate the primal above using the epigraph of the production function.

Problem 2 (Domestic-Product Primal)

$$\begin{aligned}
&\sup_{x,y,z}\left[z_1^* z_1 + z_2^* z_2 - \delta_{(y_1\star f_1)(x_1)\geq z_1} - \delta_{(y_2\star f_2)(x_2)\geq z_2} - \delta_{x_1+x_2\leq x} - \delta_{y_1+y_2\leq y}\right]\\
&\quad = \sup_{x,y,z}\left[\langle z^*, z\rangle - \delta_{\mathrm{cone}\, f_1}(x_1, z_1, -y_1) - \delta_{\mathrm{cone}\, f_2}(x_2, z_2, -y_2)\right.\\
&\qquad\left. -\delta_{\mathbf{R}_+}(x - x_1 - x_2) - \delta_{\mathbf{R}_+}(y - y_1 - y_2)\right]
\end{aligned}$$

The primal is a Karush-Kuhn-Tucker concave optimization. There are six choice variables—the quantities of capital and labor inputs and of the outputs–and two perturbation variables—the endowments of capital and labor. The two output prices are the perturbation variables for the dual.

43.2 Domestic-Income Dual

The primal is a Karush-Kuhn-Tucker concave optimization, and the Lagrangian is

$$\begin{aligned}
&L(\boldsymbol{x}, \boldsymbol{y}, z, x^*, y^*)\\
&\quad = \langle z^*, z\rangle + x^*(x - x_1 - x_2) + y^*(y - y_1 - y_2)\\
&\qquad - \delta_{\mathrm{cone}\, f_1}(x_1, z_1, -y_1) - \delta_{\mathrm{cone}\, f_2}(x_2, z_2, -y_2) + \delta_{x^*\geq 0} + \delta_{y^*\geq 0}.
\end{aligned}$$

For a domestic-product primal and a domestic-income dual, a price-quantity combination $(x^*, y^*, \boldsymbol{x}, \boldsymbol{y}, z)$ is a production equilibrium if and only if it is a saddle point of the Lagrangian. It is straightforward to show this property directly using this Lagrangian.

The objective function in the reciprocal dual is

$$\begin{aligned}
&\sup_{x,y,z} L(\boldsymbol{x}, \boldsymbol{y}, z, x^*, y^*)\\
&\quad = x^* x + y^* y + \delta_{x^*\geq 0} + \delta_{y^*\geq 0}\\
&\qquad + \sum_i \sup_{x_i, y_i, z_i}\left[z_i^* z_i - x^* x_i - y^* y_i - \delta_{\mathrm{cone}\, f_i}(x_i, z_i, -y_i)\right]
\end{aligned}$$

$$= x^*x + y^*y + \delta_{x^*\geq 0} + \delta_{y^*\geq 0} + \sum_i \delta^*_{\text{cone } f_i}(-x^*, z_i^*, y^*)$$
$$= x^*x + y^*y + \delta_{x^*\geq 0} + \delta_{y^*\geq 0} + \sum_i \delta_{\text{cone } f_{i*}}(x^*, -y^*, -z_i^*).$$

In the evaluation of the conjugate, inputs x_i and y_i and output z_i maximize profit $z_i^* z_i - x^* x_i - y^* y_i$ subject to the constraint that the production is feasible.

The domestic-income dual chooses input prices x^* and y^* to minimize the value of the resource endowments, subject to the constraint that the prices lie on or above the factor-price frontiers.

Problem 3 (Domestic-Income Dual)

$$\inf_{x^*,y^*} \left[x^*x + y^*y + \delta_{\text{cone } f_{1*}}(x^*, -y^*, -z_1^*) + \delta_{\text{cone } f_{2*}}(x^*, -y^*, -z_2^*) + \delta_{x^*\geq 0} + \delta_{y^*\geq 0}\right]$$
$$= \inf_{x^*,y^*} \left[x^*x + y^*y + \delta_{(z_1^*\star f_{1*})(x^*)\geq -y^*} + \delta_{(z_2^*\star f_{2*})(x^*)\geq -y^*} + \delta_{x^*\geq 0} + \delta_{y^*\geq 0}\right].$$

The domestic-income dual amounts to the minimization analyzed by Woodland (1977). The constraint here that for each good the input prices lie on or above the factor-price frontier is equivalent to his constraint that for each good the minimum cost of production is greater than or equal to the output price.

Both the primal and the reciprocal dual satisfy the polyhedral/relative condition, so there is no duality gap. If the optimum value is finite, then this value is attained in both problems.

Equating the primal and the dual finds Fenchel equalities necessary and sufficient for the solution of the primal and the dual:

$$0 = \delta_{\text{cone } f_i}(x_i, z_i, -y_i) + \delta^*_{\text{cone } f_i}(-x^*, z_i^*, y^*) - \langle(-x^*, z_i^*, y^*), (x_i, z_i, -y_i)\rangle \qquad (43.4)$$
$$= \delta_{(y_i\star f_i)(x_i)\geq z_i} + \delta_{(z_i^*\star f_{i*})(x^*)\geq -y^*} - (z_i^* z_i - x^* x_i - y^* y_i) \text{ for } i = 1, 2 \qquad (43.5)$$

$$0 = \delta_{x_1+x_2\leq x} + \delta_{x^*\geq 0} + x^*(x - x_1 - x_2)$$
$$0 = \delta_{y_1+y_2\leq y} + \delta_{y^*\geq 0} + y^*(y - y_1 - y_2)$$

As explained above, the first equality means that production is feasible, there is no profit opportunity, and profit is zero. The final two equalities are complementary slackness. An input in excess supply must have a zero price.

43.3 Comparative Statics

Below we carry out a traditional comparative-statics analysis, by taking the differential of necessary and sufficient conditions for solutions to the primal and the dual. We assume that the solution to the primal and the dual is unique.[1] In each sector, output and the inputs are positive. There is no specialization. The input prices are both positive. There is no excess supply of capital and labor, and all capital and labor is utilized in production. We assume that the intensive production functions are twice differentiable.

In the unique solution to the primal and the dual, we assume that sector two is more capital intensive.

Condition 1 (Capital/Labor Intensity) *The capital/labor ratio is different in the two sectors:* $k_2 > k_1$.

Equivalently, $\lambda_1 < \lambda_2$, in which λ_i is the labor share of total cost in sector i.

As output prices and input quantities are positive, there is no exceptional case in the Fenchel equality (43.4)–(43.5), so it is equivalent to the three conditions

$$(y_i \star f_i)(x_i) = z_i \tag{43.6}$$

$$(z_i^* \star f_{i*})(x^*) = -y^* \tag{43.7}$$

$$f_i(x_i/y_i) + f_{i*}(x^*/z_i^*) - (x^*/z_i^*)(x_i/y_i) = 0 \tag{43.8}$$

The first Eq. (43.6) is the production function. The second Eq. (43.7) is the factor-price frontier. For these prices, the maximum profit is zero. Substituting from the first and second into the third—a Fenchel equality—rearranges to profit equals zero.

Eight necessary and sufficient conditions for a solution to the primal and the dual form the basis for the comparative statics analysis. We have the two production functions (43.6) and the two factor-price frontiers (43.7). The marginal-product condition

$$f_i'(k_i) = \frac{x^*}{z_i^*} \tag{43.9}$$

[1] If the primal has a unique solution, then the capital/labor ratio must be different in the two sectors. If there exists a solution for which both ratios are the same, then the solution is not unique. Shifting capital and labor in this ratio from one sector to the other leaves the total value of output unchanged. Whether all production occurs in one sector or in the other sector or in both is irrelevant. Reallocation achieves the same total output value. Furthermore, an arbitrarily small increase in the price of one output would cause specialization: all inputs are devoted to producing this output, and production of the other output is zero.

is the subgradient form of the zero-profit condition (43.8). The two resource constraints are

$$x_1 + x_2 = x$$
$$y_1 + y_2 = y.$$

Equivalently, rewrite the first resource constraint in terms of the capital/labor ratio,

$$k_1 y_1 + k_2 y_2 = x.$$

For these eight equations in the eight unknowns k_i, y_i, z_i, x^*, y^*, below we calculate differential expressions that one can solve for how these unknowns change when the perturbations change. Compare the formulas obtained to Jones, who derives comparative-statics formulas in terms of input-output coefficients Jones (1965).

The differential of the production function (43.6) is

$$\frac{\mathrm{d}\,(z_i/y_i)}{z_i/y_i} = \frac{1}{z_i/y_i} f_i' \,\mathrm{d}k_i = (1-\lambda_i)\,\frac{\mathrm{d}k_i}{k_i}.$$

Average productivity is an increasing function of the capital/labor ratio.

The key relationship used below is the differential of the factor-price frontier (43.7):

$$-\mathrm{d}y^* = \mathrm{d}\,[z_i^* f_{i*}\,(x^*/z_i^*)] = f_{i*}'\,\mathrm{d}x^* + \left(f_{i*} - \frac{x^*}{z_i^*} f_{i*}'\right) \mathrm{d}z_i^* = k_i\,\mathrm{d}x^* - f_i\,(k_i)\,\mathrm{d}z_i^*,$$

substituting from the zero-profit Fenchel equality (43.8). This expression rearranges to

$$k_i\,\mathrm{d}x^* + \mathrm{d}y^* = f_i\,(k_i)\,\mathrm{d}z_i^* \tag{43.10}$$

$$(1-\lambda_i)\,\frac{\mathrm{d}x^*}{x^*} + \lambda_i \frac{\mathrm{d}y^*}{y^*} = \frac{\mathrm{d}z_i^*}{z_i^*}, \tag{43.11}$$

both envelope relationships. Given profit-maximizing production, a change in input prices changes the minimum cost by the expressions on the left-hand side, and the change in the output price changes the revenue by the expressions on the right-hand side. That maximum profit is zero means that the cost change must equal the revenue change.

The differential of the zero-profit/marginal-product condition (43.9) is

$$\mathrm{d}f_i'\,(k_i) = f_i''\,\mathrm{d}k_i = -\left(\frac{x^*}{z_i^*}\right)^2 \frac{y^*}{x^* f_i} \left(\frac{-k_i}{y^*/x^*}\frac{f_i f_i''}{f_i'^2}\right) \frac{\mathrm{d}k_i}{k_i} = -\frac{x^*}{z_i^*}\frac{\lambda_i}{s_i}\frac{\mathrm{d}k_i}{k_i},$$

which can be written as

$$\frac{\mathrm{d}k_i}{k_i} = -\frac{s_i}{\lambda_i}\frac{\mathrm{d}\,(x^*/z_i^*)}{x^*/z_i^*}.$$

Here s_i is the elasticity of substitution between capital and labor.

The differential of the capital constraint is

$$\mathrm{d}x = k_1\,\mathrm{d}y_1 + k_2\,\mathrm{d}y_2 + y_1\,\mathrm{d}k_1 + y_2\,\mathrm{d}k_2. \tag{43.12}$$

The differential of the labor constraint is

$$\mathrm{d}y = \mathrm{d}y_1 + \mathrm{d}y_2. \tag{43.13}$$

Consider first the effect of a change in factor endowments on factor prices, with output prices given.

Theorem 1 (Factor Prices and Endowments) *For small changes, factor endowments have no effect on the factor prices.*

A corollary is factor-price equalization.

Theorem 2 (Factor-Price Equalization) *For two countries facing the same world prices and having identical technology and similar endowments of capital to labor, factor prices are the same.*

These results follow from the envelope condition (43.10):

$$\begin{aligned} k_1\,\mathrm{d}x^* + \mathrm{d}y^* &= 0 \\ k_2\,\mathrm{d}x^* + \mathrm{d}y^* &= 0, \end{aligned}$$

because output prices are given. In each sector, the change in cost is zero. As $k_1 \neq k_2$, the unique solution is $\mathrm{d}x^* = 0$ and $\mathrm{d}y^* = 0$, regardless of how the factor endowment changes.

The Rybczynski theorem describes the effect of the factor endowments on production (Rybczynski 1955). It follows quickly from how factor endowments have no effect on factor prices.

Theorem 3 (Rybczynski) *If the labor endowment increases, then the output of the labor-intensive good rises more than in proportion; and the output of the capital-intensive good falls.*

The Heckscher-Ohlin theorem is a corollary of the Rybczynski theorem.

Theorem 4 (Heckscher-Ohlin) *If technology is uniform across countries, then factor endowments guide production: a country having a higher ratio of capital to labor produces relatively more of the capital-intensive good.*

Consider a change $\mathrm{d}y > 0$ in the labor endowment, holding output prices and the capital endowment fixed. Because there is no effect on input prices and no effect on either capital/labor ratio, the two endowment differentials (43.12) and (43.13) reduce to

$$k_1\,\mathrm{d}y_1 + k_2\,\mathrm{d}y_2 = 0$$
$$\mathrm{d}y_1 + \mathrm{d}y_2 = \mathrm{d}y,$$

which have the unique solution

$$\mathrm{d}y_1 = \frac{k_2}{k_2 - k_1}\mathrm{d}y > 0$$
$$\mathrm{d}y_2 = \frac{k_1}{k_1 - k_2}\mathrm{d}y < 0,$$

formulas symmetric between the two sectors. To utilize fully both inputs, production must expand in the labor-intensive industry and contract in the capital-intensive industry.

Because $\frac{k_2}{k_2 - k_1} > 1$,

$$\mathrm{d}y_1 > \mathrm{d}y.$$

And because $y_1 < y$,

$$\frac{\mathrm{d}y_1}{y_1} > \frac{\mathrm{d}y}{y},$$

Output in sector one rises more than in proportion to the labor input, which Jones terms "magnification" (Jones 1965).

The Stolper-Samuelson theorem concerns the effect of a change in the terms of trade on factor prices, for fixed factor endowments (Stolper and Samuelson 1941).

Theorem 5 (Stolper-Samuelson) *To impose a tariff on the labor-intensive good raises the real wage, expressed in terms of either good.*

Model this effect by a price change $\mathrm{d}z_1^* > 0$, with the other perturbation variables fixed. The envelope relation (43.11) says

$$(1 - \lambda_1)\frac{\mathrm{d}x^*}{x^*} + \lambda_1\frac{\mathrm{d}y^*}{y^*} = \frac{\mathrm{d}z_1^*}{z_1^*}$$
$$(1 - \lambda_2)\frac{\mathrm{d}x^*}{x^*} + \lambda_2\frac{\mathrm{d}y^*}{y^*} = 0.$$

As $1 > \lambda_1 > \lambda_2$, the unique solution is

$$\frac{dx^*}{x^*} = -\frac{\lambda_2}{\lambda_1 - \lambda_2}\frac{dz_1^*}{z_1^*} < 0$$
$$\frac{dy^*}{y^*} = \frac{1-\lambda_2}{\lambda_1 - \lambda_2}\frac{dz_1^*}{z_1^*} > \frac{dz_1^*}{z_1^*} > 0.$$

The second formula exhibits magnification (Jones 1965). These results lead readily to the following formulas (the same formula for both i):

$$\frac{d\,(y^*/z_i^*)}{y^*/z_i^*} = \frac{1-\lambda_i}{\lambda_1 - \lambda_2}\frac{dz_1^*}{z_1^*} > 0$$
$$\frac{d\,(x^*/z_i^*)}{x^*/z_i^*} = -\frac{\lambda_i}{\lambda_1 - \lambda_2}\frac{dz_1^*}{z_1^*} < 0$$
$$\frac{dk_i}{k_i} = \frac{s_i}{\lambda_1 - \lambda_2}\frac{dz_1^*}{z_1^*} > 0.$$

The first formula shows that the wage in terms of either output rises, whereas the second formula shows that the capital price in terms of either output falls. The third formula says that the capital/labor ratio rises in both sectors, explaining why the real wage rises.

References

Jones RW (1965) The structure of simple general equilibrium models. J Polit Econ 73:557–572

Rybczynski TM (1955 Nov) Factor endowments and relative commodity prices. Econ New Ser 22(88):336–341

Stolper WF, Samuelson PA (1941 Nov) Protection and real wages. Rev Econ Stud 9(1):58–73

Woodland AD (1977) A dual approach to equilibrium in the production sector in international trade theory. Canad J Econ 10(1):50–68

Part X
Consumption

Chapter 44
Hicks and Marshall Decompositions

Abstract We contrast two alternative decompositions of the calculation of the conjugate of a concave function—the "Hicks decomposition" and the "Marshall decomposition." Each decomposes the calculation into two steps. One obtains an economic interpretation by viewing the function as the utility of consumption. Economists refer to expenditure-minimizing consumption as Hicks demand (compensated demand) (Hicks, J. R. (1946). Value and capital (2nd ed.). Oxford: Clarendon) and to utility-maximizing consumption as Marshall demand (consumer demand) (Marshall, A. (1890). Principles of economics). In the Hicks decomposition, consumption minimizes the expenditure required for a given utility. In the Marshall decomposition, consumption maximizes utility subject to the budget constraint. The expenditure function E is the minimum cost of obtaining utility u. The indirect-utility function V is the maximum utility possible, subject to the budget constraint that expenditure is less than or equal to income y. Located midway between utility and its conjugate, either the expenditure function or the indirect utility function determines the conjugate. The relationship between the expenditure function and the indirect-utility function is simple: reversing the axes in the closure of the epigraph of the expenditure function obtains the epigraph of the indirect-utility function. Suppress the price from the arguments of E and V. That E and V are nondecreasing is the key to axis reversal: for a given price,

$$\delta_{\text{epi } V}(y, u) = \delta_{\text{epi}(\text{cl } E)}(u, y).$$

We prove this axis-reversal theorem via the duals of expenditure minimization and utility maximization.

The chapter "Consumption Demand" reviews the justification for modelling consumer choice via a concave utility function, and there we present the properties of the utility function for a "classical consumer." But we emphasize that the results in this chapter apply for an arbitrary proper, closed, concave function in a Euclidean

B. C. Dieffenbach, *Conjugate Duality in Economic Analysis*, Contributions to Economics,
https://doi.org/10.1007/978-3-032-21396-9_44

space. Nonetheless, it is convenient to employ the terminology of consumption demand. Consumption $\boldsymbol{x}$ has price $\boldsymbol{x}^*$. Expenditure is price times quantity $\langle \boldsymbol{x}^*, \boldsymbol{x} \rangle$, and income is y.

The expenditure function is the minimum cost of obtaining utility u. The indirect-utility function is the maximum utility possible, subject to the budget constraint that expenditure is less than or equal to income y. The relationship between the expenditure function and the indirect-utility function is simple: reversing the axes in the closure of the epigraph of the expenditure function obtains the epigraph of the indirect-utility function.

44.1 Expenditure Function and Indirect-Utility Function

Consider a proper, closed, concave function $f(\boldsymbol{x})$. One can view the calculation of the concave conjugate

$$f_*(\boldsymbol{x}^*) = \inf_{\boldsymbol{x}} [\langle \boldsymbol{x}^*, \boldsymbol{x} \rangle - f(\boldsymbol{x})] \tag{44.1}$$

as decomposed into two steps. Replace either $\langle \boldsymbol{x}^*, \boldsymbol{x} \rangle$ or $f(\boldsymbol{x})$ by its epigraph, and then carry out the minimization in two steps. The latter replacement yields the Hicks decomposition, whereas the former replacement yields the Marshall decomposition.

These decompositions apply for any proper, closed, concave function. However one obtains a simple economic interpretation by viewing f as the utility of consumption $\boldsymbol{x}$. In the chapter "Consumption Demand" we review the justification for modelling consumer choice by a concave utility function, and there we present the properties of the utility function for a "classical consumer." Axis reversal applies for a classical consumer. However the results below apply independently of this interpretation.

The expenditure function is the minimum cost of obtaining utility u.

Definition 1 (Expenditure Function)

$$E(u, \boldsymbol{x}^*) := \inf_{\boldsymbol{x}} \left[\langle \boldsymbol{x}^*, \boldsymbol{x} \rangle + \delta_{f(\boldsymbol{x}) \geq u} \right] \tag{44.2}$$

The minimization is a Karush-Kuhn-Tucker problem. Viewing u as the primal perturbation and $\boldsymbol{x}^*$ as the dual perturbation implies that E is convex in u and concave in $\boldsymbol{x}^*$. The expenditure-minimizing $\boldsymbol{x}$ is compensated demand; a change in income compensates for a change in price, to keep utility constant (Hicks 1946). There is only a substitution effect on demand but no income effect. For fixed u, let $\boldsymbol{x}_i$ denote an expenditure-minimizing consumption for price $\boldsymbol{x}_i^*$. The substitution effect

$$\langle \boldsymbol{x}_2^* - \boldsymbol{x}_1^*, \boldsymbol{x}_2 - \boldsymbol{x}_1 \rangle \leq 0;$$

a change in price changes the compensated demand in the opposite direction.

Summary 1 (Expenditure Function) The expenditure function is a convex function in utility u and a concave and positively homogeneous function in price $\boldsymbol{x}^*$.

Consumer demand maximizes utility subject to the budget constraint.[1] The indirect-utility function is the maximum utility attainable for income y and price $\boldsymbol{x}^*$.

Definition 2 (Indirect-Utility Function)

$$V(y, \boldsymbol{x}^*) := \sup_{\boldsymbol{x}} \left[f(\boldsymbol{x}) - \delta_{\langle \boldsymbol{x}^* \boldsymbol{x} \rangle \leq y} \right]. \tag{44.3}$$

The maximization is a Karush-Kuhn-Tucker problem. Viewing y as the primal perturbation reveals that V is concave in y. In contrast, the price $\boldsymbol{x}^*$ is not a perturbation, because the objective function is not jointly concave in $\boldsymbol{x}$ and $\boldsymbol{x}^*$. In general indirect utility is neither concave nor convex in price. Nevertheless, for a given income, the contour set of prices such that indirect utility is less than or equal to any particular value is convex.[2] A function having this property is called "quasi-convex."

Summary 2 (Indirect-Utility Function) Indirect utility $V(y, \boldsymbol{x}^*)$ is a concave function of income y and a quasi-convex function of the price $\boldsymbol{x}^*$. It is positively homogeneous jointly in y and $\boldsymbol{x}^*$.

The equivalence theorem follows from the Hicks and Marshall decompositions.

Theorem 1 (Equivalence) *The utility function, its conjugate, the expenditure function, and the indirect-utility function are equivalent: any one determines the other three.*

The Hicks decomposition replaces f by a maximization involving its epigraph:

$$\begin{aligned} f_*(\boldsymbol{x}^*) &= \inf_{\boldsymbol{x}} \left[\langle \boldsymbol{x}^*, \boldsymbol{x} \rangle - f(\boldsymbol{x}) \right] \\ &= \inf_{\boldsymbol{x}} \left\{ \langle \boldsymbol{x}^*, \boldsymbol{x} \rangle - \sup_u \left[u - \delta_{\mathrm{epi}\, f}(\boldsymbol{x}, u) \right] \right\} \\ &= \inf_{\boldsymbol{x}} \left\{ \langle \boldsymbol{x}^*, \boldsymbol{x} \rangle + \inf_u \left[-u + \delta_{\mathrm{epi}\, f}(\boldsymbol{x}, u) \right] \right\} \end{aligned} \tag{44.4}$$

[1] Even though economists refer to utility-maximizing consumption in Eq. (44.3) as Marshallian demand, in fact (Marshall 1890) did not put forward this exact formulation (Friedman 1949). Marshall discussed the relationship between demand and price holding other things constant, without specifying exactly what was held constant.

[2] This result is proved in the chapter "Indirect Utility."

$$
\begin{aligned}
&= \inf_u \left\{ -u + \inf_x \left[\langle x^*, x \rangle + \delta_{\text{epi}\, f}(x, u) \right] \right\} \\
&= \inf_u \left\{ -u + \inf_x \left[\langle x^*, x \rangle + \delta_{f(x) \geq u} \right] \right\} \qquad (44.5) \\
&= \inf_u [-u + E(u, x^*)].
\end{aligned}
$$

Located midway between utility and its conjugate, the expenditure function determines the conjugate.

The Marshall decomposition replaces the linear function $\langle x^*, x \rangle$ by a minimization involving its epigraph:

$$
\begin{aligned}
f_*(x^*) &= \inf_x [\langle x^*, x \rangle - f(x)] \\
&= \inf_x \left[-f(x) + \inf_y \left(y + \delta_{\langle x^*, x \rangle \leq y} \right) \right] \\
&= \inf_y \left\{ y - \sup_x \left[f(x) - \delta_{\langle x^*, x \rangle \leq y} \right] \right\} \qquad (44.6) \\
&= \inf_y [y - V(y, x^*)].
\end{aligned}
$$

Located midway between utility and its conjugate, the indirect utility function determines the conjugate.

The equivalence theorem is immediate. For example, by the Marshall decomposition, the indirect utility function determines the conjugate. By conjugate duality, the conjugate determines the utility function, which in turn determines the expenditure function.

44.2 Axis Reversal

Axis reversal connects the expenditure function and the indirect-utility function. The following analysis below pertains for a fixed price x^*. To simplify the notation, suppress the price from the arguments of E and V: the convex function $E(u)$ denotes the minimum expenditure to achieve utility u, and the concave function $V(y)$ denotes the maximum utility for income y. That E and V are nondecreasing is the key to axis reversal.

The greatest lower bound of E is also the greatest lower bound of dom V:

$$
\begin{aligned}
\inf_u E(u) &= \inf_x \left[\langle x^*, x \rangle + \delta_{\text{dom}\, f}(x) \right] \\
&= \inf_x \left[\delta_{\text{dom}\, f}(x) + \inf_y \left(y + \delta_{\langle x^*, x \rangle \leq y} \right) \right] \\
&= \inf_y \left\{ y + \inf_x \left[\delta_{\text{dom}\, f}(x) + \delta_{\langle x^*, x \rangle \leq y} \right] \right\} \\
&= \inf_y [y + \delta_{\text{dom}\, V}(y)].
\end{aligned}
$$

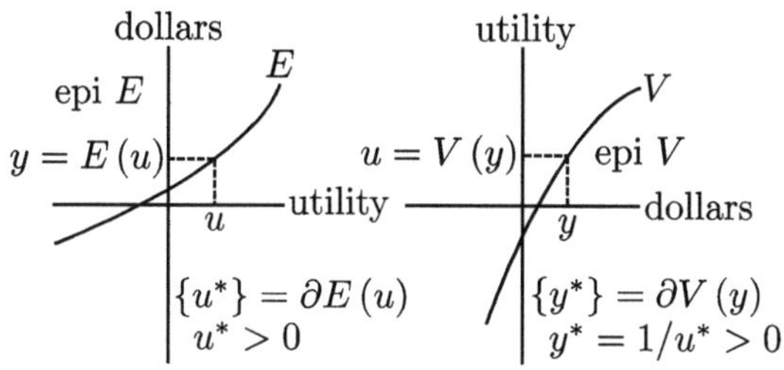

Fig. 44.1 Axis reversal

Let $\underline{y}$ denote this greatest lower bound.

The least upper bound of V is also the least upper bound of dom E:

$$\begin{aligned}\sup_y V(y) &= \sup_{\boldsymbol{x}} f(\boldsymbol{x})\\ &= \sup_{\boldsymbol{x}}\left\{\sup_u\left[u-\delta_{\text{epi } f}(\boldsymbol{x},u)\right]\right\}\\ &= \sup_u\left\{u+\sup_{\boldsymbol{x}}\left[-\delta_{\text{epi } f}(\boldsymbol{x},u)\right]\right\}\\ &= \sup_u\left[u-\inf_{\boldsymbol{x}}\delta_{\text{epi } f}(\boldsymbol{x},u)\right]\\ &= \sup_u\left[u-\delta_{\text{dom } E}(u)\right].\end{aligned}$$

Let $\bar{u}$ denote this least upper bound.

That epi V reverses the axes of epi (cl E) is the key result.

Theorem 2 (Axis Reversal) *For a given* $\boldsymbol{x}^*$,

$$\delta_{\text{epi } V}(y,u) = \delta_{\text{epi(cl } E)}(u,y).$$

The theorem refers to cl E, because the expenditure function is possibly not closed at $\bar{u}$ (it is closed for every $u < \bar{u}$). We prove the axis-reversal theorem via the duals of expenditure minimization and utility maximization.

If $y = E(u)$, then $V(y) \geq u$, because any expenditure-minimizing consumption must satisfy $f(\boldsymbol{x}) \geq u$.

The analysis of axis reversal consists of three cases—two exceptional cases and the intermediate case. Figure 44.1 illustrates the relationship. Let u^* denote the Lagrange multiplier for the Karush-Kuhn-Tucker expenditure-minimization (44.2). The case at the greatest lower bound $\underline{y}$ is the exceptional case in which $u^* = 0$. Let y^* denote the Lagrange multiplier for the Karush-Kuhn-Tucker utility-maximization (44.3). The case at the least upper bound $\bar{u}$ is the exceptional case in which $y^* = 0$.

The intermediate case is between these bounds, where $u < \overline{u}$ and $E(u) > \underline{y}$, which supposes $\overline{u} = \infty$ and $\underline{y} = -\infty$.) Here E is closed and subdifferentiable and increasing. For $y = E(u)$, then $V(y) = u$. Utility u can be attained by income y. For higher utility to be attainable ($V(y) > u$) would contradict the condition that $E(u)$ is increasing at u. Any subgradient $u^* \in \partial E(u)$ is positive, and its inverse $1/u^* = y^* \in \partial V(y)$. In this region V is the inverse of E, but elsewhere in general this relationship does not hold.

Theorem 3 (Conjugate of the Expenditure Function) *For fixed price $\boldsymbol{x}^*$, the conjugate of the expenditure function is*

$$E^*(u^*) = -(u^* \bigstar f_*)(\boldsymbol{x}^*).$$

Because E is nondecreasing, E^* is nonincreasing. We evaluate:

$$\begin{aligned}
&\sup_u [u^* u - E(u)] \\
&\quad = \sup_u \left\{u^* u - \inf_{\boldsymbol{x}} \left[\langle \boldsymbol{x}^*, \boldsymbol{x}\rangle + \delta_{f(\boldsymbol{x}) \geq u}\right]\right\} \\
&\quad = \sup_u \left\{u^* u + \sup_{\boldsymbol{x}} \left[-\langle \boldsymbol{x}^*, \boldsymbol{x}\rangle - \delta_{f(\boldsymbol{x}) \geq u}\right]\right\} \\
&\quad = \sup_{u, \boldsymbol{x}} \left[u^* u + \langle -\boldsymbol{x}^*, \boldsymbol{x}\rangle - \delta_{\operatorname{epi} f}(\boldsymbol{x}, u)\right] \\
&\quad = \delta^*_{\operatorname{epi} f}(-\boldsymbol{x}^*, u^*) \\
&\quad = -(u^* \bigstar f_*)(\boldsymbol{x}^*).
\end{aligned}$$

Theorem 4 (Conjugate of the Indirect Utility Function) *For fixed price $\boldsymbol{x}^*$, the concave conjugate of the indirect-utility function is*

$$V_*(y^*) = f_*(\boldsymbol{x}^* y^*) - \delta_{y^* \geq 0}.$$

Because V is nondecreasing, V_* is nonincreasing. We evaluate:

$$\begin{aligned}
V_*(y^*) &= \inf_y [y^* y - V(y)] \\
&= \inf_y \left\{y^* y - \sup_{\boldsymbol{x}} \left[f(\boldsymbol{x}) - \delta_{\langle \boldsymbol{x}^*, \boldsymbol{x}\rangle \leq y}\right]\right\} \\
&= \inf_{\boldsymbol{x}, w} \left[y^*(w + \langle \boldsymbol{x}^*, \boldsymbol{x}\rangle) - f(\boldsymbol{x}) + \delta_{w \geq 0}\right] \\
&\quad (w := y - \langle \boldsymbol{x}^*, \boldsymbol{x}\rangle) \\
&= \inf_{\boldsymbol{x}} [\langle \boldsymbol{x}^* y^*, \boldsymbol{x}\rangle - f(\boldsymbol{x})] - \sup_w \left[-y^* w - \delta_{w \geq 0}\right] \\
&= f_*(\boldsymbol{x}^* y^*) - \delta_{y^* \geq 0}
\end{aligned}$$

Assume first that E is proper, so conjugate duality applies to cl E.

Problem 1 (Expenditure-Minimization Dual) For perturbation u, the dual of expenditure minimization is

$$(\mathrm{cl}\, E)(u) = \sup_{u^*} [u^* u - E^*(u^*)] = \sup_{u^*} [u^* u + (u^* \bigstar f_*)(\boldsymbol{x}^*)]. \tag{44.7}$$

Problem 2 (Utility-Maximization Dual) For perturbation y, the dual of utility maximization is

$$\inf_{y^*} [y^* y - V_*(y^*)] = \inf_{y^*} \left[y^* y - f_*(\boldsymbol{x}^* y^*) + \delta_{y^* \geq 0} \right]. \tag{44.8}$$

Of course the optimum value of the dual is greater than or equal to the optimum value of the primal.

Simplifying the analysis greatly, the axis-reversal concept sidesteps the matter of whether the optimum value is attained in these primals and duals. Define the axis-reversal function

$$\delta_{\mathrm{epi}\ W}(y, u) = \delta_{\mathrm{epi}(\mathrm{cl}\ E)}(u, y).$$

Using the expenditure-minimization and utility-maximization duals, we show that $W = V$.

The subgradients $u^* \in \partial (\mathrm{cl}\, E)(u)$ constitute the solution set of the expenditure-minimization dual (44.7). Because E is nondecreasing, $u^* \geq 0$.

Consider the case in which a positive subgradient $u^* \in \partial (\mathrm{cl}\, E)(u)$. By the axis reversal, $W(y) = u$, such that $y = E(u)$. Thus $V(y) \geq W(y)$. The inverse $y^* = 1/u^*$ is a subgradient $y^* \in \partial W(y)$.

Because u^* solves the expenditure-minimization dual,

$$\begin{aligned} y &= (\mathrm{cl}\, E)(u) \\ &= u^* u - (u^* \bigstar f_*)(\boldsymbol{x}^*) \\ &= u/y^* - (1/y^*) f_*(\boldsymbol{x}^* y^*). \end{aligned}$$

Rearranging and seeing that the right-hand side equals the objective function in the utility-maximization dual implies

$$u = y^* y - f_*(\boldsymbol{x}^* y^*) \geq V(y). \tag{44.9}$$

By this inequality, $W(y) \geq V(y)$. It follows that $W(y) = V(y)$, and y^* solves the utility-maximization dual.

Figure 44.1 illustrates this case. The left-hand side graphs the expenditure function. Its slope at u is the subgradient $u^* > 0$. Because the expenditure function is nondecreasing, its epigraph is unbounded to the left, so reversing the axes obtains the epigraph of a concave function. The right-hand side graphs the indirect-utility function obtained. Its slope at y is the subgradient $y^* = 1/u^* > 0$.

Consider the exceptional case of a zero subgradient $0 \in \partial (\mathrm{cl}\, E)(u)$. This first-order condition implies that expenditure is minimized at u, and $E(u) = \underline{y}$, the greatest lower bound. If there is also a positive subgradient, then the previous analysis applies.

Figure 44.2 graphs the case in which $\tilde{u}$ is the highest utility having a zero subgradient, and there is no positive subgradient. Taking the limit as utility approaches $\tilde{u}$ from above explains the situation: $\underline{y} = E(\tilde{u})$ and $\tilde{u} = W(\underline{y}) = V(\underline{y})$. As $y^* = 1/u^* \to \infty$ the objective function in the utility-maximization dual approaches its infimum $V(\underline{y})$.

That the expenditure function is horizontal for utility less than $\tilde{u}$ squares with the possibility that consumption vector can be compensated demand but not consumer demand. A vector can perhaps minimize expenditure at $y = E(u)$ but have utility less than $V(y)$.

Figure 44.3 illustrates the other extreme, at the greatest lower bound $\overline{u}$. The boundary of $\mathrm{epi}\,(\mathrm{cl}\, E)$ is vertical at utility $\overline{u}$. The function is not subdifferentiable at $\overline{u}$. Taking the limit as utility rises to $\overline{u}$, the subgradient $u^* \to \infty$ and $y^* = 1/u^* \searrow 0$. Substituting into Eq. (44.9) and taking the limit obtains

$$V(y) = u = 0y - f_*(\mathbf{0}) = \sup_x f(\boldsymbol{x}),$$

as desired. Extra income does not increase utility, because it is already at its supremum.

Axis reversal also applies if E is improper. Because E is nondecreasing, the possibilities are limited: $(\mathrm{cl}\, E)(u) = -\infty$ for $u \le \overline{u}$ and $(\mathrm{cl}\, E)(u) = \infty$ for $u > \overline{u}$. Thus

$$\delta_{\mathrm{epi}(\mathrm{cl}\ E)}(u, y) = \delta_{u \le \overline{u}}.$$

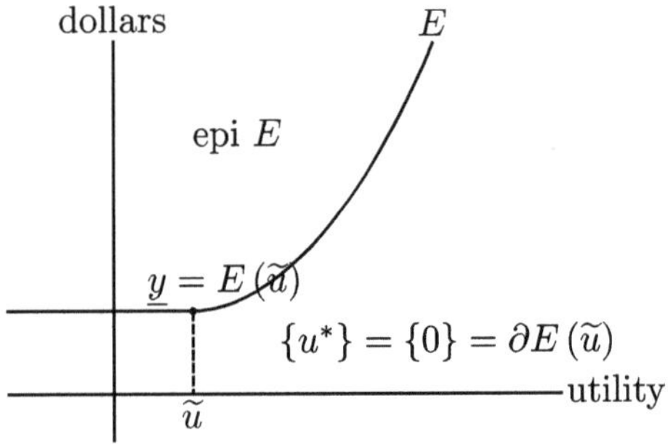

Fig. 44.2 Exceptional case

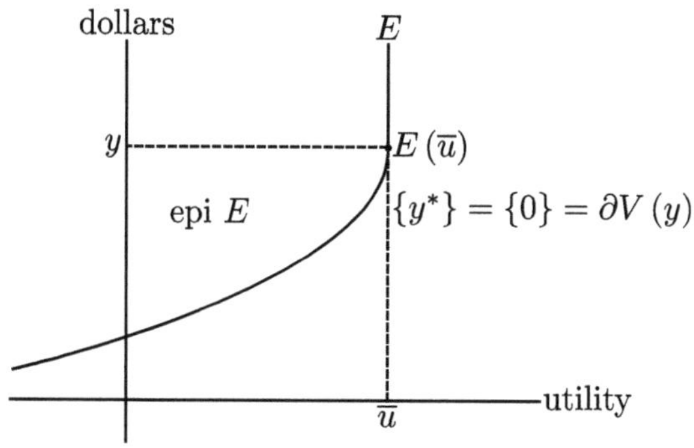

Fig. 44.3 Efficient vector

Axis reversal says

$$\delta_{\text{epi } V}(y,u) = \delta_{u \le \overline{u}},$$

so $V(y) = \overline{u}$ for all y. This result makes sense: if any expenditure, no matter how small, suffices to attain any level of utility bounded by $\overline{u}$, then any income, no matter how small, suffices to attain any level of utility bounded by $\overline{u}$.

This reasoning also applies to the case $\overline{u} = \infty$. Then $E = -\infty$ everywhere, so its epigraph is empty. Axis reversal says that the epigraph of V is also empty, meaning that the concave function $V = \infty$ everywhere. This result makes sense: if any expenditure, no matter how small, suffices to attain any level of utility, then any income, no matter how small, suffices to attain any level of utility, no matter how large.

Consider the following intuitive understanding of the formulas (44.5) and (44.6) for the concave conjugate:

$$f_*(\boldsymbol{x}^*) = -\delta^*_{\text{epi } E}(1,-1) = -\delta^*_{\text{epi } V}(-1,1).$$

When $u^* > 0$ solves the expenditure-minimization dual, then the inequality constraint is binding. If the optimum value in the primal is attained, $[u, E(u)] = [f(\boldsymbol{x}), \langle \boldsymbol{x}^*, \boldsymbol{x} \rangle]$ for any primal solution $\boldsymbol{x}$. If the optimum value is not attained, $E(u)$ is the limit of terms $\langle \boldsymbol{x}^*, \boldsymbol{x} \rangle$. The support $\delta^*_{\text{epi } E}(1,-1)$ maximizes across the terms $u - E(u) = f(\boldsymbol{x}) - \langle \boldsymbol{x}^*, \boldsymbol{x} \rangle$, so its negative is $f_*(\boldsymbol{x}^*)$. An analogous interpretation applies to $-\delta^*_{\text{epi } V}(-1,1)$.

The chapter "Consumption Demand" presents the concave utility function of a classical consumer. The setting is the Euclidean Jordan algebra $\mathbf{R}^n$, and the price $\boldsymbol{x}^* \succ \mathbf{0}$. See consumption demand as a special case of the Hicks and Marshall decompositions. Unless consumption of free goods suffices to send utility up to $\overline{u}$, the expenditure function is not constant. In this case, by axis reversal $V(y)$ is proper, so the utility-maximization primal and dual have no duality gap.

The conjugates of the expenditure function and the indirect-utility function pop up in economic applications. Seeing one of these conjugates in the dual, one infers that the optimization embodies compensated demand or consumer demand. In the efficiency analysis of resource allocation in an exchange economy, the conjugate of the expenditure function appears in the dual. By invoking axis reversal, interpret the dual via utility maximization.

References

Friedman M (1949 Dec) The Marshallian demand curve. J Polit Econ 57(6):463–495
Hicks JR (1946) Value and capital, 2nd edn. Clarendon, Oxford
Marshall A (1890) Principles of economics. Macmillan and Company, London, pp 20–22

Chapter 45
Consumption Demand

Abstract The fundamental research of Hildenbrand and Kannai justifies the use of concave utility, but what economists have not done is to develop consumer demand theory via concave utility. Because indifference curves are ordinal, the concept of the concave conjugate of utility is not part of economic thinking. Concave utility allows analysis by duality, which makes possible the derivation of results and the discovery of relationships by algebraic calculation. We specify the properties of the utility function of a "classical consumer." We contrast expenditure minimization and utility maximization via the directional derivative and the subdifferential of utility. Consider consumption $\boldsymbol{x}$ and price $\boldsymbol{v}^* > \mathbf{0}$. Utility is $f(\boldsymbol{x})$, and income/expenditure is $\langle \boldsymbol{v}^*, \boldsymbol{x} \rangle$. We study whether $\boldsymbol{x}$ is compensated demand: does it minimize expenditure among all consumption having the same utility or higher? We study whether $\boldsymbol{x}$ is consumer demand: does it maximize utility among all consumption costing the same or less? Axis reversal explains the situation. If income/expenditure is positive, then expenditure is minimized if and only if utility is maximized. If income/expenditure is zero, then necessarily expenditure is minimized, because it cannot be nonnegative. However perhaps utility is not maximized: to consume more free goods perhaps raises utility.

45.1 Concave Utility

We analyze consumer demand via a concave utility function. The fundamental research of Hildenbrand and Kannai justifies the use of concave utility, but hitherto their insight has been unappreciated by economists (Hildenbrand 1974; Kannai 1974, 1977, 1981). Hildenbrand presents a topology for preferences (Hildenbrand 1974). Under general conditions, if the preference ordering of a consumer cannot be represented by a concave utility function, then there is an ordering arbitrarily close that can be represented by concave utility. Hildenbrand and Kirman explain simply how one can construct a concave utility function from preferences (Hildenbrand and Kirman 1988, Chap. 2, appendix).

In history, initially economists saw utility as a cardinal measure permitting the interpersonal comparison of well-being. To offset this invalid point-of-view, Pareto

B. C. Dieffenbach, *Conjugate Duality in Economic Analysis*, Contributions to Economics,
https://doi.org/10.1007/978-3-032-21396-9_45

introduced ordinal utility to put forward the concept of Pareto efficiency (Pareto 2006). Hicks developed consumer demand theory via indifference curves (Hicks 1946).

Economists see consumer preferences as the fundamental concept, and Rubinstein cleverly develops consumer-demand theory directly from preferences, without resort to utility (Rubinstein 2012). Utility is not fundamental but is a theoretical construction to simplify the analysis. Debreu proves that under very general conditions one can construct a utility function from consumer preferences (Debreu 1983).

What economists have not done is to develop consumer demand theory via concave utility. Because indifference curves are ordinal, the concept of the concave conjugate of utility is not part of economic thinking. Concave utility allows analysis by duality, which makes possible the derivation of results and the discovery of relationships by algebraic calculation.

45.2 Classical Consumer

We specify the properties of the utility function of a "classical consumer" (Rubinstein 2012). Consumption $\boldsymbol{x} \in \mathbf{X}$, the Euclidean Jordan algebra $\mathbf{R}^n$. The utility function $f(\boldsymbol{x})$ is a proper, closed, concave function, for which $\operatorname{dom} f = \mathbf{X}_+$.

For a classical consumer, more of every good is better; and more of any good is no worse. We require that the utility subdifferential ∂f satisfy certain conditions, everywhere in the effective domain.

Condition 1 (Subdifferentiability) $\partial f \neq \emptyset$.

Necessarily utility is subdifferentiable in $\mathbf{X}_{++}$, but we require subdifferentiability also at a corner. This condition rules out an "infinite marginal utility" of consumption at a corner.

Condition 2 (Nonsatiation) $\mathbf{0} \notin \partial f$.

No consumption vector is an unconstrained maximum of utility.

Condition 3 (Nonnegativity) $\partial f \subset \mathbf{X}_+$.

Together with nonsatiation, this condition means that every subgradient is greater than zero. Every subgradient is nonnegative. By definition, the utility of a classical consumer satisfies these conditions, and below we analyze a classical consumer.[1]

[1] In advanced microeconomics texbooks, for simplicity it is common to assume "strong monotonicity": more of any good is better. For concave utility, $\partial f \subset \mathbf{X}_{++}$. In market equilibrium, every good has a positive price—there are no free goods. There cannot be excess supply of any good—demand equals supply for every good. Utility maximization and expenditure minimization are

The normalization $f(\mathbf{0}) = 0$ is simplifying. The conjugate f_* is nonnegative and nondecreasing, and $\operatorname{dom} f_* \subset \mathbf{X}_+$. By the Fenchel inequality

$$f(\boldsymbol{x}) + f_*(\boldsymbol{x}^*) - \langle \boldsymbol{x}^*, \boldsymbol{x} \rangle \leq 0,$$

that $f(\mathbf{0}) = 0$ implies $f_*(\boldsymbol{x}^*) \leq 0$. Also, f_* is nondecreasing, because $\langle \boldsymbol{x}^*, \boldsymbol{x} \rangle - f(\boldsymbol{x})$ is nondecreasing. Suppose that $\boldsymbol{x}^* \in \operatorname{dom} f_*$ had some component $x_i^* < 0$. Choosing $\boldsymbol{x} = -\lambda x_i^* \boldsymbol{e}_i$ for $\lambda > 0$, then

$$\begin{aligned} f_*(\boldsymbol{x}^*) &= \inf_{\boldsymbol{x}} [\langle \boldsymbol{x}^*, \boldsymbol{x} \rangle - f(\boldsymbol{x})] \\ &\leq \langle \boldsymbol{x}^*, -\lambda x_i^* \boldsymbol{e}_i \rangle - f(-\lambda x_i^* \boldsymbol{e}_i) \\ &\leq -\lambda x_i^{*2} \\ &\to -\infty \text{ as } \lambda \to \infty. \end{aligned}$$

Thus $\operatorname{dom} f^* \subset \mathbf{X}_+$. The horizon function

$$f_{*\infty}(\boldsymbol{x}^*) = -\delta^*_{\operatorname{dom} f}(-\boldsymbol{x}^*) = -\delta_{\mathbf{X}_+}(\boldsymbol{x}^*) = -\delta_{\boldsymbol{x}^* \geq \mathbf{0}}. \tag{45.1}$$

45.3 Directional Derivative and Subdifferential

We contrast expenditure minimization and utility maximization via the directional derivative and the subdifferential of utility. In line with economists' traditional exposition in terms of marginal utility, we express necessary and sufficient conditions in terms of the subdifferential. For a classical consumer, the directional derivative is proper, closed, and concave, so conjugate duality applies:

$$\begin{aligned} \delta_{\partial f}(z^*) &= -(f')_*(z^*) \\ f'(z) &= -\delta^*_{\partial f}(-z). \end{aligned}$$

Theorem 1 (Monotonicity) *The following subdifferential and directional-derivative conditions are equivalent.*

- *The subdifferential* $\partial f \subset \mathbf{X}_+, \mathbf{0} \notin \partial f$.
- *The directional-derivative* $f'(z) > 0$ *for* $z \gg \mathbf{0}$, *and* $f'(z) \geq 0$ *for* $z \geq \mathbf{0}$.

In the directional-derivative condition, the first part says that more of every good is better. The second part says that more of some goods is no worse.

---

equivalent. A compensated equilibrium is a competitive equilibrium. This book does not assume strong monotonicity.

The subdifferential condition implies the directional-derivative condition. The Fenchel equality

$$\delta_{\partial f}(z^*) + \delta^*_{\partial f}(-z) - \langle z^*, -z\rangle = 0.$$

defines a hyperplane supporting ∂f at subgradient z^* in direction $-z$. For any nonnegative z, such a hyperplane exists (if z is not nonnegative and the subdifferential is unbounded in an increasing direction, then perhaps there is no supporting hyperplane). If $z \geq \mathbf{0}$, then $f'(z) = -\delta^*_{\partial f}(-z) = \langle z^*, z\rangle \geq 0$, as $z^* \succ \mathbf{0}$. For $z \gg \mathbf{0}$, then $\langle z^*, z\rangle > 0$, so $f'(z) > 0$.

Conversely, the directional-derivative condition implies the subdifferential condition. Analyze the Fenchel inequality

$$\delta_{\partial f}(z^*) + \delta^*_{\partial f}(-z) - \langle z^*, -z\rangle \geq 0.$$

For any $z^* \in \partial f$, then $\langle z^*, z\rangle \geq -\delta^*_{\partial f}(-z) = f'(z)$, which is positive for every $z \gg \mathbf{0}$. That $z^* \succ \mathbf{0}$ follows.

The horizon cone of the subdifferential

$$(\partial f)^\infty = -(\operatorname{dom} f')^\circ. \tag{45.2}$$

The set having indicator

$$\delta_{f'(z)\geq 0}$$

is a local approximation to the contour set for consumption preferred or indifferent to $\boldsymbol{x}$. Its polar is the normal cone. The negative of the normal cone is the cone $\mathbf{R}_+\partial f \cup (\partial f)^\infty$ generated by the subdifferential.[2]

45.4 Corner

A corner is a consumption vector such that consumption is zero for one or more goods, a vector on the boundary of the effective domain of the utility function. The traditional economic analysis of a corner solution is complex and *ad hoc*, adding nonnegativity constraints for consumption. In contrast, here the complexity disappears: that consumption is nonnegative is embodied in the directional derivative and the subdifferential of utility. For a corner at which consumption is zero for goods $i \in I$,

$$\delta_{\operatorname{dom} f'}(z) = \sum_{i\in I} \delta_{\langle e_i, z\rangle \geq 0}.$$

[2] As $\partial f \neq \emptyset$, $\partial\delta_{\operatorname{dom} f} = (\partial f)^\infty$.

The horizon cone (45.2) is the negative of the cone generated by the e_i, all nonnegative linear combinations of the e_i. For $x \gg 0$, $(\partial f)^\infty = \{0\}$.

The price of consumption is $v^* > 0$. A free good is a good having price zero. For $x \geq 0$, income/expenditure $\langle v^*, x\rangle \geq 0$. Because $\langle v^*, x\rangle = 0$ requires zero consumption for any good having a positive price, the following applies.

Summary 1 (Zero Income/Expenditure) Income/expenditure $\langle v^*, x\rangle = 0$ if and only if $v^* \in (\partial f)^\infty$.

Otherwise income/expenditure is positive.

Consider a two-good, polyhedral example of consumption at a corner (Fig. 45.1). The consumption of good 1 is zero and the consumption of good 2 is positive. The derivative of utility in direction z is

$$\langle u^*, z\rangle - \delta_{\langle e_1, z\rangle \geq 0}.$$

Here $u^* \gg 0$ is the marginal utility of goods, and $\langle e_1, z\rangle = z_1$. The marginal utility of higher consumption of good 1 is u_1^*. In the direction e_1 of higher consumption of good 1, the derivative is u_1^*; in the opposite direction $-e_1$, the derivative is $-\infty$. The downward-sloping dotted line is the local approximation to the indifference curve through the corner, the direction $(u_2^*, -u_1^*)$ in which the derivative of utility is zero.

The subdifferential at the corner is

$$\partial f = \{u^*\} + e_1 \mathbf{R}_+,$$

the sum of the subdifferentials of the two terms in the directional derivative. The cone $\mathbf{R}_+\partial f \cup (\partial f)^\infty$ generated by the subdifferential is the cone generated

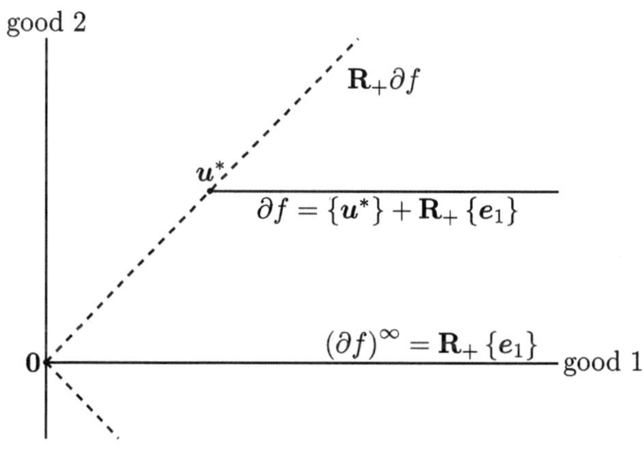

Fig. 45.1 Corner subdifferential

by $\boldsymbol{u}^*$ and $\boldsymbol{e}_1$. Here the components $\mathbf{R}_+\partial f$ and $(\partial f)^\infty$ are disjoint. The horizon cone $(\partial f)^\infty = \mathbf{R}_+\{\boldsymbol{e}_1\}$, whereas $\mathbf{R}_+\partial f$ excludes the horizontal axis, because $\mathbf{R}_+\partial f = \mathbf{R}_+(\{\boldsymbol{u}^*\} + \mathbf{R}_+\{\boldsymbol{e}_1\}) \neq \mathbf{R}_+\{\boldsymbol{u}^*\} + \mathbf{R}_+\{\boldsymbol{e}_1\}$.

45.5 Expenditure Minimization Versus Utility Maximization

We contrast expenditure minimization and utility maximization. Consider consumption $\boldsymbol{x}$ and price $\boldsymbol{v}^*$. Utility is $f(\boldsymbol{x})$, and income/expenditure is $\langle \boldsymbol{v}^*, \boldsymbol{x}\rangle$. We study whether $\boldsymbol{x}$ is compensated demand: does it minimize expenditure among all consumption having the same utility or higher? We study whether $\boldsymbol{x}$ is consumer demand: does it maximize utility among all consumption costing the same or less? If utility is maximized, then expenditure is minimized, but the converse perhaps does not hold.

Axis reversal explains the situation. If income/expenditure is positive, then expenditure is minimized if and only if utility is maximized. If income/expenditure is zero, then necessarily expenditure is minimized, because it cannot be nonnegative. However perhaps utility is not maximized: to consume more free goods perhaps raises utility.

Given $\boldsymbol{x}$, for each case we set up a primal involving the directional derivative; and the dual then involves the subdifferential. Solving the dual obtains necessary and sufficient conditions on the subdifferential for expenditure minimization and utility maximization.

45.5.1 *Expenditure Minimization*

We analyze expenditure-minimization by the following Karush-Kuhn-Tucker problem.

Problem 1 (Expenditure-Minimization Primal)

$$\inf_{z}\left[\langle \boldsymbol{v}^*, z\rangle + \delta_{f'(z)\geq 0}\right].$$

Minimize the derivative $\langle \boldsymbol{v}^*, z\rangle$ of expenditure in direction z subject to the constraint that utility stays the same or higher. The objective function is zero for the choice $z = \mathbf{0}$, and the optimum value is either zero or $-\infty$. The optimum value is zero if and only if expenditure is minimized at $\boldsymbol{x}$. By monotonicity, the Slater condition holds: $f'(z) > 0$ for some z.

Equivalently, the primal is

$$\begin{aligned}&\inf_z\left[\langle v^*,z\rangle+\delta_{f'(z)\geq 0}\right]\\&=\inf_z\left[\langle v^*,z\rangle+\delta_{-\delta^*_{\partial f}(-z)\geq 0}\right]\\&=\inf_z\left[\langle v^*,z\rangle+\delta_{\mathrm{epi}\ \delta^*_{\partial f}}(-z,0)\right]\\&=\inf_z\left[\langle v^*,z\rangle+\delta^*_{\mathrm{cone}\ \partial f}(-z,0)\right].\end{aligned}$$

We find the reciprocal dual. Rearranging the Fenchel inequality

$$\delta_{\mathrm{cone}\ \partial f}(z^*,-y^*)+\delta^*_{\mathrm{cone}\ \partial f}(-z,-y)-\langle z^*,-z\rangle-(-y^*)(-y)\geq 0$$

obtains

$$\langle z^*,z\rangle+\delta^*_{\mathrm{cone}\ \partial f}(-z,-y)\geq y^*y-\delta_{\mathrm{cone}\ \partial f}(z^*,-y^*).$$

The perturbations are $y=0$ and $z^*=v^*$. There is no duality gap.

The dual is

$$\sup_{y^*}\left[-\delta_{\mathrm{cone}\ \partial f}(v^*,-y^*)\right]=-\delta_{\mathbf{R}_+\partial f\cup(\partial f)^\infty}(v^*).$$

The choice $y^*=0$ is the exceptional case of a zero Lagrange multiplier. Expenditure is minimized at $\boldsymbol{x}$ if and only if $\boldsymbol{v}^*$ lies in the cone generated by the subdifferential. Equivalently, $\boldsymbol{v}^*\succ\mathbf{0}$ lies in the normal cone at $\boldsymbol{x}$ of the contour set of all consumption having utility greater than or equal to $f(\boldsymbol{x})$. Because $\mathbf{0}\notin\partial f$, the following summarizes the situation.

Theorem 2 (Expenditure Minimization) *Expenditure is minimized if and only if either of the following holds:*

$$y^*v^*\in\partial f\ \textit{for some}\ y^*>0; \tag{45.3}$$
$$v^*\in(\partial f)^\infty. \tag{45.4}$$

As Eq. (45.4) holds if and only if expenditure is zero, of course it implies that expenditure is minimized.

45.5.2 Utility Maximization

We analyze utility-maximization by the following Karush-Kuhn-Tucker problem.

Problem 2 (Utility-Maximization Primal)

$$\sup_z \left[f'(z) - \delta_{\langle v^*, z\rangle \leq 0}\right] = \sup_z \left[-\delta^*_{\partial f}(-z) - \delta_{\langle v^*, z\rangle \leq 0}\right].$$

Maximize the derivative of utility in direction z subject to the constraint that the directional derivative of expenditure is less than or equal to zero. The objective function is zero for the choice $z = \mathbf{0}$, and the optimum value is either zero or ∞. The optimum value is zero if and only if utility is maximized at $\boldsymbol{x}$.

In the primal, the polyhedral/relative condition holds if and only if income $\langle v^*, z\rangle$ is positive. Income is zero if and only if $v^* \in (\partial f)^\infty = -\partial\delta_{\mathrm{dom}\, f'}$. Hence $\langle v^*, z\rangle \geq 0$ for any $z \in \mathrm{dom}\, f'$, so the relative condition does not hold. If income is positive, by monotonicity there exists some $z \in \mathrm{rint}\,(\mathrm{dom}\, f')$ such that $\langle v^*, z\rangle < 0$; and the polyhedral/relative condition holds.

We find the reciprocal dual. Adding the Fenchel inequalities

$$\delta_{\partial f}(v^* y^* - z^*) + \delta^*_{\partial f}(-z) - \langle v^* y^* - z^*, -z\rangle \geq 0$$
$$\delta_{\mathbf{R}_+}(y^*) + \delta^*_{\mathbf{R}_+}(\langle v^*, z\rangle - y) - (\langle v^*, z\rangle - y)\, y^* \geq 0$$

and rearranging obtains

$$y^* y + \delta_{\partial f}(v^* y^* - z^*) + \delta_{\mathbf{R}_+}(y^*) \geq \langle z^*, z\rangle - \delta^*_{\partial f}(-z) - \delta^*_{\mathbf{R}_+}(\langle v^*, z\rangle - y).$$

The perturbations are $y = 0$ and $z^* = \mathbf{0}$. The dual is

$$\inf_{y^*} \left[\delta_{\partial f}(v^* y^*) + \delta_{\mathbf{R}_+}(y^*)\right].$$

Here y^* is a subgradient of the value function for utility maximization. Because $\mathbf{0} \notin \partial f$, the objective function is ∞ if $y^* = 0$. The optimum value of the dual is zero if and only if (45.5) holds, the traditional condition analyzed by economists.

Theorem 3 (Utility Maximization) *For positive income, the condition*

$$y^* v^* \in \partial f,\, y^* > 0, \tag{45.5}$$

is necessary and sufficient for $\boldsymbol{x}$ *to maximize utility. For zero income, the condition is sufficient but not necessary.*

The condition is sufficient for utility to be maximized at $\boldsymbol{x}$. If it holds, the optimum value of the dual is zero, so the optimum value of the primal must also be zero. When income is positive, there can be no duality gap, so the condition is necessary as well as sufficient.

When income is zero, there might be a duality gap. Perhaps the condition does not hold and the optimum value of the dual is ∞, while the optimum value of the primal is zero.

That utility-maximization implies expenditure minimization is an implication. If income is positive, then the subdifferential conditions are the same, an aspect of axis reversal. Conversely, expenditure zero does not entail utility maximization.

45.6 Examples

Figure 45.2 portrays the case $\boldsymbol{x} \gg \mathbf{0}$ and $\boldsymbol{v}^* \gg \mathbf{0}$.

Utility is differentiable at $\boldsymbol{x}$, where $\partial f = \{\boldsymbol{u}^*\}$. In the figure $\boldsymbol{v}^*$ is not proportional to $\boldsymbol{u}^*$, so (45.3) and (45.5) are violated. Using directional derivatives, we explain why utility is not maximized, and why expenditure is not minimized. The calculation is via least squares.

Utility is not maximized. Let $\lambda \boldsymbol{v}^*$ be the vector in the ray $\mathbf{R}_+ \{\boldsymbol{v}^*\}$ closest to $\boldsymbol{u}^*$: choose λ minimize $\frac{1}{2} \langle \boldsymbol{u}^* - \lambda \boldsymbol{v}^*, \boldsymbol{u}^* - \lambda \boldsymbol{v}^* \rangle$. The first-order condition is $\langle \boldsymbol{v}^*, \boldsymbol{u}^* - \lambda \boldsymbol{v}^* \rangle = 0$. Define the direction $\boldsymbol{z} := \boldsymbol{u}^* - \lambda \boldsymbol{v}^* \neq \mathbf{0}$. The derivative of expenditure in direction $\boldsymbol{z}$ is $\langle \boldsymbol{v}^*, \boldsymbol{z} \rangle = 0$, by the first-order condition. The derivative of utility in direction $\boldsymbol{z}$ is $\langle \boldsymbol{u}^*, \boldsymbol{z} \rangle = \langle \boldsymbol{u}^*, \boldsymbol{u}^* - \lambda \boldsymbol{v}^* \rangle = \langle \boldsymbol{u}^* - \lambda \boldsymbol{v}^*, \boldsymbol{u}^* - \lambda \boldsymbol{v}^* \rangle = \langle \boldsymbol{z}, \boldsymbol{z} \rangle > 0$, substituting the first-order condition.

An analogous calculation shows that expenditure is not minimized. Let $\lambda \boldsymbol{u}^*$ be the vector in the ray $\mathbf{R}_+ \{\boldsymbol{u}^*\}$ closest to $\boldsymbol{v}^*$: choosc λ minimize $\frac{1}{2} \langle \boldsymbol{v}^* - \lambda \boldsymbol{u}^*, \boldsymbol{v}^* - \lambda \boldsymbol{u}^* \rangle$. The first-order condition is $\langle \boldsymbol{u}^*, \boldsymbol{v}^* - \lambda \boldsymbol{u}^* \rangle = 0$. Define the direction $\boldsymbol{z} := -\boldsymbol{v}^* + \lambda \boldsymbol{u}^* \neq$

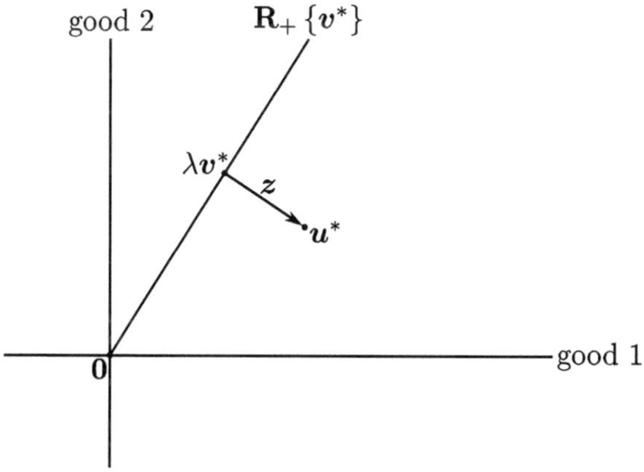

Fig. 45.2 Utility not maximized

$\mathbf{0}$. The derivative of utility in direction z is $\langle \boldsymbol{u}^*, z\rangle = 0$, by the first-order condition. The derivative of expenditure in direction z is $\langle \boldsymbol{v}^*, z\rangle = \langle \boldsymbol{v}^*, -\boldsymbol{v}^* + \lambda \boldsymbol{u}^*\rangle = -\langle -\boldsymbol{v}^* + \lambda \boldsymbol{u}^*, -\boldsymbol{v}^* + \lambda \boldsymbol{u}^*\rangle = -\langle z, z\rangle < 0$, substituting the first-order condition.

A slight modification of Fig. 45.3 portrays consumption demand at a corner. Income/expenditure is positive. Because $y^*\boldsymbol{v}^* \in \partial f$, the corner is both expenditure-minimizing and utility-maximizing. The consumer does not raise the consumption of good 1 above zero, because the extra utility obtained would be less than the utility lost by reducing the consumption of good 2.

Change the price to $\boldsymbol{v}^* \propto \boldsymbol{e}_1$. As $\boldsymbol{v}^* \propto (\partial f)^\infty$, expenditure is minimized (at zero). The sufficient condition (45.5) for utility maximization does not hold. Utility is not maximized: the derivative of utility in the vertical direction is $f'(\boldsymbol{e}_2) = -\delta^*_{\partial f}(-\boldsymbol{e}_2) = u_2^* > 0$.

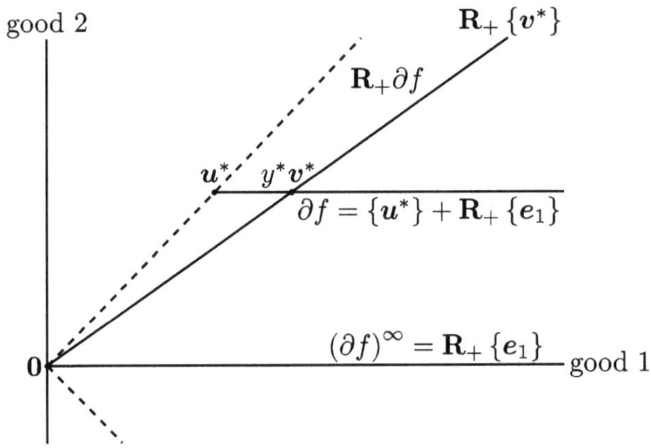

Fig. 45.3 Corner solution

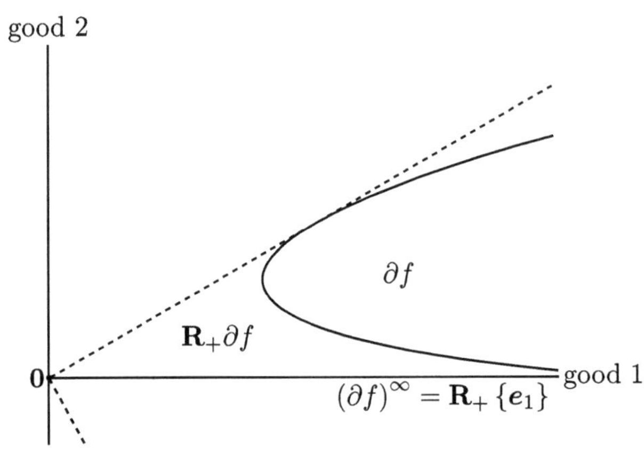

Fig. 45.4 Utility maximization duality gap

Figure 45.4 portrays consumption demand at a corner, for price $\boldsymbol{v}^* \propto \boldsymbol{e}_1$. Income/expenditure is zero. The subdifferential gradually approaches the horizontal axis but never reaches it. The cone $\mathbf{R}_+\partial f$ is the set from the dotted line sloping up and right down to the horizontal axis, but excludes the axis. The horizon cone $(\partial f)^\infty = \mathbf{R}_+\{\boldsymbol{e}_1\}$, the horizontal axis. Thus $\mathbf{R}_+\partial f$ is disjoint from $(\partial f)^\infty$. The traditional condition (45.3) does not hold; but $\boldsymbol{v}^* \propto (\partial f)^\infty$, and expenditure is minimized.

The sufficient condition (45.5) for utility maximization does not hold, and yet utility is maximized. The optimum value of the utility-maximization dual is ∞. But there is a duality gap: the optimum value of the utility-maximization primal is zero. The derivative of utility in the vertical direction is $f'(\boldsymbol{e}_2) = -\delta^*_{\partial f}(-\boldsymbol{e}_2) = 0$; utility is maximized, because the consumption of good 2 is satiated.

References

Debreu G (1983) Representation of a preference ordering by a numerical function. In: Mathematical economics: twenty papers of Gerard Debreu. Cambridge University Press, Cambridge, pp 105–110 (Originally 1952)

Hicks JR (1946) Value and capital, 2nd edn. Clarendon, Oxford

Hildenbrand W (1974) Core and equilibria of a large economy. Princeton University Press, Princeton, NJ

Hildenbrand W, Kirman AP (1988) Equilibrium analysis: variations on themes by Edgeworth and Walras, Vol 28). North-Holland, Amsterdam

Kannai Y (1974 Aug) Approximation of convex preferences. J Math Econ 1(2):101–106

Kannai Y (1977 Mar) Concavifiability and constructions of concave utility functions. J Math Econ 4(1):1–56

Kannai Y (1981) Concave utility functions: existence, constructions, and cardinality. In: Schaible S, Ziemba WT (eds) Generalized concavity in optimization and economics. Academic Press, pp 543–611

Pareto V (2006) Manuale di economia politica, con una introduzione ulla scienza sociale. Societa Editrice Libraria

Rubinstein A (2012) Lecture notes in microeconomic theory: the economic agent. Princeton University Press, Princeton, NJ (available online)

Chapter 46
Indirect Utility

Abstract By the Marshall decomposition applied to a consumer, indirect utility—the maximum utility for given income and price—is equivalent to utility; either suffices to determine the other. Consumer demand maximizes utility subject to the budget constraint set by price and income. Alternatively, see this same condition as constraining price for a given quantity and income, such that spending is less than or equal to income. A parallelogram theorem holds: quantity maximizes utility for a given price if and only if this price minimizes indirect utility for this quantity. Maximum utility equals minimum indirect utility. A parallelogram furnishes a geometric interpretation.

The consumer is classical. Consumption $\boldsymbol{x} \in \mathbf{X}$, the Euclidean Jordan algebra $\mathbf{R}^n$. The concave utility function $f(\boldsymbol{x})$ is proper, closed, and monotonic, such that $\operatorname{dom} f = \mathbf{X}_+$.

46.1 Indirect Utility

Indirect utility is the maximum utility attainable for given prices and income, the value function for utility maximization subject to the budget constraint.

Definition 1 (Indirect Utility)

$$V(\boldsymbol{v}^*, w) := \sup_{\boldsymbol{x}} \left[f(\boldsymbol{x}) - \delta_{\langle \boldsymbol{v}^*, \boldsymbol{x} \rangle \leq w} \right] \tag{46.1}$$

Indirect utility V is positively homogeneous in the price $\boldsymbol{v}^*$ and income w. For a classical consumer and nonnegative w and $\boldsymbol{v}^*$, V is increasing in w and nonincreasing in $\boldsymbol{v}^*$. If every price rises, then indirect utility falls.

B. C. Dieffenbach, *Conjugate Duality in Economic Analysis*, Contributions to Economics,
https://doi.org/10.1007/978-3-032-21396-9_46

The indirect-utility function is quasi-convex in the price. In Eq. (46.1), income w is a perturbation, so V is concave in w, for a given $\boldsymbol{v}^*$. In contrast, the price $\boldsymbol{v}^*$ is not a perturbation, because the objective function is not jointly concave in $\boldsymbol{x}$ and $\boldsymbol{v}^*$. In general indirect utility is neither concave nor convex in price. Nevertheless, for a given income, a lower contour set of prices for indirect utility is necessarily a convex set. A function having this property is called "quasi-convex."

Theorem 1 (Quasi-Convex Indirect Utility) *For a given income w, indirect utility $V(\boldsymbol{v}^*, w)$ is a quasi-convex function of the price $\boldsymbol{v}^*$.*

Since indirect utility is positively homogeneous in price and income, without loss of generality set $w = 1$, and view indirect utility $V(\boldsymbol{v}^*, 1)$ as a function only of the price $\boldsymbol{v}^*$. Let the closed, convex set C denote the upper contour set above some level of utility u; its indicator is $\delta_C(\boldsymbol{x}) = \delta_{f(\boldsymbol{x}) \geq u}$. For income $w = 1$, the lower contour set D for indirect utility for this same level of utility has indicator $\delta_D(\boldsymbol{v}^*) = \delta_{V(\boldsymbol{v}^*,1) \leq u}$.

We show that D is convex. In Fig. 46.1, the minimum income w required for $V(\boldsymbol{v}^*, w) \geq u$ is $-\delta_C^*(-\boldsymbol{v}^*) = \langle \boldsymbol{v}^*, \boldsymbol{z} \rangle$.

Indirect utility is perhaps only approached as a supremum. Hence $V(\boldsymbol{v}^*, 1) > u$ if and only if $-\delta_C^*(-\boldsymbol{v}^*) < 1$, because the marginal utility of income is positive. And $V(\boldsymbol{v}^*, 1) \leq u$ if and only if $-\delta_C^*(-\boldsymbol{v}^*) \geq 1$. Hence

$$\delta_D(\boldsymbol{v}^*) = \delta_{-\delta_C^*(-\boldsymbol{v}^*) \geq 1} = \delta_{\delta_C^*(-\boldsymbol{v}^*) \leq -1},$$

so D is convex. That indirect utility need not be convex poses no problem below, because the calculation never takes its conjugate. Figure 46.2 plots indifference curves for indirect utility. For a given income, as prices rise, indirect utility falls. Each lower contour set is convex.

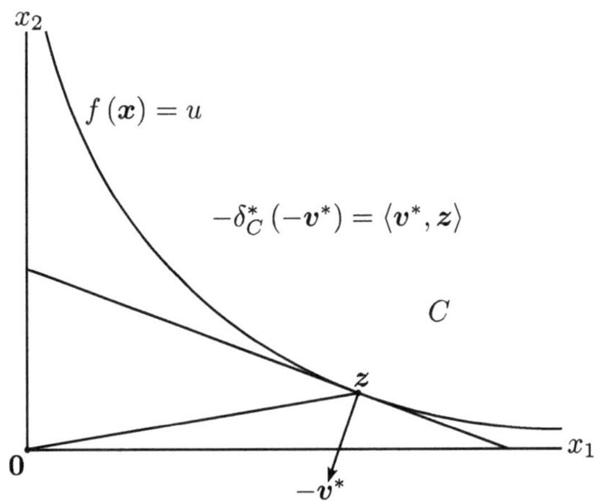

Fig. 46.1 Minimum income

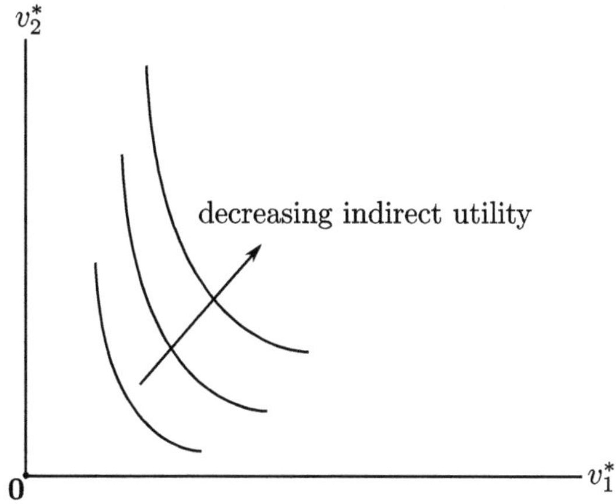

Fig. 46.2 Indirect-utility indifference curves

46.2 Parallelogram Theorem

The utility-maximization problem (46.2) chooses the quantity z to maximize utility, subject to the budget constraint set by z^*. The support to the cone representation of the price provides an alternate form for the budget constraint:

$$\delta_{\langle v^*, x\rangle \leq w} = \delta_{\text{epi } \delta^*_{\{v^*\}}}(x, w) = \delta^*_{\text{cone } \{v^*\}}(x, w).$$

Problem 1 (Utility Maximization)

$$\sup_z \left[f(z) - \delta^*_{\text{cone } \{z^*\}}(z, w) \right] \tag{46.2}$$

The indirect-utility minimization problem chooses the price z^* to minimize indirect utility, subject to the budget constraint set by z.

Problem 2 (Indirect-Utility Minimization)

$$\inf_{z^*} \left[V(z^*, w) + \delta^*_{\text{cone } \{z\}}(z^*, w) \right]. \tag{46.3}$$

For a classical consumer, any $z \in \text{dom } f$ is consumer demand for a suitable choice of price z^*. Take income w as given.

Theorem 2 (Parallelogram Theorem for Indirect Utility) *Quantity z is utility-maximizing for price z^* if and only if price z^* is indirect-utility-minimizing for quantity z. The maximum utility equals the minimum indirect utility.*

This result originates with Roy, who studies the differentiable case (Roy 1947). If the quantity z is utility maximizing for price z^*, he shows that the first-order condition for the price z^* to minimize indirect utility for quantity z holds. However the theorem does not require differentiability. Furthermore, there is no requirement that a unique quantity maximize utility, and no requirement that a unique price minimize indirect utility. The name "parallelogram theorem" derives from figure below.

We have

$$\begin{aligned} f(z) &= \delta^*_{\text{cone }\{z^*\}}(z, w) + V(z^*, w) \\ &\quad \text{(if quantity } z \text{ maximizes utility for price } z^*\text{)} \\ &\geq \inf_{v^*}\left[\delta^*_{\text{cone }\{v^*\}}(z, w) + V(v^*, w)\right] \\ &\quad \text{(possibly } v^* = z^* \text{ attains the minimum).} \end{aligned} \tag{46.4}$$

To prove the parallelogram theorem, we show that the inequality (46.4) is equality. Even though the utility maximization (46.2) and the indirect-utility minimization (46.3) do not constitute a standard primal/dual pair, the argument involves both conjugate duality and perturbation duality.

We substitute a utility-maximization dual that is an extension of the utility-maximization dual presented in "Hicks and Marshall Decompositions."

Problem 3 (Utility-Maximization Primal)

$$\sup_{x}\left[f(\boldsymbol{x}) - \delta^*_{\text{cone }\{v^*\}}(\boldsymbol{x} - \boldsymbol{y}, w)\right]$$

The perturbations have base values $\boldsymbol{y} = \boldsymbol{0}$ and income w. Unless consumption of free goods suffices to send utility up to its least upper bound, the expenditure function is not constant. In this case, by axis reversal the utility-maximization value function is proper, so the utility-maximization primal and dual have no duality gap.

Adding the Fenchel inequalities

$$-f(\boldsymbol{x}) - f_*(\boldsymbol{y}^*) + \langle \boldsymbol{y}^*, \boldsymbol{x}\rangle \geq 0 \tag{46.5}$$

$$\delta_{\text{cone }\{v^*\}}(\boldsymbol{y}^*, -w^*) + \delta^*_{\text{cone }\{v^*\}}(\boldsymbol{x} - \boldsymbol{y}, w) - \langle(\boldsymbol{y}^*, -w^*), (\boldsymbol{x} - \boldsymbol{y}, w)\rangle \geq 0 \tag{46.6}$$

and rearranging gives

$$\begin{aligned} & w^* w + \langle \boldsymbol{y}^*, \boldsymbol{y}\rangle - f_*(\boldsymbol{y}^* - \boldsymbol{x}^*) + \delta_{\text{cone }\{v^*\}}(\boldsymbol{y}^*, -w^*) \\ &\quad \geq f(\boldsymbol{x}) - \delta^*_{\text{cone }\{v^*\}}(\boldsymbol{x} - \boldsymbol{y}, w). \end{aligned}$$

The choice variables in the dual are the marginal utility $\boldsymbol{y}^*$ of goods and the marginal utility w^* of income.[1]

Problem 4 (Utility-Maximization Dual)

$$\inf_{y^*,w^*} \left[w^* w - f_* (y^*) + \delta_{\text{cone } \{v^*\}} (y^*, -w^*)\right] \tag{46.7}$$

Substituting the utility-maximization dual proves that (46.4) holds with equality:

$$\begin{aligned}
&\inf_{v^*}\left[\delta^*_{\text{cone } \{v^*\}} (z, w) + V(v^*, w)\right] \\
&= \inf_{v^*}\Big\{\delta^*_{\text{cone } \{v^*\}} (z, w) \\
&\quad + \inf_{y^*,w^*} \left[w^* w - f_* (y^*) + \delta_{\text{cone } \{v^*\}} (y^*, -w^*)\right]\Big\} \\
&\quad \text{(substituting the utility-maximization dual)} \\
&= \inf_{y^*,w^*} \{\langle y^*, z\rangle - f_* (y^*) \\
&\quad + \inf_{v^*}\left[\delta_{\text{cone } \{v^*\}} (y^*, -w^*) + \delta^*_{\text{cone } \{v^*\}} (z, w) - \langle y^*, z\rangle - (-w^*)\, w\right]\Big\} \\
&\quad \text{(reversing the order of minimization)} \\
&\geq \inf_{y^*,w^*} \left[\langle y^*, z\rangle - f_* (y^*)\right] \\
&\quad \text{(invoking the Fenchel inequality (46.6)} \\
&= f(z).
\end{aligned}$$

By the minimization inequality (46.4), all lines in this sequence equal $f(z)$, which proves the parallelogram theorem.

In Fig. 46.3, the parallelogram formed by z and z^* is furnishes a geometric interpretation.

For simplicity, income $w = 1$. The figure graphs indifference curves for utility and for indirect utility, both for utility u. Here $-z^*$ is normal to the indifference curve for utility at z. Subject to the budget constraint $\langle z^*, \boldsymbol{x}\rangle \leq 1$, the quantity z maximizes utility. And $-z$ is normal to the indifference curve for indirect utility at z^*. Subject to the budget constraint $\langle v^*, z\rangle \leq 1$, the price z^* minimizes indirect utility. The maximum utility and the minimum indirect utility are both u.

---

[1] Eliminating $\boldsymbol{y}^*$ from the this dual reduces it to the dual presented in "Hicks and Marshall Decompositions."

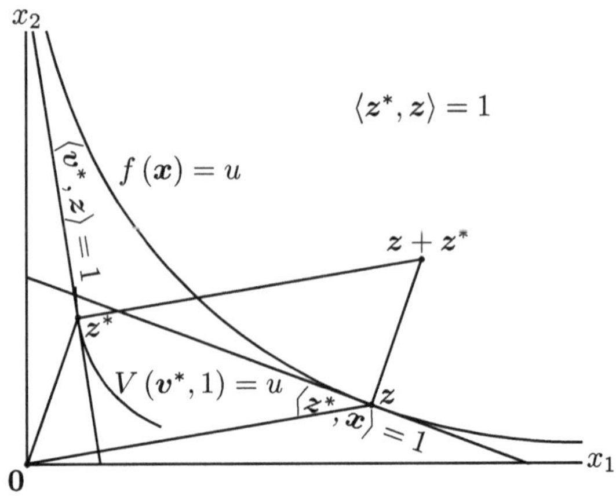

Fig. 46.3 Parallelogram theorem

Figure 46.3 has lines and curves identical to a corresponding figure in "Shepard Duality," but the labelling is different. The Shepard duality between production and cost is the geometric relationship between an isoquant and the prices such that minimum cost is one. The geometry connecting a utility indifference curve and an indirect-utility indifference curve is identical, even though the optimization is different.

Reference

Roy R (1947) La distribution du revenue entre les divers biens. Econometrica 15:205–225

Chapter 47
Benefit

Abstract Luenberger introduces and applies "benefit" for a consumer, as a novel alternative to utility. For a classical consumer having consumption $\boldsymbol{x}$ and consumption price $\boldsymbol{x}^*$, we define and analyze benefit and relate it to compensated demand. Let C denote the upper contour set for a particular indifference curve. Benefit is the maximum amount that can be subtracted without falling below the indifference curve:

$$b(\boldsymbol{x}) := \sup_y [y^* y - \delta_C(\boldsymbol{x} - \mathbf{1}y)].$$

Here $\mathbf{1}$ denotes consumption of one unit of each good. Here $y^* = 1$ is the dual perturbation. The compensated demand is the set of cost-minimizing $\boldsymbol{x}$ for price $\boldsymbol{x}^*$. Because benefit is a dollar value, it is meaningful to consider benefit minus cost. Consider a price $\boldsymbol{x}^*$:

- If $\boldsymbol{x}$ maximizes benefit minus cost and benefit is zero, then $\boldsymbol{x}$ is the compensated demand for price $\boldsymbol{x}^*$;
- Conversely, the compensated demand $\boldsymbol{x}$ maximizes benefit minus cost, and benefit is zero.

Luenberger introduces and applies "benefit" for a consumer, as an alternative to utility (Luenberger 1992, 1995). Benefit provides a novel way to model the demand of a consumer. "The benefit statements are simple, intuitive, and unified—the way economics was *meant* to be" (Luenberger 1995, addendum). Using conjugate duality and perturbation duality, we define and analyze benefit and relate it to compensated demand.

B. C. Dieffenbach, *Conjugate Duality in Economic Analysis*, Contributions to Economics,
https://doi.org/10.1007/978-3-032-21396-9_47

47.1 Expenditure Function

We study a classical consumer. We work with consumption $\boldsymbol{x}$ and consumption price $\boldsymbol{x}^*$, in the Euclidean Jordan algebra $\mathbf{X} = \mathbf{R}^m$. Let the closed, convex C denote the upper contour set for a particular indifference curve. The expenditure function $e(\boldsymbol{x}^*) = -\delta_C^*(-\boldsymbol{x}^*)$ is the minimum cost of attaining this level of preference. Here dom $e = \mathbf{X}_+$. The compensated demand is the subdifferential $\partial e(\boldsymbol{x}^*) = \partial\delta_C^*(-\boldsymbol{x}^*)$, the set of cost-minimizing $\boldsymbol{x}$ for price $\boldsymbol{x}^*$.

47.2 Benefit

Benefit is the value function in an optimization.

Problem 1 (Benefit Primal)

$$b(\boldsymbol{x}) := \sup_y [y^*y - \delta_C(\boldsymbol{x} - \mathbf{1}y)]. \tag{47.1}$$

The primal perturbation is consumption $\boldsymbol{x}$, and $y^* = 1$ is the dual perturbation. The vector $\mathbf{1}$ denotes consumption of one unit of each good. Since more of everything is better, subtracting a positive multiple of $\mathbf{1}$ makes the market basket less preferred. Benefit is the maximum amount that can be subtracted without falling below the indifference curve. Since the choice variable in the primal is a scalar, its solution $y = b(\boldsymbol{x})$ is unique. Here dom b is the entire consumption space $\mathbf{X}$.

Figure 47.1 graphs an indifference curve and its upper contour set C. For price $\boldsymbol{x}^*$, the compensated demand is $\overline{\boldsymbol{x}}$. The price $\boldsymbol{x}^*$ is normal to a hyperplane of constant cost. The cost is minimized where a hyperplane of constant cost is tangent to the indifference curve.

Like Figs. 47.1 and 47.2 graphs an indifference curve and its upper contour set C. The line sloping down and left from consumption $\boldsymbol{x}$ has direction $\mathbf{1}$ (for two goods, a line having slope one). The benefit of $\boldsymbol{x}$ is the largest y such that $\boldsymbol{x} - \mathbf{1}y$ lies in C.

The Lagrangian for the benefit primal is (here $y^* = 1$)

$$\begin{aligned} L(\boldsymbol{x}^*, y) &= \langle \boldsymbol{x}^*, \boldsymbol{x}\rangle + y^*y - \inf_z [\langle \boldsymbol{x}^*, z\rangle + \delta_C(z - \mathbf{1}y)] \\ &= \langle \boldsymbol{x}^*, \boldsymbol{x}\rangle + y^*y - \inf_w [\langle \boldsymbol{x}^*, \boldsymbol{w} + \mathbf{1}y\rangle + \delta_C(\boldsymbol{w})] \quad (\boldsymbol{w} := z - \mathbf{1}y) \\ &= \langle \boldsymbol{x}^*, \boldsymbol{x}\rangle + y^*y - \langle \boldsymbol{x}^*, \mathbf{1}y\rangle + \sup_w [\langle -\boldsymbol{x}^*, \boldsymbol{w}\rangle - \delta_C(\boldsymbol{w})] \\ &= \langle \boldsymbol{x}^*, \boldsymbol{x}\rangle + y^*y - \langle \boldsymbol{x}^*, \mathbf{1}y\rangle - e(\boldsymbol{x}^*). \end{aligned}$$

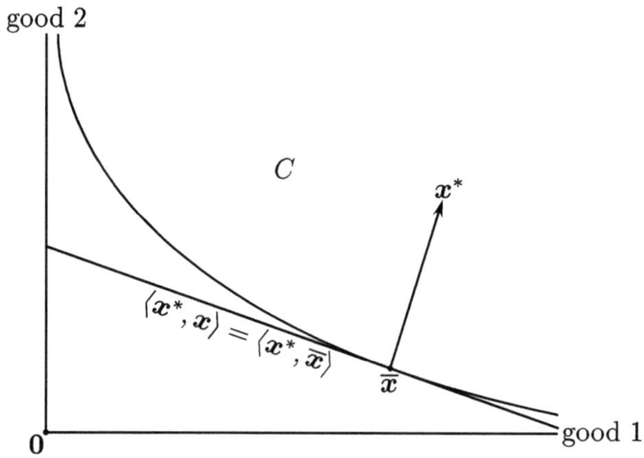

Fig. 47.1 Compensated demand

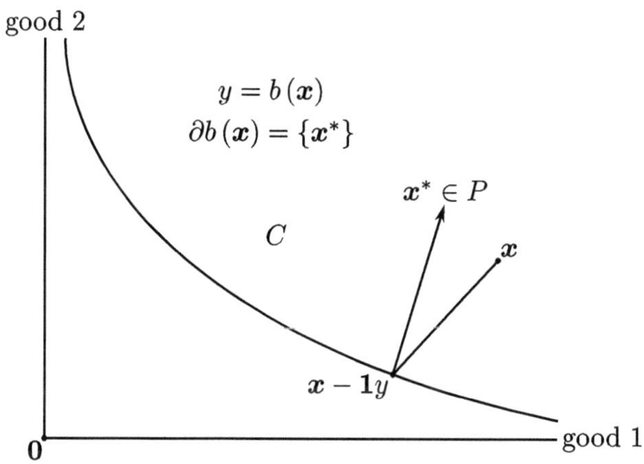

Fig. 47.2 Benefit

The reciprocal dual is

$$\begin{aligned}
&\inf_{\boldsymbol{x}^*}\left[\sup_y L(\boldsymbol{x}^*, y)\right] \\
&= \inf_{\boldsymbol{x}^*}\left\{\sup_y \left[\langle \boldsymbol{x}^*, \boldsymbol{x}\rangle + y^* y - \langle \boldsymbol{x}^*, \mathbf{1}y\rangle - e(\boldsymbol{x}^*)\right]\right\} \\
&= \inf_{\boldsymbol{x}^*}\left[\langle \boldsymbol{x}^*, \boldsymbol{x}\rangle - e(\boldsymbol{x}^*) + \delta_{\langle \boldsymbol{x}^*, \mathbf{1}\rangle = y^*}\right] \\
&= \inf_{\boldsymbol{x}^*}\left[\langle \boldsymbol{x}^*, \boldsymbol{x}\rangle - e(\boldsymbol{x}^*) - \delta_P(\boldsymbol{x}^*)\right],
\end{aligned} \tag{47.2}$$

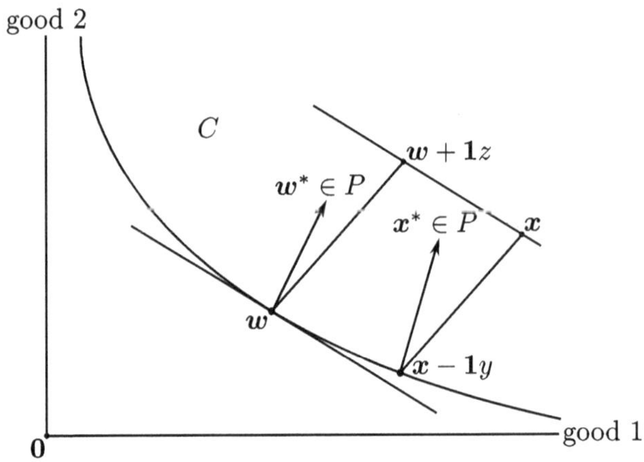

Fig. 47.3 Benefit dual

as dom $e = \mathbf{X}_+$. Hence $b_*(\boldsymbol{x}^*) = e(\boldsymbol{x}^*) - \delta_P(\boldsymbol{x}^*)$, the expenditure function restricted to $\boldsymbol{x}^* \in P$. For a classical consumer, more of everything is better, so the benefit primal satisfies the relative condition. The benefit dual satisfies the relative/polyhedral condition. There is no duality gap, and the optimum value is attained in both the primal and the dual.

Interpret $\boldsymbol{x}^* \in P$ as the price of consumption. Figure 47.3 illustrates the benefit dual. The benefit $b(\boldsymbol{x}) = y$. The solution to the dual is $\boldsymbol{x}^* \in P$, such that $-\boldsymbol{x}^*$ lies in the normal cone to C at $\boldsymbol{x} - \mathbf{1}y$. The value of the dual at $\boldsymbol{x}^*$ is

$$\langle \boldsymbol{x}^*, \boldsymbol{x} \rangle - e(\boldsymbol{x}^*) = \langle \boldsymbol{x}^*, \boldsymbol{x} \rangle - \langle \boldsymbol{x}^*, \boldsymbol{x} - \mathbf{1}y \rangle = y.$$

For some other price $\boldsymbol{w}^* \in P$, the compensated demand is $\boldsymbol{w}$, where $-\boldsymbol{w}^*$ lies in the normal cone to C. The figure plots a hyperplane through $\boldsymbol{x}$ having normal $\boldsymbol{w}^*$, which is parallel to the tangent hyperplane through $\boldsymbol{w}$. Choose z so $\boldsymbol{w} + \mathbf{1}z$ lies on the hyperplane through $\boldsymbol{x}$. The value of the dual is

$$\langle \boldsymbol{w}^*, \boldsymbol{x} \rangle - e(\boldsymbol{w}^*) = \langle \boldsymbol{w}^*, \boldsymbol{w} + \mathbf{1}z \rangle - \langle \boldsymbol{w}^*, \boldsymbol{w} \rangle = z.$$

Because the tangent hyperplane through $\boldsymbol{w}$ passes below $\boldsymbol{x} - \mathbf{1}y$, therefore $z > y$.

At a saddle point of the Lagrangian, the solution $y = b(\boldsymbol{x})$, and the solution set $\boldsymbol{x}^*$ is the subdifferential $\partial b(\boldsymbol{x})$. The first-order condition for maximization in y is

$$\langle \boldsymbol{x}^*, \mathbf{1} \rangle = 1 \text{ such that } \boldsymbol{x}^* \geq \mathbf{0}$$

Equivalently,

$$\boldsymbol{x}^* \in P. \tag{47.3}$$

The first-order condition for minimization in $\boldsymbol{x}^*$ is

$$\boldsymbol{x} - \mathbf{1}y \in \partial e(\boldsymbol{x}^*) = \partial \delta_C^*(-\boldsymbol{x}^*).$$

Inverting the subdifferential and applying the first-order condition (47.3) obtains the subdifferential, expressed via the normal cone to C.

Theorem 1 (Benefit Subdifferential) *The subdifferential $\partial b(\boldsymbol{x})$ is all $\boldsymbol{x}^* \in P$ such that $-\boldsymbol{x}^* \in \partial \delta_C[\boldsymbol{x} - \mathbf{1}b(\boldsymbol{x})]$*

Figure 47.2 plots the subgradient $\boldsymbol{x}^*$. The subdifferential is invariant to adding or subtracting multiples of $\mathbf{1}$:

$$\partial b(\boldsymbol{x}) = \partial b(\boldsymbol{x} - \mathbf{1}y),$$

for positive or negative values y. Lowering y by one raises the benefit by one but has no effect on the subdifferential. The dual can have multiple solutions, if there is a kink in the indifference curve.

Observe that benefit is a dollar value. In the benefit dual (47.2), price times quantity $\langle \boldsymbol{x}^*, \boldsymbol{x} \rangle$ and expenditure $e(\boldsymbol{x}^*)$ are both dollar amounts. In the benefit primal, y^*y is price times quantity. The term $\mathbf{1}y$ constitutes y units of every good. For price $\boldsymbol{x}^* \in P$, the cost of one unit of every good is $y^* = 1$, so the total cost is $y^*y = y$. Because benefit is a dollar value, it is meaningful to consider benefit minus cost. The conjugate of benefit is also a dollar value.

47.3 Benefit Minus Cost Maximization

Choose $\boldsymbol{x}$ to maximize benefit minus cost,.

Problem 2 (Benefit Minus Cost) For price $\boldsymbol{x}^* \in P$,

$$\sup_{\boldsymbol{x}} [b(\boldsymbol{x}) - \langle \boldsymbol{x}^*, \boldsymbol{x} \rangle] = -b_*(\boldsymbol{x}^*) = -e(\boldsymbol{x}^*).$$

The first-order condition is

$$\boldsymbol{x}^* \in \partial b(\boldsymbol{x}) = -\partial \delta_C(\boldsymbol{x} - \mathbf{1}y), \tag{47.4}$$

in which $y = b(\boldsymbol{x})$. Inverting the first-order condition gives

$$\boldsymbol{x} - \mathbf{1}y \in \partial \delta_C^*(-\boldsymbol{x}^*) = \partial e(\boldsymbol{x}^*).$$

Requiring that the benefit be zero obtains the following.

Theorem 2 (Benefit and Compensated Demand) *Consider a price $\boldsymbol{x}^* \in P$.*

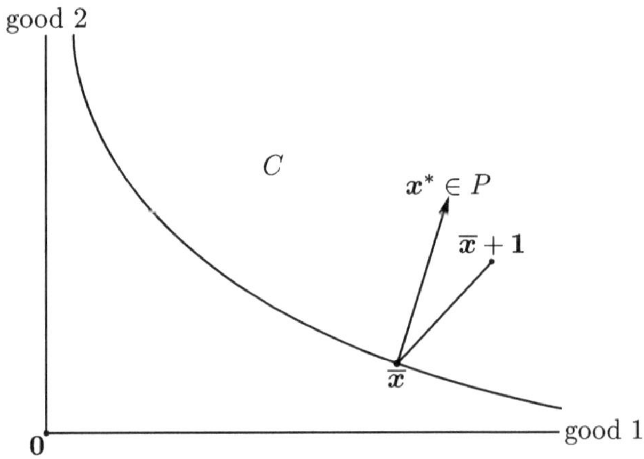

Fig. 47.4 Maximum benefit minus cost

- *If* $\mathbf{x}$ *maximizes benefit minus cost and benefit is zero, then* $\mathbf{x}$ *is the compensated demand for price* $\mathbf{x}^*$*;*
- *Conversely, the compensated demand* $\mathbf{x}$ *maximizes benefit minus cost, and benefit is zero.*

The first-order condition (47.4) for maximizing benefit minus cost is intuitive: "The marginal benefit of a good is equal to that good's price" (Luenberger 1995, addendum).

Observe that there exists a continuum of solutions maximizing benefit minus cost. Figure 47.4 illustrates the situation. One solution is the compensated demand $\overline{\mathbf{x}}$, for which the benefit $y = 0$. Raising consumption by $\mathbf{1}$ raises both benefit and cost by one, so benefit minus cost is unchanged.

In economic theory, that the normalization of utility is arbitrary is a traditional problem, preventing interpersonal comparisons of utility. In contrast, because benefit is a dollar amount, it is meaningful to add benefit for different consumers. In a later chapter, the welfare analysis of a resource allocation via benefit makes use of this property.

Benefit is not the same as utility. To increase the consumption of a single good may raise utility while leaving benefit unchanged. Economists see an analysis based on utility as the "correct" methodology. However an analysis based on benefit is simpler mathematically.

References

Luenberger DG (1992) Benefit functions and duality. J Math Econ 21(5):461–481
Luenberger DG (1995) Microeconomic theory. McGraw-Hill, New York

Part XI
Welfare Economics

Chapter 48
Resource Utilization

Abstract We study resource utilization in an exchange economy. Is an allocation of consumption to consumers weakly efficient? Could a positive amount of each good be discarded, and the remaining supply reallocated to attain the same levels of preference? If not, then the utilization of resources is weakly efficient. We model this issue by three alternative optimizations. The resource-utilization primal solves for the quantity such that this quantity of every good can be discarded, and the remaining supply is sufficient to attain the same levels of preference. From this primal, we calculate an expenditure dual (Luenberger, D. G. (1995). Microeconomic theory. New York: McGraw-Hill). The expenditure dual chooses prices to minimize the value of supply less the sum of the expenditure functions for all consumers. The resource allocation is weakly efficient if the optimum value is zero. Using this duality, we prove the first and second theorems of resource utilization, akin to the first and second theorems of welfare economics. In a compensated equilibrium, the resource utilization is weakly efficient. Conversely, an allocation in which the resource utilization is weakly efficient can be realized as a compensated equilibrium. From the expenditure dual we derive the equivalent formulation by (Debreu, G. (1983). The coefficient of resource utilization. In W. Hildenbrand (Ed.), Mathematical economics: Twenty papers of Gérard Debreu (pp. 30–49). Cambridge: Cambridge University Press. (Originally 1951)). We restate the expenditure dual as a standard positively homogeneous fractional program. We find the equivalent concave maximization and its Fenchel dual. The dual interprets the fractional program as Debreu's coefficient of resource utilization: its optimum value is the fraction of resources that is just sufficient to attain the same levels of preference. The resource allocation is weakly efficient if the value is one. Whether the resource allocation is benefit-efficient is a third formulation. The dual of the benefit-efficiency problem reduces to the expenditure dual. That the utilization of resources is weakly efficient is thus equivalent to benefit efficiency. We restate the first and second theorems of resource utilization in terms of benefit efficiency.

B. C. Dieffenbach, *Conjugate Duality in Economic Analysis*, Contributions to Economics,
https://doi.org/10.1007/978-3-032-21396-9_48

48.1 Exchange Economy

We study an exchange economy of classical consumers. The consumption space for each of the n households is $\mathbf{X} = \mathbf{R}^m$. Let

$$Y := \begin{bmatrix} \boldsymbol{y}_1 & \cdots \end{bmatrix} \in \mathbf{T}^{n \to m}$$

denote a base resource allocation, in which $\boldsymbol{y}_i \succeq \mathbf{0}$ is the base consumption of household i. Supply is aggregate consumption, $\boldsymbol{y} = Y\mathbf{1}$. Define C_i as the convex set of consumptions preferred or indifferent to $\boldsymbol{y}_i$, the upper contour set for the indifference curve through $\boldsymbol{y}_i$.

We compare this base allocation to other resource allocations

$$X := \begin{bmatrix} \boldsymbol{x}_1 & \cdots \end{bmatrix} \in \mathbf{T}^{n \to m};$$

here $\boldsymbol{x}_i$ is consumption of consumer i. Aggregate consumption $\boldsymbol{x} := X\mathbf{1}$. The resource constraint is $\boldsymbol{x} \preceq \boldsymbol{y}$.

We define compensated equilibrium for the exchange economy. For price $\boldsymbol{y}^*$, the expenditure function $e_i(\boldsymbol{y}^*) = -\delta^*_{C_i}(-\boldsymbol{y}^*)$ is the minimum cost of attaining the preference level. Compensated demand is the subdifferential $\partial e_i(\boldsymbol{y}^*)$.

Definition 1 (Compensated Equilibrium) For preference levels C_i and supply $\boldsymbol{y}$, a market allocation $(X, \boldsymbol{y}^*)$ is a compensated equilibrium if

Market Price the price $\boldsymbol{y}^* \succ \mathbf{0}$;
Compensated Demand consumption is compensated demand: $\boldsymbol{x}_i \in \partial e_i(\boldsymbol{y}^*)$;
Market Equilibrium there is complementary slackness between the excess supply $\boldsymbol{y} - X\mathbf{1}$ and the price $\boldsymbol{y}^*$.

The excess supply is nonnegative. Complementary slackness implies market equilibrium for a good having a positive price and a zero price for a good in excess supply. Since $\boldsymbol{y}^* \succ \mathbf{o}$, demand equals supply for at least one good. Below we normalize the price so $\boldsymbol{y}^* \in P$.

Defined by an optimization problem, the Scitovsky set C is the set of consumption vectors such that it is possible for each consumer to attain the specified level of preference.

Problem 1 (Scitovsky Set)

$$\delta_C(\boldsymbol{x}) := \inf_X \left[\delta_{\boldsymbol{x} = \sum_i \boldsymbol{x}_i} + \sum_i \delta_{C_i}(\boldsymbol{x}_i) \right] \tag{48.1}$$

The Scitovsky indifference curve is the lower boundary of C Scitovsky (1942).

Via perturbation duality, we show $C = \sum_i C_i$. Taking $\boldsymbol{x}$ in Eq. (48.1) as the perturbation, we find the reciprocal dual. Adding the Fenchel inequalities

$$\delta_{C_i}(\boldsymbol{x}_i) + \delta^*_{C_i}(\boldsymbol{x}^* - \boldsymbol{x}^*_i) - \langle \boldsymbol{x}^* - \boldsymbol{x}^*_i, \boldsymbol{x}_i \rangle \geq 0 \text{ for all } i$$
$$\delta_{\{0\}}\left(\boldsymbol{x} - \textstyle\sum_i \boldsymbol{x}_i\right) + \delta^*_{\{0\}}(\boldsymbol{x}^*) - \left\langle \boldsymbol{x}^*, \boldsymbol{x} - \textstyle\sum_i \boldsymbol{x}_i \right\rangle \geq 0$$

and rearranging obtains

$$\sum_i \langle \boldsymbol{x}^*_i, \boldsymbol{x}_i \rangle + \delta_{\{0\}}\left(\boldsymbol{x} - \sum_i \boldsymbol{x}_i\right) + \sum_i \delta_{C_i}(\boldsymbol{x}_i) \geq \langle \boldsymbol{x}^*, \boldsymbol{x} \rangle - \sum_i \delta^*_{C_i}(\boldsymbol{x}^* - \boldsymbol{x}^*_i) - \delta^*_{\{0\}}(\boldsymbol{x}^*).$$

The base values are $\boldsymbol{x}^*_i = \boldsymbol{0}$ and $\boldsymbol{x}$. The reciprocal dual satisfies the relative condition, so there is no duality gap. Its optimum value is

$$\delta_C(\boldsymbol{x}) = \sup_{\boldsymbol{x}^*}\left[\langle \boldsymbol{x}^*, \boldsymbol{x} \rangle - \sum_i \delta^*_{C_i}(\boldsymbol{x}^*)\right] = \sup_{\boldsymbol{x}^*}\left[\langle \boldsymbol{x}^*, \boldsymbol{x} \rangle - \delta^*_{\sum_i C_i}(\boldsymbol{x}^*)\right] = \delta_{\sum_i C_i}(\boldsymbol{x}),$$

because the sum of supports is the support to the sum. Hence $C = \sum_i C_i$.

The expenditure function for consumer i is $-\delta^*_C(-\boldsymbol{y}^*)$. The Scitovsky set fixes the aggregate expenditure function:

$$\sum_i e_i(\boldsymbol{y}^*) = \sum_i -\delta^*_{C_i}(-\boldsymbol{y}^*) = -\delta^*_C(-\boldsymbol{y}^*) := e(\boldsymbol{y}^*),$$

because the support of a sum is the sum of the supports.

48.2 Resource Utilization

Via perturbation duality, we analyze whether the allocation Y is weakly efficient. Does there exist an allocation X such that every $\boldsymbol{x}_i \in C_i$ and $\boldsymbol{x} \ll \boldsymbol{y}$? The primal (48.2) is a weak-efficiency problem: maximize $\min[\boldsymbol{y} - \boldsymbol{x}]$ subject to $\boldsymbol{x}_i \in C_i$. To maximize the smallest component of $\boldsymbol{y} - X\boldsymbol{1}$, maximize the negative of the largest component of $X\boldsymbol{1} - \boldsymbol{y}$.

Problem 2 (Resource-Utilization Primal)

$$\sup_X \left[-\delta^*_P(X\boldsymbol{1} - \boldsymbol{y}) - \sum_i \delta_{C_i}(\boldsymbol{x}_i)\right]. \tag{48.2}$$

Since the choice $X = Y$ makes the objective function zero, the optimum value is nonnegative. The resource utilization Y is weakly efficient if the optimum value is zero, but not if the optimum value is positive. For classical consumers the supremum is finite.

We calculate the reciprocal dual of the resource-utilization primal. Adding the Fenchel inequalities

$$\delta_{C_i}(\boldsymbol{x}_i) + \delta^*_{C_i}(\boldsymbol{x}_i^* - \boldsymbol{y}^*) - \langle \boldsymbol{x}_i^* - \boldsymbol{y}^*, \boldsymbol{x}_i \rangle \geq 0 \text{ for all } i \tag{48.3}$$

$$\delta_P(\boldsymbol{y}^*) + \delta^*_P(X\mathbf{1} - \boldsymbol{y}) - \langle \boldsymbol{y}^*, X\mathbf{1} - \boldsymbol{y} \rangle \geq 0 \tag{48.4}$$

and rearranging gives

$$\langle \boldsymbol{y}^*, \boldsymbol{y} \rangle + \delta_P(\boldsymbol{y}^*) + \sum_i \delta^*_{C_i}(\boldsymbol{x}_i^* - \boldsymbol{y}^*) \geq \langle X^*, X \rangle - \delta^*_P(X\mathbf{1} - \boldsymbol{y}) - \sum_i \delta_{C_i}(\boldsymbol{x}_i).$$

The perturbation base values are $\boldsymbol{y}$ and $X^* = \mathsf{0}$. Both the primal and the reciprocal dual satisfy the relative condition. There is no duality gap, and the finite optimum value is attained in both problems.

Problem 3 (Expenditure Dual)

$$\min_{\boldsymbol{y}^* \in P} \left[\langle \boldsymbol{y}^*, \boldsymbol{y} \rangle - \sum_i e_i(\boldsymbol{y}^*) \right]. \tag{48.5}$$

Luenberger defines $\langle \boldsymbol{y}^*, \boldsymbol{y} \rangle - \sum_i e_i(\boldsymbol{y}^*)$ as the *income-surplus function*, the "difference between the income generated and the income spent" —the total cost of supply less the total cost of demand (Luenberger 1995, Sect. 6.8, p. 195). If the resource utilization of allocation Y is weakly efficient, then the minimum value is zero. If it is not weakly efficient, then the minimum value is positive.

From the resource-utilization primal and the expenditure dual we prove the first and second theorems of resource utilization.

Theorem 1 (First Theorem of Resource Utilization) *In a compensated equilibrium, the resource utilization is weakly efficient.*

Market equilibrium implies just that resource utilization is weakly efficient, but not that it is efficient. There can be a free good, a good in excess supply with price zero, which precludes efficiency.

Theorem 2 (Second Theorem of Resource Utilization) *An allocation in which the resource utilization is weakly efficient is a compensated equilibrium for some price greater than zero.*

We prove the first theorem of resource utilization. Let $(X, \boldsymbol{y}^*)$ denote a compensated equilibrium, with price normalization $\boldsymbol{y}^* \in P$. We show that X solves the weak-efficiency primal and $\boldsymbol{y}^*$ solves the expenditure dual, and the optimum value is zero. It follows that the resource utilization is weakly efficient.

The two Fenchel inequalities (48.3) and (48.4) hold with equality. That consumption is compensated demand is the subdifferential form for the Fenchel factor (48.3) to be zero. That there is market equilibrium says that the Fenchel factor (48.4) is zero. Demand is less than or equal to supply, $X\mathbf{1} \preceq \boldsymbol{y}$, and equal for some good, so $\delta_P^*(X\mathbf{1} - \boldsymbol{y}) = 0$. The complementary slackness condition $\langle \boldsymbol{y}^*, X\mathbf{1} - \boldsymbol{y}\rangle = 0$ follows.

The second theorem of resource utilization is the converse. Suppose that a resource allocation Y is a weakly efficient allocation of resources. The optimum value of the resource-utilization primal and the expenditure dual is zero. The choice $X = Y$ solves the primal, because the objective function is zero. For a solution $\boldsymbol{y}^*$ to the expenditure dual, $(Y, \boldsymbol{y}^*)$ is a compensated equilibrium. That the Fenchel factor (48.3) is zero says that consumption is compensated demand. That the Fenchel factor (48.4) is zero implies market equilibrium. Necessarily $\boldsymbol{y}^* \in P$. As $X = Y$, there is market equilibrium: demand $X\mathbf{1}$ equals supply $\boldsymbol{y} = Y\mathbf{1}$.

48.3 Coefficient of Resource Utilization

One should see the formulation above as an alternative development of Debreu (1983), who measures resource utilization by the "coefficient of resource utilization." He asks, what fraction of resources could be discarded, so by reallocation the same levels of well-being could be attained? The coefficient is a fraction; a value 0.9, for example, means that ten per cent of resources could be discarded. The resource utilization is weakly efficient if the coefficient is one, but not if the coefficient is less than one.

Write the expenditure dual in terms of the Scitovsky set:

$$\min_{y^* \in P} [\langle y^*, y\rangle - e(y^*)]. \tag{48.6}$$

The objective function is nonnegative, and the minimum is zero if the resource utilization is weakly efficient. Restate the dual as a standard positively homogeneous fractional program:

$$\sup_{y^* \neq 0} \frac{e(y^*)}{\langle y^*, y\rangle + \delta_{y^* \geq 0}} = \sup_{y^* \neq 0} \frac{-\delta_C^*(-y^*)}{\delta_{\{y\}-\mathbf{X}_+}^*(y^*)}.$$

Assume $\boldsymbol{y} \gg \mathbf{0}$, so the denominator is positive for $\boldsymbol{y}^* \neq \mathbf{0}$. If the allocation is weakly efficient, then the optimum value is one; otherwise the optimum value is less than one. We show that the optimum value is the coefficient of resource utilization.

The set $\{y\} - \mathbf{X}_+$ is all consumption vectors less than or equal to $\boldsymbol{y}$. The indicator of the cone representation of its polar is

$$\delta_{\text{cone }(\{y\}-\mathbf{X}_+)^\circ}(\boldsymbol{x}^*, -z^*) = \delta_{\text{epi }\delta_{\{y\}-\mathbf{X}_+}^*}(\boldsymbol{x}^*, z^*) = \delta_{\langle \boldsymbol{x}^*, \boldsymbol{y}\rangle \leq z^*} + \delta_{\boldsymbol{x}^* \geq \mathbf{0}},$$

so

$$\delta_{(\{y\}-\mathbf{X}_+)^\circ}(\boldsymbol{x}^*) = \delta_{\langle x^*,y\rangle\le 1} + \delta_{x^*\ge 0}.$$

The equivalent concave maximization is

$$\sup_{y^*}\left[-\delta_C^*(-y^*) - \delta_{\langle y^*,y\rangle\le 1} - \delta_{y^*\ge 0}\right] = \sup_{y^*}\left[-\delta_C^*(-y^*) - \delta_{(\{y\}-\mathbf{X}_+)^\circ}(y^*)\right]. \tag{48.7}$$

Its Fenchel dual is

$$\inf_x\left[\gamma_{\{y\}-\mathbf{X}_+}(\boldsymbol{x}) + \delta_C(\boldsymbol{x})\right]. \tag{48.8}$$

The dual chooses $\boldsymbol{x}$ in the Scitovsky set to minimize the gauge $\gamma_{\{y\}-\mathbf{X}_+}(\boldsymbol{x})$, the fraction by which $\{y\} - \mathbf{X}_+$ can be shrunk toward the origin and still contain $\boldsymbol{x}$.

48.4 Benefit

Define b as the benefit function for the Scitovsky set C. Observe that the dual (48.6) is exactly the benefit dual derived in the chapter "Benefit;" its optimum value is $b(y)$. Thus the resource utilization is weakly efficient if and only if $b(y) = 0$. Always

$$\begin{aligned} b(y) &= \inf_{y^*\in P}\left[\langle y^*, y\rangle - e(y^*)\right] \\ &= \inf_{y^*\in P}\left[\left\langle y^*, \sum_i y_i\right\rangle - \sum_i e_i(y^*)\right] \\ &\ge \sum_i \inf_{y^*\in P}\left[\langle y^*, y_i\rangle - e_i(y^*)\right] \\ &= \sum_i b_i(y_i) \\ &= 0. \end{aligned}$$

In Fig. 48.1, Y is an inefficient utilization of resources, $b(y) > 0$. The dotted lines outline $\{y\}-\mathbf{X}_+$, the consumption vectors less than or equal to $\boldsymbol{y}$. The optimum choice in the dual (48.8) is the point $\boldsymbol{x} = \gamma_{\{y\}-\mathbf{X}_+}(\boldsymbol{x})\boldsymbol{y}$ on the Scitovsky indifference curve on the ray from the origin to $\boldsymbol{y}$. The coefficient of resource utilization is $|\boldsymbol{x}| / |y| = \gamma_{\{y\}-\mathbf{X}_+}(\boldsymbol{x})$. If instead Y were weakly efficient, then $b(y) = 0$ and $\boldsymbol{x} = \boldsymbol{y}$.

We show that benefit efficiency is equivalent to the weak efficiency of the utilization of resources. We study whether the allocation Y is benefit-efficient, subject to the constraint that aggregate consumption $X\mathbf{1}$ is less than or equal to $\boldsymbol{y} := Y\mathbf{1}$.

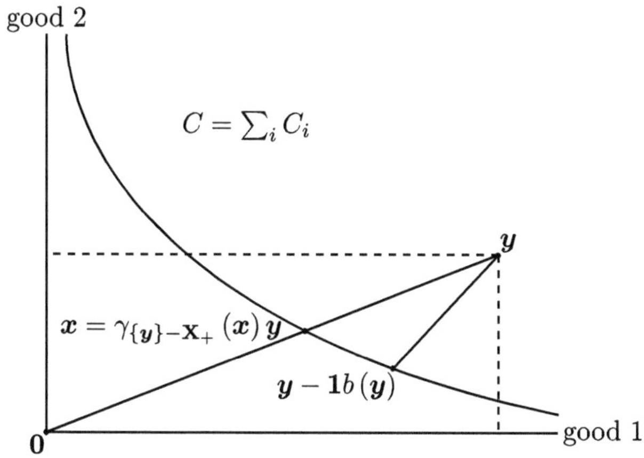

Fig. 48.1 Inefficient resource utilization

Problem 4 (Benefit-Efficiency Primal)

$$\sup_{X,z} \left[\langle X^*, X \rangle + \langle z^*, z \rangle - \delta^*_{\mathbf{R}^n_+ + \{1\}}(-z - w) - \sum_i \delta_{\text{epi } b_i}(x_i, z_i) - \delta_{\mathbf{X}_+}(y - X\mathbf{1}) \right] \quad (48.9)$$

The perturbation base values are $w = \mathbf{0}, y, X^* = 0$, and $z^* = \mathbf{0}$. The optimum value is nonnegative, because the choice $X = Y$ makes the value zero. The allocation is benefit-efficient if and only if the optimum valuc is zero.

We restate our results for resource utilization in terms of benefit.

Theorem 3 (Benefit Efficiency and Resource Utilization) *The resource allocation is benefit-efficient if and only if the resource utilization is weakly efficient.*

And we restate the first and second theorems of resource utilization in terms of benefit efficiency.

Theorem 4 (Benefit Efficiency and Compensated Equilibrium)

- *In a compensated equilibrium, the resource allocation is benefit-efficient;*
- *a benefit-efficient resource allocation is attainable as a compensated equilibrium, for some price greater than zero.*

We calculate the benefit-efficiency dual. Adding the Fenchel inequalities

$$\delta_{\text{epi } b_i}(x_i, z_i) + \delta^*_{\text{epi } b_i}(-y^* + x_i^*, w_i^* + z_i^*) - \langle -y^* + x_i^*, x_i \rangle - (w_i^* + z_i^*) z_i \geq 0 \text{ for all } i$$
$$\delta_{\mathbf{R}^n_+ + \{1\}}(w^*) + \delta^*_{\mathbf{R}^n_+ + \{1\}}(-z - w) - \langle w^*, -z - w \rangle \geq 0$$
$$\delta_{\mathbf{X}_+}(y - X\mathbf{1}) + \delta^*_{\mathbf{X}_+}(-y^*) - \langle -y^*, y - X\mathbf{1} \rangle \geq 0,$$

and rearranging obtains the reciprocal dual:

$$\langle y^*, y\rangle + \langle w^*, w\rangle + \delta_{\mathbf{R}^n_+ + \{1\}}(w^*) + \sum_i \delta^*_{\text{epi } b_i}(-y^* + x^*_i, w^*_i + z^*_i) + \delta^*_{\mathbf{X}_+}(-y^*)$$
$$\geq \langle X^*, X\rangle + \langle z^*, z\rangle - \delta^*_{\mathbf{R}^n_+ + \{1\}}(-z - w) - \sum_i \delta_{\text{epi } b_i}(x_i, z_i) - \delta_{\mathbf{X}_+}(y - X\mathbf{1}).$$

The base values are $X^* = 0, z^* = \mathbf{0}, w = \mathbf{0}$, and y. Substituting the base values obtains

$$\langle y^*, y\rangle + \delta_{\mathbf{R}^n_+ + \{1\}}(w^*) + \sum_i \delta^*_{\text{epi } b_i}(-y^*, w^*_i) + \delta_{y^* \geq 0}$$
$$\geq -\delta^*_{\mathbf{R}^n_+ + \{1\}}(-z - w) - \sum_i \delta_{\text{epi } b_i}(x_i, z_i) - \delta_{X\mathbf{1} \leq y}.$$

For classical consumers, the dual satisfies the polyhedral/relative condition, so there is no duality gap, and in the primal the optimum value is attained.

Analyzing the dual directly, we show that the choice $w^* = \mathbf{1}$ is minimizing. Substituting this value yields a reduced dual identical to expenditure dual (48.5), which establishes that benefit efficiency is equivalent to weak efficiency of the resource allocation.

In the benefit-efficiency dual, for $\delta^*_{\text{epi } b_i}(-y^*, w^*_i)$ to be finite requires $y^*/w^*_i \in P$, so $y^* > \mathbf{0}$ and the w^*_i must all be equal. Furthermore the choice $w^* = \mathbf{1}$ is minimizing:

$$\begin{aligned}
&\inf_{y^*, w^*}\left[\langle y^*, y\rangle + \delta_{\mathbf{R}^n_+ + \{1\}}(w^*) + \sum_i \delta^*_{\text{epi } b_i}(-y^*, w^*_i)\right] \\
&= \inf_{y^*, \lambda^*}\left[\langle y^*, y\rangle + \delta_{\lambda^* \geq 1} + \sum_i \delta^*_{\text{epi } b_i}(-y^*, \lambda^*)\right] \\
&= \inf_{y^*/\lambda^*, \lambda^*}\left\{\delta_{\lambda^* \geq 1} + \lambda^* \sum_i \left[\langle y^*/\lambda^*, y_i\rangle - b_{i*}(y^*/\lambda^*)\right]\right\} \qquad (48.10) \\
&= \inf_{y^*}\left[\langle y^*, y\rangle - \sum_i b_{i*}(y^*)\right] \\
&= \inf_{y^* \in P}\left[\langle y^*, y\rangle - \sum_i e_i(y^*)\right]
\end{aligned}$$

For $y^*/\lambda^* \in P$, in Eq. (48.10) $\langle y^*/\lambda^*, y_i\rangle - b_{i*}(y^*/\lambda^*)$ is nonnegative, because $b_{i*}(y^*/\lambda^*)$ is the minimum expenditure needed to achieve benefit zero. Hence the choice $\lambda^* = 1$ is minimizing.

The optimum value is also the maximum sum of benefits, subject to the resource constraint.

Problem 5 (Maximum Sum of Benefits)

$$\sup_{X}\left[\sum_{i} b_i(\boldsymbol{x}_i) - \delta_{X\mathbf{1}=\boldsymbol{y}}\right] \tag{48.11}$$

(Luenberger 1995, Chap. 6). This problem satisfies the polyhedral/relative condition, so there is no duality gap.

Although this maximum sum of benefits problem has the same optimum value as the benefit-efficiency primal, the two problems are not the same. In Eq. (48.9), each benefit is constrained to be nonnegative, whereas in (48.11) each benefit is unconstrained. The problem (48.11) has a continuum of solutions, because to transfer **1** from one consumer to a second consumer reduces the benefit of the first by one but raises the benefit of the second by one, so the sum is unchanged.

We derive (48.11) as the dual of (48.10). Adding the Fenchel inequalities

$$-b_i(\boldsymbol{x}_i) - b_{i*}(\boldsymbol{y}^*) + \langle \boldsymbol{y}^*, \boldsymbol{x}_i\rangle \geq 0 \text{ for all } i$$
$$\delta_{\mathbf{X}_+}(\boldsymbol{y} - X\mathbf{1}) + \delta^*_{\mathbf{X}_+}(-\boldsymbol{y}^*) - \langle -\boldsymbol{y}^*, \boldsymbol{y} - X\mathbf{1}\rangle \geq 0$$

and rearranging obtains

$$\langle \boldsymbol{y}^*, \boldsymbol{y}\rangle - \sum_i b_{i*}(\boldsymbol{y}^*) \geq \sum_i b_i(\boldsymbol{x}_i) - \delta_{X\mathbf{1}=\boldsymbol{y}}.$$

Thus (48.10) and (48.11) have the same optimum values.

Luenberger centers his analysis on the expenditure dual and the maximum sum of benefits problem (Luenberger 1995, Chap. 6). He explains that market equilibrium solves both problems. The price $\boldsymbol{y}^*$ minimizes the expenditure dual: "equilibrium prices minimize total surplus" (Luenberger 1995, addendum). The allocation Y maximizes the sum of benefits: "total benefit is maximized" (Luenberger 1995, addendum). He likes the simplicity of benefit: "benefit functions can be meaningfully summed across consumers to obtain a measure of aggregate benefits;" "the benefit function of the group is simply the sum of the individual benefit functions" (Luenberger 1992, p. 469).

References

Debreu G (1983) The coefficient of resource utilization. In: Hildenbrand W (ed) Mathematical economics: twenty papers of Gérard Debreu. Cambridge University Press, Cambridge, pp 30–49 (Originally 1951)

Luenberger DG (1992) Benefit functions and duality. J Math Econ 21(5):461–481

Luenberger DG (1995) Microeconomic theory. McGraw-Hill, New York

Scitovsky T (1942 Summer) A reconsideration of the theory of tariffs. Rev Econ Stud 9(2):89–110

Chapter 49
Welfare and Equilibrium

Abstract Working from the concave utility of consumers, this chapter derives first and second theorems of welfare economics for an exchange economy. General equilibrium refers to simultaneous demand and supply equilibrium in all markets. The Hicks and Marshall decompositions for utility contrast expenditure-minimizing compensated demand and utility-maximizing consumer demand. Always consumer demand is expenditure-minimizing, but compensated demand is perhaps not utility-maximizing. Compensated equilibrium is general equilibrium for compensated demand, whereas competitive equilibrium is general equilibrium for consumer demand. A competitive equilibrium is necessarily a compensated equilibrium, but not conversely. Perturbation duality leads to first and second theorems of welfare economics. Starting from a weak-efficiency primal, we derive a compensated-equilibrium dual. The primal/dual analysis yields first and second theorems of welfare economics for compensated equilibrium. Starting from an efficiency primal, we derive a competitive-equilibrium dual, resembling the compensated-equilibrium dual, but slightly different. The primal/dual analysis leads to first and second theorems of welfare economics for competitive equilibrium.

49.1 Exchange Economy

We analyze an exchange economy having m goods and n classical consumers. Consumption

$$C = \begin{bmatrix} \boldsymbol{c}_1 & \cdots \end{bmatrix} \in \mathbf{T}^{n\to m}.$$

Each consumption $\boldsymbol{c}_i > \mathbf{0}$, and aggregate consumption $C\mathbf{1} \gg \mathbf{0}$. Let $f_i(\boldsymbol{x}_i)$ denote the utility function translated by $\boldsymbol{c}_i$ and its utility; total consumption is $\boldsymbol{c}_i + \boldsymbol{x}_i$. Define

$$X := \begin{bmatrix} \boldsymbol{x}_1 & \cdots \end{bmatrix} \in \mathbf{T}^{n\to m}.$$

The resource constraint is $X\mathbf{1} \preceq \mathbf{0}$. The focus is the whether $X = 0$ is weakly efficient or efficient for utility.

B. C. Dieffenbach, *Conjugate Duality in Economic Analysis*, Contributions to Economics,
https://doi.org/10.1007/978-3-032-21396-9_49

Recall the properties of translated utility. The conjugate $f_{i*}(\boldsymbol{x}^*) \leq 0$; it is zero if and only if $\boldsymbol{x}^* \in \partial f_i(\boldsymbol{0})$. The epi-multiplication $(u_i^* \star f_{i*})(\boldsymbol{v}^*) \leq 0$.

Compensated demand minimizes the expenditure required to achieve the utility of $\boldsymbol{c}_i$. For price $\boldsymbol{v}^* \succ \boldsymbol{0}$, the expenditure-minimization dual is

$$\langle \boldsymbol{v}^*, \boldsymbol{c}_i \rangle + \sup_{u_i^*} [(u_i^* \star f_{i*})(\boldsymbol{v}^*)], \tag{49.1}$$

from the Hicks decomposition. Because the epi-multiplication is less than or equal to zero, the supremum is less than or equal to zero. Consumption $\boldsymbol{c}_i$ is expenditure-minimizing if and only if the supremum is zero.

A solution u_i^* is a subgradient of minimum expenditure with respect to utility. Rescaling $\boldsymbol{v}^*$ by a multiple $\lambda > 0$ multiplies the solution u_i^* by λ. If $u_i^* > 0$, then $\boldsymbol{c}_i$ is utility-maximizing among all consumption costing the same or less, by the Marshall decomposition. For a classical consumer, solution $u_i^* = 0$ occurs if and only expenditure is minimized at zero. From the utility-maximization dual, this zero expenditure maximizes utility if and only if

$$\lim_{\lambda \to \infty} f_{i*}(\lambda \boldsymbol{v}^*) = 0.$$

49.2 Weak Efficiency and Compensated Equilibrium

We derive the first and second theorems of welfare economics for compensated equilibrium (Arrow 1951).

Theorem 1 (First Theorem of Welfare Economics for Compensated Equilibrium) *In a compensated equilibrium, the resource allocation is weakly efficient.*

Theorem 2 (Second Theorem of Welfare Economics for Compensated Equilibrium) *If a resource allocation is weakly efficient, then it can be realized as a compensated equilibrium.*

The weak-efficiency primal tests whether the allocation C is weakly efficient for utility. Can consumption be reallocated so utility rises for every consumer?

Problem 1 (Weak-Efficiency Primal)

$$\sup_X \left[\min\left[f_i(\boldsymbol{x}_i)\right] - \delta_{X1 \leq 0}\right] = \sup_{X,\boldsymbol{y}} \left[-\delta_P^*(-\boldsymbol{y}) - \sum_i \delta_{\text{epi}\, f_i}(\boldsymbol{x}_i, y_i) - \delta_{X1 \leq 0}\right].$$

The allocation C is weakly efficient if and only if the optimum value is zero. To have $f_i(\boldsymbol{x}_i) > 0$ for all i is equivalent to $\boldsymbol{y} \gg \boldsymbol{0}$. For the choice $(X, \boldsymbol{y}) = (0, \boldsymbol{0})$, the objective function is zero, so the optimum value is greater than or equal to zero.

Problem 2 (Perturbed Weak-Efficiency Primal)

$$\sup_{X,y}\left[\langle X^*,X\rangle+\langle y^*,y\rangle-\delta_{X1\le v}-\delta_P^*(-y-u)-\sum_i\delta_{\mathrm{epi}\,f_i}(x_i,y_i)\right].$$

The perturbation base values are all zero: $X^*=0$, $y^*=\mathbf{0}$, $u=\mathbf{0}$, and $v=\mathbf{0}$.

We calculate the reciprocal dual. Adding the Fenchel inequalities

$$\begin{aligned}
\delta_{\mathbf{X}_+}(v^*)+\delta_{\mathbf{X}_+}^*(X\mathbf{1}-v)-\langle v^*,X\mathbf{1}-v\rangle&\ge 0\\
\delta_P(u^*)+\delta_P^*(-y-u)-\langle u^*,-y-u\rangle&\ge 0\\
\delta_{\mathrm{epi}\,f_i}(x_i,y_i)+\delta_{\mathrm{epi}\,f_i}^*(-v^*+x_i^*,u_i^*+y_i^*)&\\
-\langle(-v^*+x_i^*,u_i^*+y_i^*),(x_i,y_i)\rangle&\ge 0\text{ for all }i
\end{aligned}\tag{49.2}$$

and rearranging obtains

$$\begin{aligned}
&\langle v^*,v\rangle+\langle u^*,u\rangle+\delta_{\mathbf{X}_+}(v^*)+\delta_P(u^*)+\sum_i\delta_{\mathrm{epi}\,f_i}^*(-v^*+x_i^*,u_i^*+y_i^*)\\
&\quad\ge\langle X^*,X\rangle+\langle y^*,y\rangle-\delta_{\mathbf{X}_+}^*(X\mathbf{1}-v)-\delta_P^*(-y-u)-\sum_i\delta_{\mathrm{epi}\,f_i}(x_i,y_i).
\end{aligned}$$

Problem 3 (Compensated-Equilibrium Dual)

$$\inf_{u^*,v^*}\left[\delta_P(u^*)+\delta_{v^*\ge 0}-\sum_i(u_i^*\star f_{i*})(v^*)\right].\tag{49.3}$$

Both the primal and the dual satisfy the polyhedral/relative condition, so there is no duality gap, and a finite optimum value is attained.

Interpret the dual as a compensated-equilibrium problem: the triple (C,u^*,v^*) is a compensated equilibrium if and only if (u^*,v^*) solves the dual with optimum value zero. In a compensated equilibrium, compensated demand is less than or equal to supply; the price is zero for any good in excess supply. For the Fenchel factor (49.2) to be zero—complementary slackness—captures this requirement. Each term in the dual objective function is nonnegative, so the optimum value is zero if and only if every term is zero. In particular, $(u_i^*\star f_{i*})(v^*)=0$ for each consumer; consumption c_i is expenditure-minimizing for price v^*.

Our argument rests on the weak-efficiency primal and the compensated-equilibrium dual. The first theorem of welfare economics follows. If (C,u^*,v^*) is a compensated equilibrium, then the optimum value of the primal and the dual is zero, so the resource allocation C is weakly efficient. The second theorem of welfare economics also follows. If the resource allocation C is weakly efficient, that the optimum value of the primal is zero means that the optimum value of the dual is also zero. For any solution (u^*,v^*) to the dual, then (C,u^*,v^*) is a compensated equilibrium.

We fill in details for this argument. For any consumer, u_i^* is nonnegative. In a compensated equilibrium, necessarily $u_i^* > 0$ for at least one consumer. The exceptional case $u_i^* = 0$ can occur only for consumption on the boundary of effective domain of utility; only goods having a zero price are consumed—income is zero. Because $\boldsymbol{v}^* \succ \mathbf{0}$ and the aggregate endowment of every good is positive, some consumer demands a positive quantity of a good having a positive price, so $u_i^* > 0$ for this consumer. Hence a vector $\boldsymbol{u}^*$ of these subgradients must be greater than zero. Because the normalization of the price is arbitrary, when we speak of $(\boldsymbol{u}^*, \boldsymbol{v}^*)$ as a compensated equilibrium, we impose the normalization $\boldsymbol{u}^* \in P$.

Conversely, if $(\boldsymbol{u}^*, \boldsymbol{v}^*)$ solves the compensated-equilibrium dual with optimum value zero, then $(C, \boldsymbol{u}^*, \boldsymbol{v}^*)$ is a compensated equilibrium. Necessarily $\boldsymbol{v}^* \succ \mathbf{0}$, because a zero price would entail $\boldsymbol{u}^* = \mathbf{0}$.

49.3 Efficiency and Competitive Equilibrium

The efficiency primal is analogous to the weak-efficiency primal.

Problem 4 (Efficiency Primal)

$$\sup_{X,y}\left[\langle X^*, X\rangle + \langle \boldsymbol{y}^*, \boldsymbol{y}\rangle - \delta_{X\mathbf{1}\leq \boldsymbol{v}} - \delta^*_{\mathbf{X}_+ + \{1\}}(-\boldsymbol{y}-\boldsymbol{u}) - \sum_i \delta_{\text{epi } f_i}(\boldsymbol{x}_i, y_i)\right].$$

The perturbation base values are all zero: $X^* = 0$, $\boldsymbol{y}^* = \mathbf{0}$, $\boldsymbol{u} = \mathbf{0}$, and $\boldsymbol{v} = \mathbf{0}$. The allocation C is efficient if and only if the optimum value is zero. A choice $y_i > 0$ means $f_i(\boldsymbol{x}_i) > 0$. For the choice $(X, \boldsymbol{y}) = (0, \mathbf{0})$, the objective function is zero, so the optimum value is greater than or equal to zero.

We establish this result via the efficiency primal and its reciprocal dual. Adding the Fenchel inequalities

$$\begin{aligned}
&\delta_{\mathbf{X}_+}(\boldsymbol{v}^*) + \delta^*_{\mathbf{X}_+}(X\mathbf{1}-\boldsymbol{v}) - \langle \boldsymbol{v}^*, X\mathbf{1}-\boldsymbol{v}\rangle \geq 0\\
&\delta_{\mathbf{X}_+ + \{1\}}(\boldsymbol{u}^*) + \delta^*_{\mathbf{X}_+ + \{1\}}(-\boldsymbol{y}-\boldsymbol{u}) - \langle \boldsymbol{u}^*, -\boldsymbol{y}-\boldsymbol{u}\rangle \geq 0\\
&\delta_{\text{epi } f_i}(\boldsymbol{x}_i, y_i) + \delta^*_{\text{epi } f_i}(-\boldsymbol{v}^* + \boldsymbol{x}_i^*, u_i^* + y_i^*)\\
&\quad - \langle(-\boldsymbol{v}^* + \boldsymbol{x}_i^*, u_i^* + y_i^*), (\boldsymbol{x}_i, y_i)\rangle \geq 0 \text{ for all } i
\end{aligned} \tag{49.4}$$

and rearranging obtains

$$\begin{aligned}
&\langle \boldsymbol{v}^*, \boldsymbol{v}\rangle + \langle \boldsymbol{u}^*, \boldsymbol{u}\rangle + \delta_{\mathbf{X}_+}(\boldsymbol{v}^*) + \delta_P(\boldsymbol{u}^*) + \sum_i \delta^*_{\text{epi } f_i}(-\boldsymbol{v}^* + \boldsymbol{x}_i^*, u_i^* + y_i^*)\\
&\quad \geq \langle X^*, X\rangle + \langle \boldsymbol{y}^*, \boldsymbol{y}\rangle - \delta^*_{\mathbf{X}_+}(X\mathbf{1}-\boldsymbol{v}) - \delta^*_P(-\boldsymbol{y}-\boldsymbol{u}) - \sum_i \delta_{\text{epi } f_i}(\boldsymbol{x}_i, y_i).
\end{aligned}$$

Below we interpret the dual of the efficiency primal as a competitive-equilibrium problem.

Problem 5 (Competitive-Equilibrium Dual)

$$\inf_{\boldsymbol{u}^*,\boldsymbol{v}^*}\left[\delta_{\boldsymbol{u}^*\geq \boldsymbol{1}}+\delta_{\boldsymbol{v}^*\geq \boldsymbol{0}}-\sum_i \left(u_i^*\star f_{i*}\right)(\boldsymbol{v}^*)\right]. \tag{49.5}$$

The competitive-equilibrium dual satisfies the polyhedral/relative condition, so there is no duality gap. If the optimum value is finite, it is attained in the efficiency primal. However the efficiency primal does not satisfy the polyhedral/relative condition, because an efficient allocation lies on the relative boundary of the effective domain of the objective function. The optimum value is perhaps not attained in the dual but is only approached as an infimum.

The competitive-equilibrium dual (49.5) is almost the same as the compensated-equilibrium dual (49.3). In the former, the constraint $\boldsymbol{u}^* \geq \boldsymbol{1}$ captures the requirement $\boldsymbol{u}^* \gg \boldsymbol{0}$. In the latter, the constraint $\boldsymbol{u}^* \in P$ captures the requirement $\boldsymbol{u}^* > \boldsymbol{0}$. Otherwise the two duals are identical.

Nevertheless the analysis of the efficiency primal and competitive-equilibrium dual is significantly different. We summarize the situation and add details below. The resource allocation C is efficient if and only if the optimum value is zero. In contrast to the weak efficiency/compensated equilibrium case, it is not the case that $(C,\boldsymbol{u}^*,\boldsymbol{v}^*)$ is a competitive equilibrium if and only if $(\boldsymbol{u}^*,\boldsymbol{v}^*)$ solves the dual with optimum value zero.

If $(C,\boldsymbol{u}^*,\boldsymbol{v}^*)$ is a competitive equilibrium such that $\boldsymbol{u}^*$ is positive, then $(\boldsymbol{u}^*,\boldsymbol{v}^*)$ solves the dual with optimum value zero. If $(C,\boldsymbol{u}^*,\boldsymbol{v}^*)$ is a competitive equilibrium such that $\boldsymbol{u}^*$ is greater than zero but not positive, then $(\boldsymbol{u}^*,\boldsymbol{v}^*)$ does not solve the dual. However the optimum value of the dual is zero, not attained but approached as an infimum. The first theorem of welfare economics for competitive equilibrium follows.

A competitive equilibrium $(C,\boldsymbol{u}^*,\boldsymbol{v}^*)$ is efficient, because the optimum value of the dual is zero. If $\boldsymbol{u}^* \gg \boldsymbol{0}$, renormalize so $\boldsymbol{u}^* \geq \boldsymbol{1}$. Then $(\boldsymbol{u}^*,\boldsymbol{v}^*)$ solves the dual with optimum value zero. On the other hand, perhaps $\boldsymbol{u}^* > \boldsymbol{0}$ but not positive. For any $\lambda > 0$, choose

$$\begin{aligned}\widetilde{\boldsymbol{v}^*} &= \lambda \boldsymbol{v}^*\\ \widetilde{u_i^*} &= \lambda u_i^* \text{ if } u_i^* > 0\\ \widetilde{u_i^*} &= 1 \text{ if } u_i^* = 0.\end{aligned}$$

For these choices, the infimum of the dual is zero, approached as $\lambda \to \infty$. For a suitably large λ, then $\boldsymbol{u}^* \geq \boldsymbol{1}$. For a consumer such that $u_i^* > 0$, $\left(\widetilde{u_i^*}\star f_{i*}\right)\left(\widetilde{\boldsymbol{v}^*}\right) = 0$, for any $\lambda > 0$. For a consumer such that $u_i^* = 0$,

$$\lim_{\lambda\to\infty}\left(\widetilde{u_i^*}\star f_{i*}\right)\left(\widetilde{\boldsymbol{v}^*}\right) = \lim_{\lambda\to\infty} f_{i*}\left(\lambda \boldsymbol{v}^*\right) = 0,$$

the necessary and sufficient condition for utility-maximization for income zero. The optimum value of the primal/dual pair is zero, so the resource allocation is efficient.

Theorem 3 (First Theorem of Welfare Economics for Competitive Equilibrium) *A competitive equilibrium is efficient.*

Whereas we have shown the efficiency of competitive equilibrium via perturbation duality, Arrow (1951) emphasizes that efficiency holds regardless of whether utility is concave.

Conversely, we obtain only a modified second theorem of welfare economics for competitive equilibrium. If $(C, \boldsymbol{u}^*, \boldsymbol{v}^*)$ solves the competitive-equilibrium dual having optimum-value zero, then it is a competitive equilibrium. We explained previously why a solution is a compensated equilibrium. The competitive-equilibrium dual adds the requirement that each u_i^* is positive, so consumption is utility-maximizing.

Theorem 4 (Second Theorem of Welfare Economics for Competitive Equilibrium) *If the resource allocation C is efficient and there exists a solution $(\boldsymbol{u}^*, \boldsymbol{v}^*)$ to the competitive-equilibrium dual, then $(C, \boldsymbol{u}^*, \boldsymbol{v}^*)$ is a competitive equilibrium.*

This theorem is limited, because the zero optimum value of the dual is perhaps not attained but is only approached as an infimum. In this case it is possible that the efficient allocation cannot be realized as a competitive equilibrium.

Because an efficient resource allocation is weakly-efficient, it can be realized as a compensated equilibrium. If every consumer has positive income in the compensated equilibrium, expenditure-minimization implies utility maximization, so the compensated equilibrium is a competitive equilibrium. The zero optimum value in the competitive-equilibrium dual is attained. If however some consumers have zero income, then the compensated equilibrium is perhaps not a competitive equilibrium.

A two-consumer, two-good Edgeworth box example illustrates the situation. The compensated equilibrium is efficient but cannot be attained as a competitive equilibrium. The price of good one is 1, and the price of good two is 0. Consumer one has all of good one and is satiated in good two. His income is positive, and he maximizes utility. Consumer two has the remainder of good two but is not satiated. His income is zero. He minimizes expenditure but does not maximize utility.

Reference

Arrow KJ (1951) An extension of the basic theorems of classical welfare economics. In: Neyman J (ed) Proceedings of the second berkeley symposium on mathematical statistics and probability. University of California Press, Berkeley, CA, pp 507–532

Part XII
Core

Chapter 50
Core and Equilibrium

Abstract Debreu and Scarf analyze the core of a large economy, inspired by Edgeworth (Debreu, G., & Scarf, H. (1963, September). A limit theorem on the core of an economy. International Economic Review, 4 (3), 235–246; Edgeworth, F. Y. (2003 (originally 1881)). Mathematical psychics: An essay on application of mathematics to the moral sciences. In P. Newman (Ed.), F. Y. Edgeworth: Mathematical Psychics and further papers on political economy (pp. 1–174). Oxford: Oxford University Press). Consider an exchange economy in which each consumer has a utility function and an endowment of goods. An allocation of goods across the consumers is in the core if no coalition of consumers can trade their endowments among themselves so at least one member of the coalition gains and no member loses. For an economy with many consumers, the intuitive idea is that an allocation of goods lies in the core if and only if it is attainable as a general market equilibrium. For an allocation in the core, there exists a price vector such that it is a general market equilibrium. Conversely, for a general market equilibrium, the allocation is in the core. We develop their analysis via primal/dual convex optimization modelling whether an allocation is in the core of a large economy. The primal is a maximization choosing simultaneously across both allocations and coalitions, to see whether any coalition can do better. For a core allocation, no coalition can do better and the optimum value is zero. If the allocation is not in the core, then some coalition can do better, and the optimum value is positive. The dual is a general-equilibrium minimization, looking for an equilibrium price vector. For a core allocation, the dual solution solves for a price vector for which the allocation is a general equilibrium. Conversely, if the allocation is not in the core, then the dual has no solution.

We set up two primal/dual pairs modelling the core of a large economy. The situation closely resembles the chapter "Welfare and Equilibrium." The primal maximizations optimize across all possible coalitions. For the first pair, the primal is a weak-efficiency core problem, and the dual is a compensated-equilibrium problem. That the optimum value of the primal is zero is a necessary condition for the resource allocation to be in the core. The dual shows that it is necessary that the allocation

B. C. Dieffenbach, *Conjugate Duality in Economic Analysis*, Contributions to Economics,
https://doi.org/10.1007/978-3-032-21396-9_50

be attainable as a compensated equilibrium. For the second pair, the primal is an efficiency core problem, and the dual is a competitive-equilibrium problem. That the optimum value of the primal is zero is necessary and sufficient for the resource allocation to be in the core of a large economy. The dual shows that the resource allocation of a competitive equilibrium is in the core.

Consider an exchange economy with n classical consumers. Endowed initially with a vector of goods, each wants to maximize a concave utility function. An allocation of goods across the consumers is in the core if no coalition of consumers can trade their endowments among themselves so at least one member of the coalition gains and no member loses.

Debreu and Scarf extend this economy by replication: let there be k consumers of each of the n types. Each of the k consumers has the same utility function and the same endowment. Consider the core allocations for a particular k. If each utility function is strictly concave, then in any core allocation each of the k consumers of the same type has the same consumption (Debreu and Scarf 1963, Theorem 2, p. 241). Thus one can describe a core allocation by the single consumption vector for each type.

The reasoning involves two steps. If two consumers of the same type have the same utility, they must have the same consumption. If each had different consumption, then their average consumption would have higher utility.

Furthermore, all consumers of the same type must have the same utility. Suppose not, and obtain a contradiction. Select the consumer of each type who has lowest utility. Then the average consumption of the k consumers of this type has higher utility. However the sum of these average consumptions across all types equals the sum of the endowments across all types. Hence the coalition of the consumer of each type who has lowest utility could achieve these average consumptions, having higher utility, a contradiction.[1]

As k increases, more coalitions are possible, so the set of core allocations either stays the same or shrinks. Define the core allocations for a large economy as the allocations in the core for all k.

Conversely, in a general equilibrium, two consumers of the same type must have the same consumption. For each, the budget constraint is the same. For strictly concave utility, if each had different consumption, then their average consumption would have higher utility, which would contradict utility maximization.

[1] That there are k consumers of every type—the same number for each type—is fundamental to this reasoning.

50.1 Exchange Economy

Consider an allocation (C, W) of resources in an exchange economy having m goods and n types of consumer. The endowment of a type i individual is $\boldsymbol{w}_i \succeq \mathbf{0}$. Define

$$W := \begin{bmatrix} \boldsymbol{w}_1 & \cdots & \boldsymbol{w}_n \end{bmatrix} \in \mathbf{T}^{n \to m}$$
$$W\mathbf{1} \gg \mathbf{0}.$$

Consumption

$$C = \begin{bmatrix} \boldsymbol{c}_1 & \cdots & \boldsymbol{c}_n \end{bmatrix} \in \mathbf{T}^{n \to m}, \boldsymbol{c}_i \succeq \mathbf{0},$$
$$C\mathbf{1} = W\mathbf{1}.$$

The optimization compares this allocation to alternatives.

Define f_i as the translation to $\boldsymbol{c}_i$ of the utility of a consumer of type i. Recall the properties of translated utility. The conjugate $f_{i*}(\boldsymbol{x}^*) \leq 0$; $\boldsymbol{x}^* \in \partial f(\mathbf{0})$ if and only if the conjugate is zero. The epi-multiplication $(u_i^* \star f_{i*})(\boldsymbol{v}^*) \leq 0$.

The dual for the minimum expenditure required to achieve the utility of $\boldsymbol{c}_i$ or higher is

$$\langle \boldsymbol{v}^*, \boldsymbol{c}_i \rangle + \sup_{u_i} \left[(u_i^* \star f_{i*})(\boldsymbol{v}^*) \right]. \tag{50.1}$$

Because the epi-multiplication is less than or equal to zero, the supremum is less than or equal to zero. Hence $\boldsymbol{c}_i$ is expenditure-minimizing if and only if the supremum is zero. The solutions u_i^* are the subgradients of minimum expenditure with respect to utility; necessarily $u_i^* \geq 0$.

In an exchange economy, a general equilibrium $(C, W, \boldsymbol{v}^*, \boldsymbol{u}^*)$ has the following properties. The price vector $\boldsymbol{v}^* \succ \mathbf{0}$. Demand $\boldsymbol{c}_i$ must satisfy the budget constraint $\langle \boldsymbol{v}^*, \boldsymbol{a}_i \rangle \leq 0$. Expenditure must be minimized at $\boldsymbol{c}_i$ among all consumption having the same utility or more. Aggregate demand must be less than or equal to aggregate supply: $C\mathbf{1} \preceq W\mathbf{1}$. For any good in excess supply, the price must be zero.

A compensated equilibrium satisfies these conditions. It is a competitive equilibrium if in addition $\boldsymbol{c}_i$ also maximizes utility subject to the budget constraint.

In a general equilibrium, necessarily there is a subgradient $u_i^* > 0$ for at least one consumer. The exceptional case $u_i^* = 0$ can occur only for consumption on the boundary of effective domain of utility, such that only goods having a zero price are consumed. Because $\boldsymbol{v}^* \succ \mathbf{0}$ and the aggregate endowment of every good is positive, some consumer demands a positive quantity of a good having a positive price, so $u_i^* > 0$ for this consumer. Hence there is a vector $\boldsymbol{u}^*$ of subgradients greater than zero.

The normalization of $\boldsymbol{v}^*$ is arbitrary. If $\boldsymbol{u}^*$ is a subgradient vector for price $\boldsymbol{v}^*$, then $\lambda\boldsymbol{u}^*$ is a subgradient vector for price $\lambda\boldsymbol{v}^*$, for $\lambda > 0$.

50.2 Core of a Large Economy

Let $z \in \mathbf{R}^n_+$ denote a coalition, in which there are z_i consumers of type i. Aggregate consumption for type i is $\boldsymbol{c}_i z_i + \boldsymbol{x}_i$; the latter vector is net aggregate additional consumption. Define

$$X := \begin{bmatrix} \boldsymbol{x}_1 & \cdots & \boldsymbol{x}_n \end{bmatrix} \in \mathbf{T}^{n \to m}.$$

For $z_i > 0$, consumption is $\boldsymbol{c}_i + \boldsymbol{x}_i/z_i$ for a single consumer, so translated utility is $f_i(\boldsymbol{x}_i/z_i)$, which is zero if $\boldsymbol{x}_i = \mathbf{0}$.

Aggregate consumption for the coalition is $X\mathbf{1} + Cz$, and the aggregate endowment is Wz. Its resource constraint is $X\mathbf{1} + Cz \preceq Wz$. Let us define $A := C - W$ as the linear transformation of net trades from the endowment to the allocation, and the resource constraint is $X\mathbf{1} + Az \preceq \mathbf{0}$.

Definition 1 (Core of a Large Economy) The allocation (C, W) is in the core of a large economy if no coalition can trade its aggregate endowment among its members so at least one member of the coalition gains and no member loses.

Definition 2 (Weak Core of a Large Economy) The allocation (C, W) is in the weak core of a large economy if no coalition containing consumers of every type can trade its aggregate endowment among its members so that *every* member gains.

A natural efficiency problem to study the core is

$$\sup_{X,y,z} \left[-\delta^*_{\mathbf{R}^n_+ + \{1\}}(-y) - \sum_i \delta_{\text{cone}\, f_i}(\boldsymbol{x}_i, y_i, -z_i) - \delta_{X\mathbf{1}+Az \preceq \mathbf{0}}. \right] \tag{50.2}$$

If the allocation (C, W) is in the core of a large economy, the optimum value is zero. Otherwise the optimum value is ∞. The choice $(X, \boldsymbol{y}, z) = (0, \mathbf{0}, \mathbf{1})$ corresponds to (C, W), and the objective function is zero. If there is any choice $(X, \boldsymbol{y}, z)$ such that the objective function is positive, then the allocation is not in the core of a large economy: the coalition z can reallocate its aggregate endowment so that consumers of at least one type gain and no others lose. Because the objective function is positively homogeneous in the choice variables, the optimum value is ∞.

Finiteness of the objective function requires $z_i \geq 0$, for otherwise $\delta_{\text{cone}\, f_i} = \infty$. For $z_i > 0$, $(\boldsymbol{x}_i, y_i, -z_i) \in \text{cone}\, f_i$ means $f_i(\boldsymbol{x}_i/z_i) \geq y_i/z_i$. If $y_i > 0$, then the utility for a type i consumer is greater than zero (greater than the utility of $\boldsymbol{c}_i$). The problem (50.2) looks for $(X, \boldsymbol{y}, z)$ satisfying the resource constraint such that $\boldsymbol{y} \succ \mathbf{0}$.

Although literally z is a coalition only if each component is a nonnegative integer, to optimize across all $z \geq \mathbf{0}$ is an interchangeable formulation. If every component of z is a nonnegative rational number, then every component of λz is a nonnegative integer, for a suitable choice of λ. Furthermore, every irrational number is arbitrarily close to a rational number. It follows that the optimum value in Eq. (50.2) is 0 or ∞ depending on whether the allocation is or is not in the core of a large economy.

50.3 The Weak Core and Compensated Equilibrium

For an allocation to be in the core, it is necessary (but not sufficient) that it be in the weak core. A natural problem to study the weak core is the weak-efficiency problem

$$\sup_{X,y,z}\left[-\delta_P^*(-\boldsymbol{y})-\sum_i \delta_{\text{cone } f_i}(\boldsymbol{x}_i,y_i,-z_i)-\delta_{X\mathbf{1}+Az\leq\mathbf{0}}.\right] \tag{50.3}$$

The objective function is positively homogeneous, so the optimum value is either 0 or ∞. The choice $(X,\boldsymbol{y},z)=(0,\mathbf{0},\mathbf{1})$ makes the objective function zero. If $\boldsymbol{y}\gg\mathbf{0}$ for some $(X,\boldsymbol{y},z)$ satisfying the constraints, then every consumer gains, having positive utility. The allocation is not in the weak core, and the optimum value is ∞. If there is no $\boldsymbol{y}\gg\mathbf{0}$ for some $(X,\boldsymbol{y},z)$ satisfying the constraints, then the optimum value is zero, a condition necessary for the allocation to be in the weak core.

The special case $C=W$ and setting $z=\mathbf{1}$ reduces the problem (50.3) to the weak-efficiency primal in the chapter "Welfare and Equilibrium" employed to analyze Arrow's first and second theorems of compensated equilibrium. Thus the treatment here of the weak core is closely related. Taking z as an additional choice variable alters and augments the dual, but we continue to interpret it via expenditure-minimization and compensated equilibrium.

A simplification below is to add the constraint $z\in P$. If the allocation is not in the core of a large economy, this constraint makes the optimum value positive, rather than infinite. If the allocation is in the core of a large economy, then the optimum value stays at zero. Adding the constraint simplifies the analysis of the dual and its relationship to the primal.

We first analyze the weak-efficiency core primal with the added resource constraint.

Problem 1 (Weak-Efficiency Core Primal)

$$\begin{aligned}\sup_{X,y,z}\Big[&\langle X^*,X\rangle+\langle \boldsymbol{y}^*,\boldsymbol{y}\rangle+\langle z^*,z\rangle-\delta_P^*(-\boldsymbol{y}-\boldsymbol{u})-\delta_P(z-\boldsymbol{s})\\ &-\sum_i \delta_{\text{cone } f_i}(\boldsymbol{x}_i,y_i,-z_i)-\delta_{X\mathbf{1}+Az\leq\boldsymbol{v}}\Big].\end{aligned} \tag{50.4}$$

The optimum value is greater than or equal to zero, and the allocation (C,W) is in the weak core if and only if the optimum value is zero. The perturbation base values are all zero: $\boldsymbol{u}=\mathbf{0}$, $\boldsymbol{v}=\mathbf{0}$, $\boldsymbol{s}=\mathbf{0}$, $X^*=0$, $\boldsymbol{y}^*=\mathbf{0}$, $z^*=\mathbf{0}$.

We calculate the reciprocal dual. Adding the Fenchel inequalities

$$\begin{aligned}&\delta_{\text{cone } f_i}(\boldsymbol{x}_i,y_i,-z_i)+\delta^*_{\text{cone } f_i}(\boldsymbol{x}_i^*-\boldsymbol{v}^*,y_i^*+u_i^*,\langle\boldsymbol{v}^*,\boldsymbol{a}_i\rangle+s_i^*-z_i^*)\\ &\quad-\langle\boldsymbol{x}_i^*-\boldsymbol{v}^*,\boldsymbol{x}_i\rangle-(y_i^*+u_i^*)y_i-(\langle\boldsymbol{v}^*,\boldsymbol{a}_i\rangle+s_i^*-z_i^*)(-z_i)\geq 0 \text{ for all } i\end{aligned} \tag{50.5}$$

$$\begin{aligned} \delta_P(\boldsymbol{u}^*) + \delta_P^*(-\boldsymbol{y}-\boldsymbol{u}) - \langle \boldsymbol{u}^*, -\boldsymbol{y}-\boldsymbol{u}\rangle &\geq 0 \\ \delta_P(\boldsymbol{z}-\boldsymbol{s}) + \delta_P^*(\boldsymbol{s}^*) - \langle \boldsymbol{s}^*, \boldsymbol{z}-\boldsymbol{s}\rangle &\geq 0 \\ \delta_{\mathbf{R}_+^m}(\boldsymbol{v}-X\mathbf{1}-A\boldsymbol{z}) + \delta_{\mathbf{R}_+^m}^*(-\boldsymbol{v}^*) - \langle -\boldsymbol{v}^*, \boldsymbol{v}-X\mathbf{1}-A\boldsymbol{z}\rangle &\geq 0 \end{aligned} \tag{50.6}$$

and rearranging obtains

$$\begin{aligned} &\langle \boldsymbol{u}^*, \boldsymbol{u}\rangle + \langle \boldsymbol{v}^*, \boldsymbol{v}\rangle + \delta_P(\boldsymbol{u}^*) + \delta_{\mathbf{R}_+^m}^*(-\boldsymbol{v}^*) + \delta_P^*(\boldsymbol{s}^*) \\ &\quad + \sum_i \delta_{\mathrm{cone}\, f_i}^*(\boldsymbol{x}_i^* - \boldsymbol{v}^*, y_i^* + u_i^*, \langle \boldsymbol{v}^*, \boldsymbol{a}_i\rangle + s_i^* - z_i^*) \\ &\quad \geq \langle X^*, X\rangle + \langle \boldsymbol{y}^*, \boldsymbol{y}\rangle + \langle \boldsymbol{z}^*, \boldsymbol{z}\rangle \\ &\quad - \delta_P^*(-\boldsymbol{y}-\boldsymbol{u}) - \delta_P(\boldsymbol{z}-\boldsymbol{s}) - \sum_i \delta_{\mathrm{cone}\, f_i}(\boldsymbol{x}_i, y_i, -z_i) - \delta_{X\mathbf{1}+A\boldsymbol{z}\leq \boldsymbol{v}}. \end{aligned}$$

Both the primal and the reciprocal dual satisfy the polyhedral/relative condition, so there is no duality gap and the optimum value is attained in both the primal and the dual.

Substituting

$$\delta_{\mathrm{cone}\, f_i}^*(-\boldsymbol{v}^*, u_i^*, \langle \boldsymbol{v}^*, \boldsymbol{a}_i\rangle + s_i^*) \tag{50.7}$$

$$= \delta_{\mathrm{cone}\, f_{i*}}(\boldsymbol{v}^*, -\langle \boldsymbol{v}^*, \boldsymbol{a}_i\rangle - s_i^*, -u_i^*) \tag{50.8}$$

$$= \delta_{\mathrm{epi}(u_i^* \star f_{i*})}(\boldsymbol{v}^*, -\langle \boldsymbol{v}^*, \boldsymbol{a}_i\rangle - s_i^*) \tag{50.9}$$

$$= \delta_{(u_i^* \star f_{i*})(\boldsymbol{v}^*) + \langle \boldsymbol{v}^*, \boldsymbol{a}_i\rangle \geq -s_i^*} \tag{50.10}$$

and eliminating $\boldsymbol{s}^*$ obtains a reduced compensated-equilibrium core dual.

Problem 2 (Compensated-Equilibrium Core Dual)

$$\inf_{\boldsymbol{u}^*, \boldsymbol{v}^*} \left\{ \delta_P(\boldsymbol{u}^*) + \delta_{\boldsymbol{v}^* \geq 0} + \max_i \left[(u_i^* \star f_{i*})(\boldsymbol{v}^*) + \langle \boldsymbol{v}^*, \boldsymbol{a}_i\rangle \right] \right\}. \tag{50.11}$$

We interpret this dual as a compensated-equilibrium problem. Note that the dual makes no reference to coalitions or to whether the economy is large, even though these concepts underpin it.

If $(C, W, \boldsymbol{v}^*, \boldsymbol{u}^*)$ is a compensated equilibrium, then the objective function is zero. Normalize $\boldsymbol{v}^* \succ \mathbf{0}$ so $\boldsymbol{u}^* \in P$. For each consumer, u_i^* maximizes $(u_i^* \star f_{i*})(\boldsymbol{v}^*)$ at zero, by expenditure minimization. The budget constraint is satisfied, and net spending $\langle \boldsymbol{v}^*, \boldsymbol{a}_i\rangle = 0$. The minimum of the dual is zero, so the maximum of the primal is zero. The allocation is in the weak core of a large economy.

Theorem 1 (First Theorem of the Weak Core) *If $(C, W, \boldsymbol{u}^*, \boldsymbol{v}^*)$ is a compensated equilibrium, then the allocation (C, W) is in the weak core of a large economy.*

An allocation in the weak core of a large economy is also in the weak core of a smaller economy (such as an economy having only one consumer of each type), because its coalitions are only a subset of the coalitions of a large economy.

Corollary 1 *If* $(C, W, \boldsymbol{u}^*, \boldsymbol{v}^*)$ *is a compensated equilibrium, then the allocation* (C, W) *is in the weak core of the economy.*

The second theorem of the weak core is the converse of the first theorem: an allocation in the weak core can be realized as a compensated equilibrium. Consider primal and dual solutions for an allocation in the weak core of a large economy. Necessarily $\boldsymbol{v}^* \succ \boldsymbol{0}$, because a zero price would entail $\boldsymbol{u}^* = \boldsymbol{0}$. For the Fenchel inequality (50.6) to hold with equality, it must hold for all $z \in P$; by independence, $\boldsymbol{s}^* = \boldsymbol{0}$. In the derivation (50.10), it follows that

$$(u_i^* \star f_{i*})(\boldsymbol{v}^*) + \langle \boldsymbol{v}^*, \boldsymbol{a}_i \rangle \geq -s_i^* = 0 \text{ for all } i.$$

But the maximum of these is zero, so for all i

$$(u_i^* \star f_{i*})(\boldsymbol{v}^*) + \langle \boldsymbol{v}^*, \boldsymbol{a}_i \rangle = 0.$$

As $(u_i^* \star f_{i*})(\boldsymbol{v}^*) \leq 0$ and $\sum_i \langle \boldsymbol{v}^*, \boldsymbol{a}_i \rangle = 0$, for all i

$$\langle \boldsymbol{v}^*, \boldsymbol{a}_i \rangle = 0 \tag{50.12}$$

$$(u_i^* \star f_{i*})(\boldsymbol{v}^*) = 0. \tag{50.13}$$

The first equation is the budget constraint, and the second condition expresses expenditure minimization for price $\boldsymbol{v}^*$.

By Eqs. (50.12) and (50.13), the budget constraint is satisfied and expenditure is minimized at $\boldsymbol{c}_i$, so $\boldsymbol{c}_i$ is compensated demand. Aggregate demand equals supply, so $(C, W, \boldsymbol{v}^*)$ is a compensated equilibrium.

Theorem 2 (Second Theorem of the Weak Core) *If an allocation is in the weak core of a large economy, then the allocation is a compensated equilibrium for some price greater than zero.*

This theorem is the fundamental result of Debreu and Scarf (1963).

50.4 The Core and Competitive Equilibrium

We next analyze the core-efficiency primal.

Problem 3 (Core-Efficiency Primal)

$$\sup_{X,y,z}\Big[\langle X^*,X\rangle+\langle y^*,y\rangle+\langle z^*,z\rangle-\delta^*_{\mathbf{R}^n_+ +\{1\}}(-y-u)-\delta_P(z-s) \tag{50.14}$$

$$-\sum_i \delta_{\mathrm{cone}\,f_i}(x_i,y_i,-z_i)-\delta_{X1+Az\le v},\Big]. \tag{50.15}$$

The optimum value is greater than or equal to zero, and the allocation (C, W) is in the core if and only if the optimum value is zero. The perturbation base values remain $X^* = 0, y^* = \mathbf{0}, z^* = \mathbf{0}, u = \mathbf{0}, v = \mathbf{0}, s = \mathbf{0}$.

Interpret the reciprocal dual as a competitive-equilibrium problem. Adding the Fenchel inequalities

$$\begin{aligned}
\delta_{\mathrm{cone}\,f_i}(x_i,y_i,-z_i)+\delta^*_{\mathrm{cone}\,f_i}\left(x_i^*-v^*,y_i^*+u_i^*,a_i^\top v^*+s_i^*-z_i^*\right)&\\
-\langle x_i^*-v^*,x_i\rangle-(y_i^*+u_i^*)\,y_i-\left(a_i^\top v^*+s_i^*-z_i^*\right)(-z_i)&\ge 0\\
\delta_{\mathbf{R}^n_+ +\{1\}}(u^*)+\delta^*_{\mathbf{R}^n_+ +\{1\}}(-y-u)-\langle u^*,-y-u\rangle&\ge 0\\
\delta_P(z-s)+\delta_P^*(s^*)-\langle s^*,z-s\rangle&\ge 0\\
\delta_{\mathbf{R}^m_+}(v-X1-Az)+\delta^*_{\mathbf{R}^m_+}(-v^*)-\langle -v^*,v-X1-Az\rangle&\ge 0,
\end{aligned}$$

and rearranging obtains the reciprocal dual:

$$\begin{aligned}
&\langle u^*,u\rangle+\langle v^*,v\rangle+\delta_{\mathbf{R}^n_+ +\{1\}}(u^*)+\delta^*_{\mathbf{R}^m_+}(-v^*)+\delta_P^*(s^*)\\
&\quad+\sum_i \delta^*_{\mathrm{cone}\,f_i}\left(x_i^*-v^*,y_i^*+u_i^*,a_i^\top v^*+s_i^*-z_i^*\right)\\
&\quad\ge\langle X^*,X\rangle+\langle y^*,y\rangle+\langle z^*,z\rangle-\delta^*_{\mathbf{R}^n_+ +\{1\}}(-y-u)-\delta_P(z-s)\\
&\quad-\sum_i \delta_{\mathrm{cone}\,f_i}(x_i,y_i,-z_i)-\delta_{\mathbf{R}^m_+}(v-X1-Az)\,.
\end{aligned}$$

The reciprocal dual satisfies the polyhedral/relative condition, so there is no duality gap and the finite optimum value is attained the primal. As an efficiency problem, the primal fails to satisfy the polyhedral/relative condition, so the optimum value in the dual is perhaps not attained.

Invoking the derivation (50.8)–(50.9) obtains the competitive-equilibrium core dual.

Problem 4 (Competitive-Equilibrium Core Dual)

$$\inf_{u^*,v^*}\left\{\delta_{u^*\ge 1}+\delta_{v^*\ge 0}+\max_i\left[(u_i^*\star f_{i*})(v^*)+\langle v^*,a_i\rangle\right]\right\}. \tag{50.16}$$

This dual is similar to the compensated-equilibrium core dual. Effectively the compensated-equilibrium dual requires that u^* be greater than zero, whereas the competitive-equilibrium dual requires that u^* be positive.

We prove the following via the competitive-equilibrium core dual.

Theorem 3 (First Theorem of the Core) *If* $(C, W, \boldsymbol{u}^*, \boldsymbol{v}^*)$ *is a competitive equilibrium, then the allocation* (C, W) *is in the core.*

We show that a competitive equilibrium is in the core of a large economy, and so it is also in the core of a smaller economy.

Suppose that the allocation (C, W) with price $\boldsymbol{v}^*$ is a competitive equilibrium. The dual is a competitive-equilibrium problem. We explain why the optimum value of the dual is zero, and thus the optimum value of the primal is also zero. The resource allocation is in the core of a large economy and hence is in the core.

For any $\lambda > 0$, rescaling to $(\lambda \boldsymbol{u}^*, \lambda \boldsymbol{v}^*)$ is also a competitive equilibrium. For any $\lambda > 0$, choose

$$\begin{aligned}
\widetilde{\boldsymbol{v}^*} &= \lambda \boldsymbol{v}^* \\
\widetilde{u_i^*} &= \lambda u_i^* \text{ if } u_i^* > 0 \\
\widetilde{u_i^*} &= 1 \text{ if } u_i^* = 0.
\end{aligned}$$

If $\boldsymbol{u}^* \gg \boldsymbol{0}$, then $\widetilde{\boldsymbol{u}^*} \geq \boldsymbol{1}$ for a suitable large λ. Because consumption is expenditure-minimizing, the choice $\left(\widetilde{\boldsymbol{u}^*}, \widetilde{\boldsymbol{v}^*}\right)$ solves the dual with optimum value zero. If $\boldsymbol{u}^* > \boldsymbol{0}$ but is not positive, then the infimum of the dual is zero, approached as $\lambda \to \infty$. For a suitably large λ, $\widetilde{\boldsymbol{u}^*} \geq \boldsymbol{1}$. For the case $\widetilde{u_i^*} = 1$,

$$\lim_{\lambda \to \infty} \left(\widetilde{u_i^*} \star f_*\right)\left(\widetilde{\boldsymbol{v}^*}\right) = \lim_{\lambda \to \infty} f_* \left(\lambda \boldsymbol{v}^*\right) = 0,$$

the necessary and sufficient condition for utility-maximization for income zero, which implies expenditure-minimization. Consumption is expenditure-minimizing. Each epi-multiplication in the dual (50.16) is zero or approaches it. Consumption satisfies the budget constraint, so the optimum value of the dual is zero.

We prove the second theorem of the core. If $(C, W, \boldsymbol{v}^*, \boldsymbol{u}^*)$ solves the competitive-equilibrium core dual having optimum value zero, then it is a competitive equilibrium. We explained previously why a solution is a compensated equilibrium. The competitive-equilibrium core dual adds the requirement that each u_i^* is positive, so consumption is utility-maximizing.

Theorem 4 (Second Theorem of the Core) *For an allocation* (C, W) *in the core of a large economy, if the competitive-equilibrium core dual has a solution* $(\boldsymbol{u}^*, \boldsymbol{v}^*)$*, then for price* $\boldsymbol{v}^*$ *the allocation is a competitive equilibrium.*

This theorem is limited. We showed above that a competitive equilibrium perhaps does not solve the competitive-equilibrium core dual, if the zero optimum value of the dual is only approached as an infimum. And perhaps an allocation in the core of a large economy is not attainable as a competitive equilibrium, even though it is attainable as a compensated equilibrium.

References

Debreu G, Scarf H (1963 Sept) A limit theorem on the core of an economy. Int Econ Rev 4(3):235–246

Edgeworth FY (2003 [originally 1881]) Mathematical psychics: an essay on application of mathematics to the moral sciences. In: Newman P (ed) F. Y. Edgeworth: mathematical psychics and further papers on political economy. Oxford University Press, Oxford, pp 1–174

Chapter 51
Bondareva-Shapley

Abstract We study the core of a cooperative multi-person game with transferable utility. A payoff specifies the utility received by each player. By cooperating, a coalition of players can guarantee a certain utility—its *worth*—to be divided somehow among its members. The core is the set of payoffs such that no coalition can do better for itself. The worth of the grand coalition of all players is the value of the game. The primal models which payoffs are in the core. Furnishing the Bondareva-Shapley condition for the core to be non-empty, the dual finds the support to the "set of balanced weights," evaluated at the worth. Whether the core is non-empty or empty is a matter of whether the support to the set of balanced weights evaluated at the worth equals the value or is greater, respectively.

Linear inequalities define the set of payoffs in the core. Each coalition must receive its worth or more. The aggregate payoff to all players must be less than or equal to the value of the game. These linear inequalities form the primal problem.

The dual problem furnishes the Bondareva-Shapley condition for the core to be non-empty (Bondareva 1962, 1963; Shapley 1967). The dual problem finds the support to the "set of balanced weights," evaluated at the worth. This set depends only on the number of players and is independent of worth and the value. Whether the core is non-empty or empty is a matter of whether the support to the set of balanced weights evaluated at the worth equals the value or is greater, respectively.

51.1 Cooperative Game

With m players, the number of possible coalitions—the number of non-empty subsets of the players—is $n = 2^{m-1}$. We model the game via Euclidean Jordan algebras: the coalition space $\mathbf{X} = \mathbf{R}^n$ and the player space $\mathbf{Y} = \mathbf{R}^m$. The choice variable in the primal lies in the player space, and the choice variable in the dual lies in the coalition space.

B. C. Dieffenbach, *Conjugate Duality in Economic Analysis*, Contributions to Economics,
https://doi.org/10.1007/978-3-032-21396-9_51

A payoff is a vector $\boldsymbol{y}^* \in \mathbf{Y}$; the component y_i^* is the payoff to player i. A coalition j is a non-empty subset of players. Acting together, the coalition can achieve its *worth* x_j^*, which is somehow divided among the coalition. The grand coalition of all players has worth z^*, the *value* of the game. Component j in the coalition space $\mathbf{X}$ corresponds to coalition j. Let $\boldsymbol{x}^* \in \mathbf{X}$ denote the vector of the worth of each coalition.

In the player space, the idempotents are the m-tuples having components zero and one and thus correspond to the possible coalitions. For two players, the idempotents are the 2-tuples $(1,0)$, $(0,1)$, and $(1,1)$, which correspond to the possible coalitions $\{1\}$, $\{2\}$, and $\{1,2\}$. For three players, the idempotents are the 3-tuples $(1,0,0)$, $(0,1,0)$, $(0,0,1)$, $(1,1,0)$, $(1,0,1)$, $(0,1,1)$ and $(1,1,1)$, which correspond to the coalitions $\{1\}$, $\{2\}$, $\{3\}$, $\{1,2\}$, $\{1,3\}$, $\{2,3\}$, $\{1,2,3\}$. The convex hull of the idempotents is the unit cube. For one idempotent to be less than or equal to another means that the former coalition is a subset of the latter coalition.

In the coalition space, a primitive idempotent is an n-tuple having a single component one and other components zero and thus corresponds to a coalition.

51.2 Core

A payoff $\boldsymbol{y}^*$ belongs to the *core* if no coalition can do better for itself. Let $\boldsymbol{b}_j \in \mathbf{R}^m$ denote an idempotent. Coalition j can do better if $\langle \boldsymbol{y}^*, \boldsymbol{b}_j \rangle < x_j^*$, so

$$\langle \boldsymbol{y}^*, \boldsymbol{b}_j \rangle \geq x_j^*.$$

is necessary for $\boldsymbol{y}^*$ to be in the core.

Define B as a linear transformation from the coalition space to the player space:

$$\begin{aligned} B : \mathbf{X} &\to \mathbf{Y} \\ \boldsymbol{x} \mapsto \boldsymbol{y} = \textstyle\sum_j \boldsymbol{b}_j x_j &= \begin{bmatrix} \cdots & \boldsymbol{b}_j & \cdots \end{bmatrix} \boldsymbol{x}. \end{aligned}$$

The linear transformation shows the members of the coalitions: a primitive idempotent j in the coalition space maps to the idempotent $\boldsymbol{b}_j$ in the player space.

Definition 1 (Core) A payoff $\boldsymbol{y}^*$ belongs to the core if and only if

$$\begin{aligned} B^{\mathsf{T}} \boldsymbol{y}^* &\geq \boldsymbol{x}^* \\ \langle \boldsymbol{y}^*, \mathbf{1} \rangle &\leq z^*. \end{aligned}$$

No coalition can do better for itself, and the total payoff must be less than or equal to the value of the game. (For any core payoff, necessarily the second inequality must hold with equality, because the payoff to the grand coalition must be the value.)

51.3 Core Primal and Dual

We analyze these inequalities via an optimization.

Problem 1 (Core Primal)

$$\inf_{y^*}\left[\delta_{B^\top y^* \geq x^*} + \delta_{\langle y^*, 1\rangle \leq z^*}\right].$$

The primal perturbations are the worth x^* and the value z^*. Whether the optimum value is zero or infinity shows whether the core is non-empty or empty, respectively.

Finding the reciprocal dual obtains the Bondareva-Shapley theorem. Adding the Fenchel inequalities

$$\begin{aligned}
\delta_{\mathbf{X}_+}(x) + \delta^*_{\mathbf{X}_+}\left(x^* - B^\top y^*\right) - \left\langle x^* - B^\top y^*, x\right\rangle &\geq 0 \\
\delta_{\mathbf{R}_+}(z^* - \langle \mathbf{1}, y^*\rangle) + \delta^*_{\mathbf{R}_+}(z) - \langle z^* - \langle \mathbf{1}, y^*\rangle, z\rangle &\geq 0 \\
\delta_{\{0\}}(Bx + \mathbf{1}z - y) + \delta^*_{\{0\}}(y^*) - \langle y^*, Bx + \mathbf{1}z - y\rangle &\geq 0
\end{aligned}$$

and rearranging gives

$$\begin{aligned}
&\langle y^*, y\rangle + \delta^*_{\mathbf{X}_+}\left(x^* - B^\top y^*\right) + \delta_{\mathbf{R}_+}(z^* - \langle \mathbf{1}, y^*\rangle) + \delta^*_{\{0\}}(y^*) \\
&\quad \geq \langle x^*, x\rangle + \langle z^*, z\rangle - \delta_{\{0\}}(Bx + \mathbf{1}z - y) + \delta_{\mathbf{X}_+}(x) + \delta^*_{\mathbf{R}_+}(z).
\end{aligned}$$

At the dual perturbation base value $y = \mathbf{0}$, the reciprocal dual evaluates the support to a cone. The reciprocal dual satisfies the polyhedral condition, so there is no duality gap.

Define the set B of balanced weights as all nonnegative linear combinations of idempotents in the coalition space that sum to $\mathbf{1}$.

Definition 2 (Balanced Weights) The set B of balanced weights is the solution set to

$$\begin{aligned}
Bx &= \mathbf{1} \\
x &\geq \mathbf{0}.
\end{aligned}$$

As

$$\delta_{\text{cone } B}(x, z) = \delta_{Bx + \mathbf{1}z = \mathbf{0}} + \delta_{\mathbf{X}_+}(x) + \delta^*_{\mathbf{R}_+}(z),$$

the dual calculates the support to cone B.

Problem 2 (Core Dual)

$$\sup_{x,z}\left[\langle x^*, x\rangle + \langle z^*, z\rangle - \delta_{\text{cone } B}(x, z)\right].$$

The optimum value is zero if and only if $(\boldsymbol{x}^*, z^*)$ lies in the polar of cone B. Evaluating the conjugate obtains the Bondareva-Shapley result:

$$\sup_{\boldsymbol{x},z} [\langle \boldsymbol{x}^*, \boldsymbol{x}\rangle + \langle z^*, z\rangle - \delta_{\text{cone } B}(\boldsymbol{x}, z)] = \delta^*_{\text{cone } B}(\boldsymbol{x}^*, z^*)$$
$$= \delta_{\text{epi } \delta^*_B}(\boldsymbol{x}^*, z^*) = \delta_{\delta^*_B(\boldsymbol{x}^*) \le z^*}.$$

Because the grand coalition belongs to B, necessarily $\delta^*_B(\boldsymbol{x}^*) \ge z^*$.

Theorem 1 (Bondareva-Shapley) *The core is non–empty if the support to the set of balanced weights evaluated at the worth equals the value. The core is empty if the support exceeds the value.*

As the set B of balanced weights depends only on the number of players m and is independent of the worth $\boldsymbol{x}^*$ and the value z^*, it is a fundamental concept. For any $\boldsymbol{x} \in B$, its components must lie between zero and one. Because B is a bounded, polyhedral set, the support $\delta^*_B(\boldsymbol{x}^*)$ equals the maximum value $\langle \boldsymbol{x}^*, \boldsymbol{x}\rangle$ attained across its extreme points.

A partition is a set of coalitions such that each player is in one and only one coalition. Equivalently, a partition is a set of idempotents that sum to **1**, so each partition corresponds to some $\boldsymbol{x} \in B$. As each component of any $\boldsymbol{x} \in B$ must lie between zero and one, each partition is an extreme point. The number of ways to partition m players is the Bell number B_m (Bell 1934) (not to be confused with the Bernoulli number). The number increases rapidly with m: the sequence is 1, 2, 5, 15, 52, 203, 877, 4140, 21147, 115975, ...[1]

Shapley determines the extreme points for small m (Shapley 1967). For both $m = 1$ or $m = 2$, the partitions are the extreme points. For $m = 3$, there are six extreme points—the five partitions, plus a single additional extreme point that is not a partition. The latter is the vector placing weight 1/2 on each of the three two-player coalitions—{1, 2}, {1, 3}, and {2, 3}. For $m = 4$, Shapley lists the 41 extreme points, the 15 partitions plus 26 additional extreme points.

References

Bell ET (1934) Exponential numbers. Am Math Month 41:411–419

Bondareva ON (1962) Teoriia idra v igre n lits (the theory of the core in an n-person game). Vestnik Leningradskogo Universiteta, Seriia Matematika, Mekhanika i Astronomii (Bull Lenin Univ Math Mech Astron Ser 13:141–142 (Summary in English)

Bondareva ON (1963) Nekotorye primeneniia metodov linejnogo programmirovaniia k teorii kooperativnykh igr (some applications of linear programming methods to the theory of games). Problemy Kibernetiki 10:119–139

Shapley LS (1967) On balanced sets and cores. Naval Res Logist Quart 14:453–460

[1] Weisstein, Eric W. "Bell Number." From MathWorld—A Wolfram Web Resource. fntext/end

Part XIII
Econometrics

Chapter 52
Moment Space

Abstract Considerable statistical analysis and probabilistic economic modelling deals just with the first and second moments of the random variables. The full probability distribution of the variables plays no role. Moment space is a Euclidean space of random variables that embodies the first and second moments. A vector $\boldsymbol{x}$ in the space represents a random variable. The inner product $\langle \boldsymbol{y}, \boldsymbol{x} \rangle$ of random variables $\boldsymbol{y}$ and $\boldsymbol{x}$ is the expected value of their product (the second noncentral cross-moment). The mean is a linear function of $\boldsymbol{x}$. To work in moment space enables the application of convex optimization and duality theory. Moment space is convenient for modelling random variables in a problem depending only on the first and second moment. We apply moment space to econometrics and finance. Least-squares linear regression involves only the first and second moments of the random variables, so moment space is applicable. Maximum-likelihood estimation of a normal distribution involves only the first and second moments. In finance, using moment space to study mean/variance choice under uncertainty is effective. To embed the second moment in the inner product simplifies the derivation of relationships and the statement of results.

The cross-moment transformation between subspaces is the unique linear transformation $A : M \to N$ such that the inner product of $\boldsymbol{x} \in M$ and $\boldsymbol{y} \in N$ is

$$\langle \boldsymbol{y}, A\boldsymbol{x} \rangle.$$

A second interpretation is valuable. Let M and N denote the idempotents corresponding to these subspaces. Alternatively see the cross-moment transformation as the square linear transformation NM, such that the domain and range are restricted to M and N.

In probability and statistics, it is common for the same symbol to denote both the name of a random variable and its value. The situation is different in moment space. A vector $\boldsymbol{x}$ denotes a random variable named $\boldsymbol{x}$, but $\boldsymbol{x}$ is *not* the *value* of the random variable.

B. C. Dieffenbach, *Conjugate Duality in Economic Analysis*, Contributions to Economics,
https://doi.org/10.1007/978-3-032-21396-9_52

Moment space permits many different interpretations. For example, a vector might refer to a probability distribution of a scalar. It might refer to a probability distribution of a time series. It might refer to a data vector and its first and second sample moments. It might refer to random variables specified incompletely by just the first and second moments.

A specific application requires that one specify the Euclidean space by defining the sum of two random variables, multiplication by a real scalar, and the product of two random variables. One might define the sum as a random variable whose value is the sum of the values. One might define a multiple as a random variable whose value is the multiple of the value. One might define the product as a random variable whose value is the product of the values.

In a statistical application, one moment space might model the population random variables and a corresponding moment space might model the sample distribution. A common application is analog estimation (Goldberger 1968; Manski 1988, 1994). Given some function of the population moments, estimate it via the same function evaluated at the sample moments.

Chamberlain and Rothschild (1983), Chamberlain (1983), and (LeRoy and Werner 2001, ch. 17) apply moment space to finance theory. In market equilibrium two assets having the same payoff must have the same price, so one identifies the two assets as the same. Vectors $\boldsymbol{x}$ in the moment space are payoffs, and the price of a payoff is a linear function of $\boldsymbol{x}$. In general moment space involves no concept of order, of one random variable being larger than another. Consequently moment space is a Euclidean space, but perhaps not a Euclidean Jordan algebra.

52.1 Universal Space

Envision moment space as a universe of random variables, a Euclidean space $\mathbf{U}$ having high dimension.

Condition 1 (Constant Random Variable) *The universal space contains a constant random variable* $\mathbf{1}$*, which has variance* $\mathrm{Var}(\mathbf{1}) = 0$ *and expected value* $\mathrm{E}(\mathbf{1}) = 1$.

Definition 1 (Expected Value) The expected value $\mathrm{E}(\boldsymbol{x}) := \langle \mathbf{1}, \boldsymbol{x} \rangle$.

Definition 2 (Covariance) The covariance $\mathrm{Cov}(\boldsymbol{x}^*, \boldsymbol{x}) := \langle \boldsymbol{x}^*, \boldsymbol{x} \rangle - \langle \mathbf{1}, \boldsymbol{x}^* \rangle \langle \mathbf{1}, \boldsymbol{x} \rangle = \left\langle \boldsymbol{x}^*, \left(\mathsf{I} - \mathbf{1}\mathbf{1}^\top \right) \boldsymbol{x} \right\rangle$.

Standard statistical relations hold: for example, the expected value of a sum is the sum of the expected values. The constant random variable is unique, and $\langle \mathbf{1}, \mathbf{1} \rangle = 1$. The existence of a constant random variable is important in applications. For example, in a statistical application a linear regression can have an intercept. In a finance application, there can be a risk-free asset, having a constant payoff.

An application commonly involves only a subspace, itself a Euclidean space. For a subspace M, let M denote the idempotent corresponding to it. A vector $\boldsymbol{x}$ lies in the subspace if and only if $\mathsf{M}\boldsymbol{x} = \boldsymbol{x}$. By the Riesz-Fréchet theorem, because the mean is the linear function $\langle \mathbf{1}, \boldsymbol{x} \rangle$, there exists a unique mean-vector $\boldsymbol{m} \in M$ such that $\langle \boldsymbol{m}, \boldsymbol{x} \rangle$ is the mean of $\boldsymbol{x} \in M$. In fact $\boldsymbol{m} = \mathsf{M}\mathbf{1}$: for $\boldsymbol{x} \in M$,

$$\langle \mathbf{1}, \boldsymbol{x} \rangle = \langle \mathbf{1}, \mathsf{M}\boldsymbol{x} \rangle = \langle \mathsf{M}\mathbf{1}, \boldsymbol{x} \rangle.$$

If $\mathbf{1} \in M$, then $\boldsymbol{m} = \mathbf{1}$. For vectors in M, $\mathrm{Cov}(\boldsymbol{x}^*, \boldsymbol{x}) = \left\langle \boldsymbol{x}^*, \left(\mathsf{I} - \boldsymbol{m}\boldsymbol{m}^\top\right)\boldsymbol{x} \right\rangle$. If $\mathbf{1} \notin M$ and the mean-vector $\boldsymbol{m} = \mathsf{M}\mathbf{1} \neq \mathbf{0}$, then $\boldsymbol{m}$ is not constant, so

$$0 < \mathrm{Var}\,(\boldsymbol{m}) = \left\langle \boldsymbol{m}, \left(\mathsf{I} - \boldsymbol{m}\boldsymbol{m}^\top\right)\boldsymbol{m} \right\rangle = \langle \boldsymbol{m}, \boldsymbol{m} \rangle \left(1 - \langle \boldsymbol{m}, \boldsymbol{m} \rangle\right);$$

thus $0 < |\boldsymbol{m}| < 1$.

Summary 1 (Subspace Inclusion) Inequality among idempotents corresponds to subspace inclusion: $\mathsf{N} \preceq \mathsf{M}$ if and only if $N \subset M$.

A common relation in applications is to study the subspace N of M of all vectors orthogonal to some nonzero $\boldsymbol{y} \in M$; the relationship among idempotents is

$$\mathsf{N} = \mathsf{M} - \frac{\boldsymbol{y}\boldsymbol{y}^\top}{\boldsymbol{y}^\top\boldsymbol{y}},$$

the idempotent for M minus the idempotent for all multiples of $\boldsymbol{y}$. Here $\mathsf{N}\boldsymbol{y} = \mathbf{0}$, as desired. For $\boldsymbol{x} \in M$ but orthogonal to $\boldsymbol{y}$, then

$$\mathsf{N}\boldsymbol{x} = \mathsf{M}\boldsymbol{x} - \frac{\boldsymbol{y}\boldsymbol{y}^\top\boldsymbol{x}}{\boldsymbol{y}^\top\boldsymbol{y}} = \boldsymbol{x} - \mathbf{0} = \boldsymbol{x},$$

so $\boldsymbol{x} \in N$. The mean-vector $\boldsymbol{n}$ in N is

$$\boldsymbol{n} = \mathsf{N}\mathbf{1} = \mathsf{M}\mathbf{1} - \frac{\boldsymbol{y}\boldsymbol{y}^\top\mathbf{1}}{\boldsymbol{y}^\top\boldsymbol{y}} = \boldsymbol{m} - \frac{\boldsymbol{m}^\top\boldsymbol{y}}{\boldsymbol{y}^\top\boldsymbol{y}}\boldsymbol{y}.$$

52.2 Cross-Moment Transformation

Let M and N denote two subspaces of random variables, and consider vectors $\boldsymbol{x} \in M$ and $\boldsymbol{y} \in N$. The inner product between vectors $\boldsymbol{x} \in M$ and $\boldsymbol{y} \in N$ is bilinear in the vectors. By the Riesz–Fréchet theorem, there exists a unique linear transformation $\mathsf{A} : M \to N$ such that the inner product $\langle \boldsymbol{y}, \boldsymbol{x} \rangle = \langle \boldsymbol{y}, \mathsf{A}\boldsymbol{x} \rangle = \left\langle \mathsf{A}^\top \boldsymbol{y}, \boldsymbol{x} \right\rangle$. We refer to A as the "cross-moment transformation" from M to N.

Definition 3 (Cross-Moment Transformation) The cross-moment transformation $A : M \to N$ is the unique linear transformation such that the inner product of $x \in M$ and $y \in N$ is

$$\langle y, x \rangle = \langle y, Ax \rangle = \left\langle A^{\mathsf{T}} y, x \right\rangle.$$

A second interpretation is valuable. Let M and N denote the idempotents corresponding to these subspaces. As

$$\langle y, x \rangle = \langle Ny, Mx \rangle = \langle y, NMx \rangle,$$

alternatively see the cross-moment transformation as the square linear transformation $NM \in \mathbf{T}^{\mathbf{U}}$, such that the domain and range are restricted to M and N. The two concepts are closely linked. In particular both $A \in \mathbf{T}^{M \to N}$ and $NM \in \mathbf{T}^{\mathbf{U}}$ have the same ordered singular-value decomposition

$$\sum_{i}^{\operatorname{rank} A} \mu_i n_i m_i^{\mathsf{T}},$$

but having domain M versus domain $\mathbf{U}$. The $m_i \in M$ are mutually orthonormal, and the $n_i \in N$ likewise. Each set of these "canonical vectors" can be extended to an orthonormal basis, a "canonical basis." Here

$$\mu_i = \langle n_i, Am_i \rangle = \langle n_i, NMm_i \rangle = \langle Nn_i, Mm_i \rangle = \langle n_i, m_i \rangle,$$

the cross moment. Hence one can write the singular-value decomposition as

$$\sum_{i}^{\operatorname{rank} A} \langle n_i, m_i \rangle\, n_i m_i^{\mathsf{T}}.$$

If both M and N contain the constant vector $\mathbf{1}$, then the singular values are the canonical correlations between the subspaces. One can express the singular-value decomposition so that singular value $\mu_1 = 1$, and both $m_1 = \mathbf{1}$ and $n_1 = \mathbf{1}$. For $i > 1$ the μ_i are the canonical correlations between the two subspaces (Hotelling 1936). Because m_i and n_i are orthogonal to m_1 and n_1, respectively, each has mean zero; and the cross moment $\langle n_i, m_i \rangle = \langle n_i, Am_i \rangle = \mu_i$ is their correlation.

For $x \in M$ and $y \in N$, the second moment of the difference $y - x$ is

$$\langle y - x, y - x \rangle = \begin{bmatrix} -x \\ y \end{bmatrix}^{\mathsf{T}} \begin{bmatrix} \mathsf{I} & A^{\mathsf{T}} \\ A & \mathsf{I} \end{bmatrix} \begin{bmatrix} -x \\ y \end{bmatrix}. \tag{52.1}$$

The partitioned linear transformation is the identity plus a Jordan-Wielandt linear transformation. Recall that it is nonnegative if and only if every singular value is less than or equal to one. If $M \cap N = \{\mathbf{0}\}$, then the quadratic form is positive; every singular value is less than one.

The least-squares linear regression of $\boldsymbol{y}$ on M is the $\boldsymbol{x} \in M$ such that $\langle \boldsymbol{y}-\boldsymbol{x}, \boldsymbol{y}-\boldsymbol{x}\rangle$ is smallest. By the moment quadratic form (52.1), this residual moment is

$$\langle \boldsymbol{y}, \boldsymbol{y}\rangle - 2\left\langle A^{\mathsf{T}}\boldsymbol{y}, \boldsymbol{x}\right\rangle + \langle \boldsymbol{x}, \boldsymbol{x}\rangle. \tag{52.2}$$

From the first-order condition,

$$\boldsymbol{x} = A^{\mathsf{T}}\boldsymbol{y} = MN\boldsymbol{y} = M\boldsymbol{y}.$$

so the adjoint sets the regression. In an econometrics textbook, a fundamental point is that the regression of one variable on a second variable is *not* the inverse of the regression of the second variable on the first. In the formulation here, the regression of $\boldsymbol{x}$ on N is the $\boldsymbol{y}$ that minimizes the residual moment (52.2). By the first-order condition, the minimum is attained at $\boldsymbol{y} = A\boldsymbol{x} = NM\boldsymbol{x} = N\boldsymbol{x}$. The adjoint—not the inverse—gives the regression in the opposite direction.

The fitted value $\boldsymbol{x} = M\boldsymbol{y}$, and the residual $\boldsymbol{y} - \boldsymbol{x} = (\mathsf{I} - M)\boldsymbol{y}$, orthogonal to the fitted value, which implies the Pythagoras relation.

Theorem 1 (Pythagoras Theorem for Least Squares) *For the least-squares linear regression of* $\boldsymbol{y}$ *on* M,

$$|\boldsymbol{y}|^2 = |\boldsymbol{x}|^2 + |\boldsymbol{y}-\boldsymbol{x}|^2,$$

in which $\boldsymbol{x}$ *is the fitted value and* $\boldsymbol{y}-\boldsymbol{x}$ *is the residual.*

If the space of random variables contains no constant random variable, alternatively one can define the inner product as the covariance (the central second moment). For a random variable $\boldsymbol{x}$, the variance is the inner product $\langle \boldsymbol{x}, \boldsymbol{x}\rangle$, and the standard deviation is the norm $|\boldsymbol{x}|$. That the inner product is positive means that every nonzero random variable has a positive variance. The Cauchy–Schwarz inequality for the inner product amounts to the condition that any two nonzero random variables have correlation between 1 and –1. One can refer to the cross-moment transformation as the covariance transformation, and its singular values are the canonical correlations.

52.3 Coordinate-Free Versus Basis

The table compares the coordinate-free moment-space formulation with the traditional basis formulation. The latter expresses random variables as linear combinations of a basis of random variables. In applications typically the basis variables do not have variance one and are correlated. In terms of the moment-space inner product, the basis is not orthonormal. Consequently the basis expressions are complicated, whereas the coordinate-free expressions are simple.

In the table the $\boldsymbol{x}_i \in M$ and $\boldsymbol{y}_i \in N$ are basis vectors, and $\tilde{\boldsymbol{x}} \in \mathbf{R}^{\dim M}$ and $\tilde{\boldsymbol{y}} \in \mathbf{R}^{\dim N}$ are the coordinates of $\boldsymbol{x}$ and $\boldsymbol{y}$ in terms of the bases.

Coordinate-Free	Basis
	$X = \begin{bmatrix} \boldsymbol{x}_1 & \cdots & \boldsymbol{x}_m \end{bmatrix} \in \mathbf{T}^{m \to M}$
	$Y = \begin{bmatrix} \boldsymbol{y}_1 & \cdots & \boldsymbol{y}_n \end{bmatrix} \in \mathbf{T}^{n \to N}$
$\boldsymbol{x} \in M$	$X\tilde{\boldsymbol{x}}$
$\boldsymbol{y} \in N$	$Y\tilde{\boldsymbol{y}}$
M	$X\left(X^\mathsf{T}X\right)^{-1}X^\mathsf{T}$
N	$Y\left(Y^\mathsf{T}Y\right)^{-1}Y^\mathsf{T}$
$A = NM$	$Y\left(Y^\mathsf{T}Y\right)^{-1}Y^\mathsf{T}X\left(X^\mathsf{T}X\right)^{-1}X^\mathsf{T}$
$\boldsymbol{y} = A\boldsymbol{x}$	$\tilde{\boldsymbol{y}} = \left(Y^\mathsf{T}Y\right)^{-1}Y^\mathsf{T}X\tilde{\boldsymbol{x}}$
$\boldsymbol{x} = A^\mathsf{T}\boldsymbol{y}$	$\tilde{\boldsymbol{x}} = \left(X^\mathsf{T}X\right)^{-1}X^\mathsf{T}Y\tilde{\boldsymbol{y}}$

The matrix representation of A is

$$Y^{-1}AX = Y^{-1}\left[Y\left(Y^\mathsf{T}Y\right)^{-1}Y^\mathsf{T}X\left(X^\mathsf{T}X\right)^{-1}X^\mathsf{T}\right]X = \left(Y^\mathsf{T}Y\right)^{-1}Y^\mathsf{T}X.$$

The matrix representation of M is

$$X^{-1}MX = X^{-1}\left[X\left(X^\mathsf{T}X\right)^{-1}X^\mathsf{T}\right]X = \mathrm{I}.$$

When the bases are not orthonormal, the matrix representation of the adjoint A^T is not the transpose of the matrix representation of A. The complicated basis expressions are the traditional formulas for least-squares linear regression, in opposite directions.

The squared singular values of A are the positive eigenvalues of the nonnegative, self-adjoint linear transformations $A^\mathsf{T}A$ and AA^T. The matrix representation of $A^\mathsf{T}A$ is

$$\begin{aligned}
&X^{-1}A^\mathsf{T}AX \\
&= X^{-1}\left[X\left(X^\mathsf{T}X\right)^{-1}X^\mathsf{T}Y\left(Y^\mathsf{T}Y\right)^{-1}Y^\mathsf{T}\right] \\
&\quad \cdot \left[Y\left(Y^\mathsf{T}Y\right)^{-1}Y^\mathsf{T}X\left(X^\mathsf{T}X\right)^{-1}X^\mathsf{T}\right]X \\
&= \left(X^\mathsf{T}X\right)^{-1}X^\mathsf{T}Y\left(Y^\mathsf{T}Y\right)^{-1}Y^\mathsf{T}X,
\end{aligned}$$

which is symmetric if the bases are orthonormal, but in general not otherwise. Nevertheless the eigenvalues are the same nonnegative values as for the coordinate-free expressions.

References

Chamberlain G (1983) Funds, factors and diversification in arbitrage pricing models. Econometrica 51:1305–1323

Chamberlain G, Rothschild M (1983) Arbitrage, factor structure and mean variance analysis in large asset markets. Econometrica 51:1281–1304

Goldberger AS (1968) Topics in regression analysis. Macmillan, New York

Hotelling H (1936 Dec) Relations between two sets of variates. Biometrika 28(3/4):321–377

LeRoy SF, Werner J (2001) Principles of financial economics. Cambridge University Press, Cambridge

Manski CF (1988) Analog estimation methods in econometrics. Chapman and Hall, New York

Manski CF (1994) Analog estimation of econometric models. In: Engle RF, McFadden DL (eds) Handbook of econometrics, Vol 4. Elsevier, Amsterdam, pp 2559–2582

Chapter 53
Least-Squares Linear Regression

Abstract Let $\boldsymbol{y}$ denote the dependent random variable, and M is the subspace of independent variables. The least-squares linear regression of $\boldsymbol{y}$ on M is the $\boldsymbol{x} \in M$ such that the quadratic $\langle \boldsymbol{y}-\boldsymbol{x}, \boldsymbol{y}-\boldsymbol{x} \rangle$ is minimized:

$$\inf_{x}\left[\frac{1}{2}\langle \boldsymbol{y}-\boldsymbol{x}, \boldsymbol{y}-\boldsymbol{x}\rangle + \delta_M(\boldsymbol{x})\right].$$

The following three conditions are necessary and sufficient for solutions to the primal and the dual:

$$\begin{gathered} \boldsymbol{y}^* = \boldsymbol{y}-\boldsymbol{x} \\ \boldsymbol{x} \in M \\ \boldsymbol{y}^* \in M^{\perp}. \end{gathered}$$

These conditions are symmetric between the fitted value $\boldsymbol{x}$ and the residual $\boldsymbol{y}^*$.

Taking the dual of a transformation of the Fenchel dual obtains another fundamental result: the fitted values maximize the correlation with the dependent variable. We interpret least-squares linear regression as a fractional program.

We work in a moment space $\mathbf{U}$ of random variables, in which the inner product is the noncentral second cross-moment. Let $\boldsymbol{y}$ denote the dependent random variable, and M is the subspace of independent variables.

Definition 1 (Linear Regression) The linear regression of $\boldsymbol{y}$ on M is the $\boldsymbol{x} \in M$ such that $\boldsymbol{x}$ is closest to $\boldsymbol{y}$, closest by some criterion.

Definition 2 (Least-Squares Linear Regression) The least-squares linear regression of $\boldsymbol{y}$ on M is the $\boldsymbol{x} \in M$ such that the quadratic $\langle \boldsymbol{y}-\boldsymbol{x}, \boldsymbol{y}-\boldsymbol{x} \rangle$ is minimized.

Least-squares linear regression is the basic estimation technique in econometrics.

B. C. Dieffenbach, *Conjugate Duality in Economic Analysis*, Contributions to Economics,
https://doi.org/10.1007/978-3-032-21396-9_53

53.1 Coefficient Least-Squares Linear Regression

The traditional formulation of least-squares linear regression is coefficient least-squares linear regression.

Problem 1 (Coefficient Least-Squares Linear Regression)

$$\inf_{z} \frac{1}{2} \langle \boldsymbol{y} - Az, \boldsymbol{y} - Az \rangle. \tag{53.1}$$

Choose coefficients $z \in \mathbf{R}^m$ to explain the dependent variable $\boldsymbol{y}$ by independent variables $A := \begin{bmatrix} \boldsymbol{a}_1 & \cdots & \boldsymbol{a}_m \end{bmatrix}$.

The first-order condition is

$$\mathbf{0} = A^\top (\boldsymbol{y} - Az),$$

the "normal equation," which states that the residual is orthogonal (normal) to the independent variables. One solution is $z = A^+ \boldsymbol{y}$:

$$\begin{aligned} A^\top Az &= A^\top A A^+ \boldsymbol{y} \\ &= A^\top A \left[\left(A^\top A \right)^+ A^\top \right] \boldsymbol{y} \\ &= A^\top \left[A \left(A^\top A \right)^+ \right] A^\top \boldsymbol{y} \\ &= A^\top A^{\top +} A^\top \boldsymbol{y} \\ &= A^\top \boldsymbol{y}, \end{aligned}$$

the first-order condition. The general solution is

$$z \in \left\{ A^+ \boldsymbol{y} \right\} + \mathrm{N}(A).$$

A solution always exists, and it is unique if $\mathrm{N}(A) = \{\mathbf{0}\}$. The fitted value is unique:

$$\boldsymbol{x} = Az = AA^+ \boldsymbol{y} = A \left(A^\top A \right)^+ A^\top \boldsymbol{y} = M\boldsymbol{y},$$

in which $M = A \left(A^\top A \right)^+ A^\top$ is the idempotent for $\mathrm{R}(A)$.

Theorem 1 (Coefficient Least-Squares Solution) *The general solution to the coefficient least-squares linear-regression problem is*

$$z \in \left\{ A^+ \boldsymbol{y} \right\} + \mathrm{N}(A).$$

The fitted value $\boldsymbol{x} = M\boldsymbol{y}$ *is unique.*

The formula My is also called the projection of $\boldsymbol{y}$ on M. At the optimum, $\langle \boldsymbol{y}-\boldsymbol{x}, \boldsymbol{y}-\boldsymbol{x}\rangle = \langle \boldsymbol{y}, (\mathsf{I}-M)\boldsymbol{y}\rangle$.

If A^{T} is onto, then $A^{+} = \left(A^{\mathsf{T}}A\right)^{-1}A^{\mathsf{T}}$. The unique solution is $\boldsymbol{z} = \left(A^{\mathsf{T}}A\right)^{-1}A^{\mathsf{T}}\boldsymbol{y}$, the textbook formula for least-squares linear regression.

Generalized least squares and its special case weighted least squares are special cases of least squares. In the traditional problem (53.1), let the inner product embody the weighting coefficients, and the simple, coordinate-free formulas derived continue to apply for these cases. The complexity of the traditional formula for the generalized-least-squares estimate stems from expressing it via a basis of vectors that are not orthonormal.

53.2 Subspace Least-Squares Linear Regression

It is natural to make $\boldsymbol{y}$ the perturbation variable, and one could solve the problem (53.1) via extended Fenchel duality. Yet we follow an alternative path, by solving the subspace least-squares problem (53.2). The fitted values $\boldsymbol{x}$ must lie in the subspace $\mathrm{R}(A)$. Let M denote this subspace, and $M = AA^{+}$ is the idempotent linear transformation that represents M. The regression residual is $\boldsymbol{y}^{*} = \boldsymbol{y}-\boldsymbol{x}$, and the constraint requires that $\boldsymbol{x} \in M$.

Problem 2 (Subspace Least-Squares)

$$\inf_{\boldsymbol{x}}\left[\frac{1}{2}\langle \boldsymbol{y}-\boldsymbol{x}, \boldsymbol{y}-\boldsymbol{x}\rangle + \delta_{M}(\boldsymbol{x})\right]. \tag{53.2}$$

Finding the dual of this problem reveals an exact symmetry between the fitted values and the residual.

The value function is quadratic:

$$\begin{aligned}
&V(\boldsymbol{x}^{*}, \boldsymbol{y}) \\
&= \inf_{\boldsymbol{x}\in M}\left[\langle \boldsymbol{x}^{*}, \boldsymbol{x}\rangle + \frac{1}{2}\langle \boldsymbol{y}-\boldsymbol{x}, \boldsymbol{y}-\boldsymbol{x}\rangle\right] \\
&= \frac{1}{2}\langle \boldsymbol{y}, \boldsymbol{y}\rangle - \frac{1}{2}\langle \boldsymbol{y}-\boldsymbol{x}^{*}, \boldsymbol{y}-\boldsymbol{x}^{*}\rangle + \inf_{\boldsymbol{x}\in M}\frac{1}{2}\langle \boldsymbol{y}-\boldsymbol{x}^{*}-\boldsymbol{x}, \boldsymbol{y}-\boldsymbol{x}^{*}-\boldsymbol{x}\rangle \\
&= \frac{1}{2}\langle \boldsymbol{y}, \boldsymbol{y}\rangle - \frac{1}{2}\langle \boldsymbol{y}-\boldsymbol{x}^{*}, \boldsymbol{y}-\boldsymbol{x}^{*}\rangle + \frac{1}{2}\langle \boldsymbol{y}-\boldsymbol{x}^{*}, (\mathsf{I}-M)(\boldsymbol{y}-\boldsymbol{x}^{*})\rangle \\
&= \frac{1}{2}\langle \boldsymbol{y}, \boldsymbol{y}\rangle - \frac{1}{2}\langle \boldsymbol{y}-\boldsymbol{x}^{*}, M(\boldsymbol{y}-\boldsymbol{x}^{*})\rangle.
\end{aligned}$$

Theorem 2 (Value Function for Least Squares) *The value function for least squares is*

$$V(x^*, y) = \frac{1}{2}\langle y, y\rangle - \frac{1}{2}\langle y - x^*, M(y - x^*)\rangle .$$

The subdifferential with respect to x^* *is* $\{My\}$, *the fitted value* x. *The subdifferential with respect to* y *is* $\{(I - M)y\}$, *the residual* y^*.

The subdifferentials are evaluated at $x^* = \mathbf{0}$ and y.

Because problems (53.1) and (53.2) have the same value as a function of the perturbation y, their duals must be the same. We find the dual of the subspace least-squares problem. Adding the Fenchel inequalities

$$\frac{1}{2}\langle y - x, y - x\rangle + \frac{1}{2}\langle y^*, y^*\rangle - \langle y^*, y - x\rangle \geq 0 \tag{53.3}$$

$$\delta_M(x) + \delta_M^*(y^* - x^*) - \langle y^* - x^*, x\rangle \geq 0 \tag{53.4}$$

and rearranging gives

$$\langle x^*, x\rangle + \frac{1}{2}\langle y - x, y - x\rangle + \delta_M(x) \geq \langle y^*, y\rangle - \frac{1}{2}\langle y^*, y^*\rangle - \delta_M^*(y^* - x^*).$$

The perturbation base values are $x^* = \mathbf{0}$ and y. Both the primal and the dual satisfy the polyhedral/relative condition, so there is no duality gap, and the optimum is attained in both the primal and the dual.

The following three conditions are necessary and sufficient for solutions to the primal and the dual. The first Fenchel factor (53.3) is zero if and only if

$$y^* = y - x;$$

the solution to the dual is the residual. The second Fenchel factor (53.4) is zero if and only if

$$x \in M$$
$$y^* \in M^{\perp}.$$

These conditions are symmetric between the fitted value x and the residual y^*.

Rewriting the dual emphasizes this symmetry.

Problem 3 (Subspace Least-Squares Dual)

$$\sup_{y^*}\left[\frac{1}{2}\langle y, y\rangle - \frac{1}{2}\langle y - y^*, y - y^*\rangle - \delta_{M^{\perp}}(y^*)\right]. \tag{53.5}$$

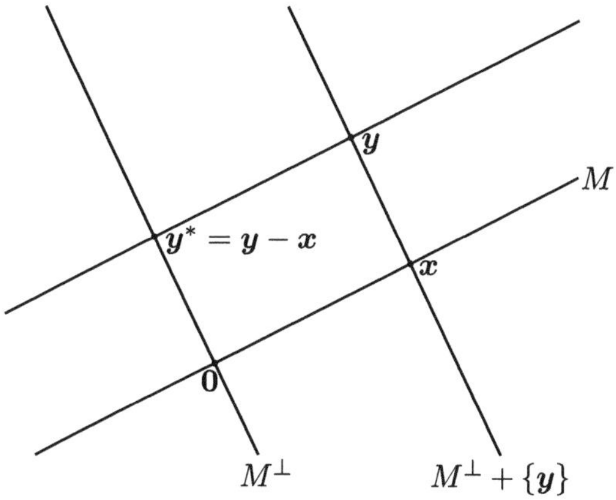

Fig. 53.1 Least-squares linear regression

The dual chooses the residual $\boldsymbol{y}^*$ as the fitted value of the least-squares regression of $\boldsymbol{y}$ on $M^{\perp}$: the residual $\boldsymbol{y}^*$ *minimizes* the quadratic function of $\boldsymbol{y} - \boldsymbol{y}^*$ [which is the fitted value in the primal (53.2)], subject to the constraint that the residual $\boldsymbol{y}^*$ is orthogonal to M. This result might seem surprising: one might expect maximization instead.

Providing a geometric interpretation, Fig. 53.1 graphs the subspace least-squares primal and dual. The solution to the primal is $\boldsymbol{x}$, and the solution to the dual is $\boldsymbol{y}^* = \boldsymbol{y} - \boldsymbol{x}$. The primal chooses $\boldsymbol{x} \in M$ to be as close to $\boldsymbol{y}$ as possible. The dual chooses $\boldsymbol{y}^* \in M^{\perp}$ to be as close to $\boldsymbol{y}$ as possible. Equivalently, $\boldsymbol{x}$ is the vector in $M^{\perp} + \{\boldsymbol{y}\}$ having minimum length. By the Pythagorean theorem, $|\boldsymbol{y}|^2 = |\boldsymbol{x}|^2 + |\boldsymbol{y}^*|^2$.

53.3 Least-Squares Fractional Program

Taking the dual of a transformation of the Fenchel dual obtains another fundamental result: the fitted values maximize the correlation with the dependent variable. We interpret least-squares linear regression as a fractional program. Recognize the following problem as a transformation of the subspace least-squares dual (53.5). Choose the fitted values $\boldsymbol{x}$ to minimize its norm, subject to the constraint that the residual $\boldsymbol{y} - \boldsymbol{x}$ is orthogonal to M. The solution remains $\boldsymbol{x} = M\boldsymbol{y}$, and the optimum value is $\sqrt{\boldsymbol{y}^{\top} M \boldsymbol{y}}$.

Problem 4 (Transformed Dual)

$$\inf_{\boldsymbol{x}} [|\boldsymbol{x}| + \delta_{M^{\perp}}(\boldsymbol{y} - \boldsymbol{x})]. \tag{53.6}$$

Assume $My \neq \mathbf{0}$. Because $\mathbf{0} \notin M^{\perp}-\{y\}$, the dual of this problem must be the concave reduction of a standard homogeneous fractional program.

We calculate the dual of (53.6), to discover the fractional program. Adding the Fenchel inequalities

$$\delta_B(\boldsymbol{y}^* - \boldsymbol{x}^*) + \delta_B^*(\boldsymbol{x}) - \langle \boldsymbol{y}^* - \boldsymbol{x}^*, \boldsymbol{x} \rangle \geq 0 \tag{53.7}$$

$$\delta_M(\boldsymbol{y}^*) + \delta_{M^\perp}(\boldsymbol{y} - \boldsymbol{x}) - \langle \boldsymbol{y}^*, \boldsymbol{y} - \boldsymbol{x} \rangle \geq 0 \tag{53.8}$$

and rearranging obtains

$$\begin{aligned}\langle \boldsymbol{x}^*, \boldsymbol{x} \rangle + |\boldsymbol{x}| + \delta_{M^\perp}(\boldsymbol{y} - \boldsymbol{x}) \\ \geq \langle \boldsymbol{y}^*, \boldsymbol{y} \rangle - \delta_M(\boldsymbol{y}^*) - \delta_B(\boldsymbol{y}^*).\end{aligned}$$

The perturbation $\boldsymbol{x}^*$ has base value zero. We know $\boldsymbol{x} = M\boldsymbol{y}$. For the Fenchel factor (53.7) to be zero requires $\boldsymbol{y}^* \in \partial \delta_B^*(\boldsymbol{x}) = \{\boldsymbol{x}/|\boldsymbol{x}|\}$

The maximization

$$\sup_{\boldsymbol{y}^*} [\langle \boldsymbol{y}^*, \boldsymbol{y} \rangle - \delta_M(\boldsymbol{y}^*) - \delta_B(\boldsymbol{y}^*)]$$

is the concave reduction of the standard homogeneous fractional program

$$\sup_{\boldsymbol{y}^* \neq 0} \left[\frac{\langle \boldsymbol{y}^*, \boldsymbol{y} \rangle - \delta_M(\boldsymbol{y}^*)}{|\boldsymbol{y}^*|} \right].$$

Both the numerator and the denominator are positively homogeneous. The denominator is positive for $\boldsymbol{y}^* \neq \mathbf{0}$. As $M\boldsymbol{y} \neq \mathbf{0}$, the numerator is positive for some $\boldsymbol{y}^*$.

Dividing by $|\boldsymbol{y}|$ obtains the following.

Theorem 3 (Least-Squares Fractional Program) *The solutions $\boldsymbol{y}^*$ to*

$$\sup_{\boldsymbol{y}^* \in M, \boldsymbol{y}^* \neq 0} \frac{\langle \boldsymbol{y}^*, \boldsymbol{y} \rangle}{|\boldsymbol{y}^*| \, |\boldsymbol{y}|} \tag{53.9}$$

are the positive multiples of the least-squares fitted value $\boldsymbol{x} = M\boldsymbol{y}$. The optimum value is

$$\sqrt{\frac{\boldsymbol{y}^\top M \boldsymbol{y}}{\boldsymbol{y}^\top \boldsymbol{y}}} = \sqrt{\frac{\boldsymbol{x}^\top \boldsymbol{x}}{\boldsymbol{y}^\top \boldsymbol{y}}} = \frac{|\boldsymbol{x}|}{|\boldsymbol{y}|}.$$

The objective function resembles the correlation of two random variables—covariance divided by the product of the standard deviations—but here the variables need not have mean zero. By the Cauchy–Schwarz inequality, the ratio is less than or equal to one.

If M contains the constant $\mathbf{1}$, a fundamental traditional result follows as a corollary. Because the residual is orthogonal to the dependent variable, the residual has mean zero, so the mean of the fitted value equals the mean of the dependent variable, $\mathrm{E}(\boldsymbol{x}) = \mathrm{E}(\boldsymbol{y})$. Assume $\mathrm{Sd}(\boldsymbol{x}) \neq 0$.

Corollary 1 (Maximum Correlation) *If* $\mathbf{1} \in M$*, then the least-squares fitted value* $\boldsymbol{x} = M\boldsymbol{y}$ *maximizes the correlation with the dependent variable:*

$$\frac{\mathrm{Sd}(\boldsymbol{x})}{\mathrm{Sd}(\boldsymbol{y})} = \max_{\boldsymbol{x} \in M, \boldsymbol{x} \neq \mathbf{0}} \mathrm{Correlation}(\boldsymbol{x}, \boldsymbol{y}).$$

The subspace of M having zero mean has idempotent $M - \mathbf{1}\mathbf{1}^{\mathsf{T}}$, and $\left(\mathsf{I} - \mathbf{1}\mathbf{1}^{\mathsf{T}}\right)\boldsymbol{y}$ is the dependent variable expressed relative to its mean. Apply the Theorem 3 to this subspace and this vector. Its least-squares regression on the subspace has fitted value

$$\left(M - \mathbf{1}\mathbf{1}^{\mathsf{T}}\right)\left(\mathsf{I} - \mathbf{1}\mathbf{1}^{\mathsf{T}}\right)\boldsymbol{y} = \left(M - \mathbf{1}\mathbf{1}^{\mathsf{T}}\right)\boldsymbol{y} = \boldsymbol{x} - \mathrm{E}(\boldsymbol{y})\,\mathbf{1} = \boldsymbol{x} - \mathrm{E}(\boldsymbol{x})\,\mathbf{1},$$

and the corollary follows. Econometricians refer to the maximum squared correlation

$$R^2 = \frac{\left\langle \boldsymbol{y}, \left(M - \mathbf{1}\mathbf{1}^{\mathsf{T}}\right)\boldsymbol{y}\right\rangle}{\left\langle \boldsymbol{y}, \left(\mathsf{I} - \mathbf{1}\mathbf{1}^{\mathsf{T}}\right)\boldsymbol{y}\right\rangle}$$

as the coefficient of determination.

Chapter 54
Multivariate Least-Squares Linear Regression

Abstract We analyze multivariate least-squares linear regression in moment space. For the multivariate dependent variable Y, choose fitted values X to minimize $\langle Y-X, Y-X\rangle$. One expresses the minimizing choice via the cross-moment transformation between the subspaces of the dependent variables and the independent variables. The multivariate least-squares linear regression is the best linear regression, in that any other choice X makes larger the matrix $(Y-X)^{\mathsf{T}}(Y-X)$ of the second moments of the residual $Y-X$. We extend the analysis to reduced-rank least-squares linear regression, in which the fitted values X are constrained to having reduced rank l. The primal embodies the reduced-rank requirement. We present primal and dual solutions such that the primal and dual values are the same. A solution X is the truncation of the singular value decomposition of $A^{\mathsf{T}}Y$ at l terms.

54.1 Moment Space

The universal space $\mathbf{U}$ is a Euclidean space of random variables in which the inner product of two vectors is their second noncentral cross-moment. Define the "dependent variable" Y as a partitioned linear transformation of n linearly independent random variables $\boldsymbol{y}_i$,

$$Y := \begin{bmatrix} \boldsymbol{y}_1 & \cdots & \boldsymbol{y}_n \end{bmatrix} \in \mathbf{T}^{n \to N}.$$

Define N as the subspace spanned by the $\boldsymbol{y}_i$; its idempotent $N = Y\left(Y^{\mathsf{T}}Y\right)^{-1}Y^{\mathsf{T}} \in \mathbf{T}^{\mathrm{U}}$. Let M denote the subspace of "independent variables," and M is its corresponding idempotent. We study a linear regression of Y on M, in which

$$X := \begin{bmatrix} \boldsymbol{x}_1 & \cdots & \boldsymbol{x}_n \end{bmatrix} \in \mathbf{T}^{n \to M}$$

denotes the fitted values of the dependent variables. A linear regression chooses X to make the residual $Y-X$ "small."

B. C. Dieffenbach, *Conjugate Duality in Economic Analysis*, Contributions to Economics,
https://doi.org/10.1007/978-3-032-21396-9_54

Definition 1 (Multivariate Least-Squares Linear Regression) The multivariate least-squares linear regression chooses $X \in \mathbf{T}^{n \to M}$ to minimize $\langle Y - X, Y - X \rangle$.

The cross-moment transformation $A \in \mathbf{T}^{M \to N}$ is the unique linear transformation such that the inner product of $\boldsymbol{x} \in M$ and $\boldsymbol{y} \in N$ is

$$\langle \boldsymbol{y}, \boldsymbol{x} \rangle = \langle \boldsymbol{y}, A\boldsymbol{x} \rangle = \left\langle A^{\mathsf{T}} \boldsymbol{y}, \boldsymbol{x} \right\rangle.$$

The cross-moment transformation $A = NM \in \mathbf{T}^{\mathrm{U}}$, such that the domain and range are restricted to M and N, respectively.

The matrix of second moments of the residual $Y - X$ is

$$(Y - X)^{\mathsf{T}} (Y - X) \in \mathbf{S}^{n}. \tag{54.1}$$

Element ij in the matrix is

$$\left\langle \boldsymbol{y}_i, \boldsymbol{y}_j \right\rangle - \left\langle \boldsymbol{y}_i, A\boldsymbol{x}_j \right\rangle - \left\langle \boldsymbol{y}_j, A\boldsymbol{x}_i \right\rangle + \left\langle \boldsymbol{x}_i, \boldsymbol{x}_j \right\rangle,$$

and the matrix is also

$$\begin{bmatrix} Y \\ -X \end{bmatrix}^{\mathsf{T}} \begin{bmatrix} \mathrm{I} & A \\ A^{\mathsf{T}} & \mathrm{I} \end{bmatrix} \begin{bmatrix} Y \\ -X \end{bmatrix} \in \mathbf{S}^{n}. \tag{54.2}$$

Choosing X to minimize

$$\langle Y, Y \rangle - \left\langle 2A^{\mathsf{T}} Y, X \right\rangle + \langle X, X \rangle$$

by solving the first-order condition yields

$$X = A^{\mathsf{T}} Y. \tag{54.3}$$

Apply least-squares linear regression to each variable: $\boldsymbol{x}_i = A^{\mathsf{T}} \boldsymbol{y}_i = MN\boldsymbol{y}_i = M\boldsymbol{y}_i$, the least-squares linear regression of $\boldsymbol{y}_i$ on M.

54.2 Best Linear Regression

The multivariate least-squares linear regression (54.3) is "best," in that it makes the matrix (54.1) of the second moments of the residuals smallest. Substituting the alternate $A^{\mathsf{T}} Y + Z$ into (54.2) obtains

$$\begin{bmatrix} Y \\ -\left(A^{\mathsf{T}} Y + Z\right) \end{bmatrix}^{\mathsf{T}} \begin{bmatrix} \mathrm{I} & A \\ A^{\mathsf{T}} & \mathrm{I} \end{bmatrix} \begin{bmatrix} Y \\ -\left(A^{\mathsf{T}} Y + Z\right) \end{bmatrix} = Y^{\mathsf{T}} \left(\mathrm{I} - AA^{\mathsf{T}}\right) Y + Z^{\mathsf{T}} Z. \tag{54.4}$$

For any nonzero Z, the matrix is larger.

Theorem 1 (Best Multivariate Linear Regression) *The multivariate least-squares linear regression is best.*

Because the determinant is an increasing function, the multivariate least-squares linear regression minimizes the determinant of the matrix of the second-moment of the residuals. The ordered singular-value decomposition

$$A = \sum_{i=1}^{\operatorname{rank} A} \lambda_i \boldsymbol{n}_i \boldsymbol{m}_i^{\mathrm{T}} = \sum_{i=1}^{\operatorname{rank} A} \langle \boldsymbol{m}_i, \boldsymbol{n}_i \rangle \boldsymbol{n}_i \boldsymbol{m}_i^{\mathrm{T}},$$

in which the singular value $\lambda_i = \langle \boldsymbol{m}_i, \boldsymbol{n}_i \rangle$.

We express the determinant of the second-moment of the residuals via the singular values. By the linear-algebra formula for the determinant of a partitioned linear transformation,

$$\left\| \begin{bmatrix} Y^{\mathrm{T}}Y & Y^{\mathrm{T}}A \\ A^{\mathrm{T}}Y & \mathrm{I} \end{bmatrix} \right\| = |\mathrm{I}| \left| Y^{\mathrm{T}}Y - Y^{\mathrm{T}}A\,(\mathrm{I})^{-1} A^{\mathrm{T}}Y \right| = \left| Y^{\mathrm{T}} \left(\mathrm{I} - AA^{\mathrm{T}} \right) Y \right|.$$

Applying the same formula in the opposite direction gives

$$\begin{aligned} \left\| \begin{bmatrix} Y^{\mathrm{T}}Y & Y^{\mathrm{T}}A \\ A^{\mathrm{T}}Y & \mathrm{I} \end{bmatrix} \right\| &= \left| Y^{\mathrm{T}}Y \right| \left| \mathrm{I} - A^{\mathrm{T}}Y \left(Y^{\mathrm{T}}Y \right)^{-1} Y^{\mathrm{T}}A \right| \\ &= \left| Y^{\mathrm{T}}Y \right| \left| \mathrm{I} - A^{\mathrm{T}}NA \right| \\ &= \left| Y^{\mathrm{T}}Y \right| \left| \mathrm{I} - A^{\mathrm{T}}A \right|. \end{aligned}$$

Thus the determinant of the second-moment of the residuals is

$$\left| Y^{\mathrm{T}} \left(\mathrm{I} - AA^{\mathrm{T}} \right) Y \right| = \left| Y^{\mathrm{T}}Y \right| \left| \mathrm{I} - A^{\mathrm{T}}A \right| = \left| Y^{\mathrm{T}}Y \right| \prod_{i=1}^{\operatorname{rank} A} (1 - \lambda_i^2).$$

The singular values of the cross-moment transformation determine the fraction of the variability in the dependent variables explained by the independent variables.

54.3 Reduced-Rank Least-Squares Linear Regression

Because the trace is also an increasing function, the multivariate least-squares linear regression also minimizes the trace of the matrix of the second-moment of the residuals. This result is obvious, as

$$\langle Y - X, Y - X \rangle = \operatorname{Tr}\,(Y - X)^{\mathrm{T}}\,(Y - X) = \sum_i (\boldsymbol{y}_i - \boldsymbol{x}_i)^{\mathrm{T}}\,(\boldsymbol{y}_i - \boldsymbol{x}_i), \tag{54.5}$$

the sum across variables of the second moments of the residuals.

Let

$$\overline{X} = A^{\mathsf{T}} Y = \sum_{i=1}^{n} \mu_i \boldsymbol{u}_i \boldsymbol{v}_i^{\mathsf{T}} \tag{54.6}$$

denote the ordered singular-value decomposition of the fitted values $\overline{X}$. By (54.4), the minimum value of the trace is

$$\begin{aligned}
&\operatorname{Tr} Y^{\mathsf{T}} \left(\mathrm{I} - AA^{\mathsf{T}}\right) Y \\
&\quad = \langle Y, Y\rangle - \operatorname{Tr} \sum_{i=1}^{n} \mu_i \boldsymbol{v}_i \boldsymbol{u}_i^{\mathsf{T}} \sum_{i=1}^{n} \mu_i \boldsymbol{u}_i \boldsymbol{v}_i^{\mathsf{T}} \\
&\quad = \langle Y, Y\rangle - \operatorname{Tr} \sum_{i=1}^{n} \mu_i^2 \boldsymbol{v}_i \boldsymbol{v}_i^{\mathsf{T}} \\
&\quad = \langle Y, Y\rangle - \sum_{i=1}^{n} \mu_i^2 .
\end{aligned}$$

Next we derive the reduced-rank least-squares linear regression, in which the fitted values X are constrained to having reduced rank:

$$\inf_{X \text{ rank } l} \frac{1}{2} \langle Y - X, Y - X \rangle . \tag{54.7}$$

An equivalent problem is

$$\inf_{X,\, L \text{ idempotent of rank } l} \left(\frac{1}{2} \langle Y - X, Y - X \rangle + \delta_{LX=X} \right).$$

Recall the quadratic function:

$$\delta_Q^* \left(A^*, -S^*\right) = \frac{1}{2} \left\langle A^{*\mathsf{T}}, S^{*+} A^{*\mathsf{T}} \right\rangle + \delta_{A^*(\mathrm{I}-S^{*+}S^*)=0} + \delta_{S^* \succeq 0}$$

Assume that L is idempotent and substitute into this formula:

$$\delta_Q^* \left(X^{\mathsf{T}}, -L\right) = \frac{1}{2} \langle X, LX \rangle + \delta_{X^{\mathsf{T}}(\mathrm{I}-L^+L)=0} + \delta_{L \succeq 0} = \frac{1}{2} \langle X, X \rangle + \delta_{LX=X}.$$

Then

$$\frac{1}{2} \langle Y - X, Y - X \rangle + \delta_{LX=X} = \frac{1}{2} \langle Y, Y \rangle - \left\langle A^{\mathsf{T}} Y, X \right\rangle + \delta_Q^* \left(X^{\mathsf{T}}, -L\right).$$

Using this formula, we analyze the following primal.

Problem 1 (Reduced-Rank Least-Squares Linear Regression)

$$\frac{1}{2}\langle Y, Y\rangle + \inf_{X,L}\left[-\langle A^{\mathsf{T}}Y, X\rangle + \delta_Q^*\left(X^{\mathsf{T}}, -L\right) - \delta_{P_l}(L)\right].$$

To obtain a convex minimization, this problem relaxes the constraint that L be idempotent to $L \in P_l$, the set having the idempotents of rank l as its extreme points; its support is the sum of the l largest eigenvalues. From the viewpoint of Euclidean Jordan algebra, the problem is linear: the quadratic function is the support to a linear inequality, and linear inequalities define P_l. Consequently there is a solution at an extreme point of the constraints. We find that there is a solution such that L is an extreme point of P_l and thus is idempotent of rank l. Consequently this solution also solves (54.7).

We find the reciprocal dual. Adding the Fenchel inequalities

$$\delta_{P_l}(L+K) + \delta_{P_l}^*(K^*) - \langle K^*, L+K\rangle \geq 0 \tag{54.8}$$

$$\delta_Q\left(-X^{*\mathsf{T}}, K^* + L^*\right) + \delta_Q^*\left(X^{\mathsf{T}}, -L\right) - \langle -X^{*\mathsf{T}}, X^{\mathsf{T}}\rangle - \langle K^* + L^*, -L\rangle \geq 0 \tag{54.9}$$

and rearranging obtains

$$\langle X^*, X\rangle + \langle L^*, L\rangle + \delta_{P_l}(L+K) + \delta_Q^*\left(X^{\mathsf{T}}, -L\right) \geq \langle K^*, K\rangle + \delta_{P_l}^*(K^*) + \delta_Q\left(-X^{*\mathsf{T}}, K^* + L^*\right).$$

The perturbation base values are $X^* = -A^{\mathsf{T}}Y = -\overline{X}$, $L^* = 0$, and $K = 0$. Subject to the constraint that K^* lies in the quadratic set, the dual chooses K^* to minimize the sum of the l largest eigenvalues,

$$\sup_{K^*}\left[-\delta_{P_l}^*(K^*) - \delta_Q\left(-X^{*\mathsf{T}}, K^*\right)\right] = -\min_{K^*}\left[\delta_{P_l}^*(K^*) + \delta_{K^* \succeq \frac{1}{2}X^*X^{*\mathsf{T}}}\right].$$

One solution is (54.10).[1]

Solution 1 (Reduced-Rank Least-Squares Linear Regression) A solution for the primal and the dual for the reduced-rank least-squares linear regression is

$$K^* = \frac{1}{2}\overline{XX}^{\mathsf{T}} = \frac{1}{2}\sum_{i=1}^{n}\mu_i^2 \boldsymbol{u}_i\boldsymbol{u}_i^{\mathsf{T}} \tag{54.10}$$

[1] If there are repeated singular values $\mu_l = \mu_{l+1}$, then there are multiple solutions. Even without repeated singular values, there can be other solutions in which K^* is slightly larger but such that the l largest eigenvalues are unchanged.

$$\delta^*_{P_l}(K^*) = \frac{1}{2}\sum_{i=1}^{l}\mu_i^2 \tag{54.11}$$

$$L = \sum_{i=1}^{l}\boldsymbol{u}_i\boldsymbol{u}_i^\top \tag{54.12}$$

$$X = L\overline{X} = \sum_{i=1}^{l}\mu_i\boldsymbol{u}_i\boldsymbol{v}_i^\top, \tag{54.13}$$

the truncation of the singular-value decomposition of $\overline{X}$ at l terms.

The optimum value of the quadratic

$$\frac{1}{2}\langle Y-X, Y-X\rangle = \frac{1}{2}\langle Y, Y\rangle - \frac{1}{2}\langle X, X\rangle = \frac{1}{2}\langle Y, Y\rangle - \frac{1}{2}\sum_{i=1}^{l}\mu_i^2,$$

the trace of the matrix of second-moments of the residual.

Expressions (54.10)–(54.13) make the primal equal the dual, so they constitute a solution. Both Fenchel factors (54.8) and (54.9) are zero. The formula (54.12) makes the Fenchel factor (54.8) zero. The Fenchel factor (54.9) is zero if and only if

$$X = L\overline{X}$$
$$\delta_{\mathbf{S}^M_+}\left(K^* - \frac{1}{2}\overline{X}\overline{X}^\top\right) + \delta^*_{\mathbf{S}^M_+}(-L) - \left\langle K^* - \frac{1}{2}\overline{X}\overline{X}^\top, -L\right\rangle = 0,$$

and the stated solution satisfies these conditions.

See the factor analysis of Eckhart and Young—the least-squares approximation of a linear transformation by another of lower rank—as the special case $M = \mathbf{R}^m$ and $N = \mathbf{R}^n$ and $Y = \mathsf{I}$ (Eckhart and Young 1936). Define $\overline{X} = A^\top$ as the matrix of cross second moments of two data sets. Then X is a least-squares approximation to this matrix giving insight into how the data sets are connected. The singular-value decomposition (54.13) is a "factor analysis" expressing the linkage of a "factor" in one data set with a "factor" in the other. The canonical vectors are the factors. If each data set includes a constant, then the singular values are the canonical correlations. An interesting economic application is Sattinger (1984).

References

Eckhart C, Young G (1936) The approximation of one matrix by another of lower rank. Psychometrika 1(3):211–218

Sattinger MJ (1984) Factor pricing in the assignment problem. Scand J Econ 86(1):17–34

Chapter 55
Instrumental Variables

Abstract Working in moment space, we develop instrumental-variables estimation via duality, in a coordinate-free way. In a structural equation, the explanatory variable causes the dependent variable. The unexplained component is the dependent variable minus the explanatory variable. An estimate of the unknown explanatory variable is a choice from the explanatory variable subspace. If one believes that the explanatory variable and the unexplained component are correlated, then estimation poses a problem, because least-squares linear regression of the dependent variable on the explanatory variable subspace will incorrectly estimate the explanatory variable. To overcome this problem, one approach is instrumental-variables estimation. One assumes that, in the population moment space, the subspace of instrumental random variables is orthogonal to the unexplained component. In the sample moment space, the estimation chooses the explanatory variable to achieve this orthogonality as well as possible. We first set up the instrumental-variables framework. The solution to a linear equation determines necessary and sufficient conditions for identification. The analysis revolves around the cross-moment transformation linking the explanatory- and instrumental-variable subspaces. For an identified equation, the explanatory variable is the dependent variable transformed by the generalized inverse of the cross-moment transformation. One also obtains the analog estimate as the solution to minimization problems. A quadratic minimization implements the generalized method of moments approach to estimating the explanatory variable. Solving this problem and its dual finds that the primal solution is the analog estimate. The two-stage least-squares estimate is also the analog estimate.

We first set up the instrumental-variables framework. Identification is the issue of whether there exists a unique structural equation satisfying specified conditions. The solution to a linear equation determines necessary and sufficient conditions for identification.

The analysis revolves around the cross-moment transformation linking the explanatory- and instrumental-variable subspaces. For an identified equation, the explanatory variable is the dependent variable transformed by the generalized inverse of the cross-moment transformation. By substituting sample moments for

B. C. Dieffenbach, *Conjugate Duality in Economic Analysis*, Contributions to Economics,
https://doi.org/10.1007/978-3-032-21396-9_55

population moments into this formula, one obtains the "analog estimate" of the structural equation (Goldberger 1968; Manski 1988, 1994).

One also obtains the analog estimate as the solution to minimization problems. A quadratic minimization implements the generalized method of moments approach to estimating the explanatory variable. Solving this problem and its dual finds that the primal solution is the analog estimate. The two-stage least-squares estimate is also the analog estimate. Seeing two-stage least squares as a special case of least-squares linear regression, we apply the standard generalized method of moments theory for the latter, to obtain asymptotic statistical inference about the estimated explanatory variable. We also obtain an asymptotic test for identification.

The first use of instrumental-variables estimation is Wright (1928) [see also Stock and Trebbi (2003, Summer)]. The first systematic treatment in economics is Reiersøl (1941, 1945). Basmann (1957) and Theil (1953) put forward the two-stage least-squares estimate.

55.1 Structural Equation

We work in a universal moment space $\mathbf{U}$ of random variables, in which the inner product of two random variables is their noncentral second moment. The population moment space models the population random variables. A corresponding data moment space models the sample random variables. Keep in mind that a vector is a name for a random variable, not a vector of observation values. We presume that as the sample size increases, the sample moments converge to the population moments.

A structural equation relates random variables to one another. Below we analyze a single structural equation. There is no distinction between endogenous and exogenous variables, and there is no reduced form. Working in the population moment space, we analyze the structural equation

$$\boldsymbol{y} = \boldsymbol{x} + \boldsymbol{e}, \tag{55.1}$$

in which the dependent variable $\boldsymbol{y}$ is the explanatory variable $\boldsymbol{x}$ plus the unexplained component $\boldsymbol{e}$. The vectors $\boldsymbol{x}$ and $\boldsymbol{e}$ are perhaps correlated.

Econometricians place a causal interpretation on a structural equation. Goldberger writes:

> By structural equation models, I refer to stochastic models in which each equation represents a causal link, rather than a mere empirical association (Goldberger 1972, p. 979).

In Eq. (55.1), the explanatory random variable $\boldsymbol{x}$ causes the dependent variable $\boldsymbol{y}$. Holding other things constant, an increase in the value of $\boldsymbol{x}$ raises the value of $\boldsymbol{y}$ by the same amount. The unexplained component is the effect of other factors on $\boldsymbol{y}$. "Holding other things constant" means that the value of the unexplained component $\boldsymbol{e}$ is held constant. Holding $\boldsymbol{e}$ constant, a change in $\boldsymbol{x}$ causes an equal change in $\boldsymbol{y}$, in the sense that the probability distribution of the difference in their values is unaltered.

The explanatory variable lies in the subspace of explanatory variables. The instrumental variables cause the explanatory variable and lie in a second subspace. The fundamental condition defining a structural equation is "exclusion": the instrumental variables do not cause the residual. A structural equation must satisfy the "identification equation," a linear equation (55.3). The structural equation is identified if a unique equation satisfies the identification equation.

Let M denote the subspace of explanatory variables, and let N denote the subspace of instrumental variables. Let M and N denote the idempotents for these subspaces. Assume that each subspace contains a constant random variable. For the special case $N = M$, the analysis reduces to least-squares linear regression.

The cross-moment transformation is fundamental.

Definition 1 (Cross-Moment Transformation) The cross-moment transformation linking N and M is

$$\mathsf{A} := \mathsf{N}\mathsf{M}.$$

Recall that one can view cross-moment transformation as either in $\mathbf{T}^{\mathrm{U}}$ or in $\mathbf{T}^{M \to N}$. From either point-of-view, the ordered singular-value decomposition is the same. The singular values of the cross-moment transformation are the canonical correlations, positive and less than or equal to one. A variable can be both an explanatory variable and an instrumental variable, so several canonical correlations can be one.

Instrumental variables cause the explanatory variable, which in turn causes the dependent variable. On the other hand, the instrumental variables do not cause the residual: the residual is orthogonal to the instrumental variables. "Exclusion" models this relationship.

Condition 1 (Exclusion) *In the population moment space*

$$\mathsf{N}(\boldsymbol{y} - \boldsymbol{x}) = \mathbf{0}. \tag{55.2}$$

Because N contains a constant random variable, the exclusion implies that the population residual has mean zero.[1]

[1] In general the residual estimated from the sample by instrumental-variables does not have sample-mean zero.

55.2 Identification

Given the dependent variable $\boldsymbol{y}$, the explanatory-variable subspace M, and the instrumental-variable subspace N, we derive necessary and sufficient conditions for exclusion to identify a unique structural equation. The exclusion condition (55.2) identifies the structural Eq. (55.1) if the identification equation

$$\begin{bmatrix} N \\ \mathsf{I} - M \end{bmatrix} \boldsymbol{x} = \begin{bmatrix} N\boldsymbol{y} \\ \mathbf{0} \end{bmatrix}. \tag{55.3}$$

has a unique solution $\boldsymbol{x}$. In this partitioned equation, the first equation is exclusion; and the second equation requires that $\boldsymbol{x}$ belong to the subspace of explanatory variables. Solving the identification equation obtains necessary and sufficient conditions for identification and derives a formula for $\boldsymbol{x}$.

The identification space is fundamental.

Definition 2 (Identification Space) The identification space $L := M + N^{\perp}$.

The subspaces M and $N^{\perp}$ need not be orthogonal. The orthogonal complement $L^{\perp} = M^{\perp} \cap N$. In a typical instrumental-variables application, the identification space is a proper subspace of the moment space, so $L^{\perp} \neq \{\mathbf{0}\}$.

Condition 2 (Existence) *The dependent variable* $\boldsymbol{y}$ *belongs to the identification space.*

There exists a solution $\boldsymbol{x}$ to the identification equation if and only if the existence condition holds. In the identification Eq. (55.3), the first equation requires $\boldsymbol{y} - \boldsymbol{x} \in N^{\perp}$; and the second equation requires $\boldsymbol{x} \in M$. Together these two requirements are equivalent to $\boldsymbol{y} \in M + N^{\perp} = L$.

We calculate the general solution to the identification equation.

Theorem 1 (Solution to the Identification Equation) *For* $\boldsymbol{y}$ *in the identification space, the general solution to the identification equation is*

$$\boldsymbol{x} \in \left\{A^{+}\boldsymbol{y}\right\} + \mathrm{N}(A) \cap M. \tag{55.4}$$

The generalized inverse

$$\begin{aligned} A^{+} &= \left(A^{\mathsf{T}}A\right)^{+} A^{\mathsf{T}} \\ &= (MNNM)^{+}\, MN \\ &= (MNM)^{+}\, MN. \end{aligned}$$

As $\boldsymbol{x} \in M$, exclusion requires

$$\begin{aligned}\mathbf{0} &= Ny - Nx \\ &= Ny - NMx \\ &= Ny - Ax,\end{aligned}$$

which has the general solution

$$x \in \{A^{+}Ny\} + \mathrm{N}(A) = \{A^{+}y\} + \mathrm{N}(A),$$

as $A^{+}N = A^{+}$. Because $x \in M$, the general solution (55.4) follows.

If the identification equation has a solution, the solution is unique if and only if the uniqueness condition holds.

Condition 3 (Uniqueness) $M \cap N^{\perp} = \{\mathbf{0}\}$.

The solution is unique if and only if the null space of the partitioned linear transformation in Eq. (55.3) is $\{\mathbf{0}\}$. The null space of the first line is $N^{\perp}$, and the null space of the second line is M, so $M \cap N^{\perp} = \{\mathbf{0}\}$ is necessary and sufficient for uniqueness.

The existence and uniqueness conditions are distinct. Existence says $y \in M + N^{\perp} = L$. The polar $L^{\perp} = M^{\perp} \cap N$. On the other hand, uniqueness requires $M \cap N^{\perp} = \{\mathbf{0}\}$, which is different.

The identification theorem pertains to the population moment space.

Theorem 2 (Identification) *Given the dependent variable* y, *the explanatory-variable subspace* M, *and the instrumental-variable subspace* N, *the exclusion restriction identifies a unique structural equation if and only if the dependent variable belongs to the identification space and uniqueness holds. The explanatory variable is*

$$x = A^{+}y. \tag{55.5}$$

Uniqueness implies this formula, because it belongs to the general solution (55.4). The uniqueness condition is equivalent to two other conditions—relevance and rank—that interpret uniqueness.

"Relevance" expresses the concept that the instrumental variables cause the explanatory variables.

Condition 4 (Relevance) *For any nonzero* $x \in M$, *its projection on* N *is nonzero:*

$$Nx \neq \mathbf{0}.$$

Relevance is equivalent to the uniqueness condition: $Nx = \mathbf{0}$ for some nonzero $x \in M$ is equivalent to some nonzero $x \in N^{\perp}$.

Another equivalent is the rank condition: rank $A = \dim M$. Always $M^{\perp} \subset \mathrm{N}(A)$. By Eq. (55.4), uniqueness requires $M^{\perp} = \mathrm{N}(A)$, so the rank condition holds. We express the rank condition in terms of the order condition and the canonical correlations between M and N.

Condition 5 (Rank)

- $\dim N \geq \dim M$;
- the canonical correlations between M and N are all positive.

The rank condition fails if any canonical correlation is zero.

Theorem 3 (Uniqueness) *The uniqueness condition, the relevance condition, and the rank condition are equivalent—any one implies the other two.*

Throughout the subsequent analysis we assume that these three conditions hold. The equivalence applies regardless of whether the existence condition holds.

55.3 Analog Estimate

A natural estimate of the explanatory variable is the analog estimate $\boldsymbol{x} = A^{+}\boldsymbol{y}$, in which one replaces a population magnitude by the corresponding sample magnitude. In a typical instrumental-variables application, the latter are random: M and M, N and N, A, $\boldsymbol{x} = \mathsf{A}^{+}\boldsymbol{y}$, etc. Under general conditions one expects that the sample magnitudes converge asymptotically to the population magnitudes. For a large sample, the sample moments are near the population moments but are not the same. Asymptotically the analog estimate is consistent (unbiased) and normally distributed. (Analog estimation makes no claim for efficiency.)

To help interpret the analog estimate, below we show that it is the unique solution to both a generalized-method-of-moments optimization and a two-stage least-squares optimization. These optimizations lead to statistical inference derived by the generalized method of moments.

Consider an identified structural equation. By the identification theorem, in the population moment space relevance holds and the dependent variable belongs to the identification space. In contrast, in the sample moment space relevance should hold; but the dependent variable does not belong to the identification space. If $\mathsf{N}\boldsymbol{x} \neq \mathbf{0}$ in the population, then in the sample one expects the same condition, because any small change keeps the vector nonzero. On the other hand, if $\boldsymbol{y} \in L$ in the population, then in the sample one expects $\boldsymbol{y} \notin L$, for a small random change in L. Equivalently, if $(\mathsf{I} - \mathsf{L})\boldsymbol{y} = \mathbf{0}$ in the population, then $(\mathsf{I} - \mathsf{L})\boldsymbol{y} \neq \mathbf{0}$ in the sample. The estimation below presumes that this situation prevails.

55.4 Generalized Method of Moments

One obtains the analog estimate as the unique solution to a generalized-method-of-moments optimization. Even though exclusion holds in the population moment space, sampling variation typically makes $\mathsf{N}(\boldsymbol{y}-\boldsymbol{x}) \neq \boldsymbol{0}$, for any $\boldsymbol{x}$ in M. The generalized method of moments chooses $\boldsymbol{x}$ to make this expression small. Specifically, choose $\boldsymbol{x}$ to minimize its second moment $\langle \mathsf{N}(\boldsymbol{y}-\boldsymbol{x}), \mathsf{N}(\boldsymbol{y}-\boldsymbol{x})\rangle$.

Problem 1 (Generalized-Method-of-Moments Primal)

$$\inf_{\boldsymbol{x}\in M} \frac{1}{2}\langle \boldsymbol{y}-\boldsymbol{x}, \mathsf{N}(\boldsymbol{y}-\boldsymbol{x})\rangle. \tag{55.6}$$

The estimation principle of instrumental variables contrasts least-squares linear regression. In the latter, the fitted value $\boldsymbol{x} = \mathsf{M}\boldsymbol{y}$ "fits the data," by choosing $\boldsymbol{x} \in M$ nearest to $\boldsymbol{y}$, in a least-squares sense. In contrast the instrumental-variables fitted value $\boldsymbol{x} = \mathsf{A}^{+}\boldsymbol{y}$ does not fit the data and does not choose $\boldsymbol{x}$ to be near $\boldsymbol{y}$. It is perhaps far from $\boldsymbol{y}$, even if the least-squares estimate is very close. Instead, the generalized-method-of-moments optimization (55.6) chooses the instrumental-variables estimate $\boldsymbol{x}$ to satisfy exclusion as well as possible.

We invoke the following conjugate to solve the generalized-method-of-moments problem.

Theorem 4 (Idempotent Quadratic Form Conjugate) *For an idempotent* N*, the conjugate*

$$\sup_{\boldsymbol{x}}\left(\langle \boldsymbol{x}^*, \boldsymbol{x}\rangle - \frac{1}{2}\langle \boldsymbol{x}, \mathsf{N}\boldsymbol{x}\rangle\right) = \frac{1}{2}\langle \boldsymbol{x}^*, \boldsymbol{x}^*\rangle + \delta_N(\boldsymbol{x}^*). \tag{55.7}$$

The first-order condition is $\mathsf{N}\boldsymbol{x} = \boldsymbol{x}^*$. A solution exists if and only if $\boldsymbol{x}^* \in N$, and then the general solution is $\boldsymbol{x} \in \{\mathsf{N}\boldsymbol{x}^*\} + N^{\perp}$. The formula (55.7) follows.

We solve the generalized-method-of-moments primal via perturbation duality, and find that the dual is a least-squares linear regression problem. Adding the Fenchel inequalities

$$\frac{1}{2}\langle \boldsymbol{y}-\boldsymbol{x}, \mathsf{N}(\boldsymbol{y}-\boldsymbol{x})\rangle + \left[\frac{1}{2}\langle \boldsymbol{y}^*, \boldsymbol{y}^*\rangle + \delta_N(\boldsymbol{y}^*)\right] - \langle \boldsymbol{y}^*, \boldsymbol{y}-\boldsymbol{x}\rangle \geq 0 \tag{55.8}$$

$$\delta_M(\boldsymbol{x}) + \delta_M^*(\boldsymbol{y}^*-\boldsymbol{x}^*) - \langle \boldsymbol{y}^*-\boldsymbol{x}^*, \boldsymbol{x}\rangle \geq 0 \tag{55.9}$$

and rearranging obtains

$$\begin{aligned}&\langle \boldsymbol{x}^*, \boldsymbol{x}\rangle + \frac{1}{2}\langle \boldsymbol{y}-\boldsymbol{x}, \mathsf{N}(\boldsymbol{y}-\boldsymbol{x})\rangle + \delta_M(\boldsymbol{x})\\ &\quad\geq \langle \boldsymbol{y}^*, \boldsymbol{y}\rangle - \left[\frac{1}{2}\langle \boldsymbol{y}^*, \boldsymbol{y}^*\rangle + \delta_N(\boldsymbol{y}^*)\right] - \delta_M^*(\boldsymbol{y}^*-\boldsymbol{x}^*).\end{aligned}$$

The perturbation base values are $x^* = \mathbf{0}$ and y. The primal and dual show immediately that the optimum value is not ∞ or $-\infty$, respectively. Both the primal and the dual satisfy the polyhedral/relative condition, so there is no duality gap and the finite optimum value is attained in both the primal and the dual.

Substituting $L^{\perp} = N \cap M^{\perp}$, rearrange the dual as a least-squares linear regression problem:

$$\begin{aligned}
&\sup_{y^*}\left\{\langle y^*, y\rangle - \left[\frac{1}{2}\langle y^*, y^*\rangle + \delta_N(y^*)\right] - \delta_M^*(y^*)\right\} \\
&= \frac{1}{2}\langle y, y\rangle + \sup_{y^*}\left[-\frac{1}{2}\langle y - y^*, y - y^*\rangle - \delta_{N\cap M^{\perp}}(y^*)\right] \\
&= \frac{1}{2}\langle y, y\rangle - \inf_{y^*}\left[\frac{1}{2}\langle y - y^*, y - y^*\rangle + \delta_{L^{\perp}}(y^*)\right].
\end{aligned}$$

Problem 2 (Generalized-Method-of-Moments Dual) The perturbation dual of the generalized method of moments primal is

$$\frac{1}{2}\langle y, y\rangle - \inf_{y^*}\left[\frac{1}{2}\langle y - y^*, y - y^*\rangle + \delta_{L^{\perp}}(y^*)\right]. \tag{55.10}$$

The inside minimization is the least-squares linear regression of y on $L^{\perp}$, with fitted-value $y^* = (\mathsf{I} - \mathsf{L})y$. The optimum value of the dual is $\frac{1}{2}\langle y, (\mathsf{I} - \mathsf{L})y\rangle = \frac{1}{2}\langle y^*, y^*\rangle$. Below this property leads to a statistical test of the null hypothesis $y \in L$.

The following are necessary and sufficient conditions for solutions to the primal and the dual:

$$y^* = \mathsf{N}(y - x) \tag{55.11}$$
$$\mathsf{M}x = x \tag{55.12}$$
$$\mathsf{M}y^* = \mathbf{0}. \tag{55.13}$$

For the Fenchel factor (55.8) to be zero is equivalent to the first equation. For the Fenchel factor (55.9) to be zero is equivalent to the final two equations. Solving these linear equations yields the analog estimate.

Theorem 5 (Generalized Method-of-Moments Estimate) *Under the uniqueness/relevance/rank condition, the generalized method-of-moments estimate is the analog estimate* $x = \mathsf{A}^{+}y$.

Eliminating y^* reduces the third equation to $\mathbf{0} = \mathsf{MN}(y - x)$. Substitution obtains

$$\mathsf{MNM}x = \mathsf{MN}y,$$

which has the general solution

$$x \in \{(MNM)^{+} MNy\} + \mathrm{N}(MNM).$$

By the uniqueness/relevance/rank condition, $\mathrm{N}(MNM) = M^{\perp}$, so the unique solution in M is $x = A^{+}y$.

A special case is exact identification.

Condition 6 (Exact Identification) *Exact identification is the case*

$$\dim N = \dim M.$$

Under exact identification, the rank of A is $\dim N$ and $AA^{+} = N$, by the uniqueness/relevance/rank condition. The analog estimate satisfies exclusion:

$$N(y - x) = N\left(y - A^{+}y\right) = N\left(y - MA^{+}y\right) = Ny - AA^{+}y = \mathbf{0}.$$

Theorem 6 (Exact Identification) *Under exact identification, the optimum value of the generalized-method-of-moments primal and dual is zero.*

55.5 Two-Stage Least Squares

The analog estimate is also the two-stage least-squares estimate.

Problem 3 (Two-Stage Least Squares)

$$\inf_{x \in M} \frac{1}{2} \langle y - Ax, y - Ax \rangle = \inf_{x \in M} \frac{1}{2} \langle y - Nx, y - Nx \rangle. \tag{55.14}$$

Theorem 7 (Two-Stage Least Squares) *The analog estimate $x = A^{+}y$ is the unique solution to the two-stage least-squares problem.*

Although the analog estimate is the unique solution to both the generalized-method-of-moments problem (55.6) and the two-stage least-squares problem (55.14), the two problems are different and do not have the same optimum value.

The formula for the analog estimate suggests this alternative least-squares linear-regression formulation. The least-squares problem

$$\inf_{x} \frac{1}{2} \langle y - Ax, y - Ax \rangle$$

has the general solution

$$x \in \{A^{+}y\} + \mathrm{N}(A).$$

By the uniqueness/relevance/rank condition, $\mathrm{N}(A) = M^{\perp}$, so the first problem in (55.14) has the unique solution $x = A^{+}y$. The second problem amounts to two-stage least squares: choose x such that its projection Nx onto the instruments is closest to y, in a least-squares sense. Because $Ax = NMx = Nx$ for any $x \in M$, the two problems in (55.14) are the same.

55.6 Statistical Inference

For the statistical inference below, we assume that a random variable refers to t observations. In the population, the observations are independent and identically distributed. In the population, the inner product of two random variables is the second noncentral moment of these variables. Given a realization of the observations, the inner product in the sample moment space is the sample noncentral second moment, the sum across all observations of one variable multiplied by the other. The moments in the sample moment space converge asymptotically to the moments in the population moment space. For a large t, the sample moments are near the population moments but are not the same. The central-limit theorem implies that sample moments are approximately normally distributed.

Below we make use of the ordered singular-value decomposition of the cross-moment transformation:

$$NM = A = \sum_{i=1}^{\dim M} \lambda_i v_i w_i^{\mathsf{T}} = V\mathbf{\Lambda}W^{\mathsf{T}},$$

in which

$$\begin{aligned} V &= \begin{bmatrix} v_1 & \cdots & v_{\dim M} \end{bmatrix} \in \mathbf{T}^{\dim M \to N} \\ W &= \begin{bmatrix} w_1 & \cdots & w_{\dim M} \end{bmatrix} \in \mathbf{T}^{\dim M \to N} \\ \mathbf{\Lambda} &= \begin{bmatrix} \lambda_1 & \mathbf{0} & 0 \\ \mathbf{0} & \ddots & \mathbf{0} \\ 0 & \mathbf{0} & \lambda_{\dim M} \end{bmatrix}. \end{aligned}$$

By the uniqueness/relevance/rank condition, there are $\dim M$ positive singular values, the canonical correlations. The w_i constitute an orthonormal basis for M, so $W^{\mathsf{T}}W = \mathrm{I}$ and $WW^{\mathsf{T}} = M$. One can extend the v_i to an orthonormal basis for N by adding vectors $v_{\dim M+1}, \ldots, v_{\dim N}$. These additional vectors constitute a basis for $L^{\perp}$. We have

$$MNM = A^{\mathsf{T}}A = \left(W\mathbf{\Lambda}V^{\mathsf{T}}\right)\left(V\mathbf{\Lambda}W^{\mathsf{T}}\right) = W\mathbf{\Lambda}^2W^{\mathsf{T}} \preceq M,$$

because each singular value is less than or equal to one.

Working from the two-stage least-squares problem (55.14), we derive asymptotic statistical inference within a standard generalized-method-of-moments framework. The relation $\boldsymbol{x} = Wz$ expresses the fitted value in terms of the regression coefficient $z \in \mathbf{R}^{\dim M}$. The two-stage least-squares problem (55.14) reduces to the least-squares linear regression

$$\inf_z \frac{1}{2} \langle y - AWz, y - AWz \rangle .$$

The asymptotic estimation variance

$$\begin{aligned} t\,\mathrm{Var}(z) &= s^2 \left[(AW)^\top AW \right]^{-1} \\ &= s^2 \left[(V\boldsymbol{\Lambda})^\top V\boldsymbol{\Lambda} \right]^{-1} \\ &= s^2 \boldsymbol{\Lambda}^{-2}. \end{aligned}$$

Here s^2 is the population variance of the residual $\boldsymbol{y}-\boldsymbol{x}$, which one can estimate by the variance of the estimated residual, $s^2 = \langle y - \boldsymbol{x}, y - \boldsymbol{x} \rangle$. Substituting $\boldsymbol{x} = Wz$ obtains

$$\begin{aligned} t\,\mathrm{Var}(\boldsymbol{x}) &= tW\mathrm{Var}(z)\, W^\top \\ &= s^2 W\boldsymbol{\Lambda}^{-2} W^\top \\ &= s^2 \left(A^\top A \right)^+ \\ &= s^2 (MNM)^+ . \end{aligned}$$

Theorem 8 (Estimation Variance) *Asymptotically the variance of the estimate $\boldsymbol{x}$ is*

$$t\,\mathrm{Var}(\boldsymbol{x}) = s^2 \left(A^\top A \right)^+ = s^2 (MNM)^+ .$$

The meaning of this expression is standard: $\mathrm{Cov}\left(\langle \boldsymbol{x}^*, \boldsymbol{x} \rangle , \langle \boldsymbol{x}^{**}, \boldsymbol{x} \rangle \right) = \langle \boldsymbol{x}^*, \mathrm{Var}(\boldsymbol{x})\boldsymbol{x}^{**} \rangle$. This variance yields statistical tests.

When $N = M$, the instrumental-variables estimate reduces to least-squares linear regression. That $(MNM)^+ \geq M$ reflects how the instrumental-variables estimate has larger variance than least squares.

We obtain a chi-square test of the null hypothesis that the fitted value $\boldsymbol{x}$ is zero: $\langle \boldsymbol{x}, [\mathrm{Var}(\boldsymbol{x})]^+ \boldsymbol{x} \rangle$ is chi-square with $\dim M$ degrees of freedom. We have

$$\frac{t}{s^2} \langle \boldsymbol{x}, [\mathrm{Var}(\boldsymbol{x})]^+ \boldsymbol{x} \rangle = \langle \boldsymbol{x}, A^\top A\boldsymbol{x} \rangle = \langle \boldsymbol{x}, MNM\boldsymbol{x} \rangle = \langle \boldsymbol{x}, N\boldsymbol{x} \rangle.$$

Asymptotically the test statistic

$$\langle \boldsymbol{x}, [\mathrm{Var}(\boldsymbol{x})]^{+} \boldsymbol{x} \rangle = \frac{1}{t} \frac{\langle \boldsymbol{x}, A^{\mathsf{T}} A \boldsymbol{x} \rangle}{\langle \boldsymbol{y} - \boldsymbol{x}, \boldsymbol{y} - \boldsymbol{x} \rangle} = \frac{1}{t} \frac{\langle \boldsymbol{x}, N \boldsymbol{x} \rangle}{\langle \boldsymbol{y} - \boldsymbol{x}, \boldsymbol{y} - \boldsymbol{x} \rangle}.$$

is chi-square with $\dim M$ degrees of freedom. A high value rejects the null hypothesis.

The special case $N = M$ is least-squares linear regression. To test the null hypothesis that the fitted value $\boldsymbol{x} = M\boldsymbol{y}$ is zero, asymptotically the test statistic

$$\langle \boldsymbol{x}, [\mathrm{Var}(\boldsymbol{x})]^{+} \boldsymbol{x} \rangle = \frac{1}{t} \frac{\langle \boldsymbol{x}, M \boldsymbol{x} \rangle}{\langle \boldsymbol{y}, (\mathsf{I} - M) \boldsymbol{y} \rangle} = \frac{1}{t} \frac{\langle \boldsymbol{y}, M \boldsymbol{y} \rangle}{\langle \boldsymbol{y}, (\mathsf{I} - M) \boldsymbol{y} \rangle} \tag{55.15}$$

is chi-square with $\dim M$ degrees of freedom. The test statistic is the ratio of the quadratic of the fitted values $M\boldsymbol{x}$ to the quadratic of the estimated residuals $(\mathsf{I} - M)\boldsymbol{y}$.

Presuming that uniqueness/relevance/rank hold, then to test whether $\boldsymbol{y} \in L$ tests whether the structural equation is identified. The generalized-method-of-moments dual reduces to a least-squares linear regression: $\boldsymbol{y}^* = (\mathsf{I} - L)\boldsymbol{y}$ is the fitted value of the regression of $\boldsymbol{y}$ on $L^{\perp}$. We want to test the null hypothesis that the population $\boldsymbol{y}^* = \boldsymbol{0}$, but the population residual $L\boldsymbol{y}$ does not have mean zero. However $\boldsymbol{x} \in L$, because $\boldsymbol{x} \in M$. Hence $\boldsymbol{y}^* = (\mathsf{I} - L)(\boldsymbol{y} - \boldsymbol{x})$, the fitted value of the regression of the residual $\boldsymbol{y} - \boldsymbol{x}$ on $L^{\perp}$. This regression has the residual $L(\boldsymbol{y} - \boldsymbol{x})$, which does have mean zero.

Invoking (55.15) obtains the first formula in (55.16). Reject the null hypothesis that the population $\boldsymbol{y}$ belongs to the identification space if the ratio is high.

Summary 1 (Residual Test for Identification) To test the null hypothesis that the population $\boldsymbol{y}$ belongs to the identification space, asymptotically the ratio

$$\frac{1}{t} \frac{\langle \boldsymbol{y} - \boldsymbol{x}, (\mathsf{I} - L)(\boldsymbol{y} - \boldsymbol{x}) \rangle}{\langle \boldsymbol{y} - \boldsymbol{x}, L(\boldsymbol{y} - \boldsymbol{x}) \rangle} = \frac{1}{t} \frac{\langle \boldsymbol{y} - \boldsymbol{x}, N(\boldsymbol{y} - \boldsymbol{x}) \rangle}{\langle \boldsymbol{y} - \boldsymbol{x}, (\mathsf{I} - N)(\boldsymbol{y} - \boldsymbol{x}) \rangle}. \tag{55.16}$$

is chi-square with $\dim L^{\perp} = \dim N - \dim M$ degrees of freedom.

By Eq. (55.11), the second formula in Eq. (55.16) is equivalent; it tests whether the least-squares linear regression of the residual $\boldsymbol{y} - \boldsymbol{x}$ on N has fitted-value zero. Under exact identification, the numerator in Eq. (55.16) is zero.[2]

The phrase "weak instruments" in the literature essentially refers to the situation in which the structural equation is not identified or almost not identified. Perhaps the population dependent variable lies near the identification space, but not in it. Or perhaps uniqueness/relevance/rank does not hold or holds only barely: one or more canonical correlations is small.

[2] The literature refers to the residual test for identification as a test for "overidentification."

References

Basmann RL (1957 Jan) A generalized classical method of linear estimation of coefficients in a structural equation. Econometrica 25(1):77–83

Goldberger AS (1968) Topics in regression analysis. Macmillan, New York

Goldberger AS (1972 Nov) Structural equation methods in the social sciences. Econometrica 40(6):979–1001

Manski CF (1988) Analog estimation methods in econometrics. Chapman and Hall, New York

Manski CF (1994) Analog estimation of econometric models. In: Engle RF, McFadden DL (eds) Handbook of econometrics, Vol 4. Elsevier, Amsterdam, pp 2559–2582

Reiersøl O (1941) Confluence analysis by means of lag moments and other methods of confluence analysis. Econometrica, 1–24

Reiersøl O (1945) Confluence analysis by means of sets of instrumental variables. Almquist and Wiksells, Stockholm

Stock JH, Trebbi F (2003 Summer) Who invented instrumental variables regression? J Econ Perspect 17(3):177–194

Theil H (1953) Estimation and simultaneous correlation in complete equation systems. The Hague

Wright PG (1928) The tariff on animal and vegetable oils. Macmillan Company

Chapter 56
Likelihood Statistical Inference

Abstract Maximum-likelihood estimation permits elegant testing of the null hypothesis that the unknown parameters satisfy a linear constraint. The context is asymptotic statistical inference when the sample size is large. The Wald test checks directly how closely the maximum-likelihood estimate satisfies the constraint. The Wilks likelihood-ratio test maximizes the likelihood subject to the constraint and checks how much it reduces the likelihood. Rao's score test looks at the derivative of the log-likelihood at the constrained maximum-likelihood estimate and checks whether this derivative is close to zero. The Aitchison and Silvey Lagrange-multiplier test checks whether the Lagrange multiplier in the constrained maximum-likelihood is close to zero. Asymptotically, these four tests are equivalent. We model the estimation in Euclidean space, in which the inner product embodies the negative of the matrix of second derivatives, in a coordinate-free formulation simplifying the mathematics. Because the sample size is large, approximate the log-likelihood by a simple quadratic in the maximum-likelihood estimate minus the vector of parameters. Each test reduces to a least-squares linear regression. The constrained maximum-likelihood estimate reduces to a least-squares linear regression, and its Lagrange-multiplier dual reduces to another least-squares linear regression. The formulas for the variance of the parameter estimates and the Lagrange multiplier are simple and show immediately the chi-square distribution of the test statistics.

Maximum-likelihood estimation permits elegant testing of the null hypothesis that the unknown parameters satisfy a linear constraint. The context is asymptotic statistical inference when the sample size is large. The Wald test checks directly how closely the maximum-likelihood estimate satisfies the constraint (Wald 1943). The likelihood-ratio test maximizes the likelihood subject to the constraint and checks how much it reduces the likelihood (Wilks 1938). The score test looks at the derivative of the log-likelihood at the constrained maximum-likelihood estimate and checks whether this derivative is close to zero (Rao 1948). The Lagrange-multiplier test checks whether the Lagrange multiplier in the constrained maximum-likelihood is close to zero (Aitchison and Silvey 1958; Silvey 1959).

B. C. Dieffenbach, *Conjugate Duality in Economic Analysis*, Contributions to Economics,
https://doi.org/10.1007/978-3-032-21396-9_56

Asymptotically these four tests are equivalent (Breusch and Pagan 1980; Engle 1984). Setting the maximum-likelihood estimation in Euclidean space provides a simple demonstration of this equivalence. The constrained maximum-likelihood estimate reduces to a least-squares linear regression, and its Lagrange-multiplier dual reduces to another least-squares linear regression. The formulas for the variance of the parameter estimates and the Lagrange multiplier are simple and show immediately the chi-square distribution of the test statistics.

The likelihood is the probability density, a function of the data and the parameters. Maximum-likelihood estimation chooses the parameters to maximize the likelihood. We consider maximization of the log-likelihood in a large sample. We assume that there are sufficient statistics, and then under general conditions the maximum-likelihood estimate is unique and has a multivariate normal distribution. We assume that the log-likelihood is twice differentiable at the maximum-likelihood estimate. The first derivative is zero, and the matrix of second derivatives is a negative, self-adjoint linear transformation.

We model the estimation in Euclidean space, in which the inner product embodies the negative of the matrix of second derivatives. Because the sample size is large, approximate the log-likelihood about the maximum-likelihood estimate by the quadratic

$$-\frac{1}{2}\langle z - \boldsymbol{x}, z - \boldsymbol{x}\rangle .$$

Here z is the maximum-likelihood estimate, and $\boldsymbol{x}$ is a vector of parameters. This coordinate-free formulation simplifies the mathematics.

Under general conditions, asymptotically, the expected value $\mathrm{E}(z)$ of the maximum-likelihood estimate is the population parameter $\boldsymbol{\zeta}$. The expected value of a linear combination $\langle z^*, z\rangle = \langle z^*, \mathrm{E}(z)\rangle$. An approximation for the variance of the maximum-likelihood estimate is the negative of the second derivative at z of the log-likelihood, so $\mathrm{Var}(z) = \mathsf{I}$. The covariance of linear combinations $\mathrm{Var}(\langle z^{**}, z\rangle, \langle z^*, z\rangle) = \langle z^{**}, z^*\rangle$.

The null hypothesis is that the population parameter $\boldsymbol{\zeta}$ satisfies the affine constraint $\boldsymbol{\zeta} \in \{\boldsymbol{y}\} + M^{\perp}$, a vector $\boldsymbol{y}$ in a subspace M plus the orthogonal complement $M^{\perp}$. Equivalently, the population parameter satisfies the constraint

$$\mathsf{M}\boldsymbol{\zeta} = \boldsymbol{y}, \tag{56.1}$$

in which M is the idempotent corresponding to M.

56.1 Wald Test

The Wald test is a simple direct test of the null hypothesis that the constraint (56.1) holds: test whether $\mathsf{M}z - \boldsymbol{y}$ is near zero (Wald 1943). Under the null hypothesis,

$\mathrm{E}(Mz - y) = \mathbf{0}$. The variance $\mathrm{Var}(Mz - y) = MIM = M$. Furthermore $Mz - y \in M$. Hence $\langle Mz - y, M(Mz - y)\rangle = \langle Mz - y, Mz - y\rangle$ is chi-square with dim M degrees of freedom.

Summary 1 (Wald Test) Under the null hypothesis (56.1), the quadratic $\langle Mz - y, Mz - y\rangle$ is chi-square with dim M degrees of freedom.

Reject the null hypothesis if $Mz - y$ is large.

56.2 Constrained Maximum Likelihood

The constrained maximum-likelihood problem maximizes the log-likelihood subject to the constraint (56.1).

Problem 1 (Constrained Maximum Likelihood) The log-likelihood ratio is the optimum value in the maximization

$$\sup_x \left[-\frac{1}{2}\langle z - x, z - x\rangle - \delta_{Mx=y}\right] = \sup_x \left[-\frac{1}{2}\langle z - x, z - x\rangle - \delta_{M^\perp}(x - y)\right]. \qquad (56.2)$$

The likelihood-ratio is the likelihood of the constrained maximum-likelihood estimate divided by the likelihood of the maximum-likelihood estimate; the ratio is less than or equal to one. The optimum value of the constrained maximum-likelihood problem is half the natural logarithm of the likelihood ratio.

A change of variables reduces the constrained maximum-likelihood problem to a least-squares linear regression:

$$\begin{aligned}&\sup_x \left[-\frac{1}{2}\langle z - x, z - x\rangle - \delta_{M^\perp}(x - y)\right]\\ &= \sup_w \left[-\frac{1}{2}\langle (z - y) - w, (z - y) - w\rangle - \delta_{M^\perp}(w)\right] \quad (w := x - y).\end{aligned}$$

Thus

$$x - y = (I - M)(z - y). \qquad (56.3)$$

is the fitted value in the least-squares linear regression of $z - y$ on $M^\perp$. Using more information reduces the variance: $\mathrm{Var}(x) = I - M \prec \mathrm{Var}(z)$. The least-squares residual is

$$z - y - (x - y) = z - x,$$

so the optimum value in (56.2) of the objective function is proportional to the quadratic of the residuals. Rearranging (56.3) obtains an alternate expression for the least-squares residual:

$$z - x = Mz - y. \tag{56.4}$$

The variance of the residual is M; if the constraint (56.1) holds, the expected value is zero.

56.3 Likelihood-Ratio Test

The concept of the likelihood-ratio test is that the likelihood of the constrained maximum-likelihood estimate $\boldsymbol{x}$ should be near the likelihood of the unconstrained maximum-likelihood estimate z. The likelihood-ratio test statistic is the quadratic $\langle z - \boldsymbol{x}, z - \boldsymbol{x}\rangle$. By Eq. (56.4), it equals the Wald test statistic.

Summary 2 (Likelihood-Ratio Test) If the constraint (56.1) holds, asymptotically –2 times the natural logarithm of the likelihood ratio is chi-square with dim M degrees of freedom

(Wilks 1938). Reject the null hypothesis if the likelihood ratio is small (if the test statistic is large).

56.4 Score Test

Definition 1 (Score) The score $z - \boldsymbol{x}$ is the derivative of the log-likelihood with respect to $\boldsymbol{x}$, evaluated at the constrained maximum-likelihood estimate

(Rao 1948). The concept is that the score should near zero if the constraint (56.1) holds. The score test statistic is $\langle z - \boldsymbol{x}, z - \boldsymbol{x}\rangle$, the same as the likelihood-ratio statistic. Reject the null hypothesis if this statistic is large.

56.5 Lagrange-Multiplier Test

A fourth test is the Lagrange-multiplier test. For perturbation $\boldsymbol{y}$, we derive the Lagrange-multiplier dual of the constrained maximum-likelihood primal. The Lagrange multiplier $\boldsymbol{y}^*$ is the fitted value of a least-squares linear regression. If the constraint (56.1) holds, then the expected value of the Lagrange multiplier is zero. The concept is that the Lagrange multiplier should be near zero under the null hypothesis. The Lagrange-multiplier test statistic is $\langle \boldsymbol{y}^*, \boldsymbol{y}^*\rangle$ and equals the Wald test statistic.

For the primal (56.2), the Lagrangian is

$$
\begin{aligned}
L(\boldsymbol{x}, \boldsymbol{y}^*) &= \langle \boldsymbol{y}^*, \boldsymbol{y} \rangle - \frac{1}{2} \langle z - \boldsymbol{x}, z - \boldsymbol{x} \rangle - \inf_{\boldsymbol{w}} [\langle \boldsymbol{y}^*, \boldsymbol{w} \rangle + \delta_{M^\perp}(\boldsymbol{x} - \boldsymbol{w})] \\
&= \langle \boldsymbol{y}^*, \boldsymbol{y} \rangle - \frac{1}{2} \langle z - \boldsymbol{x}, z - \boldsymbol{x} \rangle + \sup_{\boldsymbol{v}} [\langle \boldsymbol{y}^*, \boldsymbol{v} - \boldsymbol{x} \rangle - \delta_M^*(\boldsymbol{v})] \quad (\boldsymbol{v} := \boldsymbol{x} - \boldsymbol{w}) \\
&= \langle \boldsymbol{y}^*, \boldsymbol{y} \rangle - \frac{1}{2} \langle z - \boldsymbol{x}, z - \boldsymbol{x} \rangle - \langle \boldsymbol{y}^*, \boldsymbol{x} \rangle + \delta_M(\boldsymbol{y}^*) .
\end{aligned}
$$

The Lagrangian is quadratic in $\boldsymbol{x}$, but not in $\boldsymbol{y}^*$. The first-order condition for $\boldsymbol{x}$ to maximize the Lagrangian is

$$z - \boldsymbol{x} = \boldsymbol{y}^* .$$

The first-order condition for $\boldsymbol{y}^*$ to minimize the Lagrangian is

$$\boldsymbol{x} - \boldsymbol{y} \in M^\perp \text{ and } \boldsymbol{y}^* \in M .$$

To obtain the objective function of the dual, eliminate $\boldsymbol{x}$ from the Lagrangian by substituting the first-order condition for $\boldsymbol{x}$:

$$\sup_{\boldsymbol{x}} L(\boldsymbol{x}, \boldsymbol{y}^*) = \langle \boldsymbol{y}^*, \boldsymbol{y} + \boldsymbol{y}^* - z \rangle - \frac{1}{2} \langle \boldsymbol{y}^*, \boldsymbol{y}^* \rangle + \delta_M(\boldsymbol{y}^*) .$$

Rearrangement obtains the following.

Problem 2 (Lagrange-Multiplier Dual)

$$-\frac{1}{2} \langle z - \boldsymbol{y}, z - \boldsymbol{y} \rangle + \inf_{\boldsymbol{y}^*} \left[\frac{1}{2} \langle (z - \boldsymbol{y}) - \boldsymbol{y}^*, (z - \boldsymbol{y}) - \boldsymbol{y}^* \rangle + \delta_M(\boldsymbol{y}^*) \right] .$$

The infimum is the least-squares linear regression of z–$\boldsymbol{y}$ on M, with fitted value $\boldsymbol{y}^* \in M$:

$$\boldsymbol{y}^* = \mathsf{M}(z - \boldsymbol{y}) = \mathsf{M}z - \boldsymbol{y} . \tag{56.5}$$

Both the primal and the dual satisfy the relative/polyhedral condition.

By Eq. (56.5), then $\mathrm{E}(\boldsymbol{y}^*) = \mathbf{0}$ if the constraint (56.1) holds. The variance $\mathrm{Var}(\boldsymbol{y}^*) = \mathsf{M}$. The Lagrange multiplier is uncorrelated with the constrained estimate $\boldsymbol{x}$,

$$\mathrm{Cov}(\boldsymbol{y}^*, \boldsymbol{x}) = \mathsf{M}(\mathsf{I} - \mathsf{M}) = 0 .$$

Hence $\langle \boldsymbol{y}^*, \mathsf{M}\boldsymbol{y}^* \rangle = \langle \boldsymbol{y}^*, \boldsymbol{y}^* \rangle$ is chi-square with $\dim M$ degrees of freedom and is just the Wald test.

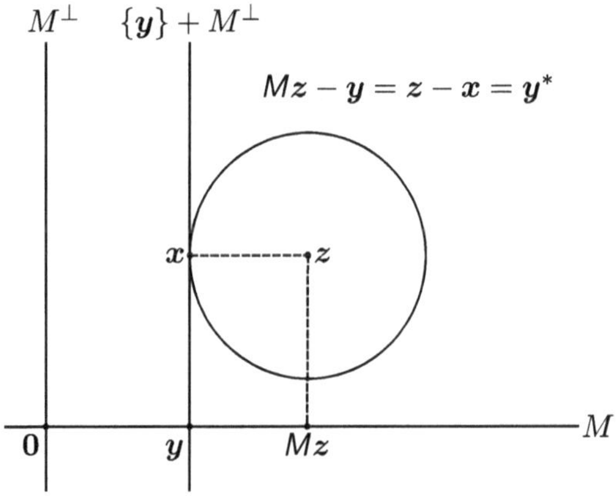

Fig. 56.1 Constrained maximum-likelihood

Figure 56.1 illustrates the constrained maximum-likelihood estimation. The horizontal axis depicts M, and the vertical axis its orthogonal complement. The parameter constraint set is the vertical set $\{y\} + M^\perp$. The circle centered on the maximum-likelihood estimate z is a contour of constant likelihood. The constrained maximum-likelihood estimate x is where the circle is tangent to the constraint set. The test statistics $Mz - y = z - x = y^*$ belong to M.

References

Aitchison J, Silvey SD (1958 Sept) Maximum-likelihood estimation of parameters subject to constraints. Ann Math Statist 29(3):813–828

Breusch TS, Pagan AR (1980 Jan) The Lagrange multiplier test and its applications to model specification in econometrics. Rev Econ Stud 47(1):239–253

Engle RF (1984) Wald, likelihood ratio, and Lagrange multiplier tests in econometrics. In: Griliches Z, Intriligator MD (eds) Handbook of econometrics, Vol 2. Elsevier Science, Amsterdam, pp 775–826

Rao CR (1948) Large sample tests of statistical hypotheses concerning several parameters with applications to problems of estimation. Proc Cambr Philos Soc 44(1):50–57

Silvey SD (1959 June) The Lagrangian multiplier test. Ann Math Statist 30(2):389–407

Wald A (1943 Nov) Tests of statistical hypotheses concerning several parameters when the number of observations is large. Trans Am Math Soc 54(3):426–482

Wilks SS (1938) The large-sample distribution of the likelihood ratio for testing composite hypotheses. Ann Math Statist 9(1):60–62

Chapter 57
Conjugate Maximum Likelihood

Abstract In some important cases, maximum-likelihood estimation amounts to evaluating a conjugate, in which the data vector of sufficient statistics is the dual perturbation for the vector of parameters. A comparative-statics relationship applies. In a large sample, the variance of the parameter estimate is the derivative of the estimate with respect to the sufficient statistics. If the parameter estimate is sensitive to the data, then the variance of the estimate is large.

The likelihood is the probability density, a function of the data and the parameters. In special but important cases, the log-likelihood can be expressed in the simple form

$$\langle \boldsymbol{x}^*, \boldsymbol{x} \rangle - f(\boldsymbol{x}) - g(\cdots), \tag{57.1}$$

set in a Euclidean space. Here $\boldsymbol{x}$ is the vector of unknown parameters, to estimate by choosing $\boldsymbol{x}$ to maximize the likelihood. The vector $\boldsymbol{x}^*$ is a data vector of sufficient statistics, paired with the parameter vector. The proper, closed, convex function f is a function of the parameters but is independent of the data. There can be an extra term g that depends on other data but is independent of the parameters. In this situation, maximum-likelihood estimation amounts to the evaluation of the conjugate

$$f^*(\boldsymbol{x}^*) = \sup_{\boldsymbol{x}} [\langle \boldsymbol{x}^*, \boldsymbol{x} \rangle - f(\boldsymbol{x})], \tag{57.2}$$

as the extra term has no effect on the estimate (The likelihood is not necessarily the probability density function of the sufficient statistic, even though one derives the likelihood from the probability density function of the full sample). A necessary condition for (57.1) is that the probability distribution belong to the natural exponential family (Morris 1982). Morris emphasizes that the choice parameters may be a transformation of the traditional parameters employed by statisticians.

B. C. Dieffenbach, *Conjugate Duality in Economic Analysis*, Contributions to Economics,
https://doi.org/10.1007/978-3-032-21396-9_57

57.1 Comparative-Statics Variance

Because there are sufficient statistics, under general conditions the maximum-likelihood estimate is unique. The estimate $\boldsymbol{x}$ is a function of the sufficient statistics $\boldsymbol{x}^*$, and assume that the partial derivative $\partial\boldsymbol{x}/\partial\boldsymbol{x}^*$ exists.

Theorem 1 (Comparative-Statics Variance) *In a large sample, an estimate of the variance of the maximum-likelihood estimate is*

$$\mathrm{Var}(\boldsymbol{x}) \approx \frac{\partial\boldsymbol{x}}{\partial\boldsymbol{x}^*}. \tag{57.3}$$

How the estimate depends on the sufficient statistics sets its variance. For example, if the estimate is sensitive to the sufficient statistics ($\partial\boldsymbol{x}/\partial\boldsymbol{x}^*$ is large), then the variance is high. To calculate the comparative-statics derivative (57.3) can be simpler than alternative ways to calculate the variance.

We derive the comparative-statics variance (57.3). The first-order condition for the maximum-likelihood estimate is

$$\boldsymbol{x}^* \in \partial f(\boldsymbol{x}). \tag{57.4}$$

By invertibility,

$$\boldsymbol{x} \in \partial f^*(\boldsymbol{x}^*).$$

The comparative-statics derivative

$$\frac{\partial\boldsymbol{x}}{\partial\boldsymbol{x}^*} = \frac{\partial^2 f^*}{\partial\boldsymbol{x}^{*2}} \in \mathbf{S}_+^{\mathbf{X}},$$

so the second partial derivatives of f^* exist.

Consider first the case $\frac{\partial x}{\partial x^*} \gg 0$. The derivative is invertible, so differentiating (57.4) above gives

$$\left(\frac{\partial\boldsymbol{x}}{\partial\boldsymbol{x}^*}\right)^{-1} = \frac{\partial^2 f}{\partial\boldsymbol{x}^2} \gg 0,$$

evaluated at the maximum-likelihood estimate. By statistical theory, under general conditions in a large sample an approximation for the variance of the maximum-likelihood estimate is

$$\mathrm{Var}(\boldsymbol{x}) \approx -\left(\frac{\partial^2 \ln L}{\partial\boldsymbol{x}^2}\right)^{-1} = \left(\frac{\partial^2 f}{\partial\boldsymbol{x}^2}\right)^{-1} = \frac{\partial\boldsymbol{x}}{\partial\boldsymbol{x}^*},$$

evaluated at the maximum-likelihood estimate.

Alternatively, perhaps $\frac{\partial x}{\partial x^*}$ is only nonnegative but not positive. By the fundamental theorem of linear algebra, $\mathbf{X} = \mathrm{R}\left(\frac{\partial x}{\partial x^*}\right) \oplus \mathrm{N}\left(\frac{\partial x}{\partial x^*}\right)$. If $\boldsymbol{y} \in \mathrm{N}\left(\frac{\partial x}{\partial x^*}\right)$, then the data have no effect on $\boldsymbol{y}$, so the variance of this vector must be zero. For the linear transformation $\frac{\partial x}{\partial x^*}$ restricted to $\mathrm{R}\left(\frac{\partial x}{\partial x^*}\right)$, the linear transformation is positive. The situation is like the case above where $\mathrm{R}\left(\frac{\partial x}{\partial x^*}\right)$ is the entire space. The second derivatives of the log-likelihood exist only in $\mathrm{R}\left(\frac{\partial x}{\partial x^*}\right)$, but in this subspace $\mathrm{Var}\,(\boldsymbol{x}) \approx \frac{\partial x}{\partial x^*}$. The comparative-statics formula (57.3) follows.

The case of $\frac{\partial x}{\partial x^*}$ singular means that the maximum-likelihood estimate of one parameter can be expressed as a function of the maximum-likelihood estimate of other parameters. This situation occurs if $\mathrm{dom}\, f$ is not the entire space, and the maximum-likelihood estimate lies on the boundary of the effective domain. An example is the conjugate maximum-likelihood estimation of linear regression coefficients analyzed in the next chapter.

Below we analyze a sample moment space $\mathbf{U}$ of data, in which a vector represents t independent observations on a variable. The inner product of two variables is the sample second noncentral moment. The sample is n data variables,

$$Y = \begin{bmatrix} \boldsymbol{y}_1 & \cdots & \boldsymbol{y}_n \end{bmatrix} \in \mathbf{T}^{n \to \mathbf{U}}. \tag{57.5}$$

The population probability distribution for each observation of these n variables is multivariate normal, with variance $V \gg 0$. The log-likelihood is proportional to t. In a large sample, the sufficient statistic $\boldsymbol{x}^*$ and $\mathrm{Var}(\boldsymbol{x}^*)$ both rise in proportion to t. In contrast, the maximum-likelihood estimate does not, and $\mathrm{Var}(\boldsymbol{x})$ is inversely proportional to t.

57.2 Mean

We apply conjugate maximum-likelihood to the estimation of the mean, for known variance. Let $\mathbf{1}$ denote the constant variable, so $Y^\top \mathbf{1}$ is the vector of the sample means. The log-likelihood is

$$\ln L = -\frac{nt}{2} \ln 2\pi - \frac{t}{2} \ln \det V - \frac{t}{2} \left\langle \left(Y - \mathbf{1}\boldsymbol{b}^\top\right)^\top, V^{-1} \left(Y - \mathbf{1}\boldsymbol{b}^\top\right)^\top \right\rangle,$$

in which Y has unknown mean $\boldsymbol{b} \in \mathbf{R}^n$ and known variance $V \in \mathbf{S}^n_{++}$. Expanding the quadratic obtains

$$\left\langle \left(Y - \mathbf{1}\boldsymbol{b}^\top\right)^\top, V^{-1} \left(Y - \mathbf{1}\boldsymbol{b}^\top\right)^\top \right\rangle = \left\langle Y^\top, V^{-1} Y^\top \right\rangle + \left\langle -2V^{-1} Y^\top \mathbf{1}, \boldsymbol{b} \right\rangle + \left\langle \boldsymbol{b}, V^{-1} \boldsymbol{b} \right\rangle$$

so the conjugate maximum-likelihood (57.3) is

$$\sup_{\boldsymbol{b}}\left(\left\langle tV^{-1}Y^{\mathsf{T}}\mathbf{1},\boldsymbol{b}\right\rangle-\frac{t}{2}\left\langle \boldsymbol{b},V^{-1}\boldsymbol{b}\right\rangle\right).$$

The sufficient statistic $\boldsymbol{b}^{*}=tV^{-1}Y^{\mathsf{T}}\mathbf{1}$. The first-order condition is

$$\mathbf{0}=tV^{-1}Y^{\mathsf{T}}\mathbf{1}-tV^{-1}\boldsymbol{b},$$

so the maximum-likelihood estimate is the sample mean

$$\boldsymbol{b}=Y^{\mathsf{T}}\mathbf{1}.$$

In a large sample, how the estimate $\boldsymbol{b}$ depends on the sufficient statistic $\boldsymbol{b}^{*}$ shows the estimated variance of the estimate. By the first-order condition, the differential

$$\mathrm{d}\boldsymbol{b}=\mathrm{d}\left(Y^{\mathsf{T}}\mathbf{1}\right)=\frac{1}{t}V\,\mathrm{d}\left(tV^{-1}Y^{\mathsf{T}}\mathbf{1}\right)=\frac{1}{t}V\,\mathrm{d}\boldsymbol{b}^{*}.$$

Theorem 2 (Variance of the Maximum-Likelihood Estimate of the Mean) *When t is large, the estimated variance of the estimate of the mean*

$$\mathrm{Var}\,(\boldsymbol{b})\approx\frac{1}{t}V.$$

(In fact, this formula is the exact variance for any t.) In this simple case,

$$\frac{\partial^{2}\ln L}{\partial\boldsymbol{b}^{2}}=-tV^{-1},$$

which yields the same result.

57.3 Variance

For known mean zero, consider the estimation of the variance V via conjugate maximum likelihood, from the sample Y. The inner product of two random variables is their sample covariance. The likelihood is

$$L\,(V)=(2\pi)^{\frac{-nt}{2}}\,(\det V)^{-\frac{t}{2}}\,\mathrm{e}^{-\frac{t}{2}\left\langle Y^{\mathsf{T}},V^{-1}Y^{\mathsf{T}}\right\rangle}.$$

The population variance is positive, so in a large sample the sample variance $Y^{\mathsf{T}}Y\gg 0$. Here

$$\left\langle Y^{\mathsf{T}},V^{-1}Y^{\mathsf{T}}\right\rangle=\mathrm{Tr}\,YV^{-1}Y^{\mathsf{T}}=\mathrm{Tr}\,Y^{\mathsf{T}}YV^{-1}=\left\langle Y^{\mathsf{T}}Y,V^{-1}\right\rangle.$$

Substituting $U := V^{-1}$ and taking the logarithm obtains a conjugate maximum-likelihood problem:

$$\max_U \ln L(U) = \max_U \frac{t}{2}\left(n \ln 2\pi - \left\langle Y^\mathsf{T} Y, U\right\rangle + \ln \det U\right).$$

As $\partial \ln \det U = \left\{U^{-1}\right\}$, the first-order condition yields the estimates

$$Y^\mathsf{T} Y = U^{-1} = V. \tag{57.6}$$

The maximum-likelihood estimate of the population variance is the sample variance.

Here $tY^\mathsf{T} Y$ has the Wishart distribution with t degrees of freedom, and its inverse has the inverse Wishart distribution with t degrees of freedom. These two distributions are not identical. But as the number of observations increases, the formulas for the variances and covariances have the same pattern. Below we derive formulas for the large-sample variances of U and V. That the formulas are the same reflects this situation.

From the conjugate maximum-likelihood problem, we calculate the estimated variance of the estimate U via comparative-statics.

Theorem 3 (Covariance of Inverse Wishart) *For large t,*

$$\operatorname{Cov}\left(U_{ij}, U_{kl}\right) \approx \frac{1}{t}\left(U_{ik}U_{jl} + U_{il}U_{jk}\right). \tag{57.7}$$

The maximization amounts to

$$\sup_U \left(-\left\langle Y^\mathsf{T} Y, U\right\rangle + \ln \det U\right) = -(\ln \det)_* \left(Y^\mathsf{T} Y\right) = -n - (\ln \det)\left(Y^\mathsf{T} Y\right).$$

The dual perturbation term is $\left\langle -\frac{t}{2} Y^\mathsf{T} Y, U\right\rangle$. Via differentials, we calculate the linear transformation relating the maximum-likelihood estimate U to the dual perturbation $-\frac{t}{2} Y^\mathsf{T} Y$. We decompose the dual perturbation into components involving the sample variances and covariances of the random variables:

$$\begin{aligned}
\left\langle Y^\mathsf{T} Y, U\right\rangle &= \operatorname{Tr} Y^\mathsf{T} Y U \\
&= \sum_{ij} \left(Y^\mathsf{T} Y\right)_{ij} U_{ij} \\
&= \sum_{ij} \boldsymbol{y}_i^\mathsf{T} \boldsymbol{y}_j U_{ij} \\
&= \sum_{i} \boldsymbol{y}_i^\mathsf{T} \boldsymbol{y}_i U_{ii} + 2 \sum_{i,j>i} \boldsymbol{y}_i^\mathsf{T} \boldsymbol{y}_j U_{ij}.
\end{aligned}$$

Because $Y^\mathsf{T} Y \gg 0$, each individual variance or covariance can be changed, while holding all the others constant.

Consider first the off-diagonal case $j > i$. We change $\boldsymbol{y}_i^\top \boldsymbol{y}_j$, while holding the variances and other covariances constant. The dual perturbation term is $\left\langle -\frac{t}{2} 2\boldsymbol{y}_i^\top \boldsymbol{y}_j, U_{kl} \right\rangle$. From the first-order condition (57.6), below we show

$$\mathrm{d}U_{kl} = \frac{1}{t}\left(U_{ik}U_{jl} + U_{il}U_{jk}\right)\mathrm{d}\left(-t\boldsymbol{y}_i^\top \boldsymbol{y}_j\right) = -\left(U_{ik}U_{jl} + U_{il}U_{jk}\right)\mathrm{d}\left(\boldsymbol{y}_i^\top \boldsymbol{y}_j\right),$$

which proves (57.7) for $i \neq j$.

Differentiating $UU^{-1} = \mathsf{I}$,

$$\begin{aligned}(\mathrm{d}U)\,U^{-1} + U\,\mathrm{d}U^{-1} &= 0\\ (\mathrm{d}U) = -U\left(\mathrm{d}U^{-1}\right)U &= -U\,\mathrm{d}\left(Y^\top Y\right)U.\end{aligned} \tag{57.8}$$

Here

$$\mathrm{d}\left(Y^\top Y\right) = \boldsymbol{e}_i\,\mathrm{d}\left(\boldsymbol{y}_i^\top \boldsymbol{y}_j\right)\boldsymbol{e}_j^\top + \boldsymbol{e}_j\,\mathrm{d}\left(\boldsymbol{y}_j^\top \boldsymbol{y}_i\right)\boldsymbol{e}_i^\top.$$

Defining $\boldsymbol{u}_k$ as the kth column of U, we have

$$\begin{aligned}&\mathrm{d}U_{kl}\\ &= -\boldsymbol{u}_k^\top\left(\mathrm{d}U^{-1}\right)\boldsymbol{u}_l\\ &= -\boldsymbol{u}_k^\top\mathrm{d}\left(Y^\top Y\right)\boldsymbol{u}_l\\ &= -\boldsymbol{u}_k^\top\left[\boldsymbol{e}_i\,\mathrm{d}\left(\boldsymbol{y}_i^\top \boldsymbol{y}_j\right)\boldsymbol{e}_j^\top + \boldsymbol{e}_j\,\mathrm{d}\left(\boldsymbol{y}_j^\top \boldsymbol{y}_i\right)\boldsymbol{e}_i^\top\right]\boldsymbol{u}_l\\ &= -\left(U_{ki}U_{lj} + U_{kj}U_{li}\right)\mathrm{d}\left(\boldsymbol{y}_i^\top \boldsymbol{y}_j\right),\end{aligned}$$

as desired.

For the diagonal case $i = j$, the dual perturbation term is $\left\langle -\frac{t}{2}\boldsymbol{y}_i^\top \boldsymbol{y}_i, U_{kl} \right\rangle$. From the first-order condition (57.6), below we show

$$\mathrm{d}U_{kl} = \frac{2}{t}U_{ik}U_{il}\,\mathrm{d}\left(-\frac{t}{2}\boldsymbol{y}_i^\top \boldsymbol{y}_i\right) = -U_{ik}U_{il}\,\mathrm{d}\left(\boldsymbol{y}_i^\top \boldsymbol{y}_i\right).$$

Here

$$\mathrm{d}\left(Y^\top Y\right) = \boldsymbol{e}_i\,\mathrm{d}\left(\boldsymbol{y}_i^\top \boldsymbol{y}_i\right)\boldsymbol{e}_i^\top.$$

We have

$$\begin{aligned}&\mathrm{d}U_{kl}\\ &= -\boldsymbol{u}_k^\top\left(\mathrm{d}U^{-1}\right)\boldsymbol{u}_l\\ &= -\boldsymbol{u}_k^\top\mathrm{d}\left(Y^\top Y\right)\boldsymbol{u}_l\\ &= -\boldsymbol{u}_k^\top\left[\boldsymbol{e}_i\,\mathrm{d}\left(\boldsymbol{y}_i^\top \boldsymbol{y}_i\right)\boldsymbol{e}_i^\top\right]\boldsymbol{u}_l\\ &= -U_{ki}U_{li}\,\mathrm{d}\left(\boldsymbol{y}_i^\top \boldsymbol{y}_i\right),\end{aligned}$$

as desired.

We derive the same formula (57.9) for the Wishart distribution from (57.7), by linearization.

Theorem 4 (Covariance of Wishart) *For large t,*

$$\mathrm{Cov}\left(V_{kl}, V_{mn}\right) \approx \frac{1}{t}\left(V_{km}V_{ln} + V_{lm}V_{kn}\right). \tag{57.9}$$

(In fact, this formula is the exact covariance for any t).

The linearization

$$\mathrm{d}V = \mathrm{d}U^{-1} = U^{-1}\,(\mathrm{d}U)\,U^{-1} = V\,(\mathrm{d}U)\,V.$$

implies

$$\begin{aligned}
\mathrm{d}V_{kl} &= \boldsymbol{v}_k^{\mathsf{T}}\,(\mathrm{d}U)\,\boldsymbol{v}_l = \sum_{gh} V_{kg}\left(\mathrm{d}U_{gh}\right)V_{lh}\\
\mathrm{d}V_{mn} &= \boldsymbol{v}_m^{\mathsf{T}}\,(\mathrm{d}U)\,\boldsymbol{v}_n = \sum_{ij} V_{mi}\left(\mathrm{d}U_{ij}\right)V_{nj}.
\end{aligned}$$

Their product

$$\mathrm{d}V_{kl}\,\mathrm{d}V_{mn} = \left[\sum_{gh} V_{kg}\left(\mathrm{d}U_{gh}\right)V_{lh}\right]\left[\sum_{ij} V_{mi}\left(\mathrm{d}U_{ij}\right)V_{nj}\right]$$

implies

$$\begin{aligned}
&\mathrm{Cov}\left(V_{kl}, V_{mn}\right)\\
&\quad= \sum_{ghij} V_{kg}V_{lh}V_{mi}V_{nj}\,\mathrm{Cov}\left(U_{gh}, U_{ij}\right)\\
&\quad= \frac{1}{t}\sum_{ghij} V_{kg}V_{lh}V_{mi}V_{nj}\left(U_{gi}U_{hj} + U_{gj}U_{hi}\right)
\end{aligned} \tag{57.10}$$

by (57.7).

That $V = U^{-1}$ simplifies this expression to (57.9). There are two terms. In Eq. (57.10), the first sum

$$\begin{aligned}
&\frac{1}{t}\sum_{ghij} V_{kg}V_{lh}V_{mi}V_{nj}U_{gi}U_{hj} \\
&= \frac{1}{t}\sum_{ij}\left(V_{mi}V_{nj}\sum_{g} V_{kg}U_{gi}\sum_{h} V_{lh}U_{hj}\right) \\
&= \frac{1}{t}\sum_{ij} V_{mi}V_{nj}\delta_{ki}\delta_{lj} \\
&= \frac{1}{t}\sum_{i} V_{mi}\delta_{ki}\sum_{j} V_{nj}\delta_{lj} \\
&= \frac{1}{t}V_{km}V_{ln}.
\end{aligned}$$

Here the Kronecker delta $\delta_{ij} = 1$ if $i = j$, $\delta_{ij} = 0$ if $i \neq j$. The second sum is analogous, and (57.9) follows.

Reference

Morris CN (1982 Mar) Natural exponential families with quadratic variance functions. Ann Statist 10(1):65–80

Chapter 58
Maximum-Likelihood Estimation of Linear Regression

Abstract We perform maximum-likelihood estimation of linear regression by perturbation duality. Transforming the parameters and deploying the quadratic function obtains a maximization concave in the parameters. The reduced dual resembles the standard least-squares linear-regression dual. To carry out a comparative-statics calculation of parameter variances and covariances, adding a choice variable obtains a conjugate maximum-likelihood problem, in which the inner product multiplies the second moments of the data by the parameters. We find the large-sample variances of the parameter estimates via the comparative-statics approach, by calculating how a change in the data affects the estimates. Each sufficient statistic can be varied independently, while holding the others constant.

58.1 Concave Likelihood

In the moment space of data, a vector refers to a random sample of t observations. The inner product is the sample second noncentral moment. The independent variables

$$X = \begin{bmatrix} \boldsymbol{x}_1 & \cdots & \boldsymbol{x}_n \end{bmatrix} \in \mathbf{T}^{n \to t}$$

and the dependent variable $\boldsymbol{y} \in \mathbf{R}^t$. Assume that there is no multicollinearity,

$$\begin{bmatrix} \boldsymbol{y}^\mathsf{T}\boldsymbol{y} & \boldsymbol{y}^\mathsf{T}X \\ X^\mathsf{T}\boldsymbol{y} & X^\mathsf{T}X \end{bmatrix} \gg 0. \tag{58.1}$$

The estimated linear regression calculates the vector in $\mathrm{R}(X)$ closest to $\boldsymbol{y}$.

The log-likelihood is

$$\ln L = -\frac{t}{2} \ln 2\pi - \frac{t}{2} \ln v - \frac{t}{2} \left\langle \boldsymbol{y} - X\boldsymbol{b}, \frac{1}{v} (\boldsymbol{y} - X\boldsymbol{b}) \right\rangle.$$

B. C. Dieffenbach, *Conjugate Duality in Economic Analysis*, Contributions to Economics,
https://doi.org/10.1007/978-3-032-21396-9_58

The distribution of the dependent variable conditional on the independent variables is normal, and different observations are independent. Choose the estimated regression coefficient $\boldsymbol{b}$ and the estimated residual variance v to maximize the likelihood. The estimated residual is $\boldsymbol{y} - X\boldsymbol{b}$.

Because $\ln v$ is concave and not convex, reparametrize the model to obtain a concave likelihood:

$$\begin{aligned} u &= v^{-1} \\ \boldsymbol{c} &= -\boldsymbol{b}u. \end{aligned}$$

The objective function is then jointly concave in u and $\boldsymbol{c}$. The inner product

$$\begin{aligned} &\left\langle \boldsymbol{y} - X\boldsymbol{b}, \frac{1}{v}(\boldsymbol{y} - X\boldsymbol{b})\right\rangle \\ &= \langle \boldsymbol{y} + X\boldsymbol{c}/u, u(\boldsymbol{y} + X\boldsymbol{c}/u)\rangle \\ &= \langle \boldsymbol{y}^{\mathsf{T}}\boldsymbol{y}, u\rangle + \langle 2X^{\mathsf{T}}\boldsymbol{y}, \boldsymbol{c}\rangle + \langle X\boldsymbol{c}, u^{-1}X\boldsymbol{c}\rangle \\ &= 2\left[\frac{1}{2}\langle \boldsymbol{y}^{\mathsf{T}}\boldsymbol{y}, u\rangle + \langle \boldsymbol{y}, X\boldsymbol{c}\rangle + \delta_Q^*(X\boldsymbol{c}, -u)\right] \end{aligned}$$

58.2 Maximum Likelihood

To maximize the likelihood, solve the following problem.

Problem 1 (Maximum-Likelihood Primal)

$$\sup_{u,\boldsymbol{c}} \left[-\frac{1}{2}\langle \boldsymbol{y}^{\mathsf{T}}\boldsymbol{y}, u\rangle - \langle X^{\mathsf{T}}\boldsymbol{y}, \boldsymbol{c}\rangle + \frac{1}{2}\ln u - \delta_Q^*(X\boldsymbol{c}, -u)\right]. \tag{58.2}$$

Adding the Fenchel inequalities

$$\delta_Q(z^*, v^*) + \delta_Q^*(-X\boldsymbol{c} - z, -u - v) - \langle z^*, -X\boldsymbol{c} - z\rangle - (v^*)(-u - v) \geq 0 \tag{58.3}$$

$$\delta_{\{0\}}\left(X^{\mathsf{T}}z^* + \boldsymbol{c}^*\right) + \delta_{\{0\}}^*(\boldsymbol{c}) - \left\langle X^{\mathsf{T}}z^* + \boldsymbol{c}^*, \boldsymbol{c}\right\rangle \geq 0 \tag{58.4}$$

$$-\frac{1}{2}\ln u - \frac{1}{2}\ln_*[2(-v^* - u^*)] + (-v^* - u^*)u \geq 0 \tag{58.5}$$

and rearranging obtains

$$\begin{aligned} &v^*v + z^*z - \frac{1}{2}\ln_*[2(-v^* - u^*)] + \delta_{X^{\mathsf{T}}z^* = \boldsymbol{c}^*} + \delta_Q(z^*, v^*) \\ &\geq u^*u + \langle \boldsymbol{c}^*, \boldsymbol{c}\rangle + \frac{1}{2}\ln u - \delta_Q^*(-X\boldsymbol{c} - z, -u - v). \end{aligned}$$

The perturbation base values are $v = 0$, $z = \mathbf{0}$, $\boldsymbol{c}^* = -X^\mathsf{T}\boldsymbol{y}$, and $u^* = -\frac{1}{2}\boldsymbol{y}^\mathsf{T}\boldsymbol{y}$.

Problem 2 (Maximum-Likelihood Dual)

$$\inf_{v^*,z^*}\left\{-\frac{1}{2}\left[1+\ln\left(-2v^*+\boldsymbol{y}^\mathsf{T}\boldsymbol{y}\right)\right]+\delta_{X^\mathsf{T}(y-z^*)=0}+\delta_{v^*\geq\frac{1}{2}\langle z^*,z^*\rangle}\right\}.$$

Because the primal satisfies the relative condition and the dual satisfies the polyhedral/relative condition, there is no duality gap; and the optimum value is attained in both the primal and the dual.

For the Fenchel factor (58.3) to be zero is equivalent to

$$v^* = \frac{1}{2}\langle z^*, z^*\rangle$$
$$z^* u = -X\boldsymbol{c},$$

so

$$z^* = X\boldsymbol{b} \tag{58.6}$$

is the fitted value of the dependent variable, and v^* is half the second moment of the fitted values.

For the Fenchel factor (58.4) to be zero means

$$X^\mathsf{T}(y - z^*) = \mathbf{0};$$

the estimated residual is orthogonal to the independent variables. Substituting (58.6) into the constraint yields

$$\boldsymbol{b} = \left(X^\mathsf{T}X\right)^{-1}X^\mathsf{T}\boldsymbol{y}.$$

For the Fenchel factor (58.5) to be zero means that the estimated variance

$$v = 1/u = -2\left(u^* + v^*\right) = \boldsymbol{y}^\mathsf{T}\boldsymbol{y} - \langle z^*, z^*\rangle,$$

the second moment of the dependent variable less the second moment of the fitted values. (If the independent variables include a constant, then the estimated variance is the variance of the estimated residuals.)

Obtain a reduced dual by eliminating v^*:

$$\inf_{z^*}\left\{-\frac{1}{2}\left[1+\ln\left(\boldsymbol{y}^\mathsf{T}\boldsymbol{y}-\langle z^*,z^*\rangle\right)\right]+\delta_{X^\mathsf{T}(y-z^*)=0}\right\}.$$

This reduced dual resembles the standard least-squares linear regression dual: *minimize* the second moment of the fitted values such that the residuals are orthogonal to the independent variables.

58.3 Conjugate Maximum Likelihood

The primal (58.2) is not a conjugate maximum-likelihood problem, because the quadratic term contains the data X. Convert this primal into a conjugate maximum-likelihood problem by adding a choice variable W:

$$\langle Xc, u^{-1}Xc\rangle = \inf_{W \in S^n} \left(\langle X^\top X, W\rangle + \delta_{S_+^{1+n}} \begin{bmatrix} u & c^\top \\ c & W \end{bmatrix} \right),$$

because $X^\top X \gg 0$ and $W = u^{-1}cc^\top$.

Problem 3 (Conjugate Maximum-Likelihood Linear Regression)

$$\sup_{\begin{bmatrix} u & c^\top \\ c & W \end{bmatrix} \in S_+^{1+n}} \left(-\frac{t}{2} \left\langle \begin{bmatrix} y^\top y & y^\top X \\ X^\top y & X^\top X \end{bmatrix}, \begin{bmatrix} u & c^\top \\ c & W \end{bmatrix} \right\rangle - \frac{t}{2} \ln 2\pi + \frac{t}{2} \ln u \right).$$

The inner product multiplies the second moments of the data by the parameters. Although it is straightforward to solve this problem via perturbation duality, we need not do so, because we already know the answer.

We find the large-sample variances of the parameter estimates via the comparative-statics approach, by calculating how a change in the data affects the estimates. The dual perturbation term

$$-\frac{t}{2} \left\langle \begin{bmatrix} y^\top y & y^\top X \\ X^\top y & X^\top X \end{bmatrix}, \begin{bmatrix} u & c^\top \\ c & W \end{bmatrix} \right\rangle = -\frac{t}{2} \left(\langle y^\top y, u\rangle + \langle 2X^\top y, c\rangle + \langle X^\top X, W\rangle \right).$$

Because there is no multicollinearity, each sufficient statistic $y^\top y$, $X^\top y$, or $X^\top X$ can be varied independently, while holding the other two constant. We linearize the estimates

$$v = \frac{1}{u} = y^\top y - y^\top X \left(X^\top X\right)^{-1} X^\top y \tag{58.7}$$

$$c = -\left(X^\top X\right)^{-1} X^\top y u. \tag{58.8}$$

First calculate how the estimate u depends on $y^\top y$, holding fixed the other sufficient statistics. By the residual variance estimate (58.7),

$$du = -u^2 \, d\left(u^{-1}\right) = -u^2 \, d\left(y^\top y\right) = \frac{2}{t} u^2 \, d\left(-\frac{t}{2} y^\top y\right).$$

Theorem 1 (Inverse of Residual) *In a large sample, the variance of the maximum-likelihood estimate u*

$$\mathrm{Var}(u) \approx \frac{2}{t}u^2.$$

Holding fixed the other sufficient statistics, calculate how the estimate $\boldsymbol{c}$ depends on $\boldsymbol{y}^\mathsf{T}\boldsymbol{y}$. Because $\boldsymbol{b}$ does not depend on $\boldsymbol{y}^\mathsf{T}\boldsymbol{y}$,

$$\begin{aligned}
\mathrm{d}\boldsymbol{c} &= -\boldsymbol{b}\,\mathrm{d}u \\
&= -\frac{2}{t}\boldsymbol{b}u^2\,\mathrm{d}\left(-\frac{t}{2}\boldsymbol{y}^\mathsf{T}\boldsymbol{y}\right) \\
&= \frac{2}{t}\boldsymbol{c}u\,\mathrm{d}\left(-\frac{t}{2}\boldsymbol{y}^\mathsf{T}\boldsymbol{y}\right).
\end{aligned}$$

Theorem 2 (Covariance) *In a large sample, the covariance of the maximum-likelihood estimates*

$$\mathrm{Cov}(\boldsymbol{c}, u) \approx \frac{2}{t}\boldsymbol{c}u. \tag{58.9}$$

Holding fixed the other sufficient statistics, calculate how u depends on $X^\mathsf{T}\boldsymbol{y}$. By the residual variance estimate (58.7),

$$\begin{aligned}
\mathrm{d}u^{-1} &= -\boldsymbol{y}^\mathsf{T}X\left(X^\mathsf{T}X\right)^{-1}\mathrm{d}\left(X^\mathsf{T}\boldsymbol{y}\right) - \mathrm{d}\left(\boldsymbol{y}^\mathsf{T}X\right)\left(X^\mathsf{T}X\right)^{-1}X^\mathsf{T}\boldsymbol{y} \\
&= -\boldsymbol{b}^\mathsf{T}\mathrm{d}\left(X^\mathsf{T}\boldsymbol{y}\right) - \mathrm{d}\left(\boldsymbol{y}^\mathsf{T}X\right)\boldsymbol{b} \\
&= -2\boldsymbol{b}^\mathsf{T}\mathrm{d}\left(X^\mathsf{T}\boldsymbol{y}\right).
\end{aligned}$$

Then

$$\mathrm{d}u = -u^2\,\mathrm{d}\left(u^{-1}\right) = -u^2\left[-2\boldsymbol{b}^\mathsf{T}\mathrm{d}\left(X^\mathsf{T}\boldsymbol{y}\right)\right] = \frac{2}{t}u\boldsymbol{c}^\mathsf{T}\mathrm{d}\left(-tX^\mathsf{T}\boldsymbol{y}\right). \tag{58.10}$$

Theorem 3 *In a large sample, the covariance of the maximum-likelihood estimates*

$$\mathrm{Cov}(u, \boldsymbol{c}) \approx \frac{2}{t}u\boldsymbol{c}^\mathsf{T}.$$

This formula confirms (58.9) above.

Holding fixed the other sufficient statistics, calculate how $\boldsymbol{c}$ depends on $X^\mathsf{T}\boldsymbol{y}$:

$$\begin{aligned}
\mathrm{d}\boldsymbol{c} &= -\left(X^\mathsf{T}X\right)^{-1}\mathrm{d}\left(X^\mathsf{T}\boldsymbol{y}\right)u - \left(X^\mathsf{T}X\right)^{-1}X^\mathsf{T}\boldsymbol{y}\,\mathrm{d}u \\
&= \frac{1}{t}\left(X^\mathsf{T}X\right)^{-1}\mathrm{d}\left(-tX^\mathsf{T}\boldsymbol{y}\right)u - \boldsymbol{b}\left[\frac{2}{t}u\boldsymbol{c}^\mathsf{T}\mathrm{d}\left(-tX^\mathsf{T}\boldsymbol{y}\right)\right] \text{ by Eq. (58.10)} \\
&= \frac{1}{t}\left[u\left(X^\mathsf{T}X\right)^{-1} + 2\boldsymbol{c}\boldsymbol{c}^\mathsf{T}\right]\mathrm{d}\left(-tX^\mathsf{T}\boldsymbol{y}\right).
\end{aligned}$$

Theorem 4 (Variance) *In a large sample, the variance of the maximum-likelihood estimate*

$$\mathrm{Var}\,(\boldsymbol{c}) \approx \frac{1}{t}\left[u\left(X^{\top}X\right)^{-1} + 2\boldsymbol{c}\boldsymbol{c}^{\top}\right].$$

Linearizing these formulas obtains the traditional formulas.

Theorem 5 (Traditional Variance and Covariance) *In a large sample, the variances*

$$\begin{aligned}\mathrm{Var}\,(v) &\approx \frac{2}{t}v^2\\ \mathrm{Var}\,(\boldsymbol{b}) &\approx \frac{1}{t}v\left(X^{\top}X\right)^{-1}\\ \mathrm{Cov}\,(v,\boldsymbol{b}) &\approx \boldsymbol{0}.\end{aligned}$$

The linearization

$$\mathrm{d}v = \mathrm{d}\left(u^{-1}\right) = \frac{1}{u^2}\mathrm{d}u \tag{58.11}$$

implies

$$\mathrm{Var}\,(v) = \frac{1}{u^4}\mathrm{Var}\,(u) = \frac{1}{u^4}\frac{2}{t}u^2 = \frac{2}{t}v^2.$$

The linearization

$$\mathrm{d}\boldsymbol{b} = \mathrm{d}\left(-\boldsymbol{c}u^{-1}\right) = -u^{-1}\mathrm{d}\boldsymbol{c} + \frac{\boldsymbol{c}}{u^2}\mathrm{d}u \tag{58.12}$$

implies

$$\begin{aligned}\mathrm{Var}\,(\boldsymbol{b}) &= \frac{1}{u^2}\mathrm{Var}\,(\boldsymbol{c}) + \frac{\boldsymbol{c}\boldsymbol{c}^{\top}}{u^4}\mathrm{Var}\,(u) - 2\frac{1}{u}\frac{\boldsymbol{c}}{u^2}\mathrm{Cov}\,(u,\boldsymbol{c})\\ &= \frac{1}{u^2}\frac{1}{t}\left[u\left(X^{\top}X\right)^{-1} + 2\boldsymbol{c}\boldsymbol{c}^{\top}\right] + \frac{\boldsymbol{c}\boldsymbol{c}^{\top}}{u^4}\frac{2}{t}u^2 - 2\frac{1}{u}\frac{\boldsymbol{c}}{u^2}\frac{2}{t}u\boldsymbol{c}^{\top}\\ &= \frac{1}{t}v\left(X^{\top}X\right)^{-1}.\end{aligned}$$

The linearizations (58.11) and (58.12) imply

$$\begin{aligned}\mathrm{Cov}\,(v,\boldsymbol{b}) &= -\frac{1}{u}\mathrm{Cov}\,(\boldsymbol{c},u) + \frac{\boldsymbol{c}}{u^4}\mathrm{Var}\,(u)\\ &= -\frac{1}{u}\frac{2}{t}u\boldsymbol{c} + \frac{\boldsymbol{c}}{u^4}\frac{2}{t}u^2\\ &= \boldsymbol{0}.\end{aligned}$$

Part XIV
Finance

Chapter 59
Stochastic Discount Factor

Abstract A stochastic discount factor generalizes the concept of discounted present value. The component of a stochastic discount factor for a state is the value per unit of probability of a one-dollar payoff in that state. The inner product of a stochastic discount factor and the payoff vector for an asset is the asset price. In a primal/dual analysis in finance theory, commonly the choice variable in the dual defines a stochastic discount factor. The law of one price states that a portfolio having payoff zero must have cost zero. Invoking the Fredholm alternative establishes a basic result: there exists a stochastic discount factor for asset markets if and only if the law of one price holds. The fundamental theorem of finance asserts that there is no opportunity for arbitrage if and only if there exists a positive stochastic discount factor. There exists a cost/mean/variance efficient portfolio if and only if there is no risk-free portfolio providing an arbitrage opportunity. Applications of efficiency duality and the Stiemke alternative obtain these results and provide insight. The payoff space is a moment space valuable in finance theory. When the law of one price holds, the "pricing payoff" is the unique payoff that is a stochastic discount factor. "Cost and mean positivity" is the condition that the inner product of the pricing payoff and the mean vector in the payoff space is positive. The cost of the mean vector is positive, and the expected value of the pricing payoff is positive. One should see positivity as a requirement for asset-market equilibrium, a condition distinct from the requirement that there should be no opportunity for arbitrage.

59.1 State Space and Asset Space

The state space $\mathbf{Y}$ is a Euclidean Jordan algebra, a moment space in which a vector $\boldsymbol{y}$ specifies a payoff y_i in state i. There are m possible states of the world, and state i occurs with positive probability π_i. The inner product of two payoffs is their second noncentral moment. Let the vector $\boldsymbol{e}_i$ denote payoff one in state i and payoff zero otherwise. Together the $\boldsymbol{e}_i$ form a natural basis for the state space. As $\langle \boldsymbol{e}_i, \boldsymbol{e}_j \rangle = \pi_i \delta_{ij}$,

B. C. Dieffenbach, *Conjugate Duality in Economic Analysis*, Contributions to Economics,
https://doi.org/10.1007/978-3-032-21396-9_59

the basis is orthogonal but not orthonormal. The payoff $\boldsymbol{y} = \sum_i y_i \boldsymbol{e}_i$ has payoff y_i in state i. The algebra has the natural partial ordering.[1]

An asset is described by its price and its payoff in the state space. There are n assets. The asset space $\mathbf{X}$ is the Euclidean Jordan algebra $\mathbf{R}^n$. A vector $\boldsymbol{x}$ in the asset space is a portfolio of quantity x_j of asset j. Its cost is $\langle \boldsymbol{v}^*, \boldsymbol{x} \rangle$, in which $\boldsymbol{v}^*$ denotes the asset-price vector. Let $\boldsymbol{a}_j \in \mathbf{Y}$ denote the payoff on asset j. Define the payoff transformation

$$A = \begin{bmatrix} \boldsymbol{a}_1 & \cdots & \boldsymbol{a}_n \end{bmatrix} \in \mathbf{T}^{\mathbf{X} \to \mathbf{Y}}.$$

The payoff on portfolio $\boldsymbol{x}$ is $A\boldsymbol{x}$. The mean vector in the state space is $\mathbf{1} := \sum_i \boldsymbol{e}_i$, which has risk-free payoff one in every state. The asset space may or may not contain a risk-free portfolio: $\mathbf{1}$ may or may not belong to $\mathrm{R}(A)$.

59.2 Stochastic Discount Factor

Generalizing the concept of discounted present value, a stochastic discount factor is a vector $\boldsymbol{y}^*$ in the state space such that $\langle \boldsymbol{y}^*, \boldsymbol{y} \rangle$ is the value of payoff $\boldsymbol{y}$. Its payoff y_i^* is the value per unit of probability of payoff one in the state. We say that a stochastic discount factor exists if it prices all assets: for each asset the price is the discounted value of its payoff, $v_j^* = \langle \boldsymbol{y}^*, \boldsymbol{a}_j \rangle$.

The law of one price is a necessary condition for market equilibrium.

Definition 1 (Law of One Price) A portfolio having payoff zero must have zero cost.

Otherwise a portfolio with payoff zero would have a negative cost, so buying the portfolio would obtain an arbitrage profit.

Theorem 1 (Law of One Price) *The law of one price holds if and only if there exists a stochastic discount factor.*

Thus the existence of a stochastic discount factor is a necessary condition for market equilibrium.

---

[1] In the state space, the Jordan product $\boldsymbol{y}^* \bullet \boldsymbol{y} = \sum_i \sqrt{\pi_i} y_i^* y_i \boldsymbol{e}_i$. The vectors $\boldsymbol{e}_i / \sqrt{\pi_i}$ are the primitive idempotents. A payoff $\boldsymbol{y}$ has eigenvalues $\sqrt{\pi_i} y_i$. The Jordan unit $\sum_i \boldsymbol{e}_i / \sqrt{\pi_i}$ is not the mean vector $\mathbf{1}$. The trace inner product $\langle \boldsymbol{y}^*, \boldsymbol{y} \rangle := \mathrm{tr}\,(\boldsymbol{y}^* \bullet \boldsymbol{y}) = \sum_i \pi_i y_i^* y_i$, the second noncentral moment of the two payoffs.

This theorem follows from the Fredholm alternative. The law of one price states that no portfolio satisfies the conditions

$$\begin{aligned} Ax &= \mathbf{0} \\ \langle v^*, x \rangle &\neq 0. \end{aligned}$$

The alternative states that

$$A^{\mathsf{T}} y^* = v^*$$

has a solution y^*, a stochastic discount factor.

59.3 Fundamental Theorem of Finance

The fundamental condition for asset-market equilibrium is that there should be no opportunity for arbitrage profit (Ross 1978). An arbitrage opportunity exists if some portfolio has a negative cost and a nonnegative payoff, or a zero cost and a payoff greater than zero.

Theorem 2 (Fundamental Theorem of Finance) *There is no opportunity for arbitrage if and only if there exists a positive stochastic discount factor*

The name "fundamental theorem of finance" was coined by Dybvig and Ross (1987).

There is no arbitrage opportunity if and only if there exists a cost/payoff efficient portfolio. Efficiency duality says that for some $y^* \gg \mathbf{0}$, the maximum

$$\max_x \left(-\langle v^*, x \rangle + \langle y^*, Ax \rangle\right)$$

is attained for some portfolio x. By the first-order condition, there exists a solution to

$$v^* = A^{\mathsf{T}} y^* \qquad (59.1)$$

$$y^* \gg \mathbf{0}, \qquad (59.2)$$

which proves the theorem.

The Stiemke alternative to Eqs. (59.1)–(59.2) provides insight, which states that

$$\begin{bmatrix} A \\ -v^{*\mathsf{T}} \end{bmatrix} x > \mathbf{0}$$

has a solution. A solution provides an arbitrage profit: the payoff is greater than zero and cost is less than or equal to zero, or the payoff is zero and cost is negative.

59.4 Cost/Mean/Variance Efficient Portfolio

Much finance theory analyses choice under uncertainty in terms of a tradeoff between mean and variance. A consumer dislikes risk, so higher variance is bad, while a higher mean is good. We study the existence of a cost/mean/variance efficient portfolio via efficiency duality. The variance $A^{\top}\left(\mathrm{i}-\mathbf{1}\mathbf{1}^{\top}\right)A$ of the payoff on portfolio $\boldsymbol{x}$ is the second noncentral moment $\langle A\boldsymbol{x}, A\boldsymbol{x}\rangle$ less the mean squared, $\left\langle \mathbf{1}^{\top}A\boldsymbol{x}, \mathbf{1}^{\top}A\boldsymbol{x}\right\rangle$. Efficiency duality says that for some $v^* > 0$ and $m^* > 0$, the maximum

$$\max_{\boldsymbol{x}}\left[-v^*\langle v^*,\boldsymbol{x}\rangle + m^*\mathbf{1}^{\top}A\boldsymbol{x} - \frac{1}{2}\left\langle \boldsymbol{x}, A^{\top}\left(\mathrm{I}-\mathbf{1}\mathbf{1}^{\top}\right)A\boldsymbol{x}\right\rangle\right]$$

is attained for some portfolio $\boldsymbol{x}$. By the first-order condition, there exists a solution to

$$\begin{aligned}\begin{bmatrix} -v^* & A^{\top}\mathbf{1}\end{bmatrix}\begin{bmatrix} v^* \\ m^*\end{bmatrix} - A^{\top}\left(\mathrm{I}-\mathbf{1}\mathbf{1}^{\top}\right)A\boldsymbol{x} &= \mathbf{0}\\ \begin{bmatrix} v^* \\ m^*\end{bmatrix} &\gg \mathbf{0},\end{aligned}$$

which falls within the generalized Stiemke alternative.

If there is no solution, then there exists a solution to the alternative

$$\begin{aligned} A^{\top}\left(\mathrm{I}-\mathbf{1}\mathbf{1}^{\top}\right)A\boldsymbol{x} &= \mathbf{0}\\ \begin{bmatrix} -v^{*\top} \\ \mathbf{1}^{\top}A\end{bmatrix}\boldsymbol{x} &> \mathbf{0}.\end{aligned}$$

The equation requires that the portfolio $\boldsymbol{x}$ is risk-free, so the mean $\mathbf{1}^{\top}A\boldsymbol{x}$ is the constant payoff. The inequality then states that there exists a risk-free portfolio providing an arbitrage opportunity: either the cost is negative and the payoff is nonnegative, or the cost is zero and the payoff is positive.

Theorem 3 (Existence of a Cost/Mean/Variance Efficient Portfolio) *There exists a cost/mean/variance efficient portfolio if and only if there is no risk-free portfolio providing an arbitrage opportunity.*

If there is no risk-free asset, then necessarily there exists a cost/mean/variance efficient portfolio.

Let us compare these three cases. The law of one price is necessary but not sufficient for the existence of a cost/mean/variance efficient portfolio. The law of one price permits a risk-free portfolio having a positive constant payoff to have a zero or negative cost.

The existence of a cost/mean/variance efficient portfolio rules out a risk-free portfolio providing an arbitrage opportunity but permits a risky portfolio providing an arbitrage opportunity. If there is no risk-free portfolio, then necessarily there exists a cost/mean/variance efficient portfolio, but there may be an opportunity for arbitrage.

59.5 Pricing Payoff

The payoff space $\mathrm{R}(A)$ is a moment space valuable in economic analysis. Because the law of one price holds, one identifies two assets having the same payoff as the same. Define the subspace $M := \mathrm{R}(A)$, and

$$M = AA^{+} = \left(AA^{+}\right)^{\top} = A^{+\top}A^{\top} = A^{\top+}A^{\top}.$$

is the corresponding idempotent. The mean vector in the payoff space is $\boldsymbol{m}^{*} = M\mathbf{1}$, which is the fitted value in the least-squares linear regression of $\mathbf{1}$ on $A^{\top}$.

We refer to $\boldsymbol{p}^{*} = A^{\top+}\boldsymbol{v}^{*}$ as the "pricing payoff," the unique stochastic discount factor in the payoff space.

Theorem 4 (Pricing Payoff) *If the law of one price holds, then there exists a unique payoff* $\boldsymbol{p}^{*} \in \mathrm{R}(A)$ *such that the cost of a portfolio having payoff* $\boldsymbol{p} \in \mathrm{R}(A)$ *is* $\langle \boldsymbol{p}^{*}, \boldsymbol{p} \rangle$.

The set of all stochastic discount factors is the solution set to

$$A^{\top}\boldsymbol{y}^{*} = \boldsymbol{v}^{*}.$$

There is a solution if and only if the law of one price holds. The solution set is then

$$\left\{A^{\top+}\boldsymbol{v}^{*}\right\} + \mathrm{N}\left(A^{\top}\right) = \left\{A^{\top+}\boldsymbol{v}^{*}\right\} + M^{\perp},$$

by the fundamental theorem of linear algebra, which proves the theorem.

Let us verify directly the formula $\boldsymbol{p}^{*} = A^{\top+}\boldsymbol{v}^{*}$. The law of one price means $\boldsymbol{v}^{*} \in \mathrm{R}\left(A^{\top}\right)$; equivalently, $\boldsymbol{v}^{*} = A^{\top}A^{\top+}\boldsymbol{v}^{*}$. The cost of portfolio $\boldsymbol{x}$

$$\langle \boldsymbol{v}^{*}, \boldsymbol{x} \rangle = \left\langle A^{\top}A^{\top+}\boldsymbol{v}^{*}, \boldsymbol{x} \right\rangle = \left\langle A^{\top+}\boldsymbol{v}^{*}, A\boldsymbol{x} \right\rangle = \langle \boldsymbol{p}^{*}, A\boldsymbol{x} \rangle,$$

the inner product of $\boldsymbol{p}^*$ and the portfolio payoff. Here $\boldsymbol{p}^* \in M$:

$$M\boldsymbol{p}^* = MA^{\mathsf{T}+}\boldsymbol{v}^* = \left(A^{\mathsf{T}+}A^{\mathsf{T}}\right)A^{\mathsf{T}+}\boldsymbol{v}^* = A^{\mathsf{T}+}\boldsymbol{v}^* = \boldsymbol{p}^*.$$

The support to the set of stochastic discount factors is the cost of a portfolio having payoff $\boldsymbol{y}$:

$$\begin{aligned}
&\delta^*_{\{\boldsymbol{p}^*\}+M^\perp}(\boldsymbol{y}) \\
&= \sup_{\boldsymbol{y}^*}\left[\langle \boldsymbol{y}^*, \boldsymbol{y}\rangle - \delta_{\{\boldsymbol{p}^*\}+M^\perp}(\boldsymbol{y}^*)\right] \\
&= \sup_{\boldsymbol{z}^*}\left[\langle \boldsymbol{p}^* + \boldsymbol{z}^*, \boldsymbol{y}\rangle - \delta_{M^\perp}(\boldsymbol{z}^*)\right] \;\; (\boldsymbol{z}^* := \boldsymbol{y}^* - \boldsymbol{p}^*) \\
&= \langle \boldsymbol{p}^*, \boldsymbol{y}\rangle + \delta_M(\boldsymbol{y}).
\end{aligned}$$

The cost is $\langle \boldsymbol{p}^*, \boldsymbol{y}\rangle$ if $\boldsymbol{y}$ is in the payoff space, and ∞ otherwise. In the payoff space, the cost is a linear function, so the pricing-payoff theorem is an instance of the Riesz-Fréchet theorem (Chamberlain and Rothschild 1983, LeRoy and Werner 2001, Chap. 17).

59.6 Cost and Mean Positivity

The inner product of the pricing payoff and the mean vector in the payoff space is

$$\langle \boldsymbol{p}^*, \boldsymbol{m}^*\rangle = \left\langle A^{\mathsf{T}+}\boldsymbol{v}^*, M\mathbf{1}\right\rangle = \left\langle A^{\mathsf{T}+}\boldsymbol{v}^*, \mathbf{1}\right\rangle = \left\langle \boldsymbol{v}^*, A^{+}\mathbf{1}\right\rangle.$$

In general, this inner product can be either positive or negative.

Definition 2 (Cost and Mean Positivity) The cost of $\boldsymbol{m}^*$ is positive and the mean of $\boldsymbol{p}^*$ is positive: $\langle \boldsymbol{p}^*, \boldsymbol{m}^*\rangle = \left\langle \boldsymbol{v}^*, A^{+}\mathbf{1}\right\rangle > 0$.

Cost and mean positivity and the absence of any arbitrage opportunity are distinct conditions; neither implies the other. Findings in subsequent chapters demonstrate that one should view cost and mean positivity as an essential condition for asset-market equilibrium.

If there is a positive stochastic discount factor, it is not necessary that $\boldsymbol{p}^* \gg \mathbf{0}$; perhaps no positive stochastic discount factor is a payoff. Figure 59.1 illustrates an example. There are two states of the world but only one asset.[2] The inner product of the payoff price $\boldsymbol{p}^*$ and the mean vector $\boldsymbol{m}^*$ is negative. The pricing payoff is $\boldsymbol{p}^*$, and the set of all stochastic discount factors is $\{\boldsymbol{p}^*\} + M^\perp$. The pricing payoff is the

[2] Markets are complete if the payoff space is the state space, $M = \mathbf{Y}$. With complete markets, absence of arbitrage implies price/mean positivity.

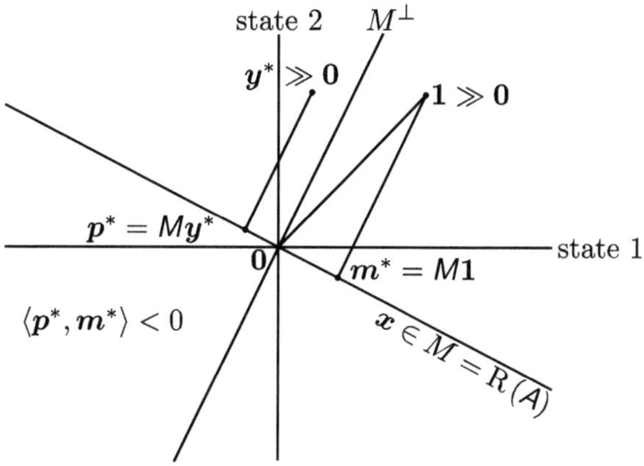

Fig. 59.1 Positivity versus no arbitrage

projection of the set of all stochastic discount factors onto M. A positive stochastic discount factor is $\boldsymbol{y}^*$, so arbitrage is not possible. The pricing payoff $\boldsymbol{p}^* = M\boldsymbol{y}^*$. The mean vector for the state space is $\mathbf{1}$, and the mean vector for the payoff space is $\boldsymbol{m}^* = M\mathbf{1}$. Here $\langle \boldsymbol{p}^*, \boldsymbol{m}^* \rangle < 0$, even though $\langle \boldsymbol{y}^*, \mathbf{1} \rangle > 0$.

References

Chamberlain G, Rothschild M (1983) Arbitrage, factor structure and mean variance analysis in large asset markets. Econometrica 51:1281–1304

Dybvig P, Ross SA (1987) Arbitrage. In: Milgate M, Eatwell J, Newman P (eds) The new Palgrave: a dictionary of economics. McMillan

LeRoy SF, Werner J (2001) Principles of financial economics. Cambridge University Press, Cambridge

Ross SA (1978) A simple approach to the valuation of risky streams. J Bus 51:453–475

Chapter 60
Expected Utility

Abstract We analyze portfolio choice by an investor who maximizes the expected utility of the payoff. The setting is state space, in which the inner product is the second non-central moment. The conjugate of the expected utility of a payoff is the expected value of the conjugate of utility evaluated at the marginal utility. The stochastic discount factors set the budget constraint. The analysis parallels and generalizes the standard theory of a consumer maximizing utility subject to a budget constraint, except that the budget constraint involves the set of stochastic discount factors, not just a single price vector. The dual chooses a particular stochastic discount factor. The pricing payoff plays no special role in expected-utility maximization. We analyze both constant relative risk aversion and constant absolute risk aversion.

60.1 von Neumann-Morgenstern Expected Utility

The von Neumann-Morgenstern utility function f is the utility from a dollar payoff.

Condition 1 (von Neumann-Morgenstern Utility) *Utility f is a proper, closed, concave function, increasing on its effective domain.*

As the payoff increases, utility increases and marginal utility (a subgradient) decreases.

We impose conditions on the effective domain of utility and its conjugate.

Condition 2 (Effective Domain of Utility) *The effective domain* $\operatorname{dom} f$ *contains* $\mathbf{R}_{++}$.

In applications, economists commonly take $\operatorname{dom} f$ to be $\mathbf{R}_{++}$, $\mathbf{R}_{+}$, or $\mathbf{R}$.

The conjugate f_* is a function of marginal utility. By Conditions 1 and 2, the conjugate is also increasing, because utility is increasing. Economic applications typically add the following condition.

B. C. Dieffenbach, *Conjugate Duality in Economic Analysis*, Contributions to Economics,
https://doi.org/10.1007/978-3-032-21396-9_60

Condition 3 (Effective Domain of the Conjugate of Utility) *The effective domain* dom f_* *is unbounded above.*

For any marginal utility sufficiently high, there is some payoff having this marginal utility.

To set the problem in state space **X** is simplifying. A payoff vector $\boldsymbol{x}$ shows the payoff x_i in the states i. The inner product of two payoff vectors is the product moment, the second non-central moment. The probability of state i is $\pi_i > 0$. Let the vector $\boldsymbol{e}_i$ denote payoff one in state i and payoff zero otherwise. Together the $\boldsymbol{e}_i$ form a natural basis for the state space. As $\langle \boldsymbol{e}_i, \boldsymbol{e}_j \rangle = \pi_i \delta_{ij}$, the basis is orthogonal but not orthonormal. The payoff $\boldsymbol{x} = \sum_i x_i \boldsymbol{e}_i$ has payoff x_i in state i, and $x_i = \boldsymbol{e}_i^\top \boldsymbol{x} / \pi_i$.

Theorem 1 (Conjugate of Expected Utility) *The conjugate of expected utility is the expected value of the conjugate of utility.*

Setting the problem in state space yields this simple relationship. The expected utility of payoff $\boldsymbol{x}$ is

$$\sum_i \pi_i f(x_i) = \sum_i \pi_i f\left(\boldsymbol{e}_i^\top \boldsymbol{x} / \pi_i\right).$$

The conjugate

$$\begin{aligned} &\inf_{\boldsymbol{x}} \left[\langle \boldsymbol{x}^*, \boldsymbol{x} \rangle - \sum_i \pi_i f\left(\boldsymbol{e}_i^\top \boldsymbol{x} / \pi_i\right) \right] \\ &= \sum_i \left\{ \pi_i \inf_{x_i} [x_i^* x_i - f(x_i)] \right\} \\ &= \sum_i \pi_i f_*(x_i^*) . \end{aligned}$$

By this derivation, it is evident that the Fenchel equality

$$\sum_i \pi_i f\left(\boldsymbol{e}_i^\top \boldsymbol{x} / \pi_i\right) + \sum_i \pi_i f_*\left(\boldsymbol{e}_i^\top \boldsymbol{x}^* / \pi_i\right) - \langle \boldsymbol{x}^*, \boldsymbol{x} \rangle = 0$$

is equivalent to the Fenchel equalities

$$f(x_i) + f_*(x_i^*) - x_i^* x_i = 0 \text{ for all } i. \tag{60.1}$$

The subdifferential interpretation $x_i^* \in \partial f(x_i)$ states that x_i^* is marginal utility for payoff x_i.

60.2 Budget Constraint

We express the budget constraint in terms of the convex set D of all stochastic discount factors $\boldsymbol{x}^* = \sum_i x_i^* \boldsymbol{e}_i$. Interpret component x_i^* as the cost of payoff $\boldsymbol{e}_i$ per unit of probability. The law of one price says that D is not empty. There are a finite number of assets, so D is polyhedral.

We stipulate conditions on D.

Condition 4 (No Arbitrage) *There is no opportunity for arbitrage.*

By the fundamental theorem of finance, this condition is equivalent to the existence of a positive stochastic discount factor.

Condition 5 (Positive Payoff) *There exists some portfolio having a positive payoff.*

By the no-arbitrage condition, this portfolio must have a positive cost. Maximum expected utility is not $-\infty$: to invest in this portfolio obtains finite expected utility. The barrier cone dom δ_D^* is the payoff space. There can be no positive vector in its polar, the horizon cone D^∞, by the positive payoff condition.

The amount to invest is wealth $z > 0$ dollars. The cost of payoff $\boldsymbol{x}$ is $\delta_D^*(\boldsymbol{x})$, so the budget constraint is

$$\delta_{\delta_D^*(x)\leq z} = \delta_{\text{epi } \delta_D^*}(\boldsymbol{x}, z) = \delta^*_{\text{cone } D}(\boldsymbol{x}, z).$$

Below we work with the cone representation of the budget constraint. The Fenchel equality for the budget constraint is

$$\delta_{\text{cone } D}(y^*, -z^*) + \delta^*_{\text{cone } D}(\boldsymbol{x}, z) - \langle (y^*, -z^*), (\boldsymbol{x}, z)\rangle = 0. \tag{60.2}$$

For $z^* > 0$, this Fenchel equality is equivalent to

$$\delta_D(y^*/z^*) + \delta_D^*(\boldsymbol{x}) - \langle y^*/z^*, \boldsymbol{x}\rangle = 0 \tag{60.3}$$

$$\delta_D^*(\boldsymbol{x}) = z. \tag{60.4}$$

60.3 Expected-Utility Maximization

The expected-utility maximization problem parallels standard consumer choice: maximize expected utility subject to the budget constraint.

Problem 1 (Expected-Utility Maximization)

$$\sup_{x} \left[\sum_i \pi_i f\left(\boldsymbol{e}_i^\top \boldsymbol{x}/\pi_i\right) - \delta^*_{\text{cone } D}(\boldsymbol{x}, z) \right].$$

Compared to standard consumer utility maximization, that the utility here is additive across states is simpler. On the other hand, the budget constraint is more complex, being the support to the cone representation of the set of stochastic discount factors, rather than just the support to the cone representation of the price vector. If asset markets are complete, then D consists of a unique stochastic discount factor—the pricing payoff—and the budget constraint reduces to the case for standard consumer choice.

We calculate the reciprocal dual. Adding the Fenchel inequalities

$$-\sum_i \pi_i f\left(e_i^\top x/\pi_i\right) - \sum_i \pi_i f_*\left[e_i^\top\left(y^* - x^*\right)/\pi_i\right] + \langle y^* - x^*, x\rangle \geq 0$$
$$\delta_{\text{cone } D}\left(y^*, -z^*\right) + \delta^*_{\text{cone } D}\left(x - y, z\right) - \langle y^*, x - y\rangle - \left(-z^*\right) z \geq 0$$

and rearranging obtains

$$\begin{aligned}\langle y^*, y\rangle + z^* z - \textstyle\sum_i \pi_i f_*\left[e_i^\top\left(y^* - x^*\right)/\pi_i\right] + \delta_{\text{cone } D}\left(y^*, -z^*\right) \\ \geq \langle x^*, x\rangle + \textstyle\sum_i \pi_i f\left(e_i^\top x/\pi_i\right) - \delta^*_{\text{cone } D}\left(x - y, z\right).\end{aligned}$$

The perturbation base values are $x^* = \mathbf{0}$, $y = \mathbf{0}$, and z.

Problem 2 (Reciprocal Dual)

$$\inf_{y^*, z^*}\left[z^* z - \sum_i \pi_i f_*\left(e_i^\top y^*/\pi_i\right) + \delta_{\text{cone } D}\left(y^*, -z^*\right)\right].$$

Minimize $z^* z$ less the expected value of the conjugate of utility, such that y^*/z^* is a stochastic discount factor.

Both the primal and the dual satisfy the polyhedral/relative condition. For the primal, that $z > 0$ implies that some $x \gg \mathbf{0}$ satisfies the budget constraint, so the polyhedral/relative condition holds. For the dual, that there exists a positive stochastic discount factor means that $y^*/z^* \gg \mathbf{0}$ for some $\left(y^*, -z^*\right) \in$ cone D. By Condition 3, choosing z^* small enough shows that the polyhedral/relative condition holds for the dual. There is no duality gap, and the finite optimum is attained in both the primal and the dual.

By the utility Fenchel equalities (60.1), y^* is the vector of marginal utilities in each state. The marginal utility of wealth $z^* > 0$.[1] The identity

$$y^* = z^*\left(y^*/z^*\right)$$

[1] Our stipulated conditions rule out the exceptional case $z^* = 0$. For $z^* = 0$, the budget-constraint Fenchel equality (60.2) is equivalent to

$$\delta_{D^\infty}\left(y^*\right) + \delta^*_{D^\infty}\left(x\right) - \langle y^*, x\rangle = 0$$
$$\delta^*_D\left(x\right) \leq z.$$

Because utility is increasing in wealth, the utility Fenchel equalities (60.1) imply that y^* is positive. However the positive-payoff condition implies that D^∞ contains no positive vector, a contradiction.

says that the marginal utility vector equals the marginal utility of wealth times the price vector, the standard requirement for consumer choice.

The pricing payoff plays no special role in expected-utility maximization. Even though D is the sum of the pricing payoff singleton and the orthogonal complement of the payoff space, the analysis here makes no use of this simple relationship. Whereas the payoff $\boldsymbol{x}$ must lie in the payoff space, in general neither the utility derived from it nor the stochastic discount factor $\boldsymbol{y}^*/z^*$ lies there. In general, different investors have different marginal utility vectors $\boldsymbol{y}^*$.

60.4 Constant Relative Risk Aversion

The relative risk aversion $-xf''(x)/f(x)$ measures the willingness to take a risk of a given size relative to wealth. We analyze the case of constant relative risk aversion (independent of wealth). For constant relative risk aversion s, positive but not one, utility is

$$\frac{x^{1-s}}{1-s},$$

in which the effective domain is $\mathbf{R}_+$ if $s < 1$ and $\mathbf{R}_{++}$ if $s > 1$. For $s = 1$, utility is logarithmic,

$$\frac{1}{2} + \ln x,$$

in which the effective domain is $\mathbf{R}_{++}$. Taking the derivative obtains the subdifferential interpretation of (60.3):

$$y_i^* = x_i^{-s}. \tag{60.5}$$

The utility is strictly concave and differentiable, so the optimum payoff is unique. When the relative risk aversion is constant (independent of wealth), then the optimum payoff is proportional to wealth.

The optimum payoff depends on D. Without knowing D, one can nevertheless show how the optimum payoff changes as wealth changes: for constant relative risk aversion, the optimum payoff is proportional to wealth.

Theorem 2 (Constant Relative Risk Aversion and Wealth) *Assume constant relative risk aversion $s > 0$. If $(\boldsymbol{x}, \boldsymbol{y}^*, z^*)$ is a solution for wealth z, then for any $\lambda > 0$, the combination $(\lambda\boldsymbol{x}, \boldsymbol{y}^*/\lambda^s, z^*/\lambda^s)$ is a solution for wealth λz.*

As wealth changes, the payoff changes in proportion, and the stochastic discount factor stays unchanged.

The proof is to verify that the Fenchel equalities necessary and sufficient for a solution continue to hold: if they hold for $(\boldsymbol{x}, \boldsymbol{y}^*, z^*)$ for wealth z, then they also hold for $(\lambda \boldsymbol{x}, \boldsymbol{y}^*/\lambda^s, z^*/\lambda^s)$ for wealth λz. By Eq. (60.5),

$$\boldsymbol{y}^*/\lambda^s = (\lambda x_i)^{-s}$$

follows. The budget-constraint conditions (60.3)–(60.4) imply

$$\begin{aligned}\delta_D\left[(\boldsymbol{y}^*/\lambda^s)/(z^*/\lambda^s)\right] + \delta_D^*(\lambda\boldsymbol{x}) - \langle(\boldsymbol{y}^*/\lambda^s)/(z^*/\lambda^s), \lambda\boldsymbol{x}\rangle = 0 \\ \delta_D^*(\lambda\boldsymbol{x}) = \lambda z,\end{aligned}$$

as desired.

60.5 Constant Absolute Risk Aversion

The absolute risk aversion $-f''(x)/f(x)$ measures the willingness to take a risk of a given absolute size. We analyze the case of constant absolute risk aversion (independent of wealth). Utility is

$$-\mathrm{e}^{-ax}/a$$

for constant absolute risk aversion $a > 0$; the effective domain is $\mathbf{R}$. Taking the derivative obtains the subdifferential interpretation of (60.3):

$$y_i^* = \mathrm{e}^{-ax_i}. \tag{60.6}$$

Assume that there exists a risk-free asset: the payoff $\mathbf{1} := \sum_i \boldsymbol{e}_i$ belongs to the payoff space. Its cost is $\delta_D^*(\mathbf{1})$, and $\delta_D^*(\boldsymbol{x} + \lambda\mathbf{1}) = \delta_D^*(\boldsymbol{x}) + \lambda\delta_D^*(\mathbf{1})$ for any λ.

For constant absolute risk aversion, the optimum is to choose a fixed amount of risk, independent of wealth. The optimum choice is a fixed risky payoff, plus investment or borrowing in the risk-free asset.

Theorem 3 (Constant Absolute Risk Aversion and Wealth) *Assume constant absolute risk aversion $a > 0$. Suppose that $(\boldsymbol{x}, \boldsymbol{y}^*, z^*)$ is a solution for wealth z. Change wealth by Δz, such that $z + \Delta z > 0$. Then the change*

$$\begin{aligned}\Delta x_i &= \Delta z/\delta_D^*(\mathbf{1}) \textit{ for all } i \\ \Delta \ln y_i^* &= -a\Delta z/\delta_D^*(\mathbf{1}) \textit{ for all } i \\ \Delta \ln z^* &= -a\Delta z/\delta_D^*(\mathbf{1})\end{aligned}$$

is a solution.

The payoff in every state changes by the same amount, as the extra wealth Δz is invested entirely in the risk-free asset.

The proof is to verify that the Fenchel equalities necessary and sufficient for a solution continue to hold. As

$$\ln y_i^* + \Delta \ln y_i^* = -a\,[x_i + \Delta z/\delta_D^*(\mathbf{1})] = -a\,(x_i + \Delta x_i),$$

therefore (60.6) holds.

The budget-constraint conditions (60.3)–(60.4) continue to hold. The stochastic discount factor is unchanged:

$$\begin{aligned}(\ln y_i^* + \Delta \ln y_i^*) - (\ln z^* + \Delta \ln z^*) \\ &= [\ln y_i^* - a\Delta z/\delta_D^*(\mathbf{1})] - [\ln z^* - a\Delta z/\delta_D^*(\mathbf{1})] \\ &= \ln y_i^* - \ln z^*.\end{aligned}$$

The cost of the payoff is

$$\delta_D^*(\boldsymbol{x} + \Delta\boldsymbol{x}) = \delta_D^*\{(\boldsymbol{x} + [\Delta z/\delta_D^*(\mathbf{1})]\,\mathbf{1})\} = \delta_D^*(\boldsymbol{x}) + \Delta z = z + \Delta z.$$

And

$$\begin{aligned}\langle(\boldsymbol{y}^* + \Delta\boldsymbol{y}^*)/(z^* + \Delta z^*), \boldsymbol{x} + \Delta\boldsymbol{x}\rangle &= \langle\boldsymbol{y}^*/z^*, \boldsymbol{x} + \Delta\boldsymbol{x}\rangle \\ &= z + \langle\boldsymbol{y}^*/z^*, \Delta\boldsymbol{x}\rangle \\ &= z + \langle\boldsymbol{y}^*/z^*, [\Delta z/\delta_D^*(\mathbf{1})]\,\mathbf{1}\rangle \\ &= z + \langle\boldsymbol{y}^*/z^*, \mathbf{1}/\delta_D^*(\mathbf{1})\rangle\,\Delta z \\ &= z + \Delta z,\end{aligned}$$

as desired.

Chapter 61
Price Bounds

Abstract The fundamental theorem of finance asserts that there is no arbitrage if and only if there exists a positive stochastic discount factor. As a corollary, one obtains upper and lower price bounds for an asset having an arbitrary payoff. Given asset payoffs and prices not allowing any arbitrage, there exists a non-empty set of positive stochastic discount factors. For an extra asset having an arbitrary payoff, to rule out arbitrage its price must be the value of the payoff discounted by one of the positive stochastic discount factors. The supremum of these prices sets an upper bound for the price, and the infimum of these prices sets a lower bound. Interpreting the bounds, the "upper-hedging" and "lower-hedging" duals furnish alternative expressions for the bounds. The upper bound is the minimum price among all larger payoffs in the payoff space, whereas the lower bound is the maximum price among smaller payoffs.

61.1 Payoffs and Pricing

The state space $\mathbf{X}$ is a Euclidean Jordan algebra, a moment space in which a vector $\boldsymbol{x}$ specifies a payoff x_i in state i. Each state occurs with positive probability; because the concern here is arbitrage, the exact value of the probability is irrelevant. The inner product of two payoffs is their second noncentral moment. The algebra has the natural partial ordering.

Let $M \subset \mathbf{X}$ denote the payoff space. Assume that the law of one price holds, so there exists a stochastic discount factor. The polyhedral set of all stochastic discount factors is $D = \{\boldsymbol{x}^*\} + M^{\perp}$, in which $\boldsymbol{x}^*$ is the pricing payoff, which need not be positive. The cost of a payoff $\boldsymbol{x}$ is $\delta_D^* (\boldsymbol{x}) = \langle \boldsymbol{x}^*, \boldsymbol{x} \rangle + \delta_M (\boldsymbol{x})$.

Definition 1 (Pricing by a Stochastic Discount Factor) A payoff $\boldsymbol{y}$ is priced by the stochastic discount factor $\boldsymbol{y}^*$ if its price z is the discounted value $\langle \boldsymbol{y}^*, \boldsymbol{y} \rangle$ of the payoff.

B. C. Dieffenbach, *Conjugate Duality in Economic Analysis*, Contributions to Economics,
https://doi.org/10.1007/978-3-032-21396-9_61

61.2 Upper and Lower Price Bounds

Let us assume that there is no arbitrage—there exists a positive stochastic discount factor. Consider an extra asset having payoff $\boldsymbol{y}$, perhaps in the payoff space, perhaps not. For this payoff, what prices z imply no arbitrage? As a corollary of the fundamental theorem of finance, to rule out arbitrage its price must be the value of the payoff discounted by one of the positive stochastic discount factors.

Summary 1 (No Arbitrage) For an asset with payoff $\boldsymbol{y}$ having price z, there is no arbitrage if and only if $z = \langle \boldsymbol{y}^*, \boldsymbol{y} \rangle$ for some positive stochastic discount factor $\boldsymbol{y}^*$.

Convex optimizations set upper and lower price bounds.

Definition 2 (Upper and Lower Price Bounds) For an asset with payoff $\boldsymbol{y}$, the upper and lower bounds for the price are

$$\bar{z} = \sup_{y^*} \left[\langle y^*, y \rangle - \delta_D (y^*) - \delta_{y^* \gg 0} \right] \tag{61.1}$$

$$\underline{z} = \inf_{y^*} \left[\langle y^*, y \rangle + \delta_D (y^*) + \delta_{y^* \gg 0} \right]. \tag{61.2}$$

The upper bound is the maximum discounted payoff such that the stochastic discount factor is positive, whereas the lower bound is the minimum discounted payoff. The upper bound is greater than or equal to the lower bound. The upper bound can be ∞, and the lower bound can be $-\infty$. If $\boldsymbol{x}^*$ is positive, then the upper and lower bounds are finite.

The fundamental theorem of finance implies the following. If $z > \bar{z}$, an arbitrage opportunity exists, because the asset is not priced by any positive stochastic discount factor. The asset is overpriced. One attains arbitrage by selling the asset short and buying some other asset. Analogously, if $z < \underline{z}$, the asset is underpriced. Buy the asset and sell some other asset short. If $\underline{z} < z < \bar{z}$, there is no arbitrage, because the asset is priced by some positive stochastic discount factor.

At the price bound, the following applies. If $z = \bar{z}$, arbitrage is possible if the maximum in (61.1) is not attained but is only approached as an infimum. If the maximum is attained, then arbitrage is not possible, because the asset is priced by some positive stochastic discount factor. The analogous situation holds for $z = \underline{z}$.

61.3 Upper-Hedging Dual

Perturbation duality interprets the price bounds. Whether there is arbitrage at the price bound depends on whether $\boldsymbol{y}$ belongs to the payoff space. If $\boldsymbol{y} \in M$, then the law of one price sets a unique price, so $\underline{z} = \bar{z}$; there is no arbitrage at the bounds. If $\boldsymbol{y} \notin M$, then $\underline{z} < \bar{z}$, and there is arbitrage at either bound.

We analyze the following primal: relaxing the constraint in (61.1) to allow a nonnegative stochastic discount factor that is not positive leaves the supremum unchanged.

Problem 1 (Maximum-Price Primal)

$$\bar{z} = \sup_{y^*}\left[\langle y^*, y\rangle - \delta_D(y^*) - \delta_{y^*\geq 0}\right] = \sup_{y^*}\left[\langle y^*, y\rangle - \delta_M^*(y^* - x^*) - \delta_{y^*\geq 0}\right].$$

We analyze the Fenchel dual. The primal and the dual are polyhedral. Because there exists a positive stochastic discount factor, the primal satisfies the polyhedral condition, so there is no duality gap. The optimum value is attained in both the primal and the dual, if the value is finite.

The Fenchel dual is the upper-hedging dual. Adding the Fenchel inequalities

$$\delta_D(y^*) + \delta_D^*(x) - \langle y^*, x\rangle \geq 0 \tag{61.3}$$

$$\delta_{\mathrm{X}_+}(y^*) + \delta_{\mathrm{X}_+}^*(y - x) - \langle y^*, y - x\rangle \geq 0 \tag{61.4}$$

and rearranging obtains

$$\delta_D^*(x) + \delta_{\mathrm{X}_+}^*(y - x) \geq \langle y^*, y\rangle - \delta_D(y^*) - \delta_{\mathrm{X}_+}(y^*).$$

Problem 2 (Upper-Hedging Dual)

$$\bar{z} = \inf_{x}\left[\delta_D^*(x) + \delta_{x\succeq y}\right] = \inf_{x}\left[\langle x^*, x\rangle + \delta_M(x) + \delta_{x\succeq y}\right].$$

The maximum price is the minimum cost among all payoffs x greater than or equal to y. A solution x is an *upper-hedging payoff.*

For the Fenchel factor (61.3) to be zero says that y^* is a stochastic discount factor and that x lies in the payoff space. If x were not in the payoff space, its cost would be ∞. As x lies in the payoff space, its cost is $\langle y^*, x\rangle$, for any stochastic discount factor.

For the Fenchel factor (61.4) to be zero is complementary slackness. If y^* solves the primal and x solves the dual, then $\langle y^*, y - x\rangle = 0$. If there exists a solution $y^* \gg \mathbf{0}$ to the primal, then $x = y$ is the unique solution to the dual. Conversely, if there exists a solution $x \succ y$ to the dual, then there is no solution $y^* \gg \mathbf{0}$ to the primal.

If $y \notin M$, then there is arbitrage, achieved by the solution x to the dual. There can be no solution $x = y$ to the dual, so there is no solution $y^* \gg \mathbf{0}$ to the primal; there is arbitrage. A solution x to the dual achieves arbitrage: $x \succ y$ but costs the same as y. Buy x and short sell y: the combined cost is zero, but the combined payoff $x - y \succ \mathbf{0}$.

If $y \in M$, then there is no arbitrage. The discounted value $\langle y^*, y\rangle$ is the same for any y^*, and $\bar{z}$ is this value. By assumption there is some positive discount factor, so there is no arbitrage. By the dual, no $x \succeq y$ costs less than $\bar{z}$. Furthermore, no $x \succ y$ costs $\bar{z}$, because then by complementary slackness no $y^* \gg \mathbf{0}$ solves the primal, a contradiction.

61.4 Two-State Example

Figure 61.1 presents a case of two states. The pricing payoff $\boldsymbol{x}^* \gg \mathbf{o}$. All payoffs lie on M. There is no opportunity for arbitrage: moving right and up along M raises the payoff but also raises the cost.

The figure compares the upper-hedging dual and the maximum-price primal. The upper-hedging dual chooses $\boldsymbol{x} \succeq \boldsymbol{y}$ in M to minimize the cost $\langle \boldsymbol{x}^*, \boldsymbol{x} \rangle$. The minimum cost is achieved at $\overline{\boldsymbol{x}}$, as any alternative must be right and up in M and costs more.

The solution to the maximum-price primal is $\overline{\boldsymbol{y}^*}$, on the vertical axis. By the decomposition

$$\begin{aligned} \boldsymbol{y}^* &= (\boldsymbol{y}^* - \boldsymbol{x}^*) + \boldsymbol{x}^* \\ \boldsymbol{y} &= (\mathsf{I} - M)\boldsymbol{y} + M\boldsymbol{y}, \end{aligned}$$

the discounted value

$$\begin{aligned} \langle \boldsymbol{y}^*, \boldsymbol{y} \rangle &= \langle (\boldsymbol{y}^* - \boldsymbol{x}^*) + \boldsymbol{x}^*, (\mathsf{I} - M)\boldsymbol{y} + M\boldsymbol{y} \rangle \\ &= \langle (\boldsymbol{y}^* - \boldsymbol{x}^*), (\mathsf{I} - M)\boldsymbol{y} \rangle + \langle \boldsymbol{x}^*, M\boldsymbol{y} \rangle \end{aligned}$$

for any choice $\boldsymbol{y}^* - \boldsymbol{x}^* \in M^{\perp}$. As the second term is independent of $\boldsymbol{y}^*$, choose nonnegative $\boldsymbol{y}^*$ to maximize the first term. In the figure $(\mathsf{I} - M)\boldsymbol{y}$ points up and left, so choose $\boldsymbol{y}^*$ as the vertical intercept of D.

There is no duality gap. The maximum value $\left\langle \overline{\boldsymbol{y}^*}, \boldsymbol{y} \right\rangle$ of the maximum-price primal equals the minimum cost $\langle \boldsymbol{x}^*, \overline{\boldsymbol{x}} \rangle$ of the upper-hedging dual:

$$\left\langle \overline{\boldsymbol{y}^*}, \boldsymbol{y} \right\rangle = \left\langle \overline{\boldsymbol{y}^*}, \overline{\boldsymbol{x}} \right\rangle = \left\langle \overline{\boldsymbol{y}^*} - \boldsymbol{x}^*, \overline{\boldsymbol{x}} \right\rangle + \langle \boldsymbol{x}^*, \overline{\boldsymbol{x}} \rangle = \langle \boldsymbol{x}^*, \overline{\boldsymbol{x}} \rangle.$$

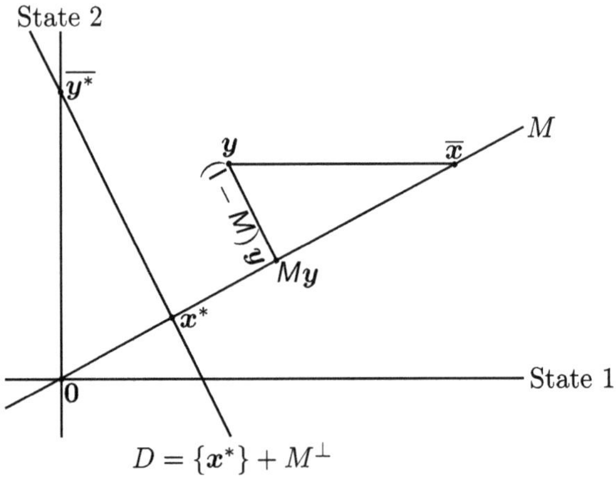

Fig. 61.1 Maximum price

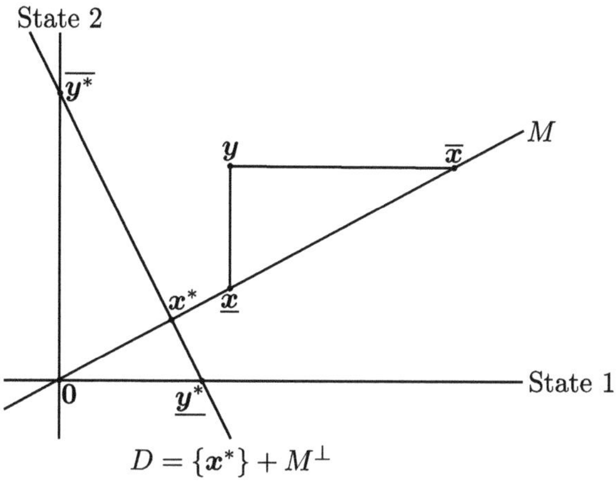

Fig. 61.2 Upper and lower bounds

The first equality is the complementary slackness: the choice $\overline{x} - y$ is zero for state two, whereas the choice $\overline{y^*}$ is zero for state one. The second equality is an identity. The third equality follows from the orthogonality of $\overline{x} \in M$ and $\overline{y^*} - x^* \in M^\perp$.

The minimum-price primal and its lower-hedging dual are analogous. The *lower-hedging dual* solves for the maximum cost among all payoffs x less than or equal to y. A solution x is a *lower-hedging payoff*.

Figure 61.2 portrays both the upper and lower price bounds. The payoff y lies above the payoff space M. For the upper price bound, the stochastic discount factor $\overline{y^*}$ solves the maximum-price primal and the payoff $\overline{x}$ solves the upper-hedging dual. Analogously, for the lower price bound, the stochastic discount factor $\underline{y^*}$ solves the minimum-price primal and the payoff $\underline{x}$ solves the lower-hedging dual.

If $y \in M$, then the upper and lower bounds are equal. At price $z = \overline{z} = \underline{z}$, no arbitrage is possible.

In the literature, price bounds have been studied in detail in the context of continuous-time portfolio choice, via Wiener-Brownian motion. See El Karoui and Quenez (1995), Föllmer and Schweizer (1991), Karatzas et al. (1991), Karatzas (1996), Karatzas and Shreve (1998), and (Roman 2007, Chap. 7).

References

El Karoui N, Quenez MC (1995 Jan) Dynamic programming and pricing of contingent claims in an incomplete market. SIAM J Contr Optim 33(1):29–66

Föllmer H, Schweizer M (1991) Hedging of contingent claims under incomplete information. In: Davis MHJ, Elliot RJ (eds) Applied stochastic analysis, Vol 5. Gordon and Breach, New York, pp 389–414

Karatzas I (1996) Lectures on the mathematics of finance, Vol 8. American Mathematical Society, Providence, RI
Karatzas I, Lehoczky J, Shreve S, Xu GL (1991) Martingale and duality methods for utility maximisation in an incomplete market. SIAM J Contr Optim 29:702–730
Karatzas I, Shreve SE (1998) Methods of mathematical finance. Springer, New York
Roman S (2007) Advanced linear algebra, 3rd edn, Vol 135. Springer

Chapter 62
Sharpe Ratio

Abstract The Sharpe ratio is the maximum value of the ratio of mean to standard deviation in a moment space of random variables. The Sharpe ratio for excess returns plays an important role in optimum portfolio choice and in the capital-asset pricing model (Sharpe, W. F. (1964, September). Capital asset prices: A theory of market equilibrium under conditions of risk. Journal of Finance, XIX (3), 425–442). We show that the Sharpe ratio is attained at the mean vector. The inverse-Sharpe-ratio theorem is that the Sharpe ratio in a subspace is the inverse of the ratio in the complementary subspace. The Hansen–Jagannathan bound in a financial market is an application of this theorem (Hansen, L. P., & Jagannathan, R. (1991, April). Implications of security market data for models of dynamic economies. Journal of Political Economy, 99 (2), 225–262). Under cost and mean positivity, the maximum ratio of mean to standard deviation among all stochastic discount factors is the inverse of the Sharpe ratio for excess returns.

62.1 Sharpe Ratio

We study a moment space in which the inner product is the second noncentral moment. The random variable $\mathbf{1}$ is constant at one. For a random variable $\boldsymbol{x}$, the mean $\mathrm{E}(\boldsymbol{x}) = \langle \mathbf{1}, \boldsymbol{x} \rangle$; the mean $\langle \mathbf{1}, \mathbf{1} \rangle = 1$. Figure 62.1 plots the least-squares linear regression of $\boldsymbol{x}$ on $\mathbf{1}$. The regression coefficient $\langle \mathbf{1}, \boldsymbol{x} \rangle / \langle \mathbf{1}, \mathbf{1} \rangle = \langle \mathbf{1}, \boldsymbol{x} \rangle = \mathrm{E}(\boldsymbol{x})$. The fitted value $\langle \mathbf{1}, \boldsymbol{x} \rangle \mathbf{1}$ has length $\mathrm{E}(\boldsymbol{x})$, and the residual $\boldsymbol{x} - \langle \mathbf{1}, \boldsymbol{x} \rangle \mathbf{1}$ has length $\mathrm{Sd}(\boldsymbol{x}) = \sqrt{\left\langle \boldsymbol{x}, \left(\mathsf{I} - \mathbf{1}\mathbf{1}^\top\right) \boldsymbol{x} \right\rangle}$.

We analyze a subspace X. Let $\boldsymbol{m}^* \neq \mathbf{0}$ denote the mean vector in X. For $\boldsymbol{x} \in X$, the mean $\mathrm{E}(\boldsymbol{x}) = \langle \boldsymbol{m}^*, \boldsymbol{x} \rangle$, and the standard deviation $\mathrm{Sd}(\boldsymbol{x}) = \sqrt{\langle \boldsymbol{x}, (\mathsf{I} - \boldsymbol{m}^* \boldsymbol{m}^{*\top}) \boldsymbol{x} \rangle}$. Assume that $\mathbf{1} \notin X$: every non-zero random variable in X has a positive variance. It follows that $1 > \langle \boldsymbol{m}^*, \boldsymbol{m}^* \rangle = \langle \boldsymbol{m}^*, \mathbf{1} \rangle > 0$.

B. C. Dieffenbach, *Conjugate Duality in Economic Analysis*, Contributions to Economics,
https://doi.org/10.1007/978-3-032-21396-9_62

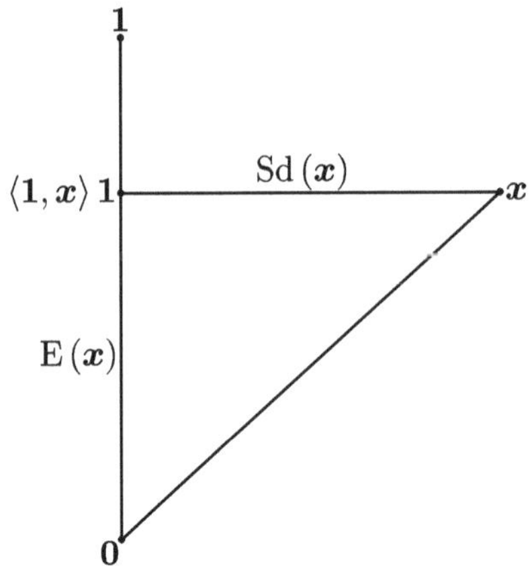

Fig. 62.1 Least-squares projection

Definition 1 (Sharpe Ratio) The Sharpe ratio is the maximum value of the ratio of mean to standard deviation.

Problem 1 (Sharpe Ratio)

$$\sup_{x \in X, x \neq 0} \frac{\mathrm{E}(\boldsymbol{x})}{\mathrm{Sd}(\boldsymbol{x})}. \tag{62.1}$$

The problem is a standard homogeneous fractional program. The denominator is positive for any nonzero $\boldsymbol{x}$, and the numerator is positive for some $\boldsymbol{x}$. The concave reduction of this fractional program is to maximize the mean such that the standard deviation is less than or equal to one.

Write the least-squares linear regression of $\boldsymbol{x}$ on $\boldsymbol{m}^*$ as

$$\boldsymbol{x} = \boldsymbol{u} + v\boldsymbol{m}^*/\mathrm{Sd}(\boldsymbol{m}^*).$$

The regression error $\boldsymbol{u}$ is orthogonal to $\boldsymbol{m}^*$ and thus has mean zero. Represent the vector by the pair $(\boldsymbol{u}, v) \in \mathbf{U} \oplus \mathbf{R}$. One calculates

$$v^* := \mathrm{E}[\boldsymbol{m}^*/\mathrm{Sd}(\boldsymbol{m}^*)] = \sqrt{\frac{\langle \boldsymbol{m}^*, \boldsymbol{m}^* \rangle}{1 - \langle \boldsymbol{m}^*, \boldsymbol{m}^* \rangle}}, \tag{62.2}$$

and the vector has mean $v^* v$ and variance $\langle \boldsymbol{u}, \boldsymbol{u} \rangle + v^2$.

We express the problem (62.1) in a Lorentz Euclidean Jordan algebra. The Sharpe ratio is the optimum value of the following standard homogeneous fractional program, the mean divided by the standard deviation.

Problem 2 (Sharpe Ratio Set in Lorentz Space)

$$\sup_{(\boldsymbol{u},w)\in\mathbf{L}^{\mathbf{U}\oplus\mathbf{R}},(\boldsymbol{u},v,w)\neq\mathbf{0}} \frac{v^*v}{w+\delta_{(\boldsymbol{u},w)\geq\mathbf{0}}}. \tag{62.3}$$

Both the numerator and the denominator are linear. To maximize the ratio, make v large and w small. The reduction of the fractional program to a concave maximization is the nonpolyhedral linear program

$$\sup_{(\boldsymbol{u},w)\in\mathbf{L}^{\mathbf{U}\oplus\mathbf{R}}} \left[v^*v-\delta_{w\leq 1}-\delta_{(\boldsymbol{u},w)\geq\mathbf{0}}\right].$$

The choice $(\boldsymbol{u}, v, w) = (\mathbf{0}, 1, 1)$ solves this problem. Thus the choice $\boldsymbol{m}^*$ maximizes the fraction (62.1), and the optimum value of the fraction is v^*.

That the Sharpe ratio is attained for the random variable $\boldsymbol{m}^*$ is intuitive. In the abstract setting analyzed here, $\boldsymbol{m}^*$ is the only structure present, so the Sharpe ratio must be attained for a random variable that is some function of $\boldsymbol{m}^*$. If the optimum were attained for a random variable not proportional to $\boldsymbol{m}^*$, this random variable would be arbitrary. Hence the optimum must be attained for any positive multiple of $\boldsymbol{m}^*$.

A more developed argument is the following. The mean of $\boldsymbol{m}^*$ is positive, so the ratio of its mean to its standard deviation is positive. Suppose that the Sharpe ratio were attained at $\boldsymbol{m}^* + \boldsymbol{u}$, in which $\boldsymbol{u}$ is a nonzero vector orthogonal to $\boldsymbol{m}^*$. By orthogonality $\boldsymbol{u}$ has mean zero and is uncorrelated with $\boldsymbol{m}^*$, but also it has a positive variance. Hence $\boldsymbol{m}^* + \boldsymbol{u}$ has the same mean as $\boldsymbol{m}^*$ but a higher variance. Thus the Sharpe ratio must be attained at $\boldsymbol{m}^*$ (Sharpe 1964).

By Eq. (62.2), the Sharpe ratio on $X^{\perp}$ as

$$\sqrt{\frac{\langle \mathbf{1}-\boldsymbol{m}^*, \mathbf{1}-\boldsymbol{m}^*\rangle}{1-\langle \mathbf{1}-\boldsymbol{m}^*, \mathbf{1}-\boldsymbol{m}^*\rangle}} = \sqrt{\frac{1-\langle \boldsymbol{m}^*, \boldsymbol{m}^*\rangle}{\langle \boldsymbol{m}^*, \boldsymbol{m}^*\rangle}} = \frac{1}{v^*},$$

the inverse of the Sharpe ratio on X.

Theorem 1 (Inverse Sharpe Ratio) *For* $0 < \langle \boldsymbol{m}^*, \boldsymbol{m}^*\rangle < 1$,

$$\max_{\boldsymbol{x}^*\in X^{\perp},\boldsymbol{x}^*\neq 0}\left[\frac{\mathrm{E}(\boldsymbol{x}^*)}{\mathrm{Sd}(\boldsymbol{x}^*)}\right]\times\max_{\boldsymbol{x}\in X,\boldsymbol{x}\neq 0}\left[\frac{\mathrm{E}(\boldsymbol{x})}{\mathrm{Sd}(\boldsymbol{x})}\right] = 1. \tag{62.4}$$

Figure 62.2 proves the inverse-Sharpe-ratio theorem by geometry. The vertical axis is the vectors proportional to $\boldsymbol{m}^*$, and the horizontal axis is the vectors proportional to $\mathbf{1} - \boldsymbol{m}^*$. Of course $\mathrm{E}(\boldsymbol{m}^*) + \mathrm{E}(\mathbf{1} - \boldsymbol{m}^*) = \mathrm{E}(\mathbf{1}) = 1$.

The line from the origin to $\mathbf{1}$ has length one, and this line divides the rectangle into two large triangles of the same dimension. Consider the upper left triangle. The projection of $\boldsymbol{m}^*$ onto $\mathbf{1}$ is $\mathrm{E}(\boldsymbol{m}^*)\,\mathbf{1}$, which is distance $\mathrm{E}(\boldsymbol{m}^*)$ from the origin. The distance from $\boldsymbol{m}^*$ to the projection is $\mathrm{Sd}(\boldsymbol{m}^*)$. Hence the area of the upper left triangle is $\frac{1}{2}\mathrm{Sd}(\boldsymbol{m}^*)$, half the base 1 times the height $\mathrm{Sd}(\boldsymbol{m}^*)$. By analogous reasoning,

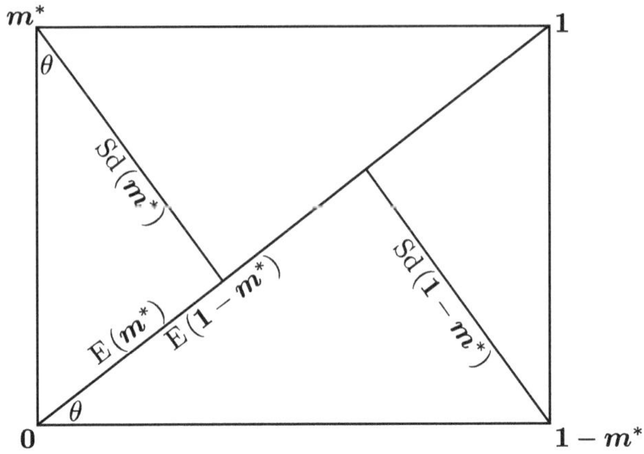

Fig. 62.2 Inverse Sharpe ratio

the area of the lower right triangle is $\frac{1}{2}\mathrm{Sd}\,(\boldsymbol{1} - \boldsymbol{m}^*)$, so both standard deviations are equal. The area of the rectangle is the standard deviation.

The smaller triangles are similar to the two large triangles. The two angles marked θ are the same. The Sharpe ratio in X is $\tan\theta$, and the ratio in $X^{\perp}$ is $\cot\theta = 1/\tan\theta$, which proves the inverse-Sharpe-ratio theorem. If $\boldsymbol{m}^*$ is near $\boldsymbol{1}$, then θ is near $\pi/2$, so the Sharpe ratio $\tan\theta$ becomes very large.

A statistical interpretation of the inverse-Sharpe-ratio theorem is the following. Consider nonzero $\boldsymbol{x} \in X$ and nonzero $\boldsymbol{x}^* \in X^{\top}$. The orthogonality condition $\langle \boldsymbol{x}^*, \boldsymbol{x} \rangle = 0$ is equivalent to

$$\mathrm{Cov}\,(\boldsymbol{x}^*, \boldsymbol{x}) + \mathrm{E}\,(\boldsymbol{x}^*)\,\mathrm{E}\,(\boldsymbol{x}) = 0.$$

Since both vectors are nonzero, each has a positive standard deviation, so this condition rearranges as

$$\frac{\mathrm{E}\,(\boldsymbol{x}^*)}{\mathrm{Sd}\,(\boldsymbol{x}^*)}\frac{\mathrm{E}\,(\boldsymbol{x})}{\mathrm{Sd}\,(\boldsymbol{x})} = -\mathrm{Correlation}\,(\boldsymbol{x}^*, \boldsymbol{x}).$$

Thus for arbitrary $\boldsymbol{x}^*$ and $\boldsymbol{x}$, the product of the mean over the standard deviation is less than or equal to one. The choices $\boldsymbol{x} = \boldsymbol{m}^*$ and $\boldsymbol{x}^* = \boldsymbol{1} - \boldsymbol{m}^*$ that maximize the Sharpe ratios sum to $\boldsymbol{1}$, and so are perfectly negatively correlated.

62.2 Hansen–Jagannathan Bound

We apply the inverse-Sharpe-ratio theorem to finance theory. Let Y denote the asset space, a subspace of the moment space. A vector $\boldsymbol{y} \in Y$ denotes an asset. Its moments are the moments of the payoff on the asset. There may or may not be a risk-free asset: Y may or may not contain $\mathbf{1}$.

Let $\boldsymbol{n}^* \in Y$ be the mean vector for the asset space. Assume $\boldsymbol{n}^* \neq \mathbf{0}$, so $0 < \langle \boldsymbol{n}^*, \boldsymbol{n}^* \rangle \leq 1$.

We assume that the law of one price holds and identify two assets having the same payoff. Let the "pricing portfolio" $\boldsymbol{p}^*$ denote the asset having the pricing payoff. An asset has mean payoff $\langle \boldsymbol{n}^*, \boldsymbol{y} \rangle$ and cost $\langle \boldsymbol{p}^*, \boldsymbol{y} \rangle$. Assume that $\boldsymbol{p}^*$ is not proportional to $\boldsymbol{n}^*$ (the asset price is not simply proportional to the mean payoff). By the Cauchy–Schwartz inequality, $\langle \boldsymbol{n}^*, \boldsymbol{n}^* \rangle \langle \boldsymbol{p}^*, \boldsymbol{p}^* \rangle - \langle \boldsymbol{n}^*, \boldsymbol{p}^* \rangle^2 > 0$.

Assume cost and mean positivity $\langle \boldsymbol{p}^*, \boldsymbol{n}^* \rangle > 0$, which we see as standard for asset-market equilibrium. If there is a risk-free asset, $\boldsymbol{n}^* = \mathbf{1}$, so $\langle \boldsymbol{p}^*, \boldsymbol{n}^* \rangle = \langle \boldsymbol{p}^*, \mathbf{1} \rangle$. The price of the risk-free asset must be positive (a necessary condition for no arbitrage). If there is no risk-free asset, cost and mean positivity is equivalent to a positive correlation between $\boldsymbol{p}^*$ and $\boldsymbol{n}^*$:

$$\begin{aligned}\operatorname{Cov}(\boldsymbol{p}^*, \boldsymbol{n}^*) &= \langle \boldsymbol{p}^*, \boldsymbol{n}^* \rangle - \operatorname{E}(\boldsymbol{p}^*) \operatorname{E}(\boldsymbol{n}^*) = \langle \boldsymbol{p}^*, \boldsymbol{n}^* \rangle - \langle \boldsymbol{p}^*, \boldsymbol{n}^* \rangle \langle \boldsymbol{n}^*, \boldsymbol{n}^* \rangle \\ &= \langle \boldsymbol{p}^*, \boldsymbol{n}^* \rangle (1 - \langle \boldsymbol{n}^*, \boldsymbol{n}^* \rangle).\end{aligned}$$

As $\langle \boldsymbol{n}^*, \boldsymbol{n}^* \rangle < 1$ if there is no risk-free asset, $\operatorname{Cov}(\boldsymbol{p}^*, \boldsymbol{n}^*)$ and $\langle \boldsymbol{p}^*, \boldsymbol{n}^* \rangle$ have the same sign.

Let X denote the subspace of excess returns, the subspace of Y such that the cost is zero: the $\boldsymbol{x} \in Y$ such that $\langle \boldsymbol{p}^*, \boldsymbol{x} \rangle = 0$. Thus $Y = X \oplus \boldsymbol{p}^* \mathbf{R}$ (all multiples of $\boldsymbol{p}^*$). Conversely, $X^\perp = Y^\perp \oplus \boldsymbol{p}^* \mathbf{R}$. The set of all stochastic discount factors is $\{\boldsymbol{p}^*\} + Y^\perp$, and $X^\perp$ is the smallest subspace containing this set.

Necessarily $\mathbf{1} \notin X$. If $\mathbf{1} \in Y$, then $\boldsymbol{n}^* = \mathbf{1}$, and its cost $\langle \boldsymbol{p}^*, \mathbf{1} \rangle > 0$, so $\mathbf{1}$ is not an excess return.

For subspace X, the mean vector

$$\boldsymbol{m}^* = \boldsymbol{n}^* - \frac{\langle \boldsymbol{n}^*, \boldsymbol{p}^* \rangle}{\langle \boldsymbol{p}^*, \boldsymbol{p}^* \rangle} \boldsymbol{p}^*, \tag{62.5}$$

the unique vector in X such that $\langle \boldsymbol{m}^*, \boldsymbol{x} \rangle = \langle \boldsymbol{n}^*, \boldsymbol{x} \rangle$ if $\langle \boldsymbol{p}^*, \boldsymbol{x} \rangle = 0$; and $\langle \boldsymbol{m}^*, \boldsymbol{p}^* \rangle = 0$. By Eq. (62.5), we calculate

$$\langle \boldsymbol{m}^*, \boldsymbol{m}^* \rangle = \langle \boldsymbol{n}^*, \boldsymbol{n}^* \rangle - \frac{\langle \boldsymbol{n}^*, \boldsymbol{p}^* \rangle^2}{\langle \boldsymbol{p}^*, \boldsymbol{p}^* \rangle},$$

which is greater than zero but less than one.

The mean vector in $X^\perp$ is

$$\mathbf{1} - \boldsymbol{m}^* = \mathbf{1} - \left[\boldsymbol{n}^* - \frac{\langle \boldsymbol{n}^*, \boldsymbol{p}^* \rangle}{\langle \boldsymbol{p}^*, \boldsymbol{p}^* \rangle} \boldsymbol{p}^* \right] = \mathbf{1} - \boldsymbol{n}^* + \frac{\langle \boldsymbol{n}^*, \boldsymbol{p}^* \rangle}{\langle \boldsymbol{p}^*, \boldsymbol{p}^* \rangle} \boldsymbol{p}^*$$

The Sharpe ratio in $X^{\perp}$ is attained for any positive multiple of the mean vector. If cost and mean positivity holds, then multiplying by $\langle \boldsymbol{p}^*, \boldsymbol{p}^* \rangle / \langle \boldsymbol{n}^*, \boldsymbol{p}^* \rangle$ shows that the ratio is also attained for the stochastic discount factor

$$\boldsymbol{p}^* + \frac{\langle \boldsymbol{p}^*, \boldsymbol{p}^* \rangle}{\langle \boldsymbol{n}^*, \boldsymbol{p}^* \rangle} (\mathbf{1} - \boldsymbol{n}^*). \tag{62.6}$$

(If cost and mean positivity does not hold, then the ratio is not attained for any stochastic discount factor.)

We have arrived at the Hansen–Jagannathan bound, as an application of the inverse-Sharpe-ratio theorem (Hansen and Jagannathan 1991). Under cost and mean positivity, the maximum ratio of mean to standard deviation among all stochastic discount factors is the inverse of the Sharpe ratio for excess returns.

Theorem 2 (Hansen–Jagannathan Bound) *Under cost and mean positivity,*

$$\max_{\boldsymbol{x}^* \in \{\boldsymbol{p}^*\} + Y^{\perp}} \left[\frac{\mathrm{E}(\boldsymbol{x}^*)}{\mathrm{Sd}(\boldsymbol{x}^*)} \right] \times \max_{\boldsymbol{x} \in X, \boldsymbol{x} \neq 0} \left[\frac{\mathrm{E}(\boldsymbol{x})}{\mathrm{Sd}(\boldsymbol{x})} \right] = 1.$$

The product is one for the choices $\boldsymbol{x} = \boldsymbol{m}^*$ in (62.5) and $\boldsymbol{x}^* = \boldsymbol{p}^* + \frac{\langle \boldsymbol{p}^*, \boldsymbol{p}^* \rangle}{\langle \boldsymbol{n}^*, \boldsymbol{p}^* \rangle} (\mathbf{1} - \boldsymbol{n}^*) \propto \mathbf{1} - \boldsymbol{m}^*$ in (62.6). If $\boldsymbol{x}^*$ is any stochastic discount factor, then the ratio $\mathrm{Sd}(\boldsymbol{x}^*) / \mathrm{E}(\boldsymbol{x}^*)$ is an upper bound for the Sharpe ratio for excess returns.

Because $\boldsymbol{p}^*$ and $\mathbf{1} - \boldsymbol{n}^*$ are orthogonal, the maximum value of $\mathrm{E}(\boldsymbol{x}^*) / \mathrm{Sd}(\boldsymbol{x}^*)$ is attained for $\boldsymbol{x}^* = \boldsymbol{p}^*$ if and only if $\boldsymbol{n}^* = \mathbf{1}$ (if and only if the asset space contains a risk-free asset). For this case the Sharpe ratio for excess returns is $\mathrm{Sd}(\boldsymbol{p}^*) / \mathrm{E}(\boldsymbol{p}^*)$, but otherwise is less.

We have

$$\begin{aligned} \frac{\mathrm{E}(\boldsymbol{p}^*)}{\mathrm{Sd}(\boldsymbol{p}^*)} \times \max_{\boldsymbol{x} \in X, \boldsymbol{x} \neq 0} \left[\frac{\mathrm{E}(\boldsymbol{x})}{\mathrm{Sd}(\boldsymbol{x})} \right] &= \frac{\mathrm{E}(\boldsymbol{p}^*)}{\mathrm{Sd}(\boldsymbol{p}^*)} \frac{\mathrm{E}(\boldsymbol{m}^*)}{\mathrm{Sd}(\boldsymbol{m}^*)} = \frac{-\mathrm{Cov}(\boldsymbol{p}^*, \boldsymbol{m}^*)}{\mathrm{Sd}(\boldsymbol{p}^*)\,\mathrm{Sd}(\boldsymbol{m}^*)} \\ &= \mathrm{Correlation}(-\boldsymbol{p}^*, \boldsymbol{m}^*), \end{aligned}$$

because $\boldsymbol{p}^*$ and $\boldsymbol{m}^*$ are orthogonal.

Corollary 1 (Pricing-Portfolio Bound) *Under cost and mean positivity,*

$$\frac{\mathrm{E}(\boldsymbol{p}^*)}{\mathrm{Sd}(\boldsymbol{p}^*)} \times \max_{\boldsymbol{x} \in X, \boldsymbol{x} \neq 0} \left[\frac{\mathrm{E}(\boldsymbol{x})}{\mathrm{Sd}(\boldsymbol{x})} \right] = \mathrm{Correlation}(-\boldsymbol{p}^*, \boldsymbol{m}^*).$$

If there is a risk-free asset, then the pricing portfolio $\boldsymbol{p}^*$ and the mean vector $\boldsymbol{m}^*$ for excess returns are perfectly negatively correlated.

References

Hansen LP, Jagannathan R (1991 Apr) Implications of security market data for models of dynamic economies. J Polit Econ 99(2):225–262

Sharpe WF (1964 Sept) Capital asset prices: a theory of market equilibrium under conditions of risk. J Fin XIX(3):425–442

Chapter 63
Efficient-Portfolios Risk-Free Asset

Abstract We solve an efficiency problem to determine the cost/mean/variance efficient portfolios for asset markets having a risk-free asset. The portfolio-choice separation theorem describes the efficient portfolios: invest a nonnegative amount in a particular risky portfolio and an arbitrary positive or negative amount in the risk-free asset. The calculations for the Sharpe ratio—the maximum ratio of mean to standard deviation—confirm Hansen–Jagannathan duality (Tobin J (1958 Feb) Liquidity preference as behavior towards risk. Rev Econ Stud XXV(2):65–86).

63.1 Moment Space of Assets

We analyze portfolio choice in the moment space of assets, a Euclidean space. The inner product of two assets is the noncentral second moment of their payoffs. There exists a risk-free asset $\mathbf{1}$, which has constant payoff of one dollar. We assume that the law of one price holds, so we identify assets having the same payoff as the same.

The asset space is the direct sum of the subspace $\mathbf{X}$ of assets having payoff mean zero and the subspace $\mathbf{R}\{\mathbf{1}\}$ of all multiples of $\mathbf{1}$. An asset $\boldsymbol{w}$ has the unique decomposition

$$\boldsymbol{w} = \boldsymbol{x} + y\mathbf{1},$$

which we denote by the pair $(\boldsymbol{x}, y) \in \mathbf{X} \oplus \mathbf{R}\{\mathbf{1}\}$. Its payoff has mean y and standard deviation $|\boldsymbol{x}|$. The mean vector is $(\mathbf{0}, 1) = \mathbf{1}$.

Let $(\boldsymbol{p}^*, q^*) = \boldsymbol{p}^* + q^*\mathbf{1}$ denote the pricing portfolio, which has the pricing payoff. We assume positivity: the inner product of the pricing portfolio and the mean vector is positive,

$$\langle(\boldsymbol{p}^*, q^*), (\mathbf{0}, 1)\rangle = q^* > 0.$$

This condition is necessary for no opportunity for arbitrage: the risk-free asset has a positive price.

B. C. Dieffenbach, *Conjugate Duality in Economic Analysis*, Contributions to Economics,
https://doi.org/10.1007/978-3-032-21396-9_63

The pricing portfolio specifies the relative price of different assets. Viewing its normalization as arbitrary, let us impose the normalization $q^* = 1$. The mean vector specifies the mean payoff on different assets, which says nothing about expected rates of return.

63.2 Excess Returns

The assets having cost zero form the subspace of excess returns. The excess return corresponding to an asset is the vector less a risk-free vector costing the same:

$$\boldsymbol{x} + y\mathbf{1} - (\langle \boldsymbol{p}^*, \boldsymbol{x}\rangle + y)\,\mathbf{1} = \boldsymbol{x} - \langle \boldsymbol{p}^*, \boldsymbol{x}\rangle\,\mathbf{1}. \tag{63.1}$$

The expected payoff of the excess return is the risk premium on the asset,

$$\langle \mathbf{1}, \boldsymbol{x} - \langle \boldsymbol{p}^*, \boldsymbol{x}\rangle\,\mathbf{1}\rangle = -\langle \boldsymbol{p}^*, \boldsymbol{x}\rangle, \tag{63.2}$$

the negative of the covariance of pricing portfolio with the payoff on the asset.

The mean vector for the subspace of excess returns is

$$\boldsymbol{m}^* = \left(\frac{1}{1 + \langle \boldsymbol{p}^*, \boldsymbol{p}^*\rangle}\right)(-\boldsymbol{p}^* + \langle \boldsymbol{p}^*, \boldsymbol{p}^*\rangle\,\mathbf{1}). \tag{63.3}$$

That the cost $\langle \boldsymbol{p}^* + \mathbf{1}, \boldsymbol{m}^*\rangle = 0$ shows that this vector is in the subspace. The mean of the excess return (63.1) is

$$\begin{aligned}
&\langle \boldsymbol{m}^*, \boldsymbol{x} - \langle \boldsymbol{p}^*, \boldsymbol{x}\rangle\,\mathbf{1}\rangle \\
&= \left(\frac{1}{1 + \langle \boldsymbol{p}^*, \boldsymbol{p}^*\rangle}\right)\langle -\boldsymbol{p}^* + \langle \boldsymbol{p}^*, \boldsymbol{p}^*\rangle\,\mathbf{1}, \boldsymbol{x} - \langle \boldsymbol{p}^*, \boldsymbol{x}\rangle\,\mathbf{1}\rangle \\
&= -\langle \boldsymbol{p}^*, \boldsymbol{x}\rangle,
\end{aligned}$$

as required by Eq. (63.2).

This case illustrates Hansen–Jagannathan duality. The mean

$$\mathrm{E}\,(\boldsymbol{m}^*) = \langle \boldsymbol{m}^*, \mathbf{1}\rangle = \frac{\langle \boldsymbol{p}^*, \boldsymbol{p}^*\rangle}{1 + \langle \boldsymbol{p}^*, \boldsymbol{p}^*\rangle},$$

which is between zero and one, and the variance

$$\mathrm{Var}\,(\boldsymbol{m}^*) = \frac{\langle \boldsymbol{p}^*, \boldsymbol{p}^*\rangle}{(1 + \langle \boldsymbol{p}^*, \boldsymbol{p}^*\rangle)^2}.$$

For the excess returns, the Sharpe ratio—the maximum ratio of mean to standard deviation—is

$$\frac{\mathrm{E}(\boldsymbol{m}^*)}{\mathrm{Sd}(\boldsymbol{m}^*)} = |\boldsymbol{p}^*|.$$

By Hansen–Jagannathan duality, the Sharpe ratio is the inverse of the Sharpe ratio for the subspace generated by all stochastic discount factors. Because the asset space contains a risk-free asset, the Sharpe ratio for this subspace is attained for the pricing portfolio:

$$\frac{\mathrm{E}(\boldsymbol{p}^* + \mathbf{1})}{\mathrm{Sd}(\boldsymbol{p}^* + \mathbf{1})} = \frac{1}{|\boldsymbol{p}^*|},$$

which is indeed the inverse.

63.3 Separation Theorem

Markowitz proposes analyzing portfolio choice as making a tradeoff of the mean and the risk of the payoff (Markowitz 1952). A simple measure of risk is the variance of the payoff. Portfolio diversification may reduce the mean, but the variance falls. An investor chooses lower variance but lower mean as risk aversion rises.

The portfolio-choice separation theorem is a version of Tobin (1958).

Theorem 1 (Portfolio-Choice Separation) *The cone of cost/mean/standard deviation efficient portfolios is*

$$\mathbf{R}_+\{-\boldsymbol{p}^*\} + \mathbf{R}\{\mathbf{1}\}. \tag{63.4}$$

The efficient portfolios are the nonnegative multiples of the risky portfolio $-\boldsymbol{p}^*$, plus any quantity of the risk-free asset. A negative quantity corresponds to borrowing risk-free. The excess-return mean vector (63.3) is an efficient portfolio. Below we derive (63.4) via an efficiency problem.

The theorem separates portfolio choice into two steps. Step one calculates the efficient risky portfolio $-\boldsymbol{p}^*$ and the cone (63.4) of efficient portfolios. At this step risk aversion plays no role, as the efficient risky portfolio is independent of risk aversion. Step two invokes risk aversion, as one chooses how much to invest in the efficient risky portfolio and in the risk-free asset.

Figure 63.1 illustrates the efficient portfolios. The upward sloping lines have slope one and plot the minimum standard deviation for a given cost and mean. The line through the origin has standard deviation zero, attained by investing or borrowing risk-free. For the lines in either direction up or down, the standard deviation is increasing. For the risk-free portfolio $\mathbf{1}$, both the cost and the mean are one. The

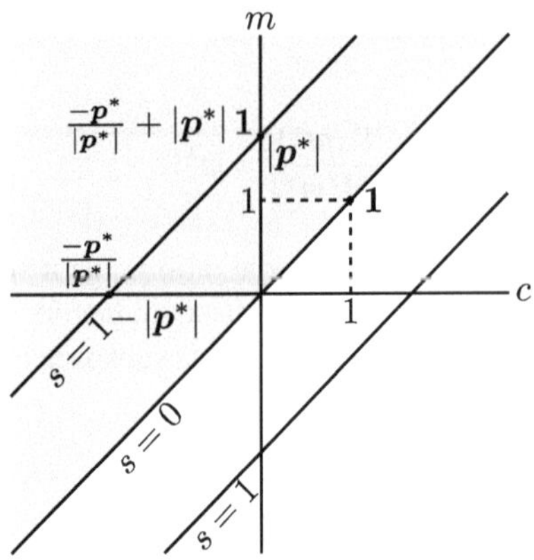

Fig. 63.1 Efficient portfolios risk-free asset

shaded area plots the efficient portfolios, the cost and mean values left and above the line through the origin, and the associated standard deviation.

The higher upward sloping line labelled $s = 1$ plots cost and mean combinations for which the minimum standard deviation is one. The portfolio $-\boldsymbol{p}^*/\,|\boldsymbol{p}^*|$ has cost $-\,|\boldsymbol{p}^*|$ and mean zero, with standard deviation one. The line intercepts the vertical axis at the Sharpe ratio $|\boldsymbol{p}^*|$. The zero-cost portfolio $-\boldsymbol{p}^*/\,|\boldsymbol{p}^*| + |\boldsymbol{p}^*|\,\mathbf{1}$ has standard deviation one and attains the Sharpe ratio.

We derive the portfolio-choice separation theorem via a cost/mean/standard deviation efficiency problem.

Problem 1 (Efficiency Primal)

$$\sup_{\boldsymbol{x},y,z}\left[-\delta^*_{\mathbf{R}_++\{1\}}\left(\langle\boldsymbol{p}^*,\boldsymbol{x}\rangle + y - c\right) - \delta^*_{\mathbf{R}_++\{1\}}\left(-y+m\right) - \delta^*_{\mathbf{R}_++\{1\}}\left(z-s\right) - \delta_{\mathbf{L}^{\mathbf{X}}_+}\left(\boldsymbol{x},z\right)\right].$$

Here $(\boldsymbol{x},z) \in \mathbf{L}^{\mathbf{X}}$. For a portfolio $(\overline{\boldsymbol{x}},\overline{y})$, the perturbation base values are $c = \langle\boldsymbol{p}^*,\overline{\boldsymbol{x}}\rangle + \overline{y}$, $m = \overline{y}$, and $s = |\overline{\boldsymbol{x}}|$. The dual perturbations $\boldsymbol{x}^*$, y^*, and z^* are all zero. The portfolio $(\overline{\boldsymbol{x}},\overline{y})$ is cost/mean/standard deviation efficient if and only if the optimum value of the efficiency primal is zero.

We find the reciprocal dual. Adding the Fenchel inequalities

$$\begin{aligned}
&\delta_{\mathbf{R}_++\{1\}}\left(c^*\right) + \delta^*_{\mathbf{R}_++\{1\}}\left(\langle\boldsymbol{p}^*,\boldsymbol{x}\rangle + y - c\right) - c^*\left(\langle\boldsymbol{p}^*,\boldsymbol{x}\rangle + y - c\right) \geq 0\\
&\delta_{\mathbf{R}_++\{1\}}\left(-m^*\right) + \delta^*_{\mathbf{R}_++\{1\}}\left(-y+m\right) - \left(-m^*\right)\left(-y+m\right) \geq 0\\
&\delta_{\mathbf{R}_++\{1\}}\left(s^*\right) + \delta^*_{\mathbf{R}_++\{1\}}\left(z-s\right) - \left(s^*\right)\left(z-s\right) \geq 0\\
&\delta_{\mathbf{L}^{\mathbf{X}}_+}\left(\boldsymbol{x},z\right) + \delta^*_{\mathbf{L}^{\mathbf{X}}_+}\left(-\boldsymbol{p}^*c^* - \boldsymbol{x}^*, -s^* + z^*\right) - \langle -\boldsymbol{p}^*c^* + \boldsymbol{x}^*, \boldsymbol{x}\rangle - \left(-s^* + z^*\right)z \geq 0\\
&\delta_{\{0\}}\left(-c^* - m^* + y^*\right) + \delta^*_{\{0\}}\left(y\right) - \left(-c^* - m^* + y^*\right)\left(y\right) \geq 0
\end{aligned}$$

and rearranging obtains

$$
\begin{aligned}
&c^*c + m^*m + s^*s + \delta_{\mathbf{R}_+ + \{1\}}(c^*) + \delta_{\mathbf{R}_+ + \{1\}}(-m^*) + \delta_{\mathbf{R}_+ + \{1\}}(s^*) \\
&\quad + \delta_{\{0\}}(-c^* - m^* + y^*) + \delta^*_{\mathbf{L}^X_+}(-\boldsymbol{p}^*c^* - \boldsymbol{x}^*, -s^* + z^*) \\
&\geq \langle \boldsymbol{x}^*, \boldsymbol{x} \rangle + y^*y + z^*z - \delta^*_{\mathbf{R}_+ + \{1\}}(\langle \boldsymbol{p}^*, \boldsymbol{x} \rangle + y - c) - \delta^*_{\mathbf{R}_+ + \{1\}}(-y + m) - \delta^*_{\mathbf{R}_+ + \{1\}}(z - s) \\
&\quad - \delta_{\mathbf{L}^X_+}(\boldsymbol{x}, z) - \delta^*_{\{0\}}(y).
\end{aligned}
$$

Problem 2 (Efficiency Dual)

$$
\begin{aligned}
\inf_{c^*, m^* s^*} &\left[c^*c + m^*m + s^*s + \delta_{c^* \geq 1} + \delta_{-m^* \geq 1} + \delta_{s^* \geq 1} \right. \\
&\left. + \delta_{m^* = -c^*} + \delta_{\mathbf{L}^X_+}(\boldsymbol{p}^*c^*, s^*) \right].
\end{aligned}
$$

The reciprocal dual satisfies the relative condition, so there is no duality gap.

A simple alternative to setting the Fenchel factors to zero is to evaluate the dual. Set $m^* = -c^*$. Because $s > 0$, $s^* = |\boldsymbol{p}^*| \, c^*$ is minimizing. Choosing c^* positive and high enough satisfies the greater than or equal to one requirements. Substituting the base values of the efficiency primal obtains

$$
\begin{aligned}
&c^*c + m^*m + s^*s \\
&= c^* (\langle \boldsymbol{p}^*, \overline{\boldsymbol{x}} \rangle + \overline{y}) - c^*\overline{y} + c^* \, |\boldsymbol{p}^*| \, |\overline{\boldsymbol{x}}| \\
&= c^* \left(|\boldsymbol{p}^*| \, |\overline{\boldsymbol{x}}| - \langle -\boldsymbol{p}^*, \overline{\boldsymbol{x}} \rangle \right) \\
&\geq 0,
\end{aligned}
$$

by the Cauchy-Swartz inequality. The value is zero if and only if $\overline{\boldsymbol{x}}$ is a nonnegative multiple of $-\boldsymbol{p}^*$, which proves the portfolio-choice separation theorem. If the portfolio is not efficient, then the optimum value is ∞, approached as $c^* \to \infty$.

References

Markowitz H (1952 Mar) Portfolio selection. J Fin VII(1):77–91

Tobin J (1958 Feb) Liquidity preference as behavior towards risk. Rev Econ Stud XXV(2):65–86

Chapter 64
Efficient Portfolios No Risk-Free Asset

Abstract Working with an asset space in which there is no risk-free asset, we solve for the portfolios that are cost/mean/variance efficient. Efficiency looks for a lower cost, a higher mean, and lower variance. Analyzing primal and dual efficiency problems obtains the separation Theorem 1. Setting the problem in Lorentz space obtains a (nonpolyhedral) linear program. Cost and mean positivity is necessary and sufficient for the existence of efficient portfolios having a positive cost. The efficient portfolios consist of the cone generated by the mean portfolio and the negative of the cost portfolio. This theorem corrects traditional thinking.

64.1 Asset Space

The asset space is a moment space of portfolios. Invoking the law of one price, identify portfolios that are perfectly correlated. Define the inner product of two portfolios as the covariance of their payoffs.

Define the mean portfolio $\boldsymbol{m}^*$ as the unique vector such that the covariance $\langle \boldsymbol{m}^*, \boldsymbol{x} \rangle$ is the mean payoff of portfolio $\boldsymbol{x}$. The portfolio $\boldsymbol{m}^* / |\boldsymbol{m}^*|$ has mean $|\boldsymbol{m}^*|$ and variance one. Recall that therefore this portfolio minimizes the variance among all portfolios having the same mean.

Analogously, define the cost portfolio $\boldsymbol{c}^*$ as the unique vector such that the covariance $\langle \boldsymbol{c}^*, \boldsymbol{x} \rangle$ is the cost of portfolio $\boldsymbol{x}$. The portfolio $\boldsymbol{c}^* / |\boldsymbol{c}^*|$ has cost $|\boldsymbol{c}^*|$ and variance one. This portfolio minimizes the variance among all portfolios having the same cost.

Condition 1 (Non-Proportionality) The mean portfolio $\boldsymbol{m}^*$ and the cost portfolio $\boldsymbol{c}^*$ are not proportional to one another.

This condition requires that neither vector is zero. Proportionality would mean that every portfolio has the same ratio of mean to cost.

B. C. Dieffenbach, *Conjugate Duality in Economic Analysis*, Contributions to Economics,
https://doi.org/10.1007/978-3-032-21396-9_64

64.2 Cost and Mean Positivity

The cost of portfolio $\boldsymbol{x}$

$$\langle \boldsymbol{c}^*, \boldsymbol{x}\rangle = \left\langle \left(\mathsf{I} + \boldsymbol{m}^*\boldsymbol{m}^{*\top}\right)^{-1} \boldsymbol{c}^*, \left(\mathsf{I} + \boldsymbol{m}^*\boldsymbol{m}^{*\top}\right)\boldsymbol{x}\right\rangle,$$

so the pricing portfolio is

$$\boldsymbol{p}^* = \left(\mathsf{I} + \boldsymbol{m}^*\boldsymbol{m}^{*\top}\right)^{-1} \boldsymbol{c}^*,$$

the portfolio such that its second noncentral moment with a portfolio is its cost. The mean of portfolio $\boldsymbol{x}$

$$\langle \boldsymbol{m}^*, \boldsymbol{x}\rangle = \left\langle \left(\mathsf{I} + \boldsymbol{m}^*\boldsymbol{m}^{*\top}\right)^{-1} \boldsymbol{m}^*, \left(\mathsf{I} + \boldsymbol{m}^*\boldsymbol{m}^{*\top}\right)\boldsymbol{x}\right\rangle,$$

so

$$\boldsymbol{e}^* = \left(\mathsf{I} + \boldsymbol{m}^*\boldsymbol{m}^{*\top}\right)^{-1} \boldsymbol{m}^* = \boldsymbol{m}^* / \left(1 + \langle \boldsymbol{m}^*, \boldsymbol{m}^*\rangle\right)$$

is the portfolio such that its second noncentral moment with a portfolio is its mean.[1]

The second noncentral moment of $\boldsymbol{p}^*$ and $\boldsymbol{e}^*$ is simultaneously both the cost of $\boldsymbol{e}^*$ and the mean of $\boldsymbol{p}^*$. An important condition in finance theory is *cost and mean positivity*: this moment is positive. Positivity is equivalent to a positive covariance between $\boldsymbol{c}^*$ and $\boldsymbol{m}^*$:

$$0 < \left\langle \boldsymbol{e}^*, \left(\mathsf{I} + \boldsymbol{m}^*\boldsymbol{m}^{*\top}\right)\boldsymbol{p}^*\right\rangle = \langle \boldsymbol{e}^*, \boldsymbol{c}^*\rangle = \langle \boldsymbol{m}^*, \boldsymbol{c}^*\rangle / \left(1 + \boldsymbol{m}^{*\top}\boldsymbol{m}^*\right).$$

Below we find that this condition is pivotal for portfolio efficiency.

64.3 Efficient Portfolios

We are looking for portfolios that are cost/mean/variance efficient; one is looking for a lower cost, a higher mean, and lower variance. The fundamental result is the following.

Theorem 1 (Portfolio-Choice Separation with no Risk-Free Asset) *The cost/mean/variance efficient portfolios consist of the closed convex cone generated by* $\boldsymbol{m}^*$ *and* $-\boldsymbol{c}^*$.

[1] The inverse $\left(\mathsf{I} + \boldsymbol{m}^*\boldsymbol{m}^{*\top}\right)^{-1} = \mathsf{I} - \boldsymbol{m}^*\boldsymbol{m}^{*\top}/1 + \langle \boldsymbol{m}^*, \boldsymbol{m}^*\rangle$.

Let us model cost/mean/standard deviation efficiency by the following nonpolyhedral linear problem, set in Lorentz space. We analyze whether a portfolio $\overline{\boldsymbol{x}}$ is efficient. The base values of the primal perturbation variables are its cost $c := \langle \boldsymbol{c}, \overline{\boldsymbol{x}} \rangle$, its mean $m := \langle \boldsymbol{m}, \overline{\boldsymbol{x}} \rangle$ and its standard deviation $s := |\overline{\boldsymbol{x}}|$. The issue is whether the portfolio and its standard deviation solve the efficiency primal.

Problem 1 (Efficient-Portfolio Primal)

$$\sup_{(x,y)\in\mathbf{L}^{\mathrm{X}}} \Big[\langle (\boldsymbol{x}^*, y^*), (\boldsymbol{x}, y)\rangle \\ - \delta^*_{\mathbf{R}_+ + \{1\}}(\langle \boldsymbol{c}^*, \boldsymbol{x}\rangle - c) - \delta^*_{\mathbf{R}_+ + \{1\}}(m - \langle \boldsymbol{m}^*, \boldsymbol{x}\rangle) \\ - \delta^*_{\mathbf{R}_+ + \{1\}}(y - s) - \delta_{(x,y)\geq \mathbf{0}}\Big].$$

The base value of the dual perturbation is $(\boldsymbol{x}^*, y^*) = (\mathbf{0}, 0)$.

We calculate the reciprocal dual. Adding the Fenchel inequalities

$$\delta_{\mathbf{R}_+ + \{1\}}(c^*) + \delta^*_{\mathbf{R}_+ + \{1\}}(\langle \boldsymbol{c}^*, \boldsymbol{x}\rangle - c) - \langle c^*, \langle \boldsymbol{c}^*, \boldsymbol{x}\rangle - c\rangle \geq 0 \tag{64.1}$$

$$\delta_{\mathbf{R}_+ + \{1\}}(-m^*) + \delta^*_{\mathbf{R}_+ + \{1\}}(m - \langle \boldsymbol{m}^*, \boldsymbol{x}\rangle) - \langle -m^*, m - \langle \boldsymbol{m}^*, \boldsymbol{x}\rangle\rangle \geq 0 \tag{64.2}$$

$$\delta_{\mathbf{R}_+ + \{1\}}(s^*) + \delta^*_{\mathbf{R}_+ + \{1\}}(y - s) - \langle s^*, y - s\rangle \geq 0 \tag{64.3}$$

$$\delta_{\mathbf{L}^+_{\mathrm{X}}}(\boldsymbol{x}, y) + \delta^*_{\mathbf{L}^+_{\mathrm{X}}}(-\boldsymbol{c}^* c^* - \boldsymbol{m}^* m^* - \boldsymbol{x}^*, -s^* - y^*) \\ - \langle (-\boldsymbol{c}^* c^* - \boldsymbol{m}^* m^* - \boldsymbol{x}^*, -s^* - y^*), (\boldsymbol{x}, y)\rangle \geq 0 \tag{64.4}$$

and rearranging gives

$$c^* c + m^* m + s^* s + \delta_{\mathbf{R}_+ + \{1\}}(c^*) + \delta_{\mathbf{R}_+ + \{1\}}(-m^*) + \delta_{\mathbf{R}_+ + \{1\}}(s^*) \\ + \delta^*_{\mathbf{L}^+_{\mathrm{X}}}(-\boldsymbol{c}^* c^* - \boldsymbol{m}^* m^*, -s^*) \\ \geq \langle (\boldsymbol{x}^*, y^*), (\boldsymbol{x}, y)\rangle - \delta^*_{\mathbf{R}_+ + \{1\}}(\langle \boldsymbol{c}^*, \boldsymbol{x}\rangle - c) - \delta^*_{\mathbf{R}_+ + \{1\}}(m - \langle \boldsymbol{m}^*, \boldsymbol{x}\rangle) \\ - \delta^*_{\mathbf{R}_+ + \{1\}}(y - s) - \delta_{\mathbf{L}^+_{\mathrm{X}}}(\boldsymbol{x}, y).$$

The reciprocal dual is a nonpolyhedral linear problem.

Problem 2 (Efficiency Dual)

$$\inf_{c^*, m^*, s^*} \Big[c^* c + m^* m + s^* s \\ + \delta_{\mathbf{L}^{\mathrm{X}}_+}(\boldsymbol{c}^* c^* + \boldsymbol{m}^* m^*, s^*) \\ + \delta_{\mathbf{R}_+ + \{1\}}(c^*) + \delta_{\mathbf{R}_+ + \{1\}}(-m^*) + \delta_{\mathbf{R}_+ + \{1\}}(s^*)\Big].$$

The dual satisfies the polyhedral/relative condition, so there is no duality gap; and the optimum value of the primal is attained. The primal does not satisfy the polyhedral/relative condition.

For the fourth Fenchel factor (64.4) to be zero is complementary slackness. The Jordan product of the arguments must be zero:

$$\mathbf{0} = -s^*\boldsymbol{x} + y\,(-\boldsymbol{c}^*c^* - \boldsymbol{m}^*m^*) \tag{64.5}$$
$$0 = \langle -\boldsymbol{c}^*c^* - \boldsymbol{m}^*m^*, \boldsymbol{x}\rangle - s^*y. \tag{64.6}$$

By Eq. (64.5),

$$\boldsymbol{x}/y = -\,(c^*/s^*)\,\boldsymbol{c}^* - (m^*/s^*)\,\boldsymbol{m}^*, \tag{64.7}$$

so $\boldsymbol{x}$ must be a positive linear combination of $-\boldsymbol{c}^*$ and $\boldsymbol{m}^*$. Substituting this relationship into (64.6) yields both

$$y = |\boldsymbol{x}|$$

and

$$s^* = |\boldsymbol{c}^*c^* + \boldsymbol{m}^*m^*|\,, \tag{64.8}$$

general properties for complementary slackness in Lorentz space.

We investigate whether $(\overline{\boldsymbol{x}}, \overline{y})$ solves the efficiency primal. If $\overline{\boldsymbol{x}}$ is a positive linear combination of $-\boldsymbol{c}^*$ and $\boldsymbol{m}^*$, we show that the optimum value of the efficiency primal is zero. Set $\overline{y} = |\overline{\boldsymbol{x}}|$. Select the dual choice variables so all Fenchel factors (64.1)–(64.4) are zero. The first three Fenchel factors (64.1)–(64.3) are zero if and only if $c^* \geq 1$, $-m^* \geq 1$, and $s^* \geq 1$, respectively. Subject to these constraints, select the dual choice variables so (64.7) holds. The left-hand side has norm one, so the right-hand side does as well, and (64.8) follows. The fourth Fenchel factor (64.4) is zero. One concludes that the separation Theorem 1 holds.

Expression (64.7) reveals the role of cost and mean positivity. As

$$\langle \boldsymbol{c}^*, \boldsymbol{x}\rangle\,/y = -\,(c^*/s^*)\,\langle \boldsymbol{c}^*, \boldsymbol{c}^*\rangle - (m^*/s^*)\,\langle \boldsymbol{c}^*, \boldsymbol{m}^*\rangle\,,$$

both terms on the right-hand side are negative if $\langle \boldsymbol{c}^*, \boldsymbol{m}^*\rangle < 0$. Every efficient portfolio has a negative cost. Cost and mean positivity is necessary and sufficient for the existence of efficient portfolios having a positive cost.

64.4 Minimum Variance

A necessary (but not sufficient) condition for efficiency is to minimize the variance among portfolios having the same cost and mean. We solve this problem, which generalizes the minimization of variance for a given cost. This minimum-variance formulation yields a graphical analysis that interprets the meaning of efficiency. The value function $V(c,m)$ is half the minimum variance.

Theorem 2 (Minimum Variance for Given Cost and Mean)

$$\begin{aligned}V(c,m) &:= \inf_{x}\left(\frac{1}{2}\langle \boldsymbol{x},\boldsymbol{x}\rangle + \delta_{\langle \boldsymbol{c}^*,\boldsymbol{x}\rangle = c} + \delta_{\langle \boldsymbol{m}^*,\boldsymbol{x}\rangle = m}\right)\\ &= \frac{1}{2}\left\langle \begin{bmatrix} c \\ m \end{bmatrix}, [\mathrm{Var}(\boldsymbol{c}^*,\boldsymbol{m}^*)]^{-1}\begin{bmatrix} c \\ m \end{bmatrix}\right\rangle.\end{aligned}$$

We derive this formula by evaluating the dual, which reduces to the evaluation of the conjugate of a quadratic function.

Because the problem has much structure, the reciprocal dual is simple. Adding the Fenchel inequalities

$$\begin{aligned}&\frac{1}{2}\langle \boldsymbol{x},\boldsymbol{x}\rangle + \frac{1}{2}\langle -\boldsymbol{c}^*c^* - \boldsymbol{m}^*m^*, -\boldsymbol{c}^*c^* - \boldsymbol{m}^*m^*\rangle \\ &\qquad + \langle -\boldsymbol{c}^*c^* - \boldsymbol{m}^*m, \boldsymbol{x}\rangle \ge 0\end{aligned} \tag{64.9}$$

$$\delta_{\{0\}}\left(\boldsymbol{c}^{*\top}\boldsymbol{x} - c\right) + \delta^*_{\{0\}}(-c^*) - \left\langle -c^*, \boldsymbol{c}^{*\top}\boldsymbol{x} - c\right\rangle \ge 0 \tag{64.10}$$

$$\delta_{\{0\}}\left(\boldsymbol{m}^{*\top}\boldsymbol{x} - m\right) + \delta^*_{\{0\}}(-m^*) - \left\langle -m^*, \boldsymbol{m}^{*\top}\boldsymbol{x} - m\right\rangle \ge 0. \tag{64.11}$$

and rearranging gives

$$\begin{aligned}&\frac{1}{2}\langle \boldsymbol{x},\boldsymbol{x}\rangle + \delta_{\{0\}}\left(\boldsymbol{c}^{*\top}\boldsymbol{x} - c\right) + \delta_{\{0\}}\left(\boldsymbol{m}^{*\top}\boldsymbol{x} - m\right)\\ &\quad \ge c^*c + m^*m - \frac{1}{2}\langle -\boldsymbol{c}^*c^* - \boldsymbol{m}^*m, -\boldsymbol{c}^*c^* - \boldsymbol{m}^*m\rangle.\end{aligned}$$

Recognize that the dual is the evaluation of the conjugate of a quadratic function.

Problem 3 (Minimum Variance for Given Cost and Mean Dual)

$$\begin{aligned}&\sup_{\begin{bmatrix} c^* \\ m^* \end{bmatrix}}\left[\left\langle \begin{bmatrix} c^* \\ m^* \end{bmatrix}, \begin{bmatrix} c \\ m \end{bmatrix}\right\rangle\right.\\ &\qquad \left. -\frac{1}{2}\left\langle \begin{bmatrix} c^* \\ m^* \end{bmatrix}, \mathrm{Var}(\boldsymbol{c}^*,\boldsymbol{m}^*)\begin{bmatrix} c^* \\ m^* \end{bmatrix}\right\rangle\right]\\ &\qquad = \frac{1}{2}\left\langle \begin{bmatrix} c \\ m \end{bmatrix}, [\mathrm{Var}(\boldsymbol{c}^*,\boldsymbol{m}^*)]^{-1}\begin{bmatrix} c \\ m \end{bmatrix}\right\rangle.\end{aligned}$$

Both the primal and the reciprocal dual satisfy the polyhedral/relative condition, so there is no duality gap, and the optimum value is attained in both the primal and the dual.

The solution to the dual is the derivative of the value function,

$$\begin{bmatrix} c^* \\ m^* \end{bmatrix} = \left(\begin{bmatrix} \boldsymbol{c}^* & \boldsymbol{m}^* \end{bmatrix}^{\mathsf{T}} \begin{bmatrix} \boldsymbol{c}^* & \boldsymbol{m}^* \end{bmatrix} \right)^{-1} \begin{bmatrix} c \\ m \end{bmatrix}.$$

When the Fenchel factor (64.9) is zero,

$$\boldsymbol{x} = -\begin{bmatrix} \boldsymbol{c}^* & \boldsymbol{m}^* \end{bmatrix} \begin{bmatrix} c^* \\ m^* \end{bmatrix}.$$

Here $\boldsymbol{x}$ is a linear combination of $\boldsymbol{c}^*$ and $\boldsymbol{m}^*$ but is not necessarily a nonnegative linear combination of $-\boldsymbol{c}^*$ and $\boldsymbol{m}^*$. Although $\boldsymbol{x}$ is variance-minimizing, it is not necessarily efficient.

A graph in (c, m)-space, Fig. 64.1 portrays the portfolio-choice separation theorem. The ellipse is the contour line along which the minimum variance is one. Consider the point on the upper right of the ellipse where the contour line is vertical. The portfolio that achieves this point is the portfolio that minimizes variance for a given cost, so portfolio must be proportional to $\boldsymbol{c}^*$. That the variance is one entails that the portfolio is $\boldsymbol{c}^* / |\boldsymbol{c}^*|$, and the cost is $|\boldsymbol{c}^*|$. Analogously, consider the point on the upper right of the ellipse where the contour line is horizontal. The portfolio that achieves this point is the portfolio that minimizes variance for a given mean, so portfolio must be proportional to $\boldsymbol{m}^*$. That the variance is one entails that the portfolio is $\boldsymbol{m}^* / |\boldsymbol{m}^*|$, and the mean is $|\boldsymbol{m}^*|$.

Figure 64.1 assumes cost and mean positivity, $\langle \boldsymbol{c}^*, \boldsymbol{m}^* \rangle > 0$. Portfolio $\boldsymbol{c}^*$ has a positive mean, and portfolio $\boldsymbol{m}^*$ has a positive cost. The principal axis of the ellipse slopes up and right. The efficient portfolios correspond to the shaded cone, generated by $-\boldsymbol{c}^*$ and $\boldsymbol{m}^*$.

Since a more efficient portfolio calls for lower cost, higher mean, and lower variance, the points along the upper left of the ellipse correspond to the efficient

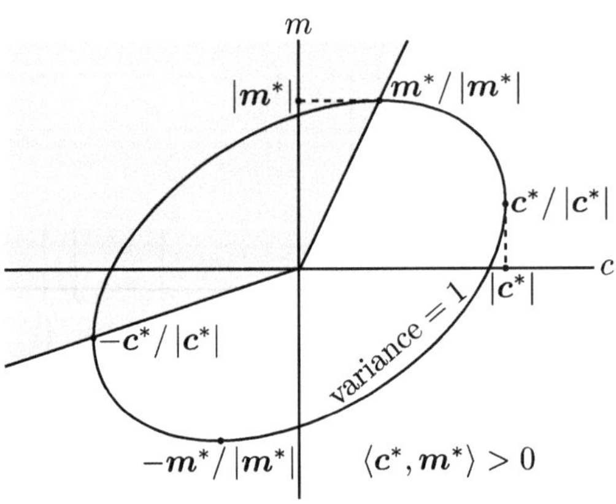

Fig. 64.1 Minimum variance given cost and mean

portfolios for this variance. Given some point on the upper left, points above and left are better, in that the cost is lower or the mean is higher; however the variance is necessarily greater. Along the upper left of the ellipse, the portfolios corresponding to the points on the right end have positive cost and positive mean. The efficient points in the middle have positive mean but negative cost. The efficient points at the left end have negative mean and negative cost. Having a negative cost is not nonsense. For example, a derivative security may involve an initial receipt of money, followed by a payment later.

The other points in Fig. 64.1 correspond to portfolios that are not efficient. For each portfolio, there exists another portfolio having a lower cost, a higher mean, and a lower variance.

This analysis corrects traditional thinking. Although the portfolio $\boldsymbol{c}^*$ minimizes the variance for a given cost, it is not efficient. The point corresponding to $\boldsymbol{c}^* / |\boldsymbol{c}^*|$, where the ellipse is vertical, is not efficient. Moving up and left in the figure obtains a lower cost, a higher mean, and a lower variance. In contrast, in traditional thinking, this portfolio is efficient, because it is mean/variance efficient for the cost $|\boldsymbol{c}^*|$.

One can describe traditional thinking via Fig. 64.2. Let r denote the correlation of $\boldsymbol{c}^*$ and $\boldsymbol{m}^*$. For cost $|\boldsymbol{c}^*|$, the parabola plots the minimum variance as a function of the mean m. The minimum variance one is achieved for portfolio $\boldsymbol{c}^* / |\boldsymbol{c}^*|$, which has mean $r\,|\boldsymbol{m}^*|$. In traditional thinking, this portfolio is efficient. And the efficient portfolios for this cost are those that attain the points up and right along the parabola. In Fig. 64.1, these portfolios attain the points vertically above the point attained by the portfolio $\boldsymbol{c}^* / |\boldsymbol{c}^*|$. Moving upward raises both the mean and the variance, while keeping the cost constant.

However these portfolios fail to be cost/mean/variance efficient, until the upward movement arrives at a portfolio proportional to $\boldsymbol{m}^*$. The efficient portfolio $\frac{1}{r}\boldsymbol{m}^* / |\boldsymbol{m}^*|$

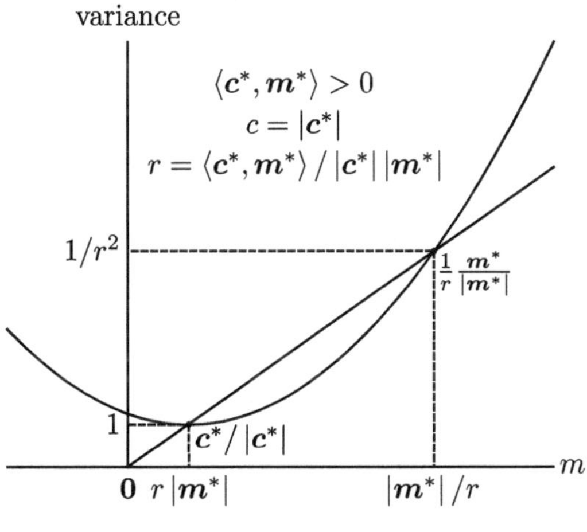

Fig. 64.2 Minimum variance

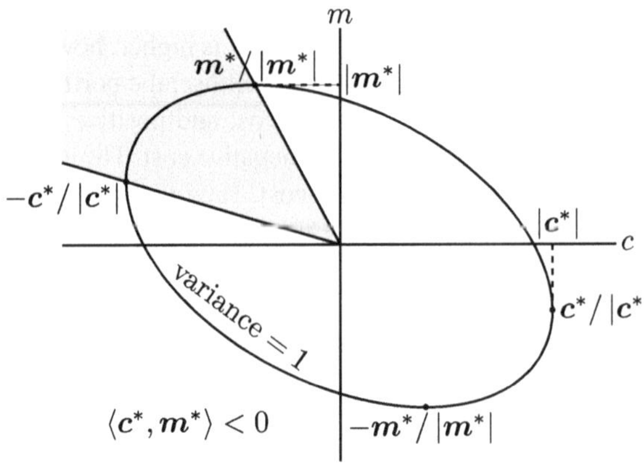

Fig. 64.3 Efficient portfolios with negativity

has mean $|\boldsymbol{m}^*|$ and variance $1/r^2$. Direct calculation shows that the ray from the origin through the minimum-variance point also passes through this point.

This result calls for a reconsideration of the theory of mean/variance portfolio choice. For example, an investor having very high risk aversion should not buy the portfolio $\boldsymbol{c}^*$. A better action is to consume more and to save less, and to invest a smaller amount in a more risky portfolio. One cannot model the portfolio choice independently of the consumption/saving choice.

Whereas so far we have assumed cost and mean positivity, now let us turn to the opposite situation: $\langle \boldsymbol{c}^*, \boldsymbol{m}^* \rangle < 0$. Figure 64.3 plots the ellipse of constant minimum variance in $(c.m)$-space. Portfolio $\boldsymbol{c}^*$ has a negative mean, and portfolio $\boldsymbol{m}^*$ has a negative cost, so the principal axis of the ellipse slopes down and right.

The efficient portfolios correspond to the shaded cone, generated by $-\boldsymbol{c}^*$ and $\boldsymbol{m}^*$. We obtain the odd result that every cost/mean/variance efficient portfolio has a positive mean but a negative cost. If each consumer held an efficient portfolio, then the market portfolio of all assets would have a negative cost.

Our interpretation is that cost and mean positivity should be seen as a required property of asset-market equilibrium.[2] A model of equilibrium should assume positivity. If there is a risk-free asset, then positivity is necessary to avoid arbitrage. If there is no risk-free asset, however, negativity is compatible with no arbitrage. Nevertheless to assume cost and mean positivity as a required property of asset market equilibrium seems essential.

---

[2] Cost and mean negativity might apply to a subset of all assets, such as a set of derivative assets.

Chapter 65
Capital-Asset Pricing

Abstract The analysis here focuses on the allocation of risk and the implication of risk for asset pricing. Working with a simple mean/variance utility, we calculate portfolio demand. We obtain the capital-asset pricing model (Sharpe, W. F. (1964, September). Capital asset prices: A theory of market equilibrium under conditions of risk. Journal of Finance, XIX (3), 425–442). By setting the aggregate portfolio demand equal to the aggregate supply (the market portfolio of all assets), we solve to obtain the equilibrium price of assets. The risk premium on an asset is proportional to the covariance of its payoff with the payoff on the market portfolio. Via primal and dual efficiency problems, we also study efficient risk sharing. How can the market portfolio be divided among investors so that the risk is shared efficiently? We find a simple necessary and sufficient condition for efficient risk sharing. The aggregate risk of the market portfolio is divided among the investors according to risk aversion. Each investor holds a fraction of the market portfolio plus some positive or negative quantity of the risk-free asset (each lends or borrows via the risk-free asset). The payoff for one investor is perfectly correlated with the payoff for another. The standard deviation of the payoff is inversely proportional to absolute risk aversion. An investor having high risk aversion holds mostly the risk-free asset, whereas an investor having low risk aversion makes a leveraged investment in the market portfolio. We obtain first and second theorems of efficient risk sharing, a variant of the first and second theorems of welfare economics. Asset-market equilibrium shares risk efficiently. Conversely, an efficient allocation of risk can be obtained as a market equilibrium, for some asset price.

65.1 Asset Space

The analysis here continues the mathematical framework of "Efficient Portfolios Risk-Free Asset." We analyze portfolio choice in the moment space of assets, a Euclidean space. The inner product of two portfolios is the noncentral second moment of their payoffs. We assume that the law of one price holds, so we identify assets hav-

B. C. Dieffenbach, *Conjugate Duality in Economic Analysis*, Contributions to Economics,
https://doi.org/10.1007/978-3-032-21396-9_65

ing the same payoff as the same. There exists a risk-free asset **1**, which has constant payoff of one dollar.

The asset space is the direct sum of the subspace **X** of portfolios having payoff mean zero and the subspace $\mathbf{R}\{\mathbf{1}\}$ of all multiples of **1**. An asset $\boldsymbol{w}$ has the unique decomposition

$$\boldsymbol{w} = \boldsymbol{x} + y\mathbf{1},$$

which we denote by the pair $(\boldsymbol{x}, y) \in \mathbf{X} \oplus \mathbf{R}\{\mathbf{1}\}$. Its payoff has mean y and standard deviation $|\boldsymbol{x}|$.

The mean vector is $(\mathbf{0}, 1) = \mathbf{1}$. This vector specifies the mean payoff on different portfolios, which says nothing about expected rates of return.

Let $(\boldsymbol{p}^*, q^*) = \boldsymbol{p}^* + q^*\mathbf{1}$ denote the pricing portfolio. Assume $\boldsymbol{p}^* \neq \mathbf{0}$. Assume cost and mean positivity: the inner product $\langle(\boldsymbol{p}^*, q^*), (\mathbf{0}, 1)\rangle = q^* > 0$. The pricing portfolio specifies the relative price of different assets. Viewing its normalization as arbitrary, impose the normalization $q^* = 1$.

65.2 Mean/Variance Portfolio Choice

Working with von Neumann–Morgenstern expected utility, Pratt and Arrow show that for small risks there is a linear tradeoff between mean and variance (Pratt 1964; Arrow 1984). Even though here risks are not small, we define utility as a linear function of mean and variance.

Definition 1 (Mean/Variance Linear Utility) The utility of portfolio $\boldsymbol{w} = (\boldsymbol{x}, y)$ is the proper, concave function

$$\mathrm{E}(\boldsymbol{w}) - \frac{a}{2}\mathrm{Var}(\boldsymbol{w}) = y - \frac{a}{2}\langle\boldsymbol{x}, \boldsymbol{x}\rangle, a > 0.$$

In the Pratt-Arrow formulation, the tradeoff between mean and variance is set by the absolute risk aversion a, which measures the willingness to take a risk of a given absolute size. We continue to refer to a as the absolute risk aversion.

The portfolio choice maximizes utility, subject to the constraint that the portfolio costs v.

Problem 1 (Portfolio Choice)

$$\sup_{(\boldsymbol{x}, y)}\left[y - \frac{a}{2}\langle\boldsymbol{x}, \boldsymbol{x}\rangle - \delta_{\langle \boldsymbol{p}^*, \boldsymbol{x}\rangle + y = v}\right].$$

Maximum utility is a conjugate:

$$\begin{aligned}&\sup_{x,y}\left[y-\frac{a}{2}\langle \boldsymbol{x},\boldsymbol{x}\rangle-\delta_{\langle \boldsymbol{p}^*,\boldsymbol{x}\rangle+y=v}\right]\\&=v+\sup_{\boldsymbol{x}}\left[-\langle \boldsymbol{p}^*,\boldsymbol{x}\rangle-\frac{a}{2}\langle \boldsymbol{x},\boldsymbol{x}\rangle\right]\\&=v+\frac{1}{2a}\langle \boldsymbol{p}^*,\boldsymbol{p}^*\rangle.\end{aligned}$$

By the first-order condition for $\boldsymbol{x}$ and the budget constraint,

$$\boldsymbol{x}=-\boldsymbol{p}^*/a \tag{65.1}$$

$$y=v-\langle \boldsymbol{p}^*,\boldsymbol{x}\rangle. \tag{65.2}$$

As $a\to\infty$, the optimum portfolio becomes the risk-free asset. A higher v raises the demand for the risk-free asset by the same amount and has no effect on the risky portfolio $\boldsymbol{x}$.

The portfolio demand (65.1)–(65.2) embodies the portfolio-choice separation theorem, which separates the choice into two steps (Tobin 1958). Step one is to determine the portfolio $-\boldsymbol{p}^*$, which is the same for every investor and is independent of risk aversion. Step two is to decide how much to invest in this portfolio (an amount inversely related to the absolute risk aversion) and how much in the risk-free asset. The portfolio demand is a cost/mean/variance efficient portfolio.

65.3 Market Equilibrium

Setting the aggregate supply of assets equal to the aggregate demand obtains the capital-asset pricing model (Sharpe 1964). Let $(\boldsymbol{p},q)$ denote the market portfolio, the aggregate supply of assets. Let a_i denote the absolute risk aversion for consumer i, and let $\boldsymbol{w}_i=(\boldsymbol{x}_i,y_i)$ denote the utility-maximizing portfolio choice. Setting supply equal to demand gives

$$\boldsymbol{p}=\sum_i \boldsymbol{x}_i=-\sum_i \boldsymbol{p}^*/a_i, \tag{65.3}$$

so the equilibrium

$$\boldsymbol{p}^*=-\frac{1}{\sum_i \frac{1}{a_i}}\boldsymbol{p}. \tag{65.4}$$

The risk premium on a portfolio is the expected value of its payoff less the payoff of a risk-free asset costing the same:

$$\langle \mathbf{1},\boldsymbol{x}+y\mathbf{1}\rangle-\langle \boldsymbol{p}^*+\mathbf{1},\boldsymbol{x}+y\mathbf{1}\rangle=-\langle \boldsymbol{p}^*,\boldsymbol{x}\rangle,$$

the negative of the covariance of $\boldsymbol{p}^*$ with the payoff. By Eq. (65.4), the risk premium is

$$\frac{1}{\sum_i \frac{1}{a_i}} \langle \boldsymbol{p}, \boldsymbol{x} \rangle ,$$

a term depending on risk aversion times the covariance of the payoff with the payoff on the market portfolio. If individuals have less risk aversion, the risk premium is less.

The risk premium derives from opportunity for diversification. If the covariance is negative, the asset is particularly useful for diversification, as it generally provides a higher payoff when other assets have a lower payoff. This property bids up its price and makes the risk premium negative. Necessarily a typical asset has a positive covariance with the market portfolio, so the risk premium is positive.[1]

65.4 Efficient Risk Sharing

Let an n-tuple $\boldsymbol{y} := (y_1, y_2, \ldots) \in \mathbf{R}^n$ and $X := \begin{bmatrix} \boldsymbol{x}_1 & \boldsymbol{x}_2 & \cdots \end{bmatrix} \in \mathbf{T}^{n \to \mathrm{X}}$ denote an allocation of assets to n individuals. The efficiency problem analyzes whether an allocation $\left(\overline{X}, \overline{\boldsymbol{y}}\right)$ is efficient. Analyzing the primal and dual furnishes a proof of the first and second theorems of efficient risk sharing, a version of the first and second theorems of welfare economics.

Problem 2 (Efficient Risk Sharing)

$$\begin{aligned} &\sup_{X,\boldsymbol{y}} \left\{ -\sum_i \delta^*_{\mathbf{R}_+ + \{1\}} \left[-u_i - y_i + \frac{a_i}{2} \langle \boldsymbol{x}_i, \boldsymbol{x}_i \rangle \right] - \delta_{(X\mathbf{1}, \boldsymbol{y}^\top \mathbf{1}) = (\boldsymbol{p}, q)} \right\} \\ &\quad = \sup_{X,\boldsymbol{y},\boldsymbol{z}} \left\{ -\sum_i \delta^*_{\mathbf{R}_+ + \{1\}} \left(-u_i - y_i + a_i z_i \right) - \sum_i \delta_{\operatorname{epi} f} \left(\boldsymbol{x}_i, z_i \right) - \delta_{(X\mathbf{1}, \boldsymbol{y}^\top \mathbf{1}) = (\boldsymbol{p}, q)} \right\}. \end{aligned}$$

---

[1] Consider the least-squares linear regression of $\boldsymbol{x} + y\mathbf{1}$ on the market portfolio $\boldsymbol{p} + q\mathbf{1}$ and the constant $\mathbf{1}$:

$$\boldsymbol{x} + y\mathbf{1} = \alpha + \beta \left(\boldsymbol{p} + q\mathbf{1} \right) + \text{error}.$$

By least-squares theory, the error has mean zero and the coefficient

$$\beta = \frac{\langle \boldsymbol{p}, \boldsymbol{x} \rangle}{\langle \boldsymbol{p}, \boldsymbol{p} \rangle},$$

so the risk premium is proportional to the beta coefficient β. If the payoff on an asset is negatively correlated with the payoff on the market portfolio, then is particularly useful for diversification, which raises its price and makes the risk premium negative.

The base value of the perturbation $(\boldsymbol{p}, q)$ is the market portfolio. The base value of the perturbation $u_i = -\overline{y}_i + \frac{a_i}{2}\langle \overline{\boldsymbol{x}_i}, \overline{\boldsymbol{x}_i} \rangle$, the negative of the utility of investor i. We have $\mathbf{1}^\top \overline{\boldsymbol{y}} = q$ and $\overline{X}\mathbf{1} = \boldsymbol{p}$.

We find the reciprocal dual. Adding the Fenchel inequalities

$$\begin{aligned}\delta_{\text{epi}\, f}(\boldsymbol{x}_i, z_i) + \delta^*_{\text{epi}\, f}(-\boldsymbol{p}^* + \boldsymbol{x}_i^*, -a_i u_i^* + z_i^*) - \langle -\boldsymbol{p}^* + \boldsymbol{x}_i^*, \boldsymbol{x}_i \rangle \\ -(-a_i u_i^* + z_i^*) z_i \geq 0 \text{ for all } i\end{aligned} \tag{65.5}$$

$$\delta_{\mathbf{R}_+ + \{1\}}(u_i^*) + \delta^*_{\mathbf{R}_+ + \{1\}}(-u_i - y_i + a_i z_i) - \langle u_i^*, -u_i - y_i + a_i z_i, \rangle \geq 0 \text{ for all } i \tag{65.6}$$

$$\delta_{\{0\}}(u_i^* - q^* + y_i^*) + \delta^*_{\{0\}}(y_i) - (u_i^* - q^* + y_i^*) y_i \geq 0 \text{ for all } i \tag{65.7}$$

$$\delta_{\{0\}}(X\mathbf{1} - \boldsymbol{p}) + \delta^*_{\{0\}}(\boldsymbol{p}^*) - \langle \boldsymbol{p}^*, X\mathbf{1} - \boldsymbol{p} \rangle \geq 0 \tag{65.8}$$

$$\delta_{\{0\}}(\boldsymbol{y}^\top \mathbf{1} - q) + \delta^*_{\{0\}}(q^*) - q^*(\boldsymbol{y}^\top \mathbf{1} - q) \geq 0 \tag{65.9}$$

and rearranging obtains

$$\begin{aligned}&\langle \boldsymbol{p}^*, \boldsymbol{p} \rangle + q^* q + \langle \boldsymbol{u}^*, \boldsymbol{u} \rangle \\ &\quad + \sum_i \delta^*_{\text{epi}\, f}(-\boldsymbol{p}^* + \boldsymbol{x}_i^*, -a_i u_i^* + z_i^*) + \sum_i \delta_{\{0\}}(u_i^* - q^* + y_i^*) + \sum_i \delta_{\mathbf{R}_+ + \{1\}}(u_i^*) \\ &\geq \langle X^*, X \rangle + \langle \boldsymbol{y}^*, \boldsymbol{y} \rangle + \langle \boldsymbol{z}^*, \boldsymbol{z} \rangle \\ &\quad - \sum_i \delta^*_{\mathbf{R}_+ + \{1\}}(-u_i - y_i + a_i z_i) - \sum_i \delta_{\text{epi}\, f}(\boldsymbol{x}_i, z_i) - \delta_{(X\mathbf{1}, \boldsymbol{y}^\top \mathbf{1}) = (\boldsymbol{p}, q)}.\end{aligned}$$

The dual perturbation base values are $X^* = 0, \boldsymbol{y}^* = \mathbf{0}, \boldsymbol{z}^* = \mathbf{0}$.

We reduce the dual, by substitution. By (65.7), as $u_i^* = q^*$, one can eliminate u_i^*. In the dual

$$\begin{aligned}&\delta^*_{\text{epi}\, f}(-\boldsymbol{p}^* + \boldsymbol{x}_i^*, -a_i u_i^* + z_i^*) \\ &= \delta^*_{\text{epi}\, f}(-\boldsymbol{p}^*, -a_i q^*) \\ &= (a_i q^* \star f^*)(-\boldsymbol{p}^*) \\ &= \frac{1}{a_i q^*} \langle \boldsymbol{p}^*, \boldsymbol{p}^* \rangle.\end{aligned}$$

We evaluate the reduced dual to find the optimum value:

$$\begin{aligned}&\inf_{(\boldsymbol{p}^*, q^*)} \left[\langle \boldsymbol{p}^*, \boldsymbol{p} \rangle + q^* q + q^* \langle \mathbf{1}, \boldsymbol{u} \rangle + \frac{1}{2} \left(\sum_i \frac{1}{a_i} \right) \frac{\langle \boldsymbol{p}^*, \boldsymbol{p}^* \rangle}{q^*} + \delta_{\mathbf{R}_+ + \{1\}}(q^*) \right] \\ &= \inf_{q^* \geq 1} q^* \left\{ q + \langle \mathbf{1}, \boldsymbol{u} \rangle - \sup_{\boldsymbol{p}^*/q^* \in \mathbf{X}} \left[\langle \boldsymbol{p}^*/q^*, -\boldsymbol{p} \rangle - \frac{1}{2} \left(\sum_i \frac{1}{a_i} \right) \langle \boldsymbol{p}^*/q^*, \boldsymbol{p}^*/q^* \rangle \right] \right\}\end{aligned} \tag{65.10}$$

$$= \inf_{q^* \geq 1} q^* \left(q + \langle \mathbf{1}, \boldsymbol{u} \rangle - \frac{1}{2} \frac{1}{\sum_i \frac{1}{a_i}} \langle \boldsymbol{p}, \boldsymbol{p} \rangle \right) \tag{65.11}$$

$$= q - \frac{1}{2} \frac{1}{\sum_i \frac{1}{a_i}} \langle \boldsymbol{p}, \boldsymbol{p} \rangle + \langle \mathbf{1}, \boldsymbol{u} \rangle. \tag{65.12}$$

Always $q^* = 1$ is minimizing. If the expression in parentheses is zero, then the optimum value is zero. If the expression is positive, the minimum value is positive, so set $q^* = 1$. The allocation is efficient if and only if the optimum value is zero. Interpret the first-order condition for the dual as aggregate supply equals aggregate demand: for $q^* = 1$, the first-order condition for the inner optimization in (65.10) is (65.3).

Consider the Fenchel equalities, after substituting $u_i^* = q^* = 1$. Fenchel factor (65.5) zero means

$$\boldsymbol{x}_i = -\boldsymbol{p}^*/a_i \text{ and } z_i = \frac{1}{2} \langle \boldsymbol{x}_i, \boldsymbol{x}_i \rangle \text{ for all } i,$$

the portfolio demand (65.1). Fenchel factor (65.6) zero means

$$y_i - a_i z_i \geq -u_i;$$

utility is greater than or equal to base utility. Fenchel factors (65.8) and (65.9) zero mean

$$X\mathbf{1} = \boldsymbol{p}$$
$$\boldsymbol{y}^\top \mathbf{1} = q,$$

aggregate demand equals supply.

We have proved the following.

Theorem 1 (Efficient Asset Allocation) *An asset allocation shares risk efficiently if and only if*

$$\boldsymbol{x}_i = \left(\frac{1}{a_i} / \sum_j \frac{1}{a_j} \right) \boldsymbol{p} \text{ for all } i.$$

For every investor $\boldsymbol{x}_i$ is some proportion of the aggregate supply $\boldsymbol{p}$, such that the proportion is less if the risk aversion is greater.

Theorem 2 (First Theorem of Efficient Risk Sharing) *In an asset-market equilibrium, risk sharing is efficient.*

An asset-market equilibrium $(\boldsymbol{p}, q, X, \boldsymbol{y}, \boldsymbol{p}^*)$ solves the efficiency primal and dual, with optimum value zero. Take the portfolio demands as the primal choice variables, and the price vector as the dual choice variable. By definition, $X\mathbf{1} = \boldsymbol{p}, \boldsymbol{y}^\top \mathbf{1} = q$, and $q^* = 1$. Set the base values $\overline{X} = X, \overline{\boldsymbol{y}} = \boldsymbol{y}$, and $\overline{\boldsymbol{u}^*} = \mathbf{1}$. These choices satisfy the primal and the dual, and the value is zero. The market equilibrium shares risk efficiently.

Theorem 3 (Second Theorem of Efficient Risk Sharing) *An asset allocation having efficient risk sharing is an asset-market equilibrium, for some price.*

Consider an allocation $(\boldsymbol{p}, q, X, \boldsymbol{y})$ having efficient risk sharing. By definition, $X\mathbf{1} = \boldsymbol{p}$ and $\boldsymbol{y}^\top \mathbf{1} = q$. The optimum value is zero. The dual solution provides the equilibrium price. Let $\boldsymbol{p}^*$, $q^* = 1$, and $\boldsymbol{u}^* = \mathbf{1}$ denote a solution to the dual. Then the allocation is an asset-market equilibrium for price $\boldsymbol{p}^*$. That the Fenchel factor (65.7) is zero means that X is the portfolio demand.

Because the utility function has its simple quadratic form, an allocation is efficient if and only if it maximizes the sum of utilities, subject to the resource constraint. Consider the maximization

$$\begin{aligned}
&\sup_{X,y} \left[\sum_i \left(y_i - \frac{a_i}{2} \langle \boldsymbol{x}_i, \boldsymbol{x}_i \rangle \right) - \delta_{(X\mathbf{1}, \boldsymbol{y}^\top \mathbf{1}) = (\boldsymbol{p}, q)} \right] \\
&= q + \sup_X \left(- \sum_i \frac{a_i}{2} \langle \boldsymbol{x}_i, \boldsymbol{x}_i \rangle - \delta_{X\mathbf{1} = \boldsymbol{p}} \right).
\end{aligned}$$

We calculate the dual of the maximization. Adding the Fenchel inequalities

$$\begin{aligned}
&\frac{a_i}{2} \langle \boldsymbol{x}_i, \boldsymbol{x}_i \rangle + \frac{1}{2a_i} \langle -\boldsymbol{p}^*, -\boldsymbol{p}^* \rangle - \langle -\boldsymbol{p}^*, \boldsymbol{x}_i \rangle \geq 0 \text{ for all } i \\
&\delta_{\{0\}} (X\mathbf{1} - \boldsymbol{p}) + \delta^*_{\{0\}} (\boldsymbol{p}^*) - \langle \boldsymbol{p}^*, X\mathbf{1} - \boldsymbol{p} \rangle \geq 0
\end{aligned}$$

and rearranging obtains

$$\begin{aligned}
&\langle \boldsymbol{p}^*, \boldsymbol{p} \rangle + \frac{1}{2} \left(\sum_i \frac{1}{a_i} \right) \langle \boldsymbol{p}^*, \boldsymbol{p}^* \rangle \\
&\geq - \sum_i \frac{a_i}{2} \langle \boldsymbol{x}_i, \boldsymbol{x}_i \rangle - \delta_{X\mathbf{1} = \boldsymbol{p}}
\end{aligned}$$

The primal satisfies the relative condition, so there is no duality gap. By the dual, the maximum sum of utilities

$$\begin{aligned}
&q + \inf_{\boldsymbol{p}^*} \left[\langle \boldsymbol{p}^*, \boldsymbol{p} \rangle + \frac{1}{2} \left(\sum_i \frac{1}{a_i} \right) \langle \boldsymbol{p}^*, \boldsymbol{p}^* \rangle \right] \\
&= q - \sup_{\boldsymbol{p}^*} \left[\langle \boldsymbol{p}^*, -\boldsymbol{p} \rangle - \frac{1}{2} \left(\sum_i \frac{1}{a_i} \right) \langle \boldsymbol{p}^*, \boldsymbol{p}^* \rangle \right] \\
&= q - \frac{1}{2} \frac{1}{\sum_i \frac{1}{a_i}} \langle \boldsymbol{p}, \boldsymbol{p} \rangle.
\end{aligned}$$

In the evaluation of the conjugate, the first-order condition is the equilibrium condition (65.4).

$$\boldsymbol{p}^* = -\frac{1}{\sum_i \frac{1}{a_i}}\boldsymbol{p}.$$

The efficiency-dual optimum value (65.12) is the maximum sum of utilities less the sum of base utilities, which must be nonnegative. The allocation is efficient if and only if it maximizes the sum of utilities.

References

Arrow KJ (1984) The theory of risk aversion. In: Individual choice under certainty and uncertainty, collected papers of Kenneth J. Arrow. Harvard University Press, Cambridge, MA, pp 147–171

Pratt JW (1964 Jan–Apr) Risk aversion in the small and in the large. Econometrica 32(1-2):122–136

Sharpe WF (1964 Sept) Capital asset prices: a theory of market equilibrium under conditions of risk. J Fin XIX(3):425–442

Tobin J (1958 Feb) Liquidity preference as behavior towards risk. Rev Econ Stud XXV(2):65–86

Chapter 66
Arbitrage Pricing Theory

Abstract Ross puts forward the *arbitrage pricing theory* (Ross, S. A. (1976, December). The arbitrage theory of capital asset pricing. Journal of Economic Theory, 13 (3), 341–360; Ross, S. A. (1977). Return, risk, and arbitrage. In I. Friend & J. L. Bicksler (Eds.), Risk and return in finance (pp. 189–218). Cambridge, MA: Ballinger). The idea is that a factor structure of asset returns leads naturally to a model of risk premia, because otherwise there would exist an opportunity for arbitrage. After summarizing the argument of Ross, we develop the idea via a moment space of portfolios, which captures the factor structure by introducing a basis. The moment space decomposes into orthogonal complements: a systematic-risk subspace and a no-systematic-risk subspace. Calculating the Sharpe ratio in each subspace obtains a formulation of the arbitrage pricing theory.

66.1 Factor Model

Assume that there exists a risk-free asset, and consider a factor model for the excess return on a set of n assets:

$$\boldsymbol{m} + \mathsf{B}\boldsymbol{f} + \boldsymbol{e}.$$

The ith component is the excess return on asset i, the payoff for a one-dollar investment in the asset financed by borrowing one dollar at the risk-free rate. The mean excess return $\boldsymbol{m} \in \mathbf{R}^n$ is the vector of risk premia.

The factor $\boldsymbol{f} \in \mathbf{R}^m$, such that

$$\begin{aligned} \mathrm{E}(\boldsymbol{f}) &= \mathbf{0} \\ \mathrm{Var}(\boldsymbol{f}) &= \mathsf{I}, \end{aligned}$$

B. C. Dieffenbach, *Conjugate Duality in Economic Analysis*, Contributions to Economics,
https://doi.org/10.1007/978-3-032-21396-9_66

The beta coefficients $B = \begin{bmatrix} \boldsymbol{b}_1 & \cdots & \boldsymbol{b}_m \end{bmatrix} \in \mathbf{T}^{m \to n}$ are also called *factor loadings*. Here $\boldsymbol{e} \in \mathbf{R}^n$, such that

$$\mathrm{E}(\boldsymbol{e}) = \mathbf{0}$$
$$\mathrm{Var}(\boldsymbol{e}) = D,$$

in which $D \in \mathbb{S}^n_+$ is a diagonal matrix, and $\boldsymbol{e}$ is uncorrelated with $\boldsymbol{f}$.

Interpret the factor model as a decomposition of risk into systematic and no-systematic risk. The arbitrage pricing theory argues that the expected excess return is mostly determined by the systematic risk $B\boldsymbol{f}$. Assume that the number of factors is small, and most of the components of each beta coefficient $\boldsymbol{b}_i$ are not near zero. The diagonal elements of D are not too large, and the number of assets is large. The term $\boldsymbol{e}$ is the no-systematic risk and can be mostly eliminated by diversification. Invoking the law of one price, one concludes that the expected excess return is approximately some linear combination of the beta coefficients (Ross 1976, 1977).

That there can be no opportunity for arbitrage is the basis for the argument. The excess return for a portfolio $\boldsymbol{y} = (y_1, \ldots, y_n) \in \mathbf{R}^n$ is

$$\boldsymbol{y}^\top \boldsymbol{m} + \boldsymbol{y}^\top B\boldsymbol{f} + \boldsymbol{y}^\top \boldsymbol{e}.$$

Suppose that the portfolio is well-diversified: most of the components of $\boldsymbol{y}$ are non-zero. By the law of large numbers,

$$\boldsymbol{y}^\top \boldsymbol{e} \approx 0;$$

diversification eliminates the no-systematic risk. If the portfolio is chosen to eliminate the systematic risk ($\boldsymbol{y}^\top B = \mathbf{0}$), then it is nearly risk-free, so the law of one price implies $\boldsymbol{y}^\top \boldsymbol{m} = 0$:

$$\boldsymbol{y}^\top B = \mathbf{0} \Rightarrow \boldsymbol{y}^\top \boldsymbol{m} = 0.$$

(Of course this argument is not valid for an arbitrary portfolio but only for a well-diversified portfolio.) Thus $\boldsymbol{m}$ is orthogonal to $\mathrm{N}\left(B^\top\right)$. By the fundamental theorem of linear algebra, $\boldsymbol{m}$ must lie in $\mathrm{R}(B)$, so

$$\boldsymbol{m} \approx B\boldsymbol{b} \tag{66.1}$$

a linear combination of the beta coefficients.

The factor structure is exact if $\boldsymbol{e} = \mathbf{0}$ ($D = 0$), and (66.1) then holds exactly. If $D \gg 0$, then any expected excess return $\boldsymbol{m}$ is consistent with the law of one price. However, Ross argues that (66.1) should hold approximately.

The factor structure determines the expected excess return. One can apply the argument to any subset of all assets that exhibits a factor structure. The argument is not based on asset demand or supply, on market equilibrium. Whether investors have high or low risk aversion does not matter for the argument. Whether investors have a short or a long time horizon in their portfolio choice does not matter. The arbitrage pricing theory is incomplete in that it furnishes no theory of the weight given to each beta coefficient for the expected excess return.

The reasoning in the capital-asset pricing model is different. One derives the risk premium by equating asset demand and supply, for all assets, and not just for some subset of assets. A separation theorem describes portfolio choice: one invests partly in the efficient portfolio of risky assets and partly in the risk-free asset, the fraction depending on risk aversion. A typical investor holds the market portfolio of all assets. The key is the beta coefficient for how the rate of return on an asset rises or falls with the rate of return on the market portfolio. The risk premium is proportional to the beta coefficient. Equivalently, it is proportional to the covariance of the asset return with the return on the market portfolio, in which the proportionality is the relative risk aversion of a typical investor.

66.2 Moment Space of Portfolios

We model the factor structure via a moment space $\mathbf{X}$ of portfolios $\boldsymbol{x}$. The inner product of two portfolios is the covariance of the excess returns. Let $\boldsymbol{x}^*$ denote the mean vector: $\langle \boldsymbol{x}^*, \boldsymbol{x} \rangle$ is the mean excess return on portfolio $\boldsymbol{x}$.

Model the factor structure via a basis

$$X = \begin{bmatrix} \boldsymbol{x}_1 & \cdots & \boldsymbol{x}_n \end{bmatrix} \in \mathbf{T}^{n \to \mathbf{X}}.$$

The portfolio $\boldsymbol{x}_i$ is a one-dollar investment in asset i financed by borrowing one dollar at the risk-free rate. The variance of the basis vectors is the matrix

$$V := X^{\mathsf{T}} X \in \mathbf{S}^n_{++}.$$

(That the basis is not orthonormal affects the basis-coefficient formulas below.)

Let $\boldsymbol{y} \in \mathbf{Y} = \mathbf{R}^n$ denote the basis coefficients for a portfolio,

$$\boldsymbol{x} = X\boldsymbol{y}.$$

The covariance between two portfolios

$$\langle \tilde{\boldsymbol{x}}, \boldsymbol{x} \rangle = \langle X\tilde{\boldsymbol{y}}, X\boldsymbol{y} \rangle = \left\langle \tilde{\boldsymbol{y}}, X^{\mathsf{T}} X\boldsymbol{y} \right\rangle = \langle \tilde{\boldsymbol{y}}, V\boldsymbol{y} \rangle$$

(so the inner product $\langle \tilde{\boldsymbol{y}}, \boldsymbol{y} \rangle$ in $\mathbf{R}^n$ is not the covariance).

The relation $\boldsymbol{x}^* = X\boldsymbol{y}^*$ defines the basis coefficients for the mean vector. The portfolio mean excess return is the covariance

$$\langle \boldsymbol{x}^*, \boldsymbol{x} \rangle = \langle X\boldsymbol{y}^*, X\boldsymbol{y} \rangle = \left\langle \boldsymbol{y}^*, X^{\mathsf{T}} X\boldsymbol{y} \right\rangle = \langle \boldsymbol{y}^*, V\boldsymbol{y} \rangle .$$

Obtain the factor structure by specifying that the variance takes the known form

$$V = BB^{\top} + D,$$

in which the matrix B has rank m, and the diagonal matrix $D \gg 0$.

66.3 Systematic and No-Systematic Risk

Decompose the portfolio space into orthogonal complements—the "systematic-risk subspace" M and the "no-systematic-risk subspace" $M^{\perp}$. Let M and $I - M$ denote the corresponding orthogonal complements. A Pythagorean theorem follows immediately.

Theorem 1 (Sharpe-Ratio Pythagorean Theorem) *The square of the Sharpe ratio is the sum of the squares of the Sharpe ratios for the systematic-risk subspace and the no-systematic-risk subspace:*

$$|\boldsymbol{x}^*|^2 = |M\boldsymbol{x}^*|^2 + |(I - M)\boldsymbol{x}^*|^2 .$$

A beta coefficient $\boldsymbol{b}_i$ is the vector of basis coefficients for the portfolio $X\boldsymbol{b}_i$. The portfolios $\boldsymbol{x}$ uncorrelated with these portfolios satisfy

$$\mathbf{0} = B^{\top}X^{\top}\boldsymbol{x} = B^{\top}X^{\top}X\boldsymbol{y} = B^{\top}V\boldsymbol{y}.$$

These portfolios constitute the no-systematic-risk subspace.

Definition 1 (No-Systematic-Risk Subspace) The no-systematic-risk subspace is $M^{\perp} := \mathrm{N}\left(B^{\top}X^{\top}\right)$.

Expressed in basis coefficients, the subspace is $\mathrm{N}\left(B^{\top}V\right)$. The fundamental theorem of linear algebra sets the orthogonal complement.

Definition 2 (Systematic-Risk Subspace) The systematic risk subspace is $M = \mathrm{R}\,(XB)$.

Portfolios in the systematic-risk subspace are uncorrelated with portfolios in the no-systematic-risk subspace. The corresponding idempotent

$$M = (XB)\,(XB)^{+} = (XB)\left[(XB)^{\top}(XB)\right]^{-1}(XB)^{\top} = XB\left(B^{\top}VB\right)^{-1}B^{\top}X^{\top}.$$

Expressed in basis coefficients, this subspace is $\mathrm{R}\,(B)$:

$$\boldsymbol{x} = XB\boldsymbol{w} \;\Leftrightarrow\; X\boldsymbol{y} = XB\boldsymbol{w} \;\Leftrightarrow \boldsymbol{y} = X^{-1}XB\boldsymbol{w} = B\boldsymbol{w}.$$

66.4 Sharpe Ratio

The Sharpe ratio for **X** is the maximum value of the expected excess return divided by the standard deviation:

$$|\boldsymbol{x}^*| = \sup_{x} \frac{\langle \boldsymbol{x}^*, \boldsymbol{x} \rangle}{|\boldsymbol{x}|}.$$

a standard homogeneous fractional program problem (standard if $\boldsymbol{x}^* \neq \mathbf{0}$). We solved this problem previously: $|\boldsymbol{x}^*|$ is the gauge of $|\boldsymbol{x}|$. A Pythagorean theorem follows immediately.

Theorem 2 (Sharpe-Ratio Pythagorean Theorem) *The square of the Sharpe ratio is the sum of the squares of the Sharpe ratios for the systematic-risk subspace and the no-systematic-risk subspace:*

$$|\boldsymbol{x}^*|^2 = |M\boldsymbol{x}^*|^2 + |(I - M)\boldsymbol{x}^*|^2 .$$

The squared Sharpe ratio for **X** is

$$\langle \boldsymbol{x}^*, \boldsymbol{x}^* \rangle = \langle X\boldsymbol{y}^*, X\boldsymbol{y}^* \rangle = \left\langle \boldsymbol{y}^*, X^\mathsf{T} X\boldsymbol{y}^* \right\rangle = \langle \boldsymbol{y}^*, V\boldsymbol{y}^* \rangle .$$

The squared Sharpe ratio for the systematic-risk subspace is

$$\begin{aligned}\langle \boldsymbol{x}^*, M\boldsymbol{x}^* \rangle &= \left\langle \boldsymbol{x}^*, XB\left(B^\mathsf{T} VB\right)^{-1} B^\mathsf{T} X^\mathsf{T} \boldsymbol{x}^* \right\rangle \\ &= \left\langle X\boldsymbol{y}^*, XB\left(B^\mathsf{T} VB\right)^{-1} B^\mathsf{T} X^\mathsf{T} X\boldsymbol{y}^* \right\rangle \\ &= \left\langle \boldsymbol{y}^*, VB\left(B^\mathsf{T} VB\right)^{-1} B^\mathsf{T} V\boldsymbol{y}^* \right\rangle .\end{aligned}$$

The squared Sharpe ratio for the no-systematic-risk subspace is the difference of these two expressions.

An optimization obtains a least-squares interpretation of the Sharpe ratio. The Sharpe ratio for the no-systematic-risk subspace is the optimum value in the following problem.

Problem 1 (No-Systematic-Risk Sharpe Ratio)

$$\sup_{x \neq 0} \frac{\langle \boldsymbol{x}^*, \boldsymbol{x} \rangle - \delta_{M^\perp}(\boldsymbol{x})}{|\boldsymbol{x}|} = \sup_{x \neq 0} \frac{\langle \boldsymbol{x}^*, \boldsymbol{x} \rangle - \delta_{\{0\}}\left(B^\mathsf{T} X^\mathsf{T} \boldsymbol{x}\right)}{|\boldsymbol{x}|}.$$

This problem is a standard homogeneous fractional program (as long as $\langle \boldsymbol{x}^*, \boldsymbol{x} \rangle \neq 0$ for some $\boldsymbol{x} \in M^\perp$).

The concave reduction is

$$\sup_{\boldsymbol{x}} \left[\langle \boldsymbol{x}^*, \boldsymbol{x} \rangle - \delta_{\{0\}} \left(B^\top X^\top \boldsymbol{x} - \boldsymbol{w} \right) - \delta_B (\boldsymbol{x}) \right],$$

and the reciprocal dual furnishes a least-squares interpretation. Adding the Fenchel inequalities

$$\begin{gathered} \delta_B (\boldsymbol{x}) + \delta_B^* (\boldsymbol{x}^* - XB\boldsymbol{w}^*) - \langle -XB\boldsymbol{w}^* + \boldsymbol{x}^*, \boldsymbol{x} \rangle \geq 0 \\ \delta_{\{0\}} \left(B^\top X^\top \boldsymbol{x} - \boldsymbol{w} \right) + \delta_{\{0\}}^* (\boldsymbol{w}^*) - \left\langle \boldsymbol{w}^*, B^\top X^\top \boldsymbol{x} - \boldsymbol{w} \right\rangle \geq 0 \end{gathered}$$

and rearranging obtains

$$\langle \boldsymbol{w}^*, \boldsymbol{w} \rangle + \delta_B^* (\boldsymbol{x}^* - XB\boldsymbol{w}^*) + \delta_{\{0\}}^* (\boldsymbol{w}^*) \geq \langle \boldsymbol{x}^*, \boldsymbol{x} \rangle - \delta_{\{0\}} \left(B^\top X^\top \boldsymbol{x} - \boldsymbol{w} \right) - \delta_B (\boldsymbol{x}).$$

The perturbation base values are the mean vector $\boldsymbol{x}^*$ and $\boldsymbol{w} = \mathbf{0}$.

The dual is

$$\inf_{\boldsymbol{w}^*} \delta_B^* (\boldsymbol{x}^* - XB\boldsymbol{w}^*) = \inf_{\boldsymbol{w}^*} |\boldsymbol{x}^* - XB\boldsymbol{w}^*|.$$

The primal satisfies the polyhedral/relative condition, and the dual satisfies the relative condition. There is no duality gap, and the optimum value is attained in both the primal and the dual. The dual solution $\boldsymbol{w}^*$ is the coefficients in the least-squares linear regression of $\boldsymbol{x}^*$ on XB. The primal solution $\boldsymbol{x}$ is the residual $\boldsymbol{x}^* – XB\boldsymbol{w}^*$ normalized to $|\boldsymbol{x}| = 1$.

The equivalent least-squares problem to solve for $\boldsymbol{w}^*$ is

$$\begin{aligned} &\inf_{\boldsymbol{w}^*} \langle \boldsymbol{x}^* - XB\boldsymbol{w}^*, \boldsymbol{x}^* - XB\boldsymbol{w}^* \rangle \\ &= \inf_{\boldsymbol{w}^*} \langle X\boldsymbol{y}^* - XB\boldsymbol{w}^*, X\boldsymbol{y}^* - XB\boldsymbol{w}^* \rangle \\ &= \inf_{\boldsymbol{w}^*} \langle \boldsymbol{y}^* - B\boldsymbol{w}^*, V(\boldsymbol{y}^* - B\boldsymbol{w}^*) \rangle. \end{aligned}$$

By the first-order condition,

$$B^\top V(\boldsymbol{y}^* - B\boldsymbol{w}^*) = \mathbf{0},$$

so

$$\boldsymbol{w}^* = \left(B^\top VB \right)^{-1} B^\top V\boldsymbol{y}^*$$

and the residual $\boldsymbol{y}^* - B\boldsymbol{w}^*$ belongs to the no-systematic-risk subspace.

66.5 Expected Excess Return

The expected excess return is the sum of the expected excess returns on the systematic-risk and no-systematic risk components. For a portfolio in the systematic-risk subspace, the expected excess return is set by a linear combination of the beta coefficients:

$$\begin{aligned}\langle \boldsymbol{x}^*, \boldsymbol{x}\rangle &= \langle \boldsymbol{x}^*, M\boldsymbol{x}\rangle \\ &= \langle M\boldsymbol{x}^*, \boldsymbol{x}\rangle \\ &= \langle MX\boldsymbol{y}^*, X\boldsymbol{y}\rangle \\ &= \left\langle \left[XB\left(B^{\mathsf{T}}VB\right)^{-1}B^{\mathsf{T}}X^{\mathsf{T}}\right]X\boldsymbol{y}^*, X\boldsymbol{y}\right\rangle \\ &= \left\langle B\left(B^{\mathsf{T}}VB\right)^{-1}B^{\mathsf{T}}V\boldsymbol{y}^*, V\boldsymbol{y}\right\rangle \\ &= \langle B\boldsymbol{w}^*, V\boldsymbol{y}\rangle ,\end{aligned}$$

the covariance between the fitted value $B\boldsymbol{w}^*$ and $\boldsymbol{y}$.[1] In the capital-asset pricing model, in contrast the expected excess return is proportional to the covariance with the excess return on the market portfolio of all assets.

For a portfolio in the no-systematic-risk subspace, the expected excess return

$$\langle \boldsymbol{y}^*, V\boldsymbol{y}\rangle = \langle \boldsymbol{y}^* - B\boldsymbol{w}^*, V\boldsymbol{y}\rangle ,$$

as $B^{\mathsf{T}}V\boldsymbol{y} = \boldsymbol{0}$. The expected excess return is the covariance of the residual $\boldsymbol{y}^* - B\boldsymbol{w}^*$ and $\boldsymbol{y}$.

The contention of the arbitrage pricing theory is that the expected excess return on a portfolio in the no-systematic-risk subspace is small. By the definition of the Sharpe ratio, an upper bound for the absolute value of the expected excess return on a portfolio in the subspace is

$$\text{Standard Deviation} \times \text{Sharpe Ratio}.$$

The expected excess return is small if either term is small. The argument of the arbitrage pricing theory is that the standard deviation of the no-systematic-risk component of a well-diversified portfolio is small. The Sharpe ratio is the standard deviation of $\boldsymbol{y}^* - B\boldsymbol{w}^*$, which is small if the factors account for most of the variability of the expected excess returns. The expected excess return is small, even if the portfolio is not well-diversified.

---

[1] If $\boldsymbol{x}^* \in M$, this same result holds for any $\boldsymbol{x} \in M$: $\langle \boldsymbol{x}^*, \boldsymbol{x}\rangle = \langle M\boldsymbol{x}^*, M\boldsymbol{x}\rangle = \langle \boldsymbol{x}^*, M\boldsymbol{x}\rangle$, so the preceding calculation applies.

References

Ross SA (1976 Dec) The arbitrage theory of capital asset pricing. J Econ Theor 13(3):341–360
Ross SA (1977) Return, risk, and arbitrage. In: Friend I, Bicksler JL (eds) Risk and return in finance. Ballinger, Cambridge, MA, pp 189–218

Part XV
Dynamics

Chapter 67
Stability

Abstract The analysis of a dynamic system is widespread in economic theory. As time passes, economic variables change. A fundamental issue is whether the system is stable. As time passes, do the economic variables converge to a steady state, either of constant values or of constant growth? Typically, economic thinking supposes stability, so to analyze whether the system is stable is fundamental. Below we investigate the stability of constant coefficients, linear difference and differential equations via the von Neumann–Morgenstern theorem of the alternative, applied in a Hermitian Euclidean Jordan algebra. The analysis connects the eigenvalues of the dynamic system with the existence of a Lyapunov linear transformation. The approach to stability by Lyapunov is to define a function that is zero at the steady state but positive elsewhere. One defines the function so away from the steady state it declines as time passes. Under general conditions, it follows that the steady state is stable. Set the analysis in the complexification $\mathbf{X}$ of a Euclidean space and the Hermitian Euclidean Jordan algebra $\mathbf{H}^{\mathbf{X}}$ of self-adjoint linear transformations on the complexification. We look for some positive linear transformation X in the complexification such that for any nonzero $\boldsymbol{x}_t$ the quadratic $\langle \boldsymbol{x}_t, X\boldsymbol{x}_t\rangle$ is positive and declines each period. The stability alternative is a restatement of the von Neumann–Morgenstern alternative derived previously. For a square linear transformation A on $\mathbf{H}^{\mathbf{X}}$, exactly one of the following has a solution:

Stable Alternative $AX \ll \mathbf{0}, X \gg \mathbf{0}$;
Not Stable Alternative $A^{\mathsf{T}}X^* \geq \mathbf{0}, X^* > \mathbf{0}$.

We apply this theorem of the alternative to study stability. For the difference equation, one choice of A applies. For the differential equation, a different choice applies. However the same two alternatives apply to either. In each case, the stable alternative states that there exists a Lyapunov transformation, so the dynamic system is stable. We interpret the not stable alternative in terms of eigenvectors and eigenvalues.

The analysis of a dynamic system is widespread in economic theory. As time passes, economic variables change. A fundamental issue is whether the system is stable. As time passes, do the economic variables converge to a steady state, either of constant

B. C. Dieffenbach, *Conjugate Duality in Economic Analysis*, Contributions to Economics,
https://doi.org/10.1007/978-3-032-21396-9_67

values or of constant growth? Typically economic thinking supposes stability, so to analyze whether the system is stable is fundamental. Below we investigate the stability of constant coefficients, linear difference and differential equations via the von Neumann–Morgenstern theorem of the alternative, applied in a Hermitian Euclidean Jordan algebra. The analysis connects the eigenvalues of the dynamic system with the existence of a Lyapunov linear transformation.

67.1 Constant Coefficients, Linear Dynamic Systems

We consider simple dynamic systems in a Euclidean space $\mathbf{X}$, in which a variable $\boldsymbol{x}_t$ evolves as time passes. A difference equation models evolution in discrete time.

Definition 1 (First-Order, Constant Coefficients, Linear Difference Equation) For $B \in \mathbf{T}^{\mathbf{X}}$,

$$\boldsymbol{x}_{t+1} = B\boldsymbol{x}_t. \tag{67.1}$$

A differential equation models evolution in continuous time.

Definition 2 (First-Order, Constant Coefficients, Linear Differential Equation) For $B \in \mathbf{T}^{\mathbf{X}}$,

$$\frac{\mathrm{d}\boldsymbol{x}_t}{\mathrm{d}t} = B\boldsymbol{x}_t. \tag{67.2}$$

Both the difference Eq. (67.1) and the differential Eq. (67.2) have the steady state $\boldsymbol{x} = \mathbf{0}$. Stability means that any solution converges to zero as time passes, regardless of the initial value $\boldsymbol{x}_0$.

Definition 3 (Stability) The difference Eq. (67.1) or the differential Eq. (67.2) is stable if $\lim_{t\to\infty} \boldsymbol{x}_t = \mathbf{0}$, regardless of the initial condition $\boldsymbol{x}_0$.

A traditional way to analyze stability is via explicit solution. The solution to the difference Eq. (67.1) is $\boldsymbol{x}_t = B^t\boldsymbol{x}_0$, and the solution to the differential equation (67.2) is $\boldsymbol{x}_t = \mathrm{e}^{Bt}\boldsymbol{x}_0$. Via the Jordan normal form of B, one finds necessary and sufficient conditions for stability, in terms of the eigenvalues of B.[1] In contrast, here we study stability via a theorem of the alternative.

The approach to stability by Lyapunov is to define a function that is zero at the steady state but positive elsewhere (Lyapunov 1949). One defines the function so away from the steady state it declines as time passes. Under general conditions, it follows that the steady state is stable.

[1] The French mathematician Camille Jordan (1838–1922) devised the Jordan normal form.

67.2 Difference Equation

Using the von Neumann–Morgenstern alternative, we present the Lyapunov formulation and the eigenvalue approach as equivalent concepts. To motivate the application of the theorem, let us first apply the Lyapunov approach to the difference Eq. (67.1).

Set the analysis in the complexification of **X** and the Hermitian Euclidean Jordan algebra $\mathbf{H}^{\mathbf{X}}$ of self-adjoint linear transformations on the complexification. For $\boldsymbol{x}_t$ in the complexification, we look for some positive $X \in \mathbf{H}^{\mathbf{X}}$ such that for any nonzero $\boldsymbol{x}_t$ the quadratic $\langle \boldsymbol{x}_t, X\boldsymbol{x}_t \rangle$ is positive and declines each period. The change from one period to the next is

$$\langle \boldsymbol{x}_{t+1}, X\boldsymbol{x}_{t+1} \rangle - \langle \boldsymbol{x}_t, X\boldsymbol{x}_t \rangle = \boldsymbol{x}_t^\top B^\top XB\boldsymbol{x}_t - \boldsymbol{x}_t^\top X\boldsymbol{x}_t = \boldsymbol{x}_t^\top \left(B^\top XB - X\right)\boldsymbol{x}_t.$$

We want this expression to be negative, for any nonzero $\boldsymbol{x}_t$.

Definition 4 (Lyapunov Transformation) For the difference Eq. (67.1), a Lyapunov transformation is some $X \gg 0$ in $\mathbf{H}^{\mathbf{X}}$ such that $B^\top XB - X \ll 0$.

The difference Eq. (67.1) is stable if there exists a Lyapunov transformation.

The stability alternative is a restatement of the von Neumann–Morgenstern alternative derived previously.

Theorem 1 (Stability Alternative) *For a square linear transformation A on $\mathbf{H}^{\mathbf{X}}$, exactly one of the following has a solution:*

Stable Alternative $AX \ll \mathbf{0}, X \gg \mathbf{0}$;
Not Stable Alternative $A^\top X^* \succeq \mathbf{0}, X^* \succ \mathbf{0}$.

We apply this theorem of the alternative to study stability. For the difference equation, one choice of A applies. For the differential equation, a different choice applies. However the same two alternatives apply to either equation. In each case, the stable alternative states that there exists a Lyapunov transformation, so the dynamic system is stable. We interpret the not stable alternative in terms of the eigenvectors and eigenvalues of B and $A^\top$.

Invoking the stability alternative Theorem 1 for the difference equation obtains the following theorem.

Theorem 2 (Difference-Equation Stability) *For the difference Eq. (67.1), the following three conditions are equivalent:*

- *the system is stable;*
- *there exists a Lyapunov transformation;*
- *every eigenvalue of B is less than one in absolute value.*

That the first two conditions are equivalent is Stein's theorem (Stein 1952).

For the difference Eq. (67.1), the linear transformation

$$\begin{aligned} A : \mathbf{H}^{\mathrm{X}} &\rightarrow \mathbf{H}^{\mathrm{X}} \\ X &\mapsto B^{\top} X B - X \end{aligned}$$

has the adjoint

$$\begin{aligned} A^{\top} : \mathbf{H}^{\mathrm{X}} &\rightarrow \mathbf{H}^{\mathrm{X}} \\ X^{*} &\mapsto B X^{*} B^{\top} - X^{*}. \end{aligned}$$

Let us verify the adjoint condition:

$$\langle X^{*}, B^{\top} X B - X \rangle = \mathrm{Tr}\left[X^{*}\left(B^{\top} X B - X\right)\right] = \mathrm{Tr}\left[\left(B X^{*} B^{\top} - X^{*}\right) X\right] = \langle B X^{*} B^{\top} - X^{*}, X \rangle.$$

The transformation is not self-adjoint.

Suppose that $B \in \mathbf{T}^{\mathrm{X}}$ has two eigenvectors and eigenvalues:

$$\begin{aligned} B\boldsymbol{x} &= \lambda \boldsymbol{x} \\ B\boldsymbol{x}^{*} &= \lambda^{*} \boldsymbol{x}^{*}, \end{aligned}$$

both normalized to length one. Then $\boldsymbol{x}^{*}\boldsymbol{x}^{\top}$ is an eigenvector of $X \mapsto BXB^{\top}$ with eigenvalue $\lambda^{*}\bar{\lambda}$:

$$B\boldsymbol{x}^{*}\boldsymbol{x}^{\top} B^{\top} = (\lambda^{*}\boldsymbol{x}^{*})\bar{\lambda}\boldsymbol{x}^{\top} = \lambda^{*}\bar{\lambda}\boldsymbol{x}^{*}\boldsymbol{x}^{\top}.$$

Every eigenvector has this form. The linear transformation $A^{\top}$ has eigenvector $\boldsymbol{x}^{*}\boldsymbol{x}^{\top}$ with eigenvalue $\lambda^{*}\bar{\lambda} - 1$.

If the stable alternative holds, then there exists a Lyapunov transformation, so the system is stable. In this case the not-stable alternative fails, so for any primitive idempotent $X^{*} = \boldsymbol{x}\boldsymbol{x}^{\top}$ it is not the case that $A^{\top} X^{*} \geq \mathbf{0}$. Thus it not the case that $|\lambda| \geq 1$ for any eigenvector $\boldsymbol{x}$ for B. Instead, every eigenvalue of B must have absolute value less than one.

If some eigenvector $\boldsymbol{x}$ of B has eigenvalue $|\lambda| \geq 1$, then system is unstable. The primitive idempotent $X^{*} = \boldsymbol{x}\boldsymbol{x}^{\top}$ satisfies $A^{\top} X^{*} \geq \mathbf{0}$, so the not-stable alternative holds.

67.3 Differential Equation

The analysis of a first-order, constant coefficients, linear differential equation is analogous. The Lyapunov approach looks for some $X \gg \mathbf{0}$ in $\mathbf{H}^{\mathrm{X}}$ such that the quadratic form $\boldsymbol{x}_t^{\top} X \boldsymbol{x}_t$ is positive, and declines as time passes, as long as $\boldsymbol{x}_t \neq 0$. We have

$$\frac{\mathrm{d}\left(\boldsymbol{x}_t^{\top} X \boldsymbol{x}_t\right)}{\mathrm{d}t} = \boldsymbol{x}_t^{\top} X \frac{\mathrm{d}\boldsymbol{x}_t}{\mathrm{d}t} + \left(\frac{\mathrm{d}\boldsymbol{x}_t}{\mathrm{d}t}\right)^{\top} X \boldsymbol{x}_t = \boldsymbol{x}_t^{\top} X B \boldsymbol{x}_t + \boldsymbol{x}_t^{\top} B^{\top} X \boldsymbol{x}_t = \boldsymbol{x}_t^{\top}\left(XB + B^{\top} X\right)\boldsymbol{x}_t,$$

We want this expression to be negative, for any nonzero $\boldsymbol{x}_t$.

Definition 5 (Lyapunov Transformation) A Lyapunov transformation for the differential equation (67.2) is some $X \gg 0$ in $\mathbf{H}^{\mathbf{X}}$ such that $XB + B^{\top}X \ll 0$.

The differential Eq. (67.2) is stable if there exists a Lyapunov transformation.

We prove the following theorem.

Theorem 3 (Differential-Equation Stability) *For the differential Eq. (67.2), the following three conditions are equivalent:*

- *the equation is stable;*
- *there exists a Lyapunov transformation;*
- *every eigenvalue of B has a negative real part.*

For the differential Eq. (67.2), the square linear transformation

$$A : \mathbf{H}^{\mathbf{X}} \to \mathbf{H}^{\mathbf{X}}$$
$$X \mapsto XB + B^{\top}X$$

has the adjoint

$$A^{\top} : \mathbf{H}^{\mathbf{X}} \to \mathbf{H}^{\mathbf{X}}$$
$$X^* \mapsto BX^* + X^*B^{\top}.$$

Let us verify the adjoint condition:

$$\langle X^*, XB + B^{\top}X\rangle = \mathrm{Tr}\left[X^*\left(XB + B^{\top}X\right)\right] = \mathrm{Tr}\left[\left(BX^* + X^*B^{\top}\right)X\right] = \langle BX^* + X^*B^{\top}, X\rangle.$$

The transformation is not self-adjoint.

Suppose that $B \in \mathbf{T}^{\mathbf{X}}$ has two eigenvectors and eigenvalues:

$$B\boldsymbol{x} = \lambda\boldsymbol{x}$$
$$B\boldsymbol{x}^* = \lambda^*\boldsymbol{x}^*,$$

both normalized to length one. Then $X^* = \boldsymbol{x}^*\boldsymbol{x}^{\top}$ is an eigenvector of $A^{\top}$ with eigenvalue $\lambda^* + \overline{\lambda}$:

$$BX^* + X^*B^{\top} = B\boldsymbol{x}^*\boldsymbol{x}^{\top} + \boldsymbol{x}^*\boldsymbol{x}^{\top}B^{\top} = \left(\lambda^* + \overline{\lambda}\right)\boldsymbol{x}^*\boldsymbol{x}^{\top}.$$

Every eigenvector has this form.

If the stable alternative holds, then there exists a Lyapunov transformation, so the system is stable. In this case the not-stable alternative fails, so for any primitive idempotent $X^* = \boldsymbol{x}\boldsymbol{x}^{\top}$ it is not the case that $A^{\top}X^* \geq \mathbf{0}$. Thus it not the case that $\lambda + \overline{\lambda} \geq 0$ for any eigenvector $\boldsymbol{x}$ for B. Every eigenvalue of B must have a negative real part.

If some eigenvector $\boldsymbol{x}$ of B has a nonnegative real part, then system is unstable. The primitive idempotent $X^* = \boldsymbol{x}\boldsymbol{x}^\top$ satisfies $A^\top X^* \geq \mathbf{0}$, so the not-stable alternative holds.

By our two theorems, the dynamic system is stable if and only if a Lyapunov transformation exists. Note that the presentation establishes this result without deriving an explicit formula for the transformation.

References

Lyapunov AM (1949) Le problème général de la stabilité du mouvement. Ann Math Stud 17. (Originally 1892)
Stein P (1952 Jan) Some general theorems on iterants. J Res Nat Bureau Stand 48(1):32–33

Chapter 68
Balanced Growth

Abstract In the von Neumann balanced growth model, an input of goods this period produces an output of goods next period. The inputs and outputs are the same goods, and there is no consumption of goods. Each period all inputs and outputs grow at the same rate. There exists a maximum rate of balanced growth, and furthermore this balanced growth is attainable as a production equilibrium.

The technology is activity analysis, with constant returns to scale. The total input of goods is Ax, and the total output is Bx, in which x denotes the vector of activity levels. We analyze the saddle-point problem

$$\min_{y^*}\left(\max_{x}\frac{\langle y^*, Bx\rangle}{\langle y^*, Ax\rangle}\right).$$

The ratio is the output value divided by the input value. The numerator and the denominator are both positive.

We establish that a saddle point $\langle x, y^*\rangle$ exists. Activity x attains the maximum feasible balanced growth, and the saddle value is the growth factor. Activity x is a production equilibrium for price y^*, in which the interest factor equals the growth factor. If the profit of an activity is negative, then the activity level is zero. If a good is in excess supply, then its price is zero. We analyze the saddle-point problem as two standard fractional programs. Together, the two zero-conjugate conditions amount to finding a Nash equilibrium for a two-person zero-sum game.

We work out the celebrated balanced growth model of von Neumann (1945–6). Writing before economists first established the existence of a general equilibrium via a fixed-point theorem, von Neumann uses saddle-point theory to establish the existence of maximum balanced-growth production equilibrium.

In the von Neumann model, an input of goods this period produces an output of goods next period. The inputs and outputs are the same goods, and there is no consumption of goods. In balanced growth, in each period all inputs and outputs

B. C. Dieffenbach, *Conjugate Duality in Economic Analysis*, Contributions to Economics,
https://doi.org/10.1007/978-3-032-21396-9_68

grow at the same rate. Von Neumann shows that there exists a maximum rate of balanced growth, and furthermore that this balanced growth is attainable as a production equilibrium.

68.1 Balanced Growth

The technology is activity analysis, with constant returns to scale. There are n activities that produce the m goods. Every activity uses a positive input of every good to produce a positive output of every good. Activity j uses positive input $\boldsymbol{a}_j \in \mathbf{R}^m_{++}$ this period to produce positive output $\boldsymbol{b}_j \in \mathbf{R}^m_{++}$ next period.[1] Define the partitioned linear transformations

$$A := \begin{bmatrix} \boldsymbol{a}_1 & \cdots & \boldsymbol{a}_n \end{bmatrix} \in \mathbf{T}^{n \to m}$$
$$B := \begin{bmatrix} \boldsymbol{b}_1 & \cdots & \boldsymbol{b}_n \end{bmatrix} \in \mathbf{T}^{n \to m}.$$

Let $\boldsymbol{x} \in \mathbf{R}^n_+$ denote the vector of activity levels. The total input of goods is $A\boldsymbol{x}$, and the total output is $B\boldsymbol{x}$. Because returns to scale are constant, for simplicity below we normalize $\boldsymbol{x} \in P^n$.

There is balanced growth with growth factor γ if the output of each good next period is γ or more times the input of the good this period: $B\boldsymbol{x} \geq \gamma A\boldsymbol{x}$, for some activity vector $\boldsymbol{x} > \mathbf{0}$. For example, $\gamma = 1.1$ means growth of 10% per period. For balanced growth, the output of each good must be at least ten per cent greater than the input of this good. Maximum balanced growth is the largest feasible growth factor.

In a balanced-growth production equilibrium, the goods price $\boldsymbol{y}^*$ is constant, the same every period. For simplicity, normalize $\boldsymbol{y}^* \in P^m$. Let ρ denote the interest factor; its inverse is the discount factor. For example, $\rho = 1.1$ means that the interest rate is 10%. In a balanced-growth production equilibrium, the interest factor is constant. Expressed as future value, the profit of activity i is output value minus input cost, $\boldsymbol{y}^{*\top}\boldsymbol{b}_i - \rho \boldsymbol{y}^{*\top}\boldsymbol{a}_i$. A production equilibrium chooses $\boldsymbol{x}$ to maximize profit, for price $\boldsymbol{y}^*$. Because returns to scale are constant, for each activity profit must be less than or equal to zero. Negative profit means that the activity level must be zero. If a good is in excess supply—output less than input times the growth factor, then its price must be zero.

[1] The assumption that $\boldsymbol{a}_j$ and $\boldsymbol{b}_j$ are both positive permits the analysis of the balanced-growth saddle point via two standard fractional programs, by ruling out division by zero. Von Neumann relaxes the assumption: $\boldsymbol{a}_j$ and $\boldsymbol{b}_j$ are both nonnegative, but their sum is positive. His formulation ensures that in each fractional program both the numerator and the denominator are not simultaneously zero.

68.2 Saddle Point

Following von Neumann, we analyze the saddle-point problem

$$\inf_{y^* \in P^m} \left(\sup_{x \in P^n} \frac{\langle y^*, Bx \rangle}{\langle y^*, Ax \rangle} \right) \tag{68.1}$$

von Neumann (1945–6). The ratio is the output value divided by the input value. The numerator and the denominator are both positive. We establish that a saddle point $\langle \boldsymbol{x}, \boldsymbol{y}^* \rangle$ exists. Activity $\boldsymbol{x}$ attains the maximum feasible balanced growth, and the saddle value is the growth factor. Activity $\boldsymbol{x}$ is a production equilibrium for price $\boldsymbol{y}^*$, in which the interest factor equals the growth factor.

We analyze the saddle-point problem as two standard fractional programs: choose $\boldsymbol{x}$ to maximize the ratio for a given $\boldsymbol{y}^*$, and choose $\boldsymbol{y}^*$ to minimize the ratio for a given $\boldsymbol{x}$. Let γ denote the saddle value. The zero-conjugate condition for the choice $\boldsymbol{x}$ is

$$0 = \max_{x \in P^n} [\langle y^*, Bx \rangle - \gamma \langle y^*, Ax \rangle],$$

and the zero-conjugate condition for the choice $\boldsymbol{y}^*$ is

$$0 = \min_{y^* \in P^m} [\langle y^*, Bx \rangle - \gamma \langle y^*, Ax \rangle].$$

Here we have written max and min, because the optimum value is attained.

68.3 Nash Equilibrium

Together, the zero-conjugate conditions for these two problems amount to finding a Nash equilibrium for a two-person zero-sum game:

$$0 = \min_{y^* \in P^m} \left[\max_{x \in P^n} \langle y^*, (B - \gamma A)\boldsymbol{x} \rangle \right].$$

The expression on the right-hand side is the Lagrangian for a two-person zero-sum game having payoff transformation $B - \gamma A$. The optimum value is increasing in γ. For small γ the value is positive, and for large γ the value is negative. For a unique γ the value is zero, so the problem (68.1) indeed has a saddle point.

The saddle value γ is the largest growth factor for balanced growth. The optimum value of the primal for the game must be zero:

$$0 = \max_{x \in P^n} \{\min [(B - \gamma A)\boldsymbol{x}]\}.$$

At the optimum $(B - \gamma A)\boldsymbol{x} \geq \mathbf{0}$, and some component must be zero. Balanced growth for a larger growth factor is not possible. The optimum value of the dual must be zero:

$$0 = \min_{y^* \in P^m} \left\{ \max \left[(B - \gamma A)^{\mathsf{T}} \boldsymbol{y}^* \right] \right\}.$$

At the optimum for every activity profit is less than or equal to zero, and for some activity profit must be zero. Thus $\rho = \gamma$ is the smallest interest factor allowing production equilibrium. That the Lagrangian is zero at a saddle point establishes that it is a production equilibrium. If the profit of an activity is negative, then the activity level is zero. If a good is in excess supply, then its price is zero.

Aubin analyzes the saddle-point problem (68.1) via a saddle-point theorem (Aubin 1998, Chap. 11). Bapat and Raghavan study maximum balanced growth directly and reduce the analysis to a two-person, zero-sum game (Bapat and Raghavan 1996). Gale formulates the analysis as two separate optimizations—a "technological expansion problem" and an "economic expansion problem" (Gale 1960, p. 311).

References

Aubin JP (1998) Optima and equilibria: an introduction to nonlinear analysis. In: Axler S, Gehring FW, Riber KA (eds) 2nd edn, Vol 140. Springer, Berlin

Bapat RB, Raghavan TES (1996). Nonnegat Mat Appl (Vol 64). Cambridge University Press, Cambridge

Gale D (1960) The theory of linear economic models. McGraw-Hill, New York

von Neumann J (1945–6) A model of general economic equilibrium. Rev Econom Stud 13(1):1–9. (Originally published in German 1937)

MIX
Papier aus verantwortungsvollen Quellen
Paper from responsible sources
FSC® C105338
www.fsc.org

If you have any concerns about our products,
you can contact us on
ProductSafety@springernature.com

In case Publisher is established outside the EU,
the EU authorized representative is:
Springer Nature Customer Service Center GmbH
Europaplatz 3, 69115 Heidelberg, Germany

Printed by Libri Plureos GmbH
in Hamburg, Germany